The Enigma of the Aerofoil

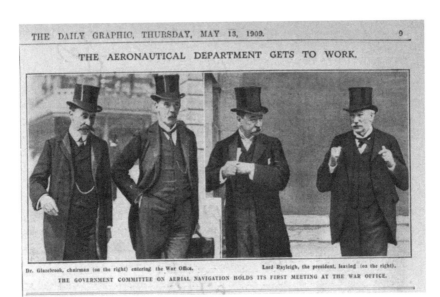

FRONTISPIECE. Before and after the first meeting of the Advisory Committee for Aeronautics. The meeting took place on May 12, 1909, with Rayleigh assuming the chair at 2:30 p.m. The caption identifies Glazebrook and Rayleigh, but who were the other two gentlemen? I tentatively identify the man next to Rayleigh as H. R. A. Mallock and Glazebrook's companion as F. J. Selby. From the *Daily Graphic*, May 13, 1909. (By permission of the Trustees of the National Library of Scotland)

The Enigma of
the Aerofoil

Rival Theories in Aerodynamics, 1909–1930

DAVID BLOOR

The University of Chicago Press
Chicago and London

David Bloor is professor emeritus in the Science Studies Unit at the University of Edinburgh. He is the author of *Knowledge and Social Imagery* and coauthor of *Scientific Knowledge: A Sociological Analysis*, both published by the University of Chicago Press.

The University of Chicago Press, Chicago 60637
The University of Chicago Press, Ltd., London
© 2011 by The University of Chicago
All rights reserved. Published 2011.
Printed in the United States of America

20 19 18 17 16 15 14 13 12 11   1 2 3 4 5

ISBN-13: 978-0-226-06094-1 (cloth)
ISBN-13: 978-0-226-06095-8 (paper)
ISBN-10: 0-226-06094-2 (cloth)
ISBN-10: 0-226-06095-0 (paper)

Library of Congress Cataloging-in-Publication Data

Bloor, David.
    The enigma of the aerofoil : rival theories in aerodynamics, 1909–1930 / David Bloor.
        p.   cm.
    Includes bibliographical references and index.
    ISBN-13: 978-0-226-06094-1 (cloth : alk. paper)
    ISBN-10: 0-226-06094-2 (cloth : alk. paper)
    ISBN-13: 978-0-226-06095-8 (pbk. : alk. paper)
    ISBN-10: 0-226-06095-0 (pbk. : alk. paper)   1. Aerodynamics—History.
2. Prandtl, Ludwig, 1875-1953.   I. Title.
TL570.B566 2011
629.132'3—dc22

                                                                    2011013059

Contents

Illustrations

Acknowledgments

In April 1997 Peter Galison and Alex Roland organized the conference "Atmospheric Flight in the Twentieth Century," which was held at the Dibner Institute in Cambridge, Massachusetts. By a stroke of good fortune, and the generosity of the Dibner Institute, I was able to attend the meeting. My role was to act as an outside commentator. I was deeply impressed by the high quality of all of the papers that were presented, though I confess I was somewhat daunted by the technical expertise of the contributors. The conference opened my eyes to a field of work, the history of aeronautics, that was new to me but which proved immediately attractive.[1]

One paper in the conference that caught my attention dealt with early British research in aerodynamics and the way in which, in Britain, the gulf between science and technology was bridged. The paper was titled "The Wind Tunnel and the Emergence of Aeronautical Research in Britain."[2] After the conference its author, Dr. Takehiko Hashimoto, kindly sent me the unpublished Ph.D. thesis on which his paper had been based.[3] Dr. Hashimoto's main concern was with the role of those important individuals who act as mediators, middlemen, and "translators" between mathematicians and engineers. By comparing the development of British and American aerodynamics (and their respective responses to German aerodynamics after World War I), he reached the gratifying conclusion that the British had been somewhat more successful in this process of mediation than had the Americans. I say "gratifying" because I am British, and the British frequently take a pessimistic attitude toward their own technological capabilities and tend to assume that other countries always do things better. I did not pursue the theme of the mediator or middleman, but it was this work that prompted me to do the research presented here. Although we paint a somewhat different picture of certain people who feature in both of our studies, I express my indebtedness to Dr. Hashimoto and my appreciation of his work.

I began by following up some of Dr. Hashimoto's references in the Public Record Office in London and soon found a set of research questions of my own that I wanted to answer, as well as evidence that there was material available with which to pursue them. My questions were these: In the early days of aviation, that is, in the early 1900s, there were rival accounts of how an aircraft wing provides "lift." One account was supported by British experts, while the other was mainly developed by German experts. This was well known to historians working in the field.[4] These two theories of lift were also featured, though not in technical detail, in Dr. Hashimoto's account.[5] But I wanted to know (1) why the rivalry arose, (2) what sustained it for almost twenty years, and (3) how it was resolved. These questions were not addressed in Dr. Hashimoto's work, nor had they been convincingly answered in any of the broader historical literature in the field. The present book sets out the conclusions that I eventually reached on these three questions.

My kind colleagues in the Science Studies Unit at the University of Edinburgh bore the disruptions caused by my research-related comings and goings with understanding and good humor. I am all too aware that my activities must have added to their own already considerable work load. Relief from teaching and administrative duties during crucial parts of the research was made possible by the Economic and Social Research Council (ESRC). I thank the Council for its financial support in the form of a project grant ESRC Res 000-23-0088. Grants specifically designed to offset the costs of publication came from two further sources: Trinity College, Cambridge, and the Royal Society of London. I thank the Master and Fellows of Trinity for their generosity, and I also express my appreciation for the continued support of the Royal Society, in these financially straitened times, for work in the history of science.

The argument of my book involves a detailed comparison between British and German aerodynamic work, and this subject would have proven impossible to study without a number of lengthy visits to the Max-Planck-Institut für Wissenschaftsgeschichte in Berlin. I must record my deep gratitude to Lorraine Daston and Hans-Jörg Rheinberger, the directors of Abteilung II and Abteilung III, respectively, and to Ursula Klein and Otto Sibum, who were directors of two of the independent research groups in the Institute. Their warm welcome and great generosity will never be forgotten, nor will the stimulus provided by the research environment they all worked so hard, and so successfully, to create. I also express particular thanks to Urs Schoepflin, the Institute librarian, and his dedicated team. They met my endless stream of requests and queries with unfailing professionalism, kindness, and scholarly understanding. Special mention must be made of one member of the library team, Monika Sommerer, who, in the final phases of writing the book, kindly began the work of approaching copyright holders for permission to reproduce the photographs and diagrams that illustrate my narrative.

One of the first things I did in Berlin was to make working translations of the main German technical papers that were relevant to the analysis. (By a "working translation" I mean something adequate for my own use rather than for public consumption.) Here I thank Marc Staudacher, a resourceful teacher of German and a professional translator, who spent many hours with me going over my attempts in order to check them and to explain points of grammar and meaning that were eluding me.

In developing the British side of the story I am indebted to the Royal Aeronautical Society in London for access to their unique collection of early aeronautical literature. I am deeply grateful to Brian Riddle, the librarian, who put this material, as well as his profound knowledge of the field, at my disposal. It was also through the good offices of Brian Riddle that I was able to make contact with Dr. Audrey Glauert of Clare Hall, Cambridge. Dr. Glauert generously made available to me material relating to her father and mother, both of whom played an important role in the development of aerodynamics and therefore feature prominently in my book. I hope I have been able to put that material to good use. The opportunity to talk with someone directly connected with the historical actors and episodes I was describing was a moving experience, and I express my gratitude to Dr. Glauert for her hospitality and kindness.

From its inception I have discussed my research project with Walter Vincenti of the University of Stanford. I have benefited immeasurably from numerous and lengthy conversations drawing on his firsthand experience of aerodynamic research. His patience in discussing the arguments of the early technical papers and his willingness to read and comment so carefully on the first drafts of many of the chapters of this book have been invaluable to me in learning to find my way in this new field. It has been a privilege to be able to put my questions and problems to him and to be the recipient of his expert and thoughtful answers. Donald MacKenzie read and commented on a number of early draft chapters; later, drafts of the complete book were read by Barry Barnes, Celia Bloor, Michael Eckert, Jon Harwood, and Horst Nowacki. Not only their encouragement but also their critical comments have been invaluable, and I have made extensive alterations as a result of their suggestions. The responsibility for the defects that remain can only be laid at my doorstep.

In addition I have accumulated many other debts of gratitude for the help I have received in the course of the research—guidance to the literature and new sources, help in approaching and gaining access to archives, and numerous conversations on historiographical, methodological, and philosophical questions. I hope the following persons will forgive me if I do not mention individually their many and varied acts of kindness and generosity that, nevertheless, I so clearly remember. My sincere thanks to Andrew Barker, Jed Buchwald, Dianna Buchwald, Harry Collins, Ivan Crozier, Olivier Darrigol, David Edgerton, Heinz Fuetterer, Zae-Young Ghim, Judith Goodstein, Ivor Grattan-Guinness, John Henry, Dieter Hoffmann, Christoph Hoffmann, Marion Kazemi, Kevin Knox, Martin Kusch, Wolfgang Lefevre, David Musker, Jürgen Renn, Simon Schaffer, Suman Seth, Steven Shapin, Skuli Sigurdsen, Richard Staley, Nelson Studart, Steve Sturdy, Thomas Sturm, Annette Vogt, Andrew Warwick, and Richard Webb.

I have used material from the following archives and express my thanks to the archivists for permission to consult their holdings: Archives of the California Institute of Technology (Kármán); Archiv zur Geschichte der Max-Planck-Gesellschaft (Prandtl); Churchill Archive Centre, Cambridge (Farren); Einstein Papers at Caltech (Einstein and Frank); Göttingen Archive of the Deutsche Gesellschaft für Luft-und Raumfahrt (Prandtl); Library of the University of Cambridge (Tripos exam papers); National Library of Scotland (Haldane); Public Record Office (minutes of the ARC); Royal Aeronautical Society (Lanchester and Grey); Royal Air Force Museum, Hendon (Melvill Jones); St. John's College, Cambridge (Jeffreys and Love); Trinity College, Cambridge (Taylor and Thomson); University of Coventry (Lanchester); and University of Edinburgh (A. R. Low).

The provenance of all photographic images and diagrams from published and unpublished sources is indicated in the caption along with an acknowledgment of copyright and permission to reproduce the material. In a few cases it proved impossible, despite every effort, to make contact with the holders of the copyright.

Finally I must mention my greatest debt. Throughout the research and the writing of this book I have benefited from the unstinting help of my wife. The book is dedicated to her. It is as good as I can make it, but it still seems little to give in return. I proffer it with the sentiment *Wenig, aber mit Liebe.*

Introduction: The Question to Be Answered

> 'Tis evident, that all the sciences have a relation, greater or less, to human nature; and that however wide any of them may seem to run from it, they still return back by one passage or another.
>
> DAVID HUME, *A Treatise of Human Nature* (1739–40)[1]

Why do aircraft fly? How do the wings support the weight of the machine and its occupants? Even the most jaded passengers in the overcrowded airliners of the present day may experience some moments of wonder — or doubt — as the machine that is to transport them lifts itself off the runway. Because the action of the air on the wing cannot be seen, it is not easy to form an idea of what is happening. Some physical processes are at work that must generate powerful forces, but the nature of these processes, and the laws they obey, are not open to casual inspection. If the passengers looking out of the window really want an explanation of how a wing works, they must do what any lay person has to do and ask the experts. Unfortunately the answers that the experts will give are likely to be highly technical. It will take patience by both parties if communication is not to break down. But given goodwill on both sides, the experts should be able to find some simplified formulations that will be useful to the nonexperts, and the nonexperts should be able to deepen their grasp of the problem.

In this book I discuss the question of why airplanes fly, but I approach the problem in a slightly unusual way. I describe the history behind the technical answer to the question about the cause of "lift," that is, the lifting force on the wing. I analyze the path by which the experts, after much disagreement, arrived at the account they would now give. I am therefore not simply asserting that airplanes fly for this or that reason; I am asserting that they were *understood* to fly for this or that reason. I am interested in the fact that different and rival understandings were developed by different persons and in different places. I cannot speak as a professional in the field of aerodynamics; nor is my position exactly that of a layperson. I speak as a historian and sociologist of science who is poised between these categories.[2]

What are the specific questions that I am addressing and to which I hope to offer convincing answers? To identify them I first need to give some background. The practical problem of building machines that can be flown, that is, the problem of "mechanical" or "artificial" flight, was solved in the final years of the nineteenth century and the early years of the twentieth century. In the 1890s Otto Lilienthal in Germany successfully built and flew what we today would call hang gliders. From 1903 to 1905 the Wright brothers in the United States showed that sustained and controlled powered flight was possible and practical. What had long been called the "secret" of flight was now no longer a secret. But not all of the secret was revealed. Some parts of it remained hidden, and indeed, some parts are still hidden today. The practical successes of the pioneer aviators still left unanswered the question of how a wing generated the lift forces that were necessary for flight. The pioneers mostly worked by trial and error. Some had experimented with models and taken measurements of lift and drag (the air resistance opposing the motion), but the measurements were sparse and unreliable.[3] No deeper theoretical understanding had prompted or significantly informed the early successes of the pioneers, nor had theory kept pace with the growth of practical understanding. The action of the air on the wing remained an enigma.

A division of labor quickly established itself. Practical constructors continued with their trial-and-error methods, while scientists and engineers began to study the nature of the airflow and the relation between the flow and the forces that it would generate. For this purpose the scientists and engineers did not just perform experiments and build the requisite pieces of apparatus, such as wind channels. They also exploited the resources of a branch of applied mathematics that was usually called hydrodynamics. The name "hydrodynamics" makes it sound as if the theory was confined to the flow of water, but in reality it was a mathematical description that, with varying degrees of approximation, was applied to "fluids" in general, including air. Thus was born the new science of aerodynamics. The birth was accompanied by much travail. One problem was that the mathematical theory of fluid flow was immensely difficult. The need to work with this theory effectively excluded the participation of all but the most mathematically sophisticated persons, and this did not go down well with the practical constructors. The mathematical analysis also depended for its starting point on a range of assumptions and hypotheses, about both the nature of the air and the more or less invisible pattern of the flow of air over, under, and around the wing. Only when the flow was known and specified could the forces on the wing be calculated. Assumptions had to be made. The unavoidable need to base their investigations on a set of assumptions proved to be deeply divisive. Different groups of experts adopted different assumptions and, for reasons I explain, stuck to them.

The first part of this historical story, the practical achievement of controlled flight, has been extensively discussed by historians. Pioneers, such as the Wright brothers, have been well served, and the attention given to them is both proper and understandable.[4] The second part of the history, the development of the science of aerodynamics, is somewhat less developed as a historical theme, though a number of outstanding works have been written and published on the subject in recent years.[5] The present book is a contribution to this developing field in the history of science and technology.

In the early years of aviation there were two, rival theories that were intended to explain the origin and nature of the lift of a wing. They may be called, respectively, the discontinuity theory and the circulatory (or vortex) theory. The names derive from the particular character of the postulated flow of air around the wing. (I should mention that the circulatory theory is, in effect, the one that is accepted today.) My aim is to give a detailed account of how the advocates of the two theories developed their ideas and how they oriented themselves to, and engaged with, the empirical facts about flight. To do this I found that I also needed to understand how they oriented themselves to, and engaged with, one another. I show that these two dimensions cannot be kept separate. This is why I have prefaced the work with the quotation from the famous Edinburgh historian and sociologist David Hume. The more one studies the technical details of the scientific work, the more evident it becomes that the social dimension of the activity is deeply implicated in these details. The more closely one analyses the technical reasoning, the more evident it becomes that the force of reason is a social force. The historical story that I have to tell about the emerging understanding of lift is, therefore, at one and the same time both a scientific and a sociological story. To understand the course taken by the science it is necessary to understand the role played by the social context, and to appreciate the role played by the social context it is necessary to deconstruct the technical and mathematical arguments.

In principle none of this should occasion surprise. Scientists and engineers do not operate as independent agents but as members of a group. They cannot achieve their status as scientists and engineers without being educated, and education is the transmission of a body of culture through the exercise of authority. Education is socialization.[6] Scientists and engineers see themselves as contributing to a certain discipline, as being members of certain institutions, as having loyalties to this laboratory or that tradition, as being students of A or rivals of B. Their activities would be impossible unless behavior were coordinated and concerted. For this the individuals concerned must be responsive to one another and in constant interaction. Their knowledge is necessarily shared knowledge, though, in its overall effects, the process of sharing can be divisive as well as unifying. The sharing is always what Hume would call a "confined" sharing.

All too frequently, when scientific and technical achievements become objects of commentary, analysis, or celebration, these simple truths are obscured. Academic culture is saturated with individualistic prejudices, which encourage us to trivialize the implications of the truth that science is a collective enterprise and that knowledge is a collective accomplishment. Philosophers of science actively encourage historians to distinguish between, on the one side, "cognitive," "epistemic," or "rational" factors and, on the other side, "social" factors. They enjoin the sociologist to "disentangle" scientific reasoning from "social influences" and to distinguish what is truly "internal" to science from what is truly "external."[7] These recommendations are treated as if they were preconditions of mental hygiene and based on self-evident truths. Historians and sociologists of science know better. They know that the problem of cognitive order *is* the problem of social order.[8] These are not two things, even two things that are closely connected; they are one thing described from different points of view. The division of a historical narrative into "the cognitive" and "the social," or "the rational" and "the social," is wholly artificial. It is methodologically lazy and epistemologically naïve.

I shall now briefly sketch the overall structure of the events I describe in this volume. Of the two theories of lift that I mentioned, one of them, the discontinuity theory, was mainly developed in Britain. It was based on work by the eminent mathematical physicist Lord Rayleigh. The other, the circulatory theory, was mainly developed in Germany. It is associated primarily with the German engineer Ludwig Prandtl, although it had originally been proposed by the English engineer Frederick Lanchester. It rapidly became clear that the discontinuity theory was badly flawed because it only predicted about half of the observed amount of lift. At this point, shortly before the outbreak of World War I (or what the British call the Great War) in 1914, the British awareness of failure might have reasonably led them to turn their attention to the other theory, the theory of circulation. They did not do this. They knew about the theory but they dismissed it. At Cambridge, G. I. Taylor, for example, treated the discontinuity theory as a mathematical curiosity, but he also found Lanchester's theory of circulation equally unacceptable. The reasons he gave to support this judgment were important and widely shared. Meanwhile the Germans embraced the idea of circulation and developed it in mathematical detail. The British also knew of this German reaction but still did not take the theory of circulation seriously. It was not until after the war ended in 1918 that the British began to take note. They found that the Germans had developed a mathematically expressed, empirically supported, and practically useful account of lift. Even then the British had serious reservations. The negative response had nothing to do with mere anti-German feeling. The British scientific experts were patriots, but, unlike some in the world of aviation, they were not bigots. Why then were they so reluctant to take the theory of circulation seriously? This is the main question addressed in the book.[9]

There are already candidate answers to this question in the literature, but they are answers of a different kind to the one I offer. The neglect of Lanchester's work became something of a scandal in the 1920s and 1930s, so it was natural that explanations and justifications were manufactured to account for it. Sir Richard Glazebrook, the head of the National Physical Laboratory, played an important role in British aviation during these years and was the source of one of the standard excuses, namely, that Lanchester did not present his ideas with sufficient mathematical clarity. Well into the midcentury, British experts in aerodynamics, who, along with Glazebrook, shared responsibility for the neglect of Lanchester's ideas, were scratching their heads and wondering how they could have allowed themselves to get into this position. Clarity or no clarity, they had turned their backs on the right theory of lift and had become bogged down with the wrong one.

The retrospective accounts and excuses that have been given have been both fragmentary and feeble, though Lanchester's biographer, P. W. Kingsford, writing in 1960, still went along with a version of Glazebrook's excuse.[10] Other existing accounts merely tend to embellish the basic excuse by invoking the personal idiosyncrasies of the leading actors. The problem is analyzed as a clash of personalities. It is true that some of those involved had strong characters as well as powerful intellects, and some of them could pass as colorful personalities. All this will become apparent in what follows. The psychology of those involved is clearly an integral part of the historical story, but such accounts miss the very thing that I want to emphasize and that I believe is essential for a proper analysis, namely, the interconnection of the sociological and technical dimensions. Only an account that is technically informed, and sensitive to the social processes built into the technical content of the aerodynamic work, will make sense of the history. I want to show that the real reasons for the resistance to the vortex or circulatory theory of lift were deep and interesting, but not really embarrassing at all.

Although I have posed the question of why the British resisted the theory of circulation, I do not believe it can be answered in isolation from the question of why the Germans embraced it. Both reactions should be seen as equally problematic. The historical record shows that the same type of causes were at work in both British and German aerodynamics. In both cases the actors drew on the resources of their local culture and elaborated them in ways that were typical of their milieu and were encouraged by the institutions of which they were active members. Of course, the cultures and the institutions were subtly different. My explanation of the German behavior is thus of the same kind as my explanation of the British. The same variables are involved, but the variables have different values. Seen in this way the explanation possesses a methodological characteristic that has been dubbed "symmetry." Because the point continues to be misunderstood, I should perhaps emphasize the words "same kind." I am not saying that the very same causes were at work but that the same *kinds* of cause were in operation. Symmetry, in this sense, is now widely (though not universally) accepted as a methodological virtue in much historical and sociological work. Conversely, it is widely rejected as an error, or treated as a triviality, by philosophers. I hope that seeing the symmetry principle in operation will help convey its meaning more effectively than merely trying to capture it in verbal formulas or justify it by abstract argument.

The overall plan of the book is as follows. In chapter 1 I start my account of the early British work in aerodynamics with the foundation of the controversial Advisory Committee for Aeronautics in 1909. The committee was presided over by Rayleigh. The frontispiece, taken from the *Daily Graphic* of May 13, 1909, shows some of the leading members of the committee striding purposefully into the War Office for their first meeting, and then emerging afterward looking somewhat more relaxed. The minutes of that important meeting are in the Public Record Office and reveal what they talked about in the interval between those two pictures.[11] It is a matter of central concern throughout this book. Chapter 2 lays the foundation for understanding the two competing theories of lift by sketching the basic ideas of hydrodynamics and the idealized, mathematical apparatus that was used to describe the flow of air. A nontechnical summary is provided at the end of the chapter. In chapter 3, I introduce the discontinuity theory of lift and describe the British research program on lift and the frustrations that were encountered. Chapter 4 is devoted to the circulatory or vortex theory and describes the hostile reception accorded to Lanchester among British experts. I pay particular attention to the reasons that were advanced to justify the rejection. In chapter 5, I identify and contrast two different intellectual traditions that were brought to bear on the theory of lift. One of them was grounded in the mathematical physics cultivated in Britain and preeminently represented by the graduates of the Cambridge Mathematical Tripos. The other tradition, called *technische Mechanik*, or "technical mechanics," was developed in the German technical colleges and was integral to Prandtl's work on wing theory. Chapters 6 and 7 provide an account of the German development and extension of the circulation theory as worked out in Munich, Göttingen, Berlin, and Aachen. In chapters 8 and 9 there is a description of the British postwar response, which took the form of a period of intense experimentation; it also gave rise to some remarkable and revealing theoretical confrontations. What, exactly, did the experiments prove? The British did not find it easy to agree on the answer.

The divergence between British and German approaches was effectively ended in 1926 with the publication, by Cambridge University Press, of a textbook that became a classic statement of the circulation theory. The book was Hermann Glauert's *The Elements of Aerofoil and Airscrew Theory*.[12] Glauert, an Englishman of German extraction, was a brilliant Cambridge mathematician who, in the 1920s, broke ranks and became a determined advocate of the circulation theory. As the title of Glauert's book indicates, he did not just work on the theory of the aircraft wing, but he also addressed the theory of the propeller. This is a natural generalization. The cross section of a propeller has the form of an aerofoil, and a propeller can be thought of as a rapidly rotating wing. The "lift" of this "wing" becomes the thrust of the propeller, which overcomes the air resistance, or "drag," as the aircraft moves through the air. Glauert's book also dealt with the theory of the flow of air in the wind channel itself, that is, the device used to test both wings and propellers. This aspect of the overall theory was needed to ensure that aerodynamic experiments and tests were correctly interpreted. As always in science, experiments are made to test theories, but theories are needed to understand the experiments.[13] The discussions of propellers and wind channels in Glauert's book are important and deserve further historical study, but, on grounds of practicality, I set aside both the aerodynamics of the propeller and the methodology of wind-channel tests in order to concentrate exclusively on the story of the wing itself.[14]

In the final chapter, chapter 10, I survey the course of the argument and consider objections to my analysis, particularly those that are bound to arise from its sociological character. I use the case study to challenge some of the negative and inaccurate stereotypes that still surround the sociology of scientific and technological knowledge. I also ask what lessons can be drawn from this episode in the history of aerodynamics. Does it carry a pessimistic message about British academic traditions and elitism? What does it tell us about the difference between Göttingen and Cambridge or between engineers and physicists? Finally, I ask what light the history of aerodynamics casts on the fraught arguments between historians, philosophers, and sociologists of science concerning relativism.[15] Does the success of aviation show that relativism must be false? I believe that, by drawing on this case study, some clear answers can be given to these questions, and they are the opposite of what may be expected.

During the writing of this book I had the great advantage of being able to make use of Andrew Warwick's *Masters of Theory: Cambridge and the Rise of Mathematical Physics.*[16] Although historians of British science had previously accorded significance to the tradition of intense mathematical training that was characteristic of late Victorian and Edwardian Cambridge, Warwick took this argument to a new level. By adopting a fresh standpoint he compellingly demonstrated the constitutive and positive role played by this pedagogic tradition in electromagnetic theory and the fundamental physics of the ether in the early 1900s.[17]

For me, one of the intriguing things about Warwick's book is that the actors in his story are, in a number of cases, also the actors in my story. What is more, his account of the resistance that some Cambridge mathematicians displayed to Einstein's work runs in parallel with my story of the resistance to Prandtl's work. Like Warwick I found that their mathematical training could exert a significant hold over the minds of Cambridge experts as they formulated their research problems. In many ways the study that I present here can be seen as corroborating the picture developed in Warwick's book. Of course, shifting the area of investigation from the history of electromagnetism to the history of fluid mechanics throws up differences between the two studies, and not surprisingly there is some divergence in our conclusions. Whereas Warwick's attention is mainly (though not exclusively) devoted to the British scene, my aim, from the outset, is that of comparing the British and German approaches to aerodynamics. Furthermore, on the British side, I follow the actors in my story as they move out of the cloisters of their Cambridge colleges into a wider world of politics, economics, aviation technology, and war. If Warwick studied Cambridge mathematicians as masters of theory, I ask how they acquitted themselves as servants of practice.

Mathematicians versus Practical Men:
The Founding of the Advisory
Committee for Aeronautics

In the meantime every aeroplane is to be regarded as a collection of unsolved mathematical problems; and it would have been quite easy for these problems to have been
solved years ago, before the first aeroplane flew.

        G. H. BRYAN, "*Researches in Aeronautical Mathematics*" (1916)[1]

The successful aeroplane, like many other pieces of mechanism, is a huge mass of
compromise.

        HOWARD T. WRIGHT, "*Aeroplanes from an Engineer's Point of View*" (1912)[2]

The Advisory Committee for Aeronautics (the ACA) was founded in 1909.
This Whitehall committee provided the scientific expertise that guided British research in aeronautics in the crucial years up to, and during, the Great
War of 1914–18. From the outset the ACA was, and was intended to be, the
brains in the body of British aeronautics.[3] It offered to the emerging field of
aviation the expertise of some of the country's leading scientists and engineers. In 1919 it was renamed the Aeronautical Research Committee, and in
this form the committee, and its successors, continued to perform its guiding role for many years. After 1909 the institutional structure of aeronautical research in Britain soon came to command respect abroad. When the
United States government began to organize its own national research effort
in aviation in 1915, it used the Advisory Committee as its model.[4] The resulting American National Advisory Committee for Aeronautics, the NACA, was
later turned into NASA, the National Aeronautics and Space Administration.
The British structure, however, was abolished by the Thatcher administration
in 1980, some seventy years after its inception.[5]

    If the Advisory Committee for Aeronautics was meant to offer the best,
there were some in Britain, especially in the early years, who argued that, in
fact, it gave the worst. For these critics the ACA held back the field of British aeronautics and encouraged the wrong tendencies. The reason for these
strongly divergent opinions was that aviation in general, and aeronautical science in particular, fell across some of the many cultural fault-lines running
through British society. These fault lines were capable of unleashing powerful

and destructive forces. From the moment of its inception the Advisory Committee was subject to the fraught relations, and conflicting interests, that divided those in government from those in industry; the representatives of the state from those seeking profit in the market place; the university-based academic scientist from the entrepreneur-engineer; the "mathematician" and "theorist" from the "practical man." Throughout its entire life these structural tensions dominated the context in which the ACA had to work.[6]

"I Was at Cambridge on Saturday"

The political pressures that originally prompted Herbert Asquith's Liberal administration to set up the Advisory Committee for Aeronautics can be epitomized by the reaction to the first cross-channel flight from France to England, made by Louis Blériot on July 25, 1909. Newspaper headlines declared that Britain was no longer an island. Blériot's heroic feat was greeted with sporting cheers, but the depressing military implications of the flight were evident. The nation's basic line of defense had been breached. The channel was no longer a moat that made the island an impregnable fortress. Blériot's flight dramatically confirmed the warnings that had been voiced since the inception of "aerial navigation." These reactions have been described in detail by the historian Alfred Gollin, who documents the atmosphere of alarm and the fear of invasion that gripped the country during the early years of the century, particularly with regard to the emerging power of Germany.[7] There was anxiety, assiduously cultivated by the press, that Britain was falling behind in the race to exploit the military potential of the new flying machines: the airship and the airplane. The anxiety was expressed in newspaper reports of mysterious (and almost certainly nonexistent) Zeppelins lurking in the night skies over Ipswich and Cardiff.[8]

The government, represented by Richard Burdon Haldane (fig. 1.1), the secretary of state for war, did not participate in this sort of unseemly clamor. Haldane was a patrician and highly intellectual figure who combined his politics with philosophical writing and a successful legal career. Educated at the universities of Edinburgh and Göttingen, he was fluent in German, translated Schopenhauer, and had a passion for Hegel.[9] To the fury of his critics the portly Haldane proceeded at his own steady pace. Much preoccupied with the long-overdue reform and rationalization of Britain's major institutions, from the army to the universities, Haldane always insisted that a cautious and "scientific" approach was needed.[10] Critics of the government policy on aeronautics called for the immediate purchase of foreign machines. Airships could be bought from France, and aircraft were on offer from the American

FIGURE 1.1. Richard Burdon Haldane (1856–1928). Haldane, the secretary of state for war, was responsible for the creation of the Advisory Committee for Aeronautics. He was the object of a campaign of hate and was hounded out of office. Photograph by Elliott and Fry, in Haldane 1929.

Wright brothers, who had been the first to master powered flight. Haldane thought that Britain should go more cautiously, even if it meant going more slowly. He was disinclined to rely on the results of mere trial-and-error methods developed by others. In fact he looked down on those who proceeded in a merely empirical manner, devoid of guiding principles to broaden and deepen their understanding.[11]

Haldane met the Wright brothers in May 1909 when they came to Europe touting for government contracts. An editorial in *Flight*, on May 8, hinted at inside information and expressed confidence in the outcome of the meeting: "On Monday, Messers Wilbur and Orville Wright paid a visit to Mr. Haldane, and, while naturally it is needful and fitting to preserve secrecy as regards official matters, it may be taken as assured that our Government will duly acquire Wright aeroplanes and the famous American brothers will themselves instruct the first pupils in England."[12] In fact it was not assured. Despite much pressure and lobbying, Haldane declined to do a commercial deal with the Wrights. Rather, the minister concluded, Britain should follow the German example—or what he took to be the German example. The National Physical Laboratory at Teddington, outside London, had been founded in 1900 on the model of Helmholtz's great, government-funded institute of physics in Berlin, the Physikalisch-Technische Reichsanstalt, and this was the pattern that Haldane wanted to see developed.[13] The government must locate the best scientists that were available and set them to work on the fundamental problems of flight. The pioneers had got their machines into the air, but how and why they flew remained obscure. The working of a wing, for example, the secret of its lift, remained an unsolved problem, as did the basis of stability and control. Furthermore, it was unclear whether the future lay with heavier-than-air flight or airships. Like many others, Haldane was more impressed by airships, but scientists must address these issues in all their generality and then, with scientific theory leading practice, the best technology would be able to progress on sound principles. As Haldane put it, "the newspapers and the contractors keep clamouring for action first and thought afterwards, whereas the energy which is directed by reflection is the energy which really gives the most rapid and stable results."[14] The issue of the Wright brothers and their rebuff deserves comment. To Haldane's critics the refusal to buy these aircraft seemed a gross error of judgment on the government's part. In fact it was grounded in a defensible line of reasoning. The Wright machine was known to be clumsy and unstable. It could only take off along specially constructed rails, and the subsequent flight demanded great skill and cease-less intervention by the pilot. As a British test pilot put it a few years later, in a report to the ACA on the flying qualities of different machines, it needed an "equilibrist of the first order" to keep the Wright machine in the air.[15] (This judgment is corroborated by modern aerodynamic research conducted on the Wright machine.)[16] The pilots of such aircraft would (1) require extensive training, (2) become exhausted on long flights, and (3) be so preoccupied that they could hardly perform any military task such as map reading, recon-naissance, or photography. The need was for aircraft that were easy to fly and would leave their pilots with spare mental and physical capacity. The British government's view was that power in the air would go to the nation that pos-sessed *stable* aircraft.[17]

Even before he met the Wrights, Haldane had sanctioned secret tests to be carried out at Blair Atholl, in the Scottish highlands, on a machine de-signed by a British inventor, J. W. Dunne.[18] Dunne was a friend of H. G. Wells and later in life became well known for his metaphysical speculations on the nature of time.[19] After his early military career had been terminated by ill health, Dunne turned to aviation and won the confidence of the super-intendent of the Army Balloon Factory at Farnborough. Dunne's airplane, unlike the Wrights', was meant to be stable and, to achieve this he used a novel, swept-wing configuration. The tests, however, which took place in the summer of 1907 and 1908, were a failure, and the machine did not maintain sustained flight.[20] A retrospective report of the episode in the journal *Aeronautics* contained the fanciful claim that, after an indiscrete mention of the trials in the press, the Scottish estate where they took place was alive with foreigners who, it was implied, must have been German secret agents. "In two days the place was buzzing with Teutons." Fortunately, the article continued, loyal local citizenry misdirected the unwanted foreign visitors so that the nation's secrets remained secure.[21] *Flight* even hinted that some of these alleged spies had been disposed of by the Scotch gillies, who acted as lookouts for the trials.[22] Haldane's worst suspicions about empirics and inventors, and everything to do with them, were confirmed by such goings on.[23] Dunne and his supporters were dismissed. It was time to bring in the scientists and develop a serious policy. Haldane had no intention of being deflected from this course just because the Wrights turned up in London.

Haldane had already laid out his ideas of a sound policy at the first meeting of a new subcommittee of the powerful Committee of Imperial Defence on December 1, 1908.[24] The prime minister had formed the subcommittee to report on three questions: (1) the military problem that aerial navigation posed to the country, (2) the naval and military advantages of airships and airplanes, and (3) the amount of money that should be spent and where that money should go. The chairman, Lord Esher, invited Haldane to open the proceedings. Haldane said it was important to have the navy and the army working together on these issues in order to provide the preconditions for real progress. Haldane meant by this the preconditions for developing a genuine, scientific understanding of aerial navigation and the problems it posed. He went on: "I was at Cambridge on Saturday, and I spent Sunday talking over some of these questions with Sir George Darwin, the mathematician. Some of them there have given a good deal of attention to this matter, and what strikes them—certainly what has struck me—is the little attempt which has been made, at any rate as far as the War Office is concerned, to answer these questions." Nobody in the navy, he said, would think of building ships without testing models in water, but if there was ever a need for model work it was in aeronautics. Darwin had told him that the French had experimental establishments using artificial currents of air. In reply to a direct question, Darwin had also told him that there was a great deal of mathematical work that needed to be done. Haldane therefore asked the subcommittee to consider appointing a further committee of experts to advise them on technical questions. The advisory body might have "somebody presiding over it like Lord Rayleigh or Lord Justice Fletcher Moulton, or Sir George Darwin." This, concluded Haldane, was "a very important preliminary to any real progress."

Esher's committee went on to take evidence from a number of expert witnesses, such as the aviator the Hon. C. S. Rolls and, to appease Churchill, the bombastic businessman, arms dealer, and aviation enthusiast Hiram Maxim.[25] The Esher committee did not succeed in bringing the interests of the army and navy into alignment and became bogged down in complicated differences over policy. Overall, it backed airships over airplanes and even recommended stopping research on heavier-than-air machines, although later this policy was quietly dropped.[26] In the course of the protracted discussions, Haldane was challenged by the navy representative over the desirability of his proposed committee of scientists. Should not scientists be on tap rather than on top? Eventually Haldane got his way but, perhaps as a result of this challenge, made sure that his projected advisory committee of experts would report directly to the prime minister.

The Structure of the Committee

The administrative structure that crystallized in Haldane's mind was for a committee of ten or eleven, involving persons of the highest scientific talent, to address technical problems presented to them by the Admiralty and War Office. Unlike the proposed committee itself, these two old-established bodies would be responsible for commissioning and even constructing military airships and aircraft. The committee would analyze and define the scientific and technical problems encountered by these constructive branches of the military and would pass them on to the National Physical Laboratory (the NPL). The laboratory, which was based at Teddington just outside London, was to have a new department specializing in aeronautical experiments. This department would produce the answers to the questions posed by the Advisory Committee. Financially, the committee would be accountable not to the War Office or Admiralty but to the Treasury.

The structure that emerged conformed to this plan except for the addition of one more unit. In 1911 the former Balloon Factory at Farnborough, belonging to the army and the home of Dunne and his supporters, was turned into the Aircraft Factory and then (in 1912) into the Royal Aircraft Factory (the RAF).[27] After the Dunne episode it had been decided to drop aircraft research at Farnborough, but this resolution was now rescinded. It was thus determined, after some indecision on Haldane's part, that new aircraft were to be designed by the government itself and built at its behest by private manufacturers.[28]

An organizational chart of Haldane's arrangement would therefore take the form shown in figure 1.2. Problems passed from left to right on the chart,

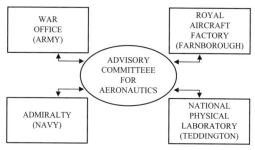

FIGURE 1.2. The Advisory Committee for Aeronautics and its institutional context. The Advisory Committee was founded in 1909 and reported directly to the prime minister.

from the Admiralty and War Office through the ACA to the National Physical Laboratory and the Royal Aircraft Factory. After experiments and tests had been completed, according to a schedule agreed on with the ACA, information and answers were passed back, from right to left on the chart, in the form of confidential technical reports. After these were discussed and agreed on by the ACA, and any required amendments had been made, the outcome was to be published in the form of a numbered series called Reports and Memoranda—a series that, over the years, ran into thousands and was to become famous for its depth and scientific authority. Each year the Advisory Committee presented an annual report containing an overview of its activities to which was attached, as a technical appendix, a selection of the more important Memoranda.

With the passage of time, and the increased workload imposed on the ACA, the original committee was broken down into a number of subcommittees to which further experts were recruited from the universities, Farnborough, and Teddington. Thus there was an Aerodynamics Sub-Committee, an Accidents Sub-Committee, an Engine Sub-Committee, a Meteorological Sub-Committee, and so on. Sometimes the subcommittees were further broken down into panels, such as the Fluid Motion Panel, which was part of the Aerodynamics Sub-Committee. Such a structure may seem complicated and bureaucratic, but viewed with the benefit of hindsight, it proved highly effective.

The Personnel of the ACA

Writing in 1920, R. T. Glazebrook, the director of the National Physical Laboratory, recalled the events of 1909 and the inception of the Advisory Committee.[29] Mr. Haldane, he said, "appealed to Lord Rayleigh and myself to know if we could help at the National Physical Laboratory. A scheme of work was suggested, and at a meeting at the Admiralty at which Mr. McKenna, then First Lord, and Mr. Haldane were present, the details were agreed upon" (435). Haldane had made the acquaintance of John William Strutt (Lord Rayleigh) while they had worked together on an earlier committee, the Explosives Committee, developing an improved and less corrosive propellant for the artillery.[30] Rayleigh was a world-renowned physicist with formidable mathematical powers. He had published on aeronautical themes and had made fundamental contributions to fluid mechanics, the branch of physics that might explain how the flow of air over a wing could keep the machine in the air.[31] Rayleigh was to become the president of the Advisory Committee for Aeronautics.

Given that Rayleigh was already the president of the National Physical Laboratory, so Glazebrook (fig. 1.3), as the director of the NPL, was an obvious choice to work under him as the chairman of the Advisory Committee for Aeronautics. What of the other members? There were, of course, representatives of the Admiralty and the War Office. These were Major General Sir Charles Hadden for the army and Captain (later Admiral) R. H. S. Bacon

FIGURE 1.3. Richard Tetley Glazebrook (1854–1935). Glazebrook was chairman of the Advisory Committee for Aeronautics from its inception in 1909. A long-standing colleague of Rayleigh's, he was also the head of the National Physical Laboratory. (By permission of the Royal Society of London)

for the navy. (Haldane knew them both from the Committee of Imperial Defence.) Bacon was soon to be joined (and then replaced) by Captain Murray Seuter. Mervin O'Gorman, an engineer, joined the committee after his appointment as the new head of the Balloon Factory and then the Aircraft Factory.[32] The remaining six members were Sir George Greenhill (mathematician), Dr. W. Napier Shaw (physicist and meteorologist), Horace Darwin (brother of Sir George Darwin and the founder of the Cambridge Scientific Instrument Company), H. R. A. Mallock (physicist), Prof. J. R. Petavel (engineer), and F. W. Lanchester (an engineer who had published a pioneering book on aerodynamics).[33] The secretary of the committee was F. J. Selby, who had got to know Glazebrook when he went up to Trinity in 1888 and took classes from Glazebrook in physics and mathematics. In 1903 Selby joined the staff of the National Physical Laboratory and acted as Glazebrook's personal secretary.[34] The assistant secretary was J. L. Nayler, again of Cambridge and the NPL, who would coauthor a number of the early experimental reports.

Over fifty years after the founding of the ACA, Nayler gave a talk in which he recalled some of the personalities involved.[35] He brought out clearly the closely knit character of the core group of scientists and the intellectual tradition to which they belonged. Glazebrook and Shaw, he said, had been assistants to Rayleigh in Cambridge, when Rayleigh had taken over the Cavendish laboratory after the death of Clerk Maxwell. Glazebrook and Shaw had written textbooks of practical physics together, though "they both started as mathematicians."[36] Mallock had also worked as an assistant to Rayleigh.[37] Nayler went on: "Seven out of the twelve were Fellows of the Royal Society and another became a Fellow later on, three were serving officers, two were heads of aeronautical establishments. Six, including the Secretary, began as wranglers, and five were trained engineers" (1045). The term "wrangler" is an old Cambridge label for a student who came at the top of the result list in the university's highly competitive mathematical examination called the Mathematical Tripos. The senior wrangler was at the very top of the list, followed by the second wrangler, and so on. Nayler notes that Rayleigh was a senior wrangler in 1865. Glazebrook and Shaw "were wranglers in the same year, 1876," while Greenhill was "a second wrangler and Smith Prizeman" (1045). The award of the Smith's Prize was an opportunity for the two or three top scorers in the Tripos to meet in a final contest in the battle for mathematical supremacy.

Nayler's talk was given at Cambridge, which may explain some of the care taken to delineate the connections between the Advisory Committee members and the Mathematical Tripos. But we should not miss the significance or the specificity of the message. It would be difficult to overstate the centrality of the Mathematical Tripos to the scientific life of Cambridge at the end of the nineteenth and the beginning of the twentieth centuries. Andrew Warwick's impressive study *Masters of Theory: Cambridge and the Rise of Mathematical Physics* leaves no doubt about the intensity and brilliance of the Tripos tradition.[38] In many areas of science in Cambridge, success in the Tripos was a precondition of scientific and academic preferment. The order of merit was published in the *Times*, and the senior wrangler of the year achieved celebrity status and, invariably, the offer of a college fellowship. Such success could only be achieved by intense coaching from one of a select band of brilliant but demanding tutors—men such William Hopkins, Edward John Routh, William Besant, R. R. Webb, and Robert Herman.[39] Perhaps the greatest of all the coaches was Routh, whose caliber as a mathematician can be judged from the fact that, as a student, he had pushed James Clerk Maxwell into second place in the Tripos of 1854. Rayleigh had been coached by Routh, while Greenhill had been coached by Besant.

The top wranglers in turn would then become the examiners and the coaches for the next generation of students facing the rigors of the Senate House examinations. Lower-placed wranglers would often become schoolmasters and prepare their charges for mathematical scholarships to Cambridge, thus ensuring that each generation of students was better prepared than the last. This system increased the competition and forced up the standards still further. So extraordinarily high did these standards become that examiners would use their current research and their latest discoveries as the basis for their questions. The most celebrated example was a result that subsequently played an important role in the mathematical underpinning of aerodynamics. The mathematical equation, usually called Stokes' theorem, relating circulation and vorticity, first appeared in print in the Smith's Prize examination of 1854. The candidates were required to prove the theorem. Stated in words, the theorem is that "the circulation round the edge of any finite surface is equal to the sum of the circulations round the boundaries of the infinitely small elements into which the surface may be divided."[40] (The meaning of the technical terms involved, such as "circulation," are explained when the circulation theory of lift is introduced in chap. 4.)

The Tripos system certainly had its critics. It placed ambitious students under great strain, and many young minds found the demands too great. Critics also argued that the questions were too difficult for all but the best, and to rectify the problem various reforms and modifications were introduced over the years. The last order of merit list was published in 1909; thereafter, candidates were placed in classes and the names listed alphabetically within the classes. There were also those who argued that the difficulty of the

questions was because they were artificial and contrived so that their answers
called for the mere mastery of technical tricks that had little educational value.
This line taken in 1906 by G. H. Hardy, although his claims should be put in
context. They were made in the course of arguments about the reform of the
syllabus.[41] Hardy was pressing the case for pure mathematics and rigorous
foundations as against the applied mathematics and mathematical physics of
the traditional Tripos.[42] It was the more traditional form of the Tripos that
informed the training of Rayleigh, Glazebrook, Shaw, Greenhill, and others,
such as Horace Lamb, who later joined the Advisory Committee. Lamb had
been coached by Routh and was second wrangler in 1872.

The scientific character of the Advisory Committee for Aeronautics, on
which Haldane placed so much emphasis, was clearly weighted toward the
Cambridge tradition of mathematical physics. In this connection, recall the
three names that Haldane mentioned as possible leaders of the committee:
Rayleigh, Moulton, and Darwin. Nayler noted that Rayleigh had been a senior wrangler, and to this one may add that Darwin, who had discussed matters with Haldane on the Saturday before the Esher committee had met, was
himself second wrangler in 1868.[43] What about Lord Justice Moulton? He
looks the odd one out. In fact John Fletcher Moulton was senior wrangler in
1868, soundly beating George Darwin into second place and achieving higher
marks than any previous Tripos candidate.[44] Rayleigh, Moulton, and Darwin
had all been coached by Routh. It would be wrong to say that the ACA was, or
was meant to be, simply a committee of wranglers. Lanchester was no mathematician, nor were the military men, but it would not be an exaggeration to
say that Cambridge wranglers were a powerful, and perhaps predominant,
presence. Whether by accident or design the scientific culture of the Advisory Committee was, to a significant degree, the culture of the Mathematical
Tripos.[45]

The Research Agenda

What was the scheme of work that was drawn up in the meeting at the Admiralty mentioned by Glazebrook? Some indication was given in the House
of Commons when questions were taken on the founding of the committee. More details, however, can be gleaned from a document titled "Preliminary Draft for Programme of Possible Experimental Work," dated June 1,
1909, and used as the basis for an interim report during the first year of the
committee's activities.[46] The program of work was divided into six sections:
I, "General Questions on Aerodynamics"; II, "Questions Especially Relating
to Aeroplanes"; III, "Propeller Experiments"; IV, "Motors"; V, "Questions

Especially Related to Airships"; and VI, "Meteorology." For our purposes, only the first two sections are of interest; within them, some fourteen distinct topics were identified. They repay scrutiny and call for some comment. The list of topics was as follows:

I. General Questions on Aerodynamics
   1. Determination of the vertical and horizontal components of the forces on inclined planes in a horizontal current of air, especially for small angles of inclination to the current.
   2. Determination of surface friction on plates exposed to currents of air.
   3. Centre of pressure for inclined planes.
   4. Distribution of pressure on inclined planes.
   5. Pressure components, distribution of pressure for curved surfaces of various forms.
   6. Resistance to motion of bodies of different shapes = long and short cylinders &c.
   7. Combination of planes: effect on pressure components of various arrangements of two or more planes.

II. Questions Especially Relating to Aeroplanes
   8. Resistance components of aeroplane models.
   9. Resistance of struts and connections.
   10. Resistance of different stabilising planes both horizontal and vertical.
   11. Problems connected with stability.
       (i) Mathematical investigation of stability.
       (ii) The stability of aero curves of different section and plane.
       (iii) Effect of stabilising planes.
       (iv) Effect of sudden action.
       (v) Effect of gusts of wind.
       (vi) Investigations as to stability of models for different dispositions of weight etc.
   12. Materials for aircraft construction.
   13. Consideration of different forms of aeroplane: monoplane, biplane, etc.
   14. Other forms of heavier than air machines, helicopter, etc.

The list conveys the range of problems confronting the committee but also something of its priorities and preferences. The emphasis to be placed on stability stands out clearly in the degree of definition accorded to the problem, which is carefully divided into six subsections. Likewise, the scientific style of the approach is clear. Experimentally, a significant amount of the work was to be done with models while, theoretically, the complexity of the real flying machine was replaced by simplified concepts such as planes and cylinders and centers of pressure.[47] The operation of an aircraft was being assimilated to the abstract categories familiar to the committee from their Tripos textbooks on mechanics and hydrodynamics.

The Reception of the ACA

Before we look at the early Reports and Memoranda generated by this research program, something should be said about the public and professional reception that was given to Haldane's new committee. If there was a cautious welcome in some quarters, elsewhere bitter disappointment was expressed that the commercial manufacturers of aircraft and the pioneers of flight (who were usually the same men) were not represented. To these critics the ACA was just a committee of professors, not of producers. Even the inclusion of Lanchester did not satisfy the critics. He had written books on airplanes, but these were dismissed as theoretical works. He had not built airplanes, only motorcars (and some of the critics didn't even like his cars). Where were the names of Britain's aeronautical pioneers, such as Handley Page, Fairey, Roe, Rolls, Short, or Grahame-White?[48]

To prepare the ground for the prime minister's announcement of the Advisory Committee, Haldane had written on May 4, 1909, to the newspaper magnate Lord Northcliffe, who had been agitating for government action. "We have," Haldane said, "at last elaborated our plans for the foundation of a system of Aerial Navigation." The government had created "a real scientific Department of State" for its study. In his reply of May 9, Northcliffe was dismissive. He gracelessly declared that the composition of the committee "is one of the most lamentable things I have read in connection with our national organisation." He conceded that Rayleigh was a good choice as chairman, but "the Committee should certainly include the names of some of the now numerous English practical exponents." As for Lanchester, he was known to be critical of the Wright machine,[49] about which Northcliffe was enthusiastic, and was "the same Mr Lanchester, I understand, who is responsible for . . . one of the most complicated motor cars we have ever had." This was Northcliffe writing to a member of the Cabinet; when corresponding with his political cronies he simply referred to the Advisory Committee as Haldane's "collection of primeval men."[50]

In reply to Northcliffe (on May 18) Haldane said that the advice he had been given had convinced him "of what I was very ready to be convinced, that here as in other things we English are far behind in scientific knowledge. The men you mention are not scientific men nor are they competent to work out great principles: they are very able constructors and men of business. But in this big affair much more than that is needed."[51] Predictably, this response failed to mollify Northcliffe and his friends, such as J. L. Garvin of the *Observer* ("too many theorists"; May 7). Nor were they alone in their negative response. The Tory *Morning Post* of May 7 had declared that "too much value has been attached to the purely theoretical side, while no evidence is forthcoming that the practical side will be advanced at all." In an interview for the *Post*, the aviation and motoring enthusiast Lord Montague said that "the Commission is composed of theoretical and official people as distinct from practical men. . . . I do not recognize the name of any man on it of actual practical experience."[52] The journal *Flight* joined in and got its revenge for its failed prediction over the Wright brothers' contract: "It is a bad system to encumber enterprise by establishing 'Boards of Opinion.' The opinions of the practical men who are doing the work are worth more to the nation than those of a miscellaneous collection of scientists."[53]

A more positive response to the committee was to be found in a short article in the pages of *Nature* on May 13, 1909.[54] It was by the brilliant and opinionated mathematician George Hartley Bryan, himself a wrangler and a former fellow of Peterhouse College.[55] Bryan welcomed the creation of the Advisory Committee, saying: "It is clear . . . that mathematical and physical investigations are to receive a large share of attention, and that the mere building of aeroplanes and experience in manipulating them are not to interfere with the less enticing and no less important work of finding out the fundamental principles underlying their construction" (313). The problem of stability, he noted with satisfaction, had been singled out for attention, though the "mathematics of this problem are pretty complicated" (313). Bryan was not surprised that newspapers were complaining that the committee was too theoretical in its orientation and that the "practical man" was not properly represented. The real problem, said Bryan, was not too much theory but too many publications that contained equations and algebraic symbols written by people who did not understand mathematics. "Indeed, in many cases it is the 'practical man' who revels in the excessive use and abuse of formulae, and the mathematician and physicist who would like to bring themselves in touch with practical problems are consequently deterred from reading such literature" (314). There was an urgent need, Bryan concluded, "for a clear division of labour between the practical man and the physicist" (314). The failure to create such a division, he argued, had already cost England the loss of its chemical and optical industries, and France had a long head start in automobiles. Now at last there was a chance to make up the ground in aeronautics.

Bryan's reaction was just the kind that Haldane would have been hoping for, though the reference to the "mere building of aeroplanes" was hardly politic. These two initial responses—that of Northcliffe and his allies and that of Bryan—indicate the tension surrounding the ACA. They also serve to introduce some of the labels that were used at the time to signalize the different and opposed parties. The term "practical man" does not refer to a cloth-capped artisan but primarily to engineers and entrepreneurs, and included, for example, the Hon. Charles Stewart Rolls. The label was a badge of honor intended to mark the contrast with university academics, civil servants, and others with no direct involvement in market processes.[56] I follow out some of the further expressions and consequences of this social divide.

The "Reptile Aeronautical Press"

Two related complaints came together in the arguments that were mobilized on behalf of the "practical man." First, it was said that the designers of airplanes did not need mathematical knowledge and that persons who did possess such knowledge were mere theorists who were ill equipped to deal with real problems. The National Physical Laboratory, it was said, was staffed by mathematicians and theorists rather than engineers with practical experience. Second, those who were in the employ of the government led cushioned and subsidized lives that protected them from the bracing rigors of the market. This complaint included the staff at the NPL but was particularly directed at the Royal Aircraft Factory.

Within a year of the ACA's founding, the aeronautical press was asking for evidence that the committee's work was bearing fruit. The anonymous writer of an editorial in *Aeronautics* in 1910 lamented that "it is too late to renew criticism of the composition of the Committee or of the limitations placed on its work; it will be sufficient once again to place on record our opinion that in both respects the Committee is bound largely to be a failure so far as results of immediate practical value are concerned."[57] In an article on aeronautical research in the *Aeroplane* of August 31, 1911, P. K. Turner, a regular contributor, said it was time that theory and practice were brought together. Was this not why institutions such as the National Physical Laboratory had been set up? "But it appears that the workers at these institutions, like the monks of old, are growing fat and useless; and of all the shameful wastes perpetuated in our alleged civilisation, the worst, in my eyes, is an equipped factory, laboratory, or office, where owing to the incompetence of those in charge or the laziness of their subordinates or both, or vice versa, nothing is done."[58] The issues came to a head at the military air trials of 1912 held on Salisbury Plain. Private contractors, British and foreign, were invited to enter their aircraft into a competition to see which ones best met the performance criteria laid down by the military. The competition was organized by O'Gorman, though the terms of the competition precluded the government Factory from formally entering its own designs. Although not an official competitor, a Factory model, a tractor biplane called the BE2, was put through its paces during the trials and informally faced the same tests as the others. The designation BE came from a system of classification developed by O'Gorman. The E stood for "experimental"; the B for "Blériot-type" and referred to the position of the engine at the front of the aircraft. Pusher aircraft with the engine behind the pilot were given the designation F after the pioneer Henri Farnam. The BE2 was designed by Geoffrey de Havilland, who had been taken onto the staff of the Factory after successfully building some aircraft of his own. The BE2's performance manifestly outclassed that of the entries from private firms. It was reasonably stable, made good speed, and de Havilland even set a new altitude record with the BE2 during the trials. The official winner was a machine entered by Cody, but the War Office proceeded to ignore the competition and focused its interest on the superior machine, even though it had been precluded from official entry. It offered contracts to private constructors to build not their own machines, but twelve of the government-designed BE2s. In an improved form this machine became, for a number of years, the mainstay of the Royal Flying Corps. Figure 1.4 shows a side elevation of the BE2A.

For the government's critics the policy of contracting out government designs constituted an outrage. It was denounced as an attack on private enterprise that strangled all design initiative.[59] But private enterprise had put up a miserable showing at the air trials. As one eyewitness noted: "Of the seventeen British aeroplanes that were nominally in evidence, at least seven of the newer

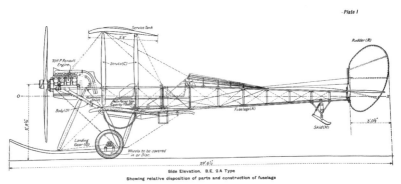

FIGURE 1.4. The BE2 was developed by the Royal Aircraft Factory and played a central but controversial role in the British war effort. The highly stable BE2 has been called one of the most interesting airplanes ever built. This side elevation is from Cowley and Levy 1918.

makes were either unfinished or untested on the opening day, and thus some
of the very firms for whose benefit the trials had, in a measure, been organized, spoiled their own chances in competition with the older constructors
who, for the most part, had entered well-tried models."[60] An editorial in the
Aero for September 1912 admitted: "It is undoubtedly a fact that the majority
of our home manufacturers have not gained in reputation through partici-
pating in the military trials."[61] The same point was conceded by an editorial
in *Flight* in which it was acknowledged that the BE2 was "one of the best flyers ever produced." Of the firms that were granted a contract to produce the
BE2, the editorial continued, "not everyone could have as readily justified a
similar demand for its own machines on demonstrated merits in the Military
trials."[62] The War Office decision was reasonable; if anything it was more ac-
commodating of the sensibilities of the private manufacturers than it should
have been. Looking back from 1917, an editorial in the journal *Aeronautics* acknowledged that "before the War there was in the whole country not a single
decently organised aircraft manufacturing firm."[63] For example, the Handley
Page Company had accepted orders to produce five of the twelve BE2s, but
it failed to deliver even this small number. Only three of the machines had
been delivered by 1914.[64] If the editorial in *Aeronautics* is to be believed, the
situation at Handley Page was the rule rather than the exception; not a single
manufacturer of the BE2s made its delivery on time. The editorial went on:
"We are not blind to the faults of the Royal Aircraft Factory, which are of a
nature which seems inseparable from any state-owned institution. Nor do
we ignore the fact that the Factory was bitterly detested and thoroughly dis-
trusted by the industry at large. But truth compels us to recognise the fact
that the industry was chiefly responsible for its own grievances" (185). The
writer of this passage was probably J. H. Ledeboer, the editor of *Aeronautics.*
The self-critical tone was hardly typical of most of the polemics unleashed
against the Royal Aircraft Factory, though the assumption that government
institutions would be inferior to those of the business world certainly was.

In Parliament the aircraft manufacturers had the support of, among others, William Joynson-Hicks and Arthur Hamilton Lee, both Conservative
MPs, and Noel Pemberton Billing, an independent MP and founder of the
Supermarine company. Billing had conducted his theatrical campaign for
election on the basis of his commitment to, and knowledge of, aviation and
all matters relating to it. The various critics of the government did not always
agree with one another, but their combined voice was loud and persistent.
Week after week, and year after year in the pages of the aeronautical journals,
in Parliament, and in the right-wing press, they directed their anger and con-
tempt at the Advisory Committee, the National Physical Laboratory, and the

Royal Aircraft Factory. Every setback, every accident, every tragedy was used as a stick with which to beat the government and as proof of the inferiority of government design and construction compared with private enterprise.

The BE2, and everyone associated with it, became the objects of a campaign of denigration. As so often happened in the early years of aviation, accidents occurred, and a number of persons flying the BE2 were tragically killed. One was Lt. Desmond Arthur, whose BE2 broke up in the air at Montrose at 7:30 a.m. on May 27, 1913. Lt. Arthur was a friend of C. G. Grey, the editor of the *Aeroplane*, and the loss fed Grey's state of permanent anger against the government.[65] Another victim was E. T. "Teddy" Busk, a brilliant engineering graduate from Cambridge who had joined the Royal Aircraft Factory in the summer of 1912. Busk had been conducting a program of experiments designed to improve the stability of the BE2 when the machine he was piloting caught fire.[66] This accident, and the other fatalities, provided the critics with the excuse for which they were looking. "The Victims of Science" was the headline in the *Aeroplane* of March 19, 1914.[67] Grey (see fig. 1.5) exploited the opportunity to the full. He argued that it was the scientific approach to airplane design that had killed these unfortunate men. He wrote: "I submit

FIGURE 1.5. C. G. Grey, editor of the *Aeroplane* and vehement critic of the Advisory Committee for Aeronautics. (By permission of the Royal Aeronautical Society Library)

that if the Department of Military Aeronautics will hold an enquiry into the design and construction of Mark BE2 biplanes and will take the evidence of workshop foreman and practical constructors—apart from the scientists and theoreticians—among contractors who are building the BEs they will obtain sufficient criticism to condemn almost every distinctive feature of the BE— provided always they can guarantee that in the event of the practical men speaking their minds they will not jeopardise their firm's chances of obtaining further orders" (320).

Grey was an accomplished polemicist and he took care to cover himself lest the criticisms he was confidently predicting were not forthcoming. He implied that this could only mean that sinister, government forces were suppressing them. Having secured his line of retreat, Grey then asserted that the deaths had been caused by criminal negligence and he knew who the criminals were: "Those responsible are the people, if you please, who have 'the best brains in the world,' and through whom aeroplane design is to excel. These are the people who base their calculations on the theories of the armchair airmen of the National Physical Laboratory" (321). When the Aeronautical Society opened a subscription to honor Busk, Grey accused it of exploiting the young man's death. He had an unpleasant talent for criticizing others for what he was doing himself.[68]

Political attacks on the aeronautical establishment became even more intense after the start of the war in 1914. The summer of 1915 saw the Fokker Scourge. Anthony Fokker, a Dutch designer working for the Germans, had developed a forward-firing machine gun, synchronized to fire through the propeller disc. He fitted it to an otherwise undistinguished monoplane, and the new arrangement marked the emergence of the specialized "fighter aircraft." It gave the Germans a marked advantage and, for a while, increased the losses of British pilots and machines. The BE2, whose stability compromised its maneuverability, was no match for the Fokker *Eindecker*—not, at least, when the Fokker was flown by the particularly skilled pilots to whom it was selectively assigned. By any standards this issue was one of importance for a country at war. Rhetorically, however, it became another opportunity to voice the interests of the aircraft manufacturers. In the House of Commons, Pemberton Billing denounced the government and military authorities as "murderers." His claim was that if the young men of the Royal Flying Corps had been given machines designed and built by private firms rather than government agencies, they would be alive today.

It would be a study in its own right to trace all the twists and turns of the protracted, political campaign conducted by Billing, Grey, and the manufacturers, and it would be no easy matter to decide, in every case, which complaints had substance and which were unscrupulous exaggerations and self-serving falsehoods. Given the seriousness of Billing's allegations, and the place in which they were made, it was inevitable that official inquiries had to be launched. Two issues had to be unraveled: (1) was the Royal Flying Corps conducting its military business properly? and (2) was the Royal Aircraft Factory dealing improperly with the private manufacturers? The Burbridge Committee addressed the first problem and the Bailhache Committee the second. During the course of these inquiries, Pemberton Billing's behavior became so eccentric and evasive that even former supporters began to back away. *Flight*, which had previously welcomed his election, decided that the talk of the "deadly Fokker" was a gross exaggeration and suggested that there must be some ulterior motive.[69] Soon the editorials were dismissing his indictments as "irresponsible" and "sensational" rather than "the measured views of a man in earnest for the welfare of his country."[70]

The official inquiries could find no basis in fact for Pemberton Billing's accusations against the Royal Flying Corps, but he and the other critics effectively "won" the argument against the Royal Aircraft Factory.[71] The report conceded that there had been inefficiencies. The tepid defense of the Factory meant that the interests represented by the critics ultimately prevailed.[72] *Flight*, which had frequently been supportive of the Advisory Committee, the National Physical Laboratory, and the Factory, now concluded that the rights of the manufacturers had indeed been encroached upon.[73] The editorial column proudly affirmed the principle that private enterprise was always superior to government, and then promptly asked for government subsidy for the aircraft industry.[74] The government acceded to the critical pressure of the manufacturers and the Aircraft Factory was turned exclusively toward research rather than design. O'Gorman was removed, and his team of designers and engineers dispersed into the private sector. An impartial assessment of the rights and wrongs of the issue would, however, have to note that, before the restriction on its activities, the personnel of the Royal Aircraft Factory had produced one of the most outstanding fighter aircraft of the war, Henry Folland's SE5. (In O'Gorman's nomenclature the S stood for "scout" and the E for "experimental.") Folland and his colleagues had skillfully balanced the competing demands of stability and maneuverability to produce one of the most formidable fighting machines of the war and an aircraft that was a match for anything its pilots might meet.[75]

The critics also "won" in that, by 1915, they had managed to hound Haldane out of office.[76] Haldane was denounced in the right-wing press as a pro-German sympathizer. He was said to have opposed and delayed the dispatch of the British Expeditionary Force to France in 1914 and to have known of the

German war plans without informing his Cabinet colleagues. There was not a shred of truth in any of these allegations. Indeed, it was only thanks to Haldane's earlier army reforms that the country had a viable expeditionary force at all. The charges even alluded to a secret wife in Germany and to Haldane being an illegitimate half brother of the kaiser. It was ludicrous and vile but it worked, and the "reptile aeronautical press," as O'Gorman justifiably called it, played its part in the affair with enthusiasm.[77] The episode must count as one of the most disreputable in twentieth-century British politics.[78]

During the Great War enormous social pressure was placed on men to contribute to the war effort and not to shirk their patriotic duty to lay down their lives on the field of battle. Pacifists and critics of the war were reviled. This practice was routine in the aeronautical press.[79] Grey was happy to mobilize the hatred of "trench-dodgers" and use it against those who, instead of being at the front, were working at Farnborough and Teddington. He reprinted an article from the *Times* (under the title "The Farnborough 'Funk-Hole'") asking why the fit young men seen coming in and out of the Aircraft Factory were not in France.[80] The theme was taken up again when reviewing the Advisory Committee's report for 1917. As usual, said Grey, the report is devoted to the glorification of the National Physical Laboratory, though he noted with approval that two "practical men of proven merit" had been brought on to the Engine Sub-Committee and the Light Alloys Sub-Committee.[81] He was, however, censorious of those who were performing basic, hydrodynamic experiments, for example, those involving water tanks. Making play with the word "tank," Grey declared that "if some of these able-bodied young men were to take a course of experimental work in motor-tanks at the front they would confer greater benefits on their native land" (315). Meanwhile Grey was corresponding with Winston Churchill to plead for his own exemption from the inconvenience of wartime obligations. Judging from surviving letters, now in the archives of the Royal Aeronautical Society, Churchill, though brief and formal in his responses, duly obliged, and through his intervention Grey got the exemptions for which he had asked.[82] *Dulce et decorum est pro patria mori.*

Theory and Experiment

I now move from the context to the content of the Advisory Committee's work to see how it carried out the research program it had originally set itself. Mr. Asquith assured the House of Commons on May 20, 1909, that the new committee would pursue the problems of aeronautics "by the application of both theoretical and experimental methods of research."[83] No significance

should be attached to the word order, placing theory before experiment, because both found vigorous expression, although the relation between theory
and experiment assumed very different forms in the different areas of the
committee's work.

Important tests on full-scale aircraft were carried out at Farnborough,
but the main arena in which theory and experiment confronted one another
was the wind channel (and sometimes the water channel) in which fluid flow
over model wings and model aircraft could be observed and measured. The
National Physical Laboratory already had a small water channel, and even
a small vertical air channel, but the first task of the ACA was to oversee the
construction of a better and more modern horizontal air channel to match
those already known to be in use in Paris and Göttingen. By the end of the
first year they were able to report on their plans to build a 4 × 4 × 20-foot
channel with a draught of nearly 50 feet per second produced by a fan of 6 feet
in diameter.[84]

Difficulty was experienced getting a steady flow, but by keeping the velocity down to 30 feet per second, the flow was found to be "satisfactorily uniform." The measuring apparatus for registering the aerodynamic forces on
various plates and models was also ready. It was now possible to measure the
force component perpendicular to the flow (the lift) and that in the direction
of the air current (the "drag" or "drift"). The apparatus could also be set up
to determine centers of pressure, and the model could be adjusted to be at
any angle with the current without stopping the flow.[85]

How was the apparatus to be used? Would it be employed to study the
behavior of wings and other models in a purely empirical manner to build up
an inductive knowledge of the regularities in their behavior? Or would it be
used in a theory-testing manner for work that started not with the observable
facts but with some theoretical conjecture? If the latter, what theories would
be tested and where would they be found? The answer is that both strategies
were present in the empirical work. Many of the measurements on model
wings involved the highly empirical, and essentially inductive, engineering
method of "parameter variation," that is, systematically altering one factor at
a time.[86] For example, in one of the studies of a model biplane, the procedure
involved keeping the sections, spans, chords, and the distance between the
wings constant while altering the angle of stagger in order to try to isolate
its effect on lift.[87] But there were also bodies of important and sophisticated
theoretical work waiting to be explored. The provenance of this theoretical
work lay almost exclusively in the achievements of Cambridge mathematical physics. Predictably, the orientation toward the fundamental, theoretical
problems of aerodynamics was swept aside in 1914 by the demands of the

war, which gave precedence to short-term, practical investigations. Before the cataclysm, however, in the period between 1909 and 1914, theory testing provided the focus for much of the research.

The theories in question concerned two general areas: (1) stability and control and (2) lift and drag. They therefore lay in two quite distinct areas of physics—one being grounded in rigid-body mechanics, the other in fluid dynamics. I consider them in turn, beginning, in this chapter, with the work on stability and, in the next chapter, moving to the fluid dynamics underlying the theory of lift and drag.

Stability and Routh's Discriminant

As the minutes of their first meeting show, G. H. Bryan had been in touch with the Advisory Committee and, though not a member, was considered central to their effort to understand stability.[88] Bryan (see fig. 1.6) was a versatile applied mathematician who wrote on thermodynamics and fluid dynamics but had become interested in aeronautics through contact with Sir Hiram

FIGURE 1.6. George Hartley Bryan (1864–1928), a British pioneer in the analysis of aircraft stability. Bryan applied the mathematical techniques that had been developed by his Cambridge coach Edward Routh. (By permission of the Royal Society of London)

Maxim and the pioneer glider flyer Percy Pilcher.[89] Bryan had been publishing calculations on stability since 1904 and was, without doubt, the leading British authority in the field.[90] It was failure to understand stability, he argued, that led to so many fatal accidents.[91] Lanchester had written on stability, but in Bryan's eyes, and judged by Tripos standards, this work "certainly appears wanting in rigour."[92] Lanchester's approach was original, conceded Bryan, and he avoided the errors that had vitiated many other attempts, but he did not deduce his conclusions from clearly stated assumptions. Describing how he had arrived at his own, highly mathematical, analysis Bryan recalled that "about the year 1903 I noticed that if a glider or other body is moving in a resisting medium, such as air, in a vertical plane with respect to which it is symmetrical, the small oscillations about steady motion in that plane are determined by a biquadratic equation; and Prof. Love directed my attention to the condition of stability given by Routh."[93] A quadratic equation has the form $ax^2 + bx + c = 0$, whereas a biquadratic of the kind referred to by Bryan has a term in x^4 and takes the form $ax^4 + bx^3 + cx^2 + dx + e = 0$. When Bryan said that he "noticed" that the oscillations of a glider were determined by a biquadratic equation, he did not mean that he drew this conclusion simply by looking at a model glider in flight. He meant that he noticed this mathematical fact in the course of using Newton's laws to write down the general equations of motion of a body, such as an airplane, moving with a specified velocity and subject to specified forces such as gravity, lift, and drag.

Bryan posed the following question: If an airplane was in steady flight and hence in dynamic equilibrium, and was then subject to small, disturbing forces, such as a gust of wind or a sudden alteration of the control surfaces, what would happen? Would the disturbance die away or would it get bigger and bigger? If the disturbance died away, the machine would count as stable; if the result was that the disturbance became amplified and disruptive, then the machine would count as unstable. He treated the airplane as a rigid body subject to forces of acceleration and rotation. Given the force of gravity and the aerodynamic forces to which it was subject, how did this mechanical system respond? What sort of longitudinal or lateral oscillations would follow from the disturbance? At this point Prof. Love stepped in. A. E. H. Love, a student at St. John's, was second wrangler in 1885 and first Smith's prizeman in 1887. He later became professor of mathematics at Oxford.[94] Love apparently reminded Bryan that the techniques and concepts he needed to answer his question about stability had already been worked out by Routh, who had been Bryan's old Cambridge coach. E. J. Routh's Adams Prize essay of 1877 and his textbook, *Dynamics of a System of Rigid Bodies*, contained a general analysis of stability for mechanical systems. Both of these works had shown the importance of a mathematical device that came to be known as Routh's discriminant, an expression whose negative or positive value indicated the stability or instability of the system under analysis.[95]

Following Routh's methods, and citing Routh's results, Bryan was able to reduce the problem of the stability of an aircraft subject to small disturbances to the behavior of an equation of the general form

$$A\lambda^4 + B\lambda^3 + C\lambda^2 + D\lambda + E = 0,$$

where λ was the symbol for the modulus of decay or the strength of the damping tendency on the oscillations that were being investigated. (This equation in λ was the "biquadratic" that Bryan "noticed.") The coefficients A, B, C, etc., in Bryan's equation were complicated mathematical expressions involving terms that were called "resistance derivatives" and "rotary derivatives." These described the rate of change of the various forces, and their leverage on the aircraft, relative to its varying conditions of speed and orientation. The values of the derivatives, and hence the values of the coefficients A, B, etc., depend on the details of the particular machine. They could not be calculated from first principles but could be given numerical values on the basis of appropriate measurements made on models in a wind channel.

An examination of the four roots, that is, the values of λ that satisfy the equation, would determine whether the machine were stable. As Bryan put it, "the small oscillations . . . are determined by an equation of the fourth degree, so the conditions for stable steady motion are those obtained by Routh."[96] Routh had discovered the general result that the stability of an oscillating system required that the coefficients A, B, C, D, and E should all be positive and that the quantity $BCD - AD^2 - EB^2$ should also be positive. This latter expression was called Routh's discriminant. Abstract though it was, it cast light on design features that unwittingly rendered many aircraft dangerous to fly and prone to accidents. The proper mathematical understanding of an aircraft in terms of this equation, argued Bryan, could diminish the risks. In his 1904 paper he had recommended that mathematical investigations should be carried out on any "aerial machines that may be designed or constructed" (115) before they take to the air. Like Haldane, he had no reservations about asserting the priority of theory over practice.[97] Bryan's studies culminated in 1911 in a treatise titled *Stability in Aviation*.[98]

Rayleigh used to say that when he hit a hard mathematical problem he would pick up pen and paper, call to mind his old coach, and "write it out for Routh."[99] This may also have been Bryan's procedure. That he too was writing it out for Routh is suggested by the way he echoed the title of Routh's book when he projected a second volume to follow from his own 1911 book.

Bryan intended to call the combined, two-volume work *The Rigid Dynamics of Aeroplane Motions*. The aim was to carry the analysis into much more difficult problems, such as that of circling and helical flight, which would generate an equation with terms involving λ to the power eight.[100] Whatever the underlying psychological processes, however, there can be no doubt that the skills honed in the Tripos classes and coaching rooms of Cambridge were about to be given a new application, and one whose potential importance would be inestimable.

If he were given the right empirical data about an aircraft, Bryan was in a position to make predictions about its stability. Now the question became: Were those predictions correct? It was not evident, a priori, that even Bryan's sophisticated mathematics would capture the complex reality of the behavior of a real aircraft. At a discussion at the Aeronautical Society, Greenhill, with considerable experience in ballistics to back up his words, expressed his concern that gyroscopic effects such as those from the engine and propeller had been neglected. "I must confess it alarms me," he said in response to an exposition of the theory by E. H. Harper, a co-worker of Bryan's, "that w, p, q, have no influence on u, v, r, especially with gyroscopic influence," where the first three letters referred to rotations around the axes of the aircraft and the latter three to velocities of translation along those axes. Greenhill could not resist a further dig at Bryan by adding that of course the pioneers of flight "could not wait for the solution of a differential equation or its determinantel quartic." Greenhill's reservations could only have emboldened the "practical men" in the audience, who also suspected that all manner of simplifications must have been introduced into the calculations. Bryan's colleague and representative was questioned closely by Handley Page and others. What about the tangential forces on the wings? Would this approach be of help designing a new machine rather than comparing two given machines?[101]

Such suspicions were shared by the reviewers of *Stability in Aviation* in the scientific press. The review in *Nature* was signed W.H.W.[102] The writer was clearly impressed by the book but drew attention to the problematic relation between mathematics and reality, and to Bryan's uncompromising attitude. The reviewer quoted the following passage, observing dryly, "it strikes the keynote of the book itself." In this book, said Bryan,

> attention is concentrated on the mathematical aspect of the problem for several reasons. In the first place, there is no obvious alternative between developing the mathematical theory fairly thoroughly and leaving it altogether alone; any attempt at a *via media* would probably lead to erroneous conclusions. In the second place, the formulae arrived at, even in the simplest cases, are such that it is difficult to see how they could be established without a

mathematical theory. In the third place, there is probably no lack of competent workers in the practical and experimental side of aviation, and under these conditions it is evident that the balance between theory and practice can be improved by throwing as much weight as possible on the mathematical side of the scale.[103]

Bryan's position, first stated in his 1904 paper, was that even if the analysis was wrong, provided it was not too wrong, it would provide a "basis of comparison" and the means for interpreting experimental results "in their true light" (100). As for the problem created for his theory by gyroscopic effects, of the kind that worried Greenhill, Bryan took the view that the fault was with reality not with his theory: "surely it may be left to practical men to get rid of these objectionable influences by proper balancing."[104] This attitude was precisely what worried W.H.W.

W.H.W. was probably Sir William H. White, FRS, an expert in naval architecture. If so, then the reviewer and Bryan had crossed swords before. At a heated meeting of the British Association in 1910, White had taken Bryan to task for insisting that mathematicians and practical men should stick to their own, separate spheres of activity. The report of this confrontation, as given in the *Aero*, is worth quoting:

> The advocacy of watertight compartments, so to speak, drew from Sir William White a strong protest against drawing any such sharp demarcation, for he conceived the existence of an engineer who was a mathematician and a mathematician who was an engineer. Sir William White was also somewhat severe on a suggestion made by Dr. Bryan that had the mathematical problems been sufficiently studied many, if not all, of the unfortunate fatal accidents to flying men would have been avoided, and that the practical man's refusal to work on these lines rendered the accidents the results of foolhardiness rather than bravery.[105]

Another reviewer of Bryan's book, this time in the *Mathematical Gazette*, went into the presuppositions behind the analysis of stability in some detail and remarked:

> the author is obliged to make a series of assumptions—that the air resistance on the planes are linear functions of the small changes in linear and angular velocities; that in steady motion they are proportional to the square of the velocity; that they are normal to the planes; that they are proportional to $\sin \alpha$; that the angle of attack α is small; that the pressure on an element of a narrow plane is independent of the motion of neighbouring elements, etc. Methods of approximation are also at times employed to simplify the algebra. The cumulative effect of small inaccuracies in each assumption may be considerable.[106]

Only experiment would reveal if the approximations were cumulative and failed to cancel out. If this were so, then the predictions would fail, however elegant the mathematics and however pure its Tripos pedigree.

A young scientist called Leonard Bairstow—a product of London University rather than Cambridge—led the wind-channel work at the NPL that gave empirical content to Bryan's equations. Working with Nayler and Bennett Melvill Jones, a Cambridge engineering graduate, Bairstow provided the data needed to attach values to the coefficients in the equations and hence to check on the viability of the assumptions behind the calculations.[107] The measurements were delicate, involving the timing of oscillations on models of complete aircraft supported on a spindle, and damped by a spring, when they were exposed, respectively, to still and moving streams of air. As might be expected with difficult experiments, there were problems behind the scenes that were not always apparent in the published reports. As the aircraft designer J. D. North pointed out, "torsional oscillations in the spindle connecting the model with the indicating or recording apparatus" was a disturbing factor and gave rise to "varying results with different moments of inertia of the apparatus."[108] Despite these complications, Bairstow's experiments seemed to show the models in the wind channel behaving in the manner predicted from Bryan's equations. There was a gratifying coordination between experiment and theory.

Because both the experiments and the theory concerned small disturbances, the results necessarily had their limitations, and the scope of the agreement between fact and theory was still open to discussion. Bairstow vigorously defended the work on stability by insisting that some, at least, of the limitations were "more apparent than real." Consider, for example, "the necessity for assuming infinitesimally small disturbances from the path of flight." A similar assumption had to be made, said Bairstow, invoking one of the classic achievements of mechanics, when setting up the differential equation for the motion of a simple pendulum. But the solution can then be "applied to oscillations of finite magnitude, without sacrificing any great proportion of accuracy."[109] The appeal to infinitesimal motions does not vitiate the empirical significance of the inquiry. Rhetorically this was a powerful comparison, and the move from infinitesimal to small, finite disturbances can be justified by the analogy. Cautious persons, however, would note that this argument still left the move from small finite disturbances to large finite disturbances unaccounted for. The inference from the stability of an aircraft under small disturbing forces to its stability when confronted with larger forces therefore remained problematic. Bairstow's colleague Melvill Jones, who worked on control during slow flying and stalling, and who was strongly supportive of the stability research program, nevertheless acknowledged that Bryan's equations became inapplicable under these circumstances.[110] Some experts also remained troubled by the points raised by the reviewers—that the forces and couples were assumed to depend on linear and angular velocities but not on accelerations.[111]

The most visible symbol of the British preoccupation with the problem of stability was the excellent BE2, the machine subject to so much hatred in the aviation press. Even here it could not be asserted that de Havilland's original machine had been stable because it had been designed according to Bryan's equations. It had not. The aircraft had been the result of good judgment and had then been further improved and, in the form of the BE2C, rendered inherently stable by subsequent trial and error. This result had been achieved not just by mathematics and the wind tunnel, but also through the dangerous flying experiments of Busk and his colleagues. Furthermore, whatever may be true of the relation between Bryan's equations and small-scale models, J. D. North continued to argue that the relation between these equations and full-size aircraft remained problematic. Speaking specifically of the longitudinal damping of the BE2, North said that it was "the only rotary derivative deduced from quantitative results" and insisted that even then it had "not shown good agreement with the estimated figures."[112] Added to reservations of this kind was a more general issue. Stability was only one of the competing virtues that might be desirable in a design. Maneuverability was another, and often incompatible, demand. Contrary to the critics, Bairstow always maintained that, scientifically and technically, the BE2C was "one of the most interesting aeroplanes ever built."[113] Despite the confidence of men such as Bairstow, however, the politics of stability would not be resolved in the laboratory.

The undoubted achievement of a greater understanding of stability was sufficient to impress at least one practical man, Archibald Reith Low, of Vickers (see fig. 1.7). Low was himself a pilot and had designed the Vickers "Gun Bus" of 1913, a machine that earned the accolade of being the first purely military aircraft.[114] Low had been to an evening lecture given by Bairstow to the Aeronautical Society on January 21, 1914. Bairstow reported on the NPL's stability work and illustrated the findings with model gliders.[115] In the discussion after the demonstration Low, who had previously expressed reservations about Bryan, affirmed his enthusiastic conversion. The NPL work, he said, "constituted a triumphant vindication of Professor Bryan and the Advisory Committee." He promised (generously if not perhaps entirely seriously) to spend the next couple of years digesting the theory of small oscillations and learning about Routh's discriminant. Low recalled that, despite the advances

FIGURE 1.7. Grahame-White type 10 aero-charabanc, 1912. A. R. Low is seated third from the left; J. D. North is fourth from the right. (By permission of the Royal Aeronautical Society Library)

he had made, Bryan had been laughed out of a British Association meeting by "so called 'practical' engineers." Low also expressed the hope that the "ignorant agitation" in the press would be stopped by the dawning realization on the part of those responsible that "there were problems in aviation that they had not begun to be able to understand."[116]

Certainly the scientists directly responsible for mastering the problem of stability were in no doubt about the value of their achievements—even if there was more work yet to be done. It was clear to them that mathematicians could now contribute to the design of inherently stable aircraft (and they were beginning to convince at least some practical men). In 1915 the Aeronautical Society awarded Bryan their Gold Medal and, if the subsequent history of aeronautics is to be the judge, the honor was well deserved: Bryan's equations are still used.[117] Understandably, Bryan was deeply grateful to those who had rendered his theory applicable. In a letter of February 21, 1916, he said it was an "extraordinary feat" that Bairstow and E. T. Busk and their colleagues had got inherent stability "into a sufficiently practical form to be incorporated into military aeroplanes." But, he went on, in the present wartime conditions it was necessary for everyone to keep working in both pure and applied research. In the prewar days, "Reissner and Bader were running us pretty hard on the mathematical side," so no one could be complacent. He ended with a warning: "the Germans are probably putting their best brains into improving their aeroplanes."[118] Bryan was right, and his sentiments did not fall on deaf ears. As Greenhill had said in 1914 in the pages of *Nature*, this was a "Mathematical War.[119] Despite the scoffing directed at mathematicians, the exponents of scientific aerodynamics were proud of their contribution to the understanding of stability and the progress that had been made. "It cannot be regarded otherwise," said W. L. Cowley and H. Levy, two of the leading experts at the National Physical Laboratory, "than in the light of a signal triumph for mathematical science."[120]

The Air as an Ideal Fluid:
Classical Hydrodynamics and the
Foundations of Aerodynamics

The following investigations proceed on the assumption that the fluid with which we deal may be treated as practically continuous and homogeneous in structure; i.e. we assume that the properties of the smallest portions into which we can conceive them to be divided are the same as those of the substance in bulk.

HORACE LAMB, *Treatise on the Mathematical Theory of the Motion of Fluids* (1879)[1]

Let me now prepare the ground for an account of the theory of lift and drag. The disputes over the correct analysis of lift and drag provide the central topic of this book. It was here that the scientists and engineers who addressed the new problems of aerodynamics called upon the highly mathematical techniques of what used to be called, simply, "hydrodynamics." The modern label, which better captures the true generality of the subject, is "fluid dynamics." Fluid dynamics provided the intellectual resources that were common to both the British and German work on lift and drag, although the stance toward that common heritage was often very different in the two cases. It is vital to have a secure sense of what the two groups of experts were disagreeing about. The present chapter is a description of this common heritage and these shared resources. It is meant to provide background and orientation. In it I do my best to explain the basic concepts in simple terms, though this hardly does justice to the ideas and techniques that are mentioned. I sketch some of the initial, mathematical steps that went into their construction in order to convey something of the style and feel of the work. At the end of the chapter, I summarize the main points in nonmathematical terms.

Two members of the British Advisory Committee for Aeronautics—Lord Rayleigh and Sir George Greenhill—made important contributions to the field of hydrodynamics in the 1870s and 1880s. The numerous references to papers and results by Rayleigh and Greenhill in the standard textbooks of hydrodynamics of that time, for example, Lamb's *Hydrodynamics*, attest to their prominence in the field.[2] Rayleigh had arrived at some classical results, which are described later in this chapter, and Greenhill had written the authoritative article on hydrodynamics in the eleventh edition of the *Encyclopaedia Britannica*. Significantly the encyclopedia had two lengthy and detailed entries that dealt with fluid flow. One was the article titled "Hydro-mechanics" written by Greenhill; the other article was titled "Hydraulics" and was written by a distinguished engineer.[3] The former presentation was filled with mathematics, while the latter was filled with descriptions and dia-grams of turbine machinery. The reason it was felt necessary to recognize this division of labor in drawing up the encyclopedia is relevant to my story and will become clear in what follows.

The Theory of Ideal Fluids

Physicists, chemists, physiologists, and engineers are all interested in air, and each group studies it from the perspective of its own discipline. In the history of each discipline there is a strand that represents the history of the chang-ing conceptions of the nature of air adopted by its practitioners. Sometimes aerodynamics is counted as a branch of physics and sometimes as a branch of engineering, but however it is classified, it is evident that it involved a determined attempt to relate the flow of air to the basic principles of me-chanics. The most important of these are the laws of motion first delineated by Newton, for example, the law that force equals mass times acceleration. The complexity of the air's behavior, however, means that there is no unique way to connect the flow to the fundamental laws of Newtonian mechan-ics. How the relation is to be articulated depends on the model of air that is used.

Newton himself treated fluids in different ways at different times. When he was thinking about the pressure of the air in a container, he conjectured, for the purposes of calculation, that air was made up of static particles that repelled one another by a force that varied inversely with distance.[4] This con-cept was a guess that explained some of the known facts, but it was a con-ception of the nature of air and gas that physicists later abandoned. In its place they adopted what is called the kinetic theory of gases in which it was assumed that a gas is made up of small, rapidly and randomly moving par-ticles. According to the kinetic theory, as developed by James Clerk Maxwell and others, gas pressure is not the effect of repulsion between the molecules of the gas but is identified with the repeated impact of the molecules on the walls of the container.[5]

When Newton was thinking of a flowing fluid impinging on the surface of an obstacle, he did not use his repulsion model but spoke, for mathematical purposes, simply of a "rare medium" and treated the fluid as made up of a lot of point masses or isolated particles that do not interact with one another.[6] The fluid medium was treated as if it were like a lot of tiny hailstones (though this was not Newton's comparison). Again, the model is not to be identified with the later kinetic theory of gases. The hailstone model, too, dropped by the wayside, though, as we shall see, in certain quarters it still played some role in early aerodynamics. The concept of a fluid that proved most influential in hydrodynamics was different from either of the ideas used by Newton as well as being different from the kinetic theory of gases. The model that came to dominate hydrodynamics, and aerodynamics, was first developed in the eighteenth century by mathematicians such as d'Alembert, Lagrange, the two Bernoullis (father and son), and Euler. They thought of the air as a continuous medium.[7]

Because the aim was to be realistic, the hypothetical, continuous-fluid picture had to be endowed with, or shown to explain, as many of the actual properties of real fluids as possible. Thus air has density so the continuous fluid must also possess density. Density is usually represented by the Greek letter rho, written ρ. Empirically, density is defined as the ratio of mass (M) to volume (V), which holds for some finite volume. The number that results, $\rho = M/V$, represents an average which holds for that volume at that moment. To apply the concept to a theoretically continuous fluid requires the assumption that it makes sense to speak not merely of an average density but of the density at a *point* in the fluid, that is, the ratio of mass to volume as the volume under consideration shrinks to zero. If the air is actually made up of distinct molecules, then, strictly, the density will be zero in the space between the molecules and nonzero within the molecules, and neither of these values would qualify as values of the density of the *fluid*. This dilemma did not appear to be a problem in practice, but it is a reminder that the relation between physical models of the air based on particles and physical models based on a continuum may, under some circumstances, prove problematic.[8]

Air is also compressible. The same mass can occupy different volumes at different pressures. For many of the purposes of aerodynamics, however, it can be assumed that the density stays the same. This is because (perhaps counter to intuition) the pressure changes involved in flight turn out to be small. The fluid continuum can then be treated as "incompressible." This approximation only becomes false when speeds approach the speed of sound, which is around 760 miles an hour. In the early days of aviation, when aircraft flew at about 70 miles an hour, compressibility was no problem for wing theory. Things were different for propellers. The tips of propellers moved at a much higher speed, and here compressibility effects began to make themselves felt, but that part of the story I put aside.[9]

Another important attribute of a fluid is its viscosity, which refers to the sluggishness with which the fluid flows. If a body of fluid is thought of as made up of layers, then the viscosity can be said to arise from the internal friction between these layers. Pitch and treacle are highly viscous fluids, whereas water is not very viscous. Viscosity can be measured by experimental arrangements involving the flow through narrow tubes. The results are summarized in terms of a coefficient of viscosity, which is usually represented by the Greek letter mu, written μ. A highly viscous fluid will be given a high value of μ; a fluid with small viscosity will have a correspondingly small value of μ. Air only has a very slight viscosity. At the extreme, if there were a fluid that was completely free of viscosity, it would be necessary to write $\mu = 0$. In reality no such wholly inviscid fluids exist, but if the fiction of zero viscosity is combined with the fiction of total incompressibility, this concept can be taken to specify what might be called a "perfect" fluid or an "ideal" fluid.

The single most important fact to know about the historical development of wing theory and the aerodynamics of lift is that its mathematical basis lay in the theory of perfect fluids, that is, in a theory in which viscosity was apparently ignored and assumed to be zero. The assumption that air can be treated as an ideal fluid was the cause of much argument, doubt, and frustration, which becomes apparent in subsequent chapters, but its central, historical role is beyond dispute. What turned out to be the most striking developments in aerodynamics (as well as some failed attempts) depended on the idea that viscosity and compressibility were effectively zero. The attractions of this assumption were twofold. First, it seemed highly plausible, and second, it produced an enormous simplification in the mathematical task of describing the flow of a fluid. The exercise produced a set of partial differential equations that determined the velocity and pressure of the fluid, provided that the starting conditions of the flow and the solid boundaries that constrain it are specified. The equations were developed by imagining a small volume of fluid, called a fluid element, and identifying the forces on it. The forces derive from pressure imbalances on the surfaces of the fluid element.

Fluid elements, it must be stressed, are mathematical abstractions rather than material constituents of the fluid. They are not to be equated with the molecules that interest chemists and physicists or the particles that feature in the kinetic theory of gases. The equations of flow do not refer to the hidden, inner constitution of fluids. The reality that is described by the differential equations that govern fluid motion concerns the *macro*behavior of fluids rather than their *micro*structure. The abstract character of a fluid element is evident from the way it is typically represented by a small rectangle. The simple geometry of the representation derives from the mathematical techniques that are being brought to bear on the flow. These are the techniques of the differential and integral calculus.[10] The concept of a fluid element is the means by which these techniques can be used to gain a purchase on reality. The differential equations that were the outcome are called the Euler equations. They can be said to describe in a strict way the flow of an ideal fluid, but the hope was that they would also describe, albeit in an approximate way, the flow of a real fluid, air, whose viscosity is small but not actually zero.

To give a feel for the style of thinking that went into the classical hydrodynamics of ideal fluids (and, later, into aerodynamics), I shall give a simple, textbook derivation of the Euler equations. It is the kind of derivation that was wholly familiar to many of the actors in my story, and certainly to those who worked in and for the Advisory Committee for Aeronautics. The discussion in the next section is therefore slightly more technical. It is based on the treatment given in one of the standard works of early British aerodynamics, namely, W. H. Cowley and H. Levy's *Aeronautics in Theory and Experiment* that was published in 1918.[11] Both Cowley and Levy worked at the National Physical Laboratory. Levy had graduated from Edinburgh in 1911, visited Göttingen on a scholarship, and had then worked with Love in Oxford. During the Great War he had been commissioned in the Royal Flying Corps but was seconded to the NPL. As a left-wing activist who wanted to unionize his fellow scientists, his relations with Glazebrook were not of the easiest. After the war Levy left to join the mathematics staff at Imperial College, where he was eventually awarded a chair.[12] Cowley stayed at the NPL and worked on problems of drag reduction with R. J. Mitchell, who was designing the racing seaplanes that won the Schneider Trophy for Britain in 1929 and 1931.[13]

The Equation of Continuity (and Some Conventions)

The equations of motion for an ideal fluid were derived from two basic statements. In a sense, said Cowley and Levy, "these two statements and all that they involve are a definition of the nature of the fluid." Furthermore, "all deductions regarding its behavior can only be a recasting of these [statements] into a new but equivalent form" (37). The two statements were as follows: (1) The mass of fluid in any region remains constant. This is called the condition of "continuity." And (2) the motion of every fluid element is consistent with Newton's Laws of Motion. To spell out the first of these principles, consider a small volume through which the fluid flows. This is represented in cross section in figure 2.1 by the rectangle *ABCD*, with sides labeled *dx* and *dy*. Before going further I should say something about the conventions used in such diagrams in hydrodynamics. The *d* and δ (delta) symbols indicate that the lengths are not just "small" in a commonsense manner of speaking but have been, or will be, made "infinitely small" in the course of the mathematical reasoning. They will be subject to a "limiting process" in which it is assumed that they can be made ever smaller without the shrinkage demanding any significant changes in the pattern of reasoning (which is essentially what Lamb meant in the passage quoted at the head of the chapter). The reason why the volume can be represented by an area is because the volume is assumed to be of unit depth, so the number 1, representing the depth, is present but can be suppressed. Diagrams of this kind thus amount to a two-dimensional cross section of the flow, and the situation portrayed is routinely referred to as a two-dimensional flow.[14]

Two-dimensional flow diagrams do not allow any representation of what happens at the edges of the figure other than those shown in the cross section. The parts of the object that go into, and out of, the page are not shown. In describing a situation in this truncated way, the mathematician assumes that nothing significant happens at the edges that are not represented. In reality this is not true, so all discussions of two-dimensional flow are by their nature simplifications. These simplifications will become especially significant when the object under discussion is a wing and the cross section takes on the shape of an aerofoil. In the literature on aerodynamics the simplified, two-dimensional diagram of the flow is then often called a diagram of the flow around an "infinite" wing. This usage can be disconcerting, but it is simply a way of saying that the flow shown in the picture is representative of what goes on in the central parts of the wing. The wingtips are assumed to be

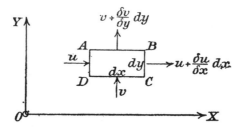

FIGURE 2.1. Small control volume used to arrive at the equation of continuity. Fluid flowing into the volume equals fluid flowing out. From Cowley and Levy 1918, 37.

sufficiently far from the action that they do not interfere in any way and can be ignored. The literature on hydrodynamics and aerodynamics is full of references to infinity. The word "infinity" can nearly always be read as meaning either "so far away that it causes no disturbance" or "so far away that it can be considered to be undisturbed."

Having dealt with these terminological matters, I now come back to the small volume represented in figure 2.1.[15] The fluid is assumed to be incompressible with constant density ρ, so in a given time interval the mass of fluid flowing into the volume always equals the mass flowing out. In the x-direction the speed of flow into the volume is designated by u and the speed of outflow by $\left(u+\dfrac{\partial u}{\partial x}\delta x\right)$. The symbol $\partial u/\partial x$ means "the rate of change of u with x," so the expression in parentheses refers to the original speed plus the change of speed. The change can be positive or negative. The mass entering the control volume per unit time is $\rho u \delta y$ and the mass leaving is $\rho\left(u+\dfrac{\partial u}{\partial x}\delta x\right)\delta y$. The same procedure is repeated for the flow in the y-direction. In each case the quantity of fluid entering and leaving the control volume is obtained by multiplying the speeds by the density of the fluid and dimensions of the face crossed by the flow. Mathematically, the condition of continuity is then expressed by summing all these quantities and equating the sum to zero. Fluid in must equal fluid out, with zero shortfall. When the expression for this zero sum is simplified, the equation that results takes the form

$$\frac{\partial u}{\partial x}+\frac{\partial v}{\partial y}=0.$$

This expression states, in a mathematical form, the condition of continuity.

The Euler Equations

The second principle states that fluid elements obey Newton's second law of motion, $F = M \times A$, that is, force equals mass times acceleration. Figure 2.2 is a picture of a fluid element whose sides are parallel with the x- and y-axes. As in the previous discussion, the flow is assumed to be "two-dimensional," so the z-axis plays no role in the analysis. The element is again assumed to be of unit depth and can be represented by a rectangle $ABCD$ (whose "depth" $\delta z = 1$). Thus the mass of a small, rectangular fluid element with sides δx and δy is $\rho \delta x \delta y$. In aerodynamics the effect on the air of external forces, such as gravity, can be neglected. The resultant force on the fluid element in the x-direction comes from difference between the pressures on the two faces AD

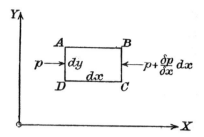

FIGURE 2.2. Small fluid element showing pressure on the faces AD and BC. The pressure difference is the force producing the acceleration along the x-axis. From Cowley and Levy 1918, 37.

and BC. Let the pressure on AD be p. The pressure on BC is then $p + \frac{\partial p}{\partial x}\delta x$, that is, the original pressure plus the change in pressure over the length of the element. The resultant thrust on the element is

$$p\delta y - \left(p + \frac{\partial p}{\partial x}\delta x\right)\delta y = -\frac{\partial p}{\partial x}\delta x \delta y.$$

This is the force causing the acceleration. Newton's second law then takes the form

$$-\frac{\partial p}{\partial x}\delta x \delta y = (\rho \delta x \delta y) \times accelaration.$$

What is the acceleration in the x-direction? The desired term is du/dt, the rate of change of the velocity u with time t. The velocity u is a function of three variables: x, y, and t. The acceleration is given by a process of differentiation involving all three variables which is called "differentiation following the motion of the fluid."[16] Thus,

$$\frac{du}{dt} = \frac{\partial u}{\partial x}\frac{\partial x}{\partial t} + \frac{\partial u}{\partial y}\frac{\partial y}{\partial t} + \frac{\partial u}{\partial t}\frac{\partial t}{\partial t}$$

$$= \frac{\partial u}{\partial x}\frac{\partial x}{\partial t} + \frac{\partial u}{\partial y}\frac{\partial y}{\partial t} + \frac{\partial u}{\partial t}.$$

This expression can be simplified. If the restriction is introduced that the flow is steady, then $\frac{\partial u}{\partial t} = 0$. Also, by definition, $\frac{\partial x}{\partial t} = u$ and $\frac{\partial y}{\partial t} = v$. Substituting these terms in the expression for Newton's law yields

$$-\frac{\partial p}{\partial x} = \rho\left(u\frac{\partial u}{\partial x} + v\frac{\partial u}{\partial y}\right).$$

Similar reasoning gives the equation for the y-direction:

$$-\frac{\partial p}{\partial y} = \rho\left(u\frac{\partial v}{\partial x} + v\frac{\partial v}{\partial y}\right).$$

These are the Euler equations for the two-dimensional steady flow of an ideal fluid. A mathematician will identify them as nonlinear, partial differential equations that relate the pressure p and the velocity components u and v. Along with the equation of continuity they constitute the fundamental equations of motion of an ideal fluid. They completely determine the motion. The integration, or solution, of the equations will, however, involve arbitrary functions and constants, and these require knowledge of (1) the initial conditions of the motion and (2) the position of any fixed boundaries. These two specifications are called the boundary conditions.

Stream Functions and Streamlines

To apply and solve the Euler equations, mathematicians had to introduce various techniques to relate them to specific flow problems. "As they stand," said Cowley and Levy, "these equations are not very suitable for solution" (39). They need to be fleshed out. This was done by means of a variety of auxiliary concepts such as source, sink, vortex, and stream function. The general connotations of the labels "source," "sink," and "vortex" will be evident, and their mathematical idealization refines, but does not essentially alter, the everyday meaning of the word. A vortex is like a whirlwind around a central point. A source is a geometrical point at which fluid is created at a certain rate, and a sink is a geometrical point at which it disappears and is destroyed. The words "stream function," however, do not have any obvious counterpart in common usage. In this section I describe briefly what they mean.

Imagine a coordinate system of x- and y-axes that is to be used to describe a flow of fluid. The value of the stream function at some point P is given by the amount of fluid that flows in unit time across a line drawn from the origin to P. To specify this quantity is to specify the value of the stream function. In hydrodynamics this value is usually designated by the Greek letter psi, ψ. Altering the position of the origin only alters the value of ψ by the same constant amount at all points in the flow. It follows from the definition that such a function has a simple relation to the velocity components of the flow, and this is the utility of the stream function. If u is the speed along the x-axis at P and v is the speed of flow along the y-axis, it can be shown that

$$u = -\frac{\partial \psi}{\partial y}, \text{ and}$$

$$v = \frac{\partial \psi}{\partial x}.$$

Given the stream function, a process of differentiation gives the velocity components. Here is a simple example. The stream function for a uniform flow of speed U along the x-axis is

$$\psi = -Uy = -Ur\sin\theta.$$

The first expression is in Cartesian coordinates and the second is in polar coordinates, giving the value of ψ at the point (r, θ). In Cartesian coordinates, differentiating ψ with respect to x gives the correct answer $v = 0$, meaning that the flow has zero velocity along the y-axis. Differentiating with respect to y gives the speed $u = U$ along the x-axis. Notice that putting $\psi = c$, a constant, gives a straight line parallel with the x-axis. Such a line can be called a streamline of the flow. Later in the discussion it will become evident that, for all its simplicity, this flow plays a basic role in hydrodynamic reasoning. Logically, it provides the foundation of the edifice.

I have referred to a streamline of this basic flow, but what is a streamline? In everyday language the words connote speed. Modern aircraft are "streamlined," whereas aircraft in the period of the old Advisory Committee for Aeronautics, with their struts and protruding engines and undercarriage, were certainly not. This usage, and the idea of low-resistance, streamlined bodies, was already well established in early aerodynamics, even if it could not be realized in the construction of flying machines.[17] The technical meaning of the term "streamline" in hydrodynamics, though related to this popular meaning, is more specific. A streamline drawn through a point in a fluid flow is a line that conforms to the direction of motion of the fluid element that is located at that point at that moment in time. A moment later the point may be occupied by another fluid element with a different velocity. The picture becomes much clearer if the flow is steady so that the speed and direction of the flow at a given point are constant over time. When the flow is steady, then streamlines will coincide with the path taken by the fluid element. Looking at the streamlines will give a picture of what the fluid elements are doing. Streamlines also indicate something about the speed of the flow. For steady incompressible flow they come closer together as the flow speeds up and become wider apart as the flow slows down.[18]

How does the mathematician identify streamlines in order to draw a diagram of a flow? The answer is by reference to the stream function. Once in possession of an expression ψ for the steam function of the flow, the mathematician generates a series of curves by putting $\psi = c$, a constant, and giving

the constant a sequence of values c_1, c_2, c_3, etc. The curves are conventionally plotted at equal intervals. *These are the streamlines.* As a simple example,
recall the stream function for the uniform flow parallel to the x-axis—the
basic flow. The formula for the stream function was $\psi = -Uy$. Putting $\psi =$
(say) 0, 1, 2, 3, etc. gives the straight, horizontal lines $y = 0$, $y = -1/U$, $y = -2/U$,
$y = -3/U$, etc. Notice that the greater the speed U, the smaller the gap between
the lines. Because, by definition, a fluid element will not cross over a streamline, then any streamline can be selected and interpreted as a solid boundary
without this in any way changing the picture of the flow. (It is sometimes said
that the fluid bounded by a streamline can be suddenly "frozen" or "solidified" without altering the rest of the flow.) In the present case the line $\psi = 0$
can be selected for this treatment. The flow then becomes (that is, can now
be *regarded as*) the uniform flow of an infinite ideal fluid along a flat, smooth
wall located on the x-axis.

Other, more complicated, flows call for other, more complicated, formulas for the stream function. For example, there are stream functions for the
flow around point sources and point sinks and for vortices. The streamlines
of sources and sinks radiate away from, or toward, their center point while
the streamlines of a simple vortex are concentric circles. By the expedient
of adding the stream functions, the flow can be found for combinations of
sources, sinks, and vortices. Shortly I shall give the stream function and the
streamlines for another, particularly important flow; for the moment, however, the point to retain is that a streamline is specified by setting the stream
function equal to a constant $\psi = c$.

Irrotational Flow and Laplace's Equation

The motion of a fluid element involves three different kinds of change:
(1) translation, (2) strain, and (3) rotation. Translation involves change of
position of the element, strain involves a deformation of the shape of the
element, and rotation involves a change of angular orientation of the element. Rotation may seem to be an intuitively clear idea because the image
that comes to mind is the rotation of a rigid body in which the fluid element
is pictured as if it behaves like, say, a spinning ball. Sometimes fluid elements
are indeed represented as spinning balls. Although shape is not really crucial,
the picture of a sphere is sometimes invoked when explaining the striking
result that a fluid element in an ideally inviscid fluid can never be made to
rotate if it is not already rotating, nor can it be stopped from rotating if it
is already in rotation. The rotation of an ideal fluid element can neither be
created nor destroyed by, for example, the motion of a solid body that is

immersed in, and surrounded by, a fluid. The argument is that, in a perfect fluid, neither the surrounding fluid nor such a moving body can exert any traction on the smooth surface of the element in order to change its existing state of rotatory motion. It will be evident that, in light of this result, the origin of rotation becomes something of a mystery.[19]

Cowley and Levy, however, do not avail themselves of an intuitive picture of fluid elements as rotating spheres of fluid. They opt for the more austere technical definition. Technically, the rotation of a fluid element (in two-dimensional flow) is defined as the average angular velocity of any two infinitesimal linear elements within the fluid element that are instantaneously perpendicular to one another. Mathematically this definition is expressed in the formula

$$rotation = \frac{1}{2}\left(\frac{\partial v}{\partial x} - \frac{\partial u}{\partial y}\right).$$

The virtue of the technical definition is that commonsense comparisons tend to omit the possibility that the angular velocity of the two linear elements might cancel out so that, under some circumstances, rotation can be equal to zero by virtue of the deformation of the fluid element.[20] A flow in which the quantity in the brackets in the previous formula is zero is called an irrotational flow.

Methodologically, the important point about the rotation of a fluid element is that by neglecting it, and restricting attention to irrotational flow, the mathematics is greatly simplified. Why is this? A glimpse into the reasons can be gained by taking another look at the stream function discussed in the previous section. Consider the following expression involving the stream function ψ. The expression is arrived at by differentiating ψ twice with respect to x and twice with respect to y and adding the result. Thus,

$$\frac{\partial^2 \psi}{\partial x^2} + \frac{\partial^2 \psi}{\partial y^2}.$$

It will be recalled that differentiating ψ once yields the velocity components of the flow and that the x and y components of the fluid velocity at a point are given by

$$u = -\frac{\partial \psi}{\partial y} \quad \text{and}$$

$$v = \frac{\partial \psi}{\partial x}.$$

Substituting these definitions of the velocity components in the expression under consideration gives

$$\frac{\partial^2 \psi}{\partial x^2} + \frac{\partial^2 \psi}{\partial y^2} = +\frac{\partial}{\partial x}\left(+\frac{\partial \psi}{\partial x}\right) - \frac{\partial}{\partial y}\left(-\frac{\partial \psi}{\partial y}\right)$$

$$= \frac{\partial v}{\partial x} - \frac{\partial u}{\partial y}.$$

The result of the substitution is precisely the expression that was used in the technical definition of the term "rotation." It is in fact twice the rotation. If the rotation is zero, that is, if the flow is irrotational, then this term must be zero, and so, therefore, is the expression cited at the outset of the discussion. In other words, if the flow is irrotational, then the stream function ψ is governed by the equation

$$\frac{\partial^2 \psi}{\partial x^2} + \frac{\partial^2 \psi}{\partial y^2} = 0.$$

This equation is called Laplace's equation. Although the equation itself may look far from simple, it is not difficult to appreciate that it is simpler than if the right-hand side were equated to some complicated function of x and y rather than to zero. Irrotational flow is thus a (relatively) simplified form of flow governed by Laplace's equation.

Laplace's equation is one of the most significant differential equations in the history of mathematical physics.[21] The equation is often written as $\nabla^2 \psi = 0$.[22] The restriction to "irrotational" flow, which it signifies, not only simplified the mathematics, but it brought out the analogies between fluid flow and the results that had emerged or were emerging in other fields. Irrotational flow obeyed simple mathematical laws that were similar to those in areas such as the theory of gravitational force, the theory of heat, the theory of elasticity, and the theory of magnetism and electricity. Maxwell used the analogy, and Laplace's equation, to shed light on the hydrodynamics of the flow of fluid through an orifice and the *vena contracta*, that is, the contraction shown by the jet of fluid a short distance from the orifice.[23] Because of the electrical analogies, irrotational flows used to be called "electrical" flows. The interplay between hydrodynamics and the theory of electric phenomena was not only suggestive theoretically, but it was also exploited in the laboratory. In the interwar years it provided the basis of a laboratory-bench technique used by E. F. Relf at the National Physical Laboratory for graphically plotting the streamlines of the flow around objects with complicated shapes, such as aerofoils.[24] The resulting representation was, of course, a representation of the flow as it would take place if the air were an ideal fluid.[25]

Bernoulli's Equation

The Euler equations permit the deduction of an important result known as Bernoulli's law.[26] Stated simply, Bernoulli's law implies that the pressure of the fluid increases when the velocity decreases, and vice versa. There are technical restrictions imposed on its application, but the law has many practical uses in aerodynamics. It lies at the basis of an important measuring instrument used for determining the speed of flow of a real fluid such as air. The instrument is called a Pitot-static probe and is used, for example, in wind tunnels to establish the speed of flow. Furthermore, every aircraft is equipped with this device in order to determine the speed of flight. The instrument registers pressures, but it yields information about velocities in virtue of the relation given by Bernoulli's law.

Stated quantitatively, the version of Bernoulli's law to which I have referred is

$$p + \frac{1}{2}\rho V^2 = H,$$

where H is a constant called Bernoulli's constant. The formula only strictly applies to the steady, irrotational motion of an ideal fluid. It refers, in the first instance, to a single streamline and relates the pressure and velocity at any point on the streamline to the value of H that characterizes the streamline. In aeronautics all the streamlines can be taken to originate from a region of constant pressure and velocity, and then all of the streamlines have the same value of H. The Bernoulli constant has the same value for all parts of the flow, and its value can be established for the entire flow if it is known for any given point in the fluid. The first term in the equation, p, is called the static pressure. The second term is called the dynamic pressure, and their sum, H, the Bernoulli constant, is the sum of the static and dynamic pressures and is therefore called the total pressure. The formula indicates that as the velocity V increases at some point in the flow, the static pressure p goes down at that point because the two quantities, p and $(1/2)\,\rho V^2$, must always add up to the same value. Furthermore, by knowing the density ρ and the value of p and H, we can calculate V, the speed of the flow. This is evident because the formula can be rearranged and restated as

$$V = \sqrt{\frac{2(H-p)}{\rho}}.$$

Both the static pressure (p) and the total pressure (H) of a flow can be measured. Figure 2.3 shows a simple arrangement of tubes and manometers that

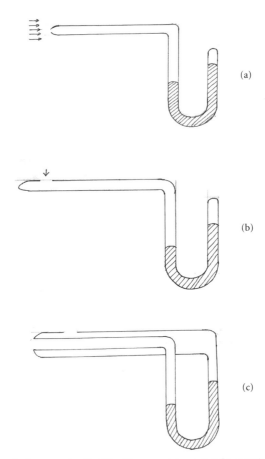

FIGURE 2.3. A Pitot tube for measuring "total head" pressure (*a*), a static tube or static head for measuring the static pressure (*b*), and a Pitot-static probe combining a Pitot tube and a static head (*c*). The
pressure difference indicated by the manometer in (*c*) gives a measure of the speed of flow. This device is
the basis of the air-speed indicator in an aircraft cockpit.

would yield measures of these quantities. The total pressure measurement (*a*)
uses an open-ended tube. The static pressure measurement (*b*) uses a closed
tube with a small hole in its side. The side hole is called the static tap. Both
tubes are connected to their respective manometers. The third part of the
figure (*c*) indicates how the two measuring devices can be unified to form a
Pitot-static probe. In the combined instrument, the single manometer measures the pressure difference $(H - p)$ needed to establish the velocity.

Measuring the speed of flow by means of a Pitot-static probe can be accurate to about 0.1 percent, but it has a slow response rate and demands care
and suitable conditions. The formula contains a term for the density of the

air, and density varies with altitude, a fact of importance when the instrument is used in an aircraft to measure speed. Furthermore, the Pitot probe itself can disturb the flow it is used to measure. Small faults such as a burr around the mouth of the static tap, or a misalignment of the probe, as well as turbulence in the flow, can significantly affect the readings.[27] Conditions such as the formation of ice in and around the inlet holes can also falsify the instrument readings of an aircraft, and for this reason such devices are usually equipped with electrical heating elements. The formula underlying the use of the Pitot-static probe, which I have given, only applies to airflows that can be considered as incompressible and has to be modified to allow for compression effects for high-speed subsonic flight. Yet further modifications are needed to correct for the presence of shock waves at the nose of the Pitot tube as the speed of sound is approached.[28]

Lines of Equal Potential

Suppose the mathematician has managed, by good fortune or guesswork, to write down the stream function for a steady flow of fluid under certain boundary conditions. By equating the stream function to a sequence of constants, a family of streamlines can be drawn and a picture of the flow can be exhibited. Now suppose that, guided by the streamlines, the mathematician draws another family of curves. These new curves are to be drawn so that they always cut across the streamlines at right angles. A network of orthogonal lines is built up. If the first set of lines were the streamlines of the flow, what are these new lines that have been drawn so that they are always at right angles to them?

They are called potential lines. They are in fact another way of implicitly representing the velocity distribution of a flow. Their immediate interest is that the potential lines of a given flow can always be reinterpreted as the streamlines of a new flow, while the old streamlines become the potential lines of the new flow. Streamlines and potential lines can be interchanged, provided that appropriate changes are made to the boundary conditions of the flow. This possibility of interchange can be interpreted to mean that, just as there exists a stream function, so there must exist another, closely related function ready to perform the same role with regard to the lines of potential that ψ played with regard to streamlines. This function is called the potential function, and it is conventionally designated by the Greek letter phi, ϕ. The role of the potential function may be illustrated by the uniform flow along the x-axis, where the axis can be taken as a solid boundary. This flow is the one discussed earlier whose stream function is $\psi = -Uy$. The streamlines are horizontal lines parallel with the x-axis, so the potential lines are vertical lines parallel with the y-axis. Now switch the potential lines and the streamlines, that is, switch the two families of curves given by ψ = constant and ϕ = constant. The streamlines are now vertical and parallel with the y-axis, which can be treated as a boundary to the new flow. The horizontal lines parallel with the x-axis are the new potential lines.

The intimate relationship between potential lines and the streamlines finds expression in the mathematics of irrotational flow. Because the two families of curves are orthogonal, it is possible to write the equations for the velocity components u and v of a given flow either in terms of the stream function that applies to the flow or in terms of the potential function that applies to it. The result gives rise to the following relationships between ϕ and ψ:

$$u = -\frac{\partial \psi}{\partial y} = -\frac{\partial \phi}{\partial x} \text{ and}$$

$$v = \frac{\partial \psi}{\partial x} = -\frac{\partial \phi}{\partial y}.$$

It follows immediately from these equations that the potential function ϕ obeys Laplace's equation, just as the stream function does when representing an irrotational flow. One useful mathematical property of solutions to Laplace's equation is that they are additive. If ψ_1 is a solution and ψ_2 is a solution, then $\psi_3 = \psi_1 + \psi_2$ is also a solution. Stream functions can be added. Again, the point can be illustrated by reference to the simplest possible cases. The flow of speed U along the x-axis ($\psi_1 = -Uy$) can be combined with, say, a flow of the same speed U but along the y-axis (that is, the flow arrived at by switching the streamlines and the potential lines of the original flow so that $\psi_2 = Ux$), and the result is another flow that moves diagonally and whose stream function is $\psi_3 = \psi_1 + \psi_2$. In this way complicated flows can be constructed out of simple flows.

The Indirect Method and Complex Variables

After introducing Laplace's equation, Cowley and Levy made the following observations about its centrality to the mathematics of ideal-fluid flow: "The real key to the solution of any problem in the irrotational motion of a non-viscous fluid lies in the determination of the appropriate expression that satisfies this equation and at the same time gives the requisite boundaries to the fluid" (44). If this is the "real key," then where and how is the key to be found? I have just indicated that some flows can be built up by combining the stream functions of existing flows, but how does the process get started? How were these stream functions arrived at? Apart from the simplest possible of all flows, the primordial straight-line, steady flow, how does the mathematician determine which expressions satisfy Laplace's equation and meet the requisite boundary conditions? Cowley and Levy acknowledge that it is not easy and draw attention to the expedients that have been used to cope with the difficulty.

One expedient is called the indirect method. Instead of beginning by stating a problem (for example, what is the mathematical description of the flow around such-and-such a given object?) the researcher starts with some known piece of mathematics and asks what flow it might be used to describe. Various mathematical functions are investigated to determine which boundaries might be fitted to them. The difference between the direct and indirect approach is like that between the carpenter who wants to put up a shelf and looks for a suitable piece of wood, and the carpenter who finds an interesting piece of wood lying around and looks for an opportunity to use it. As Cowley and Levy put it, "it will be clear that this indirect method of attack does not furnish a method of obtaining the solution of any proposed problem but rather furnishes the solution from which the problem is obtained" (51–52). This method works because there is a rich field of possible candidates. There is (so to speak) a lot of interesting wood lying around. In this and the next section, I show why there are so many solutions in search of problems. I then look at some more direct lines of attack on the problem of arriving at a mathematical description of a desired flow. The survey will reveal what is, at first sight, an almost uncanny relationship between pure mathematics and the physical world. This feature is one of the most intriguing in classical hydrodynamics.

A body of mathematics called the theory of functions of a complex variable places a large body of material at the disposal of the student of hydrodynamics. Complex variables are really just pairs of numbers which represent the coordinates (x, y) of a point relative to a standard coordinate system. But instead of a number pair (x, y), the coordinates are expressed in the form $z = x + iy$, where $i = \sqrt{-1}$. Conventionally the symbol x is called the "real" part of the complex number z, and y is called the "imaginary" part. The symbol i is sometimes called an imaginary number.[29] Complex numbers obey all the usual rules for manipulating numbers apart from needing the extra rule that $i^2 = -1$. A function of a complex variable is some combination of complex numbers z whose value is arrived at by adding or multiplying the variable z or subjecting it to some other mathematical operation. Thus, to take an example that will play a prominent role in the story, $f(z) = z + 1/z$ is a function of a complex variable. Insert a value of z into the formula, perform the requisite operations (finding the reciprocal and adding), and out comes the value of the function $f(z)$. Classical hydrodynamics was able to develop as it did because of the fortunate and remarkable fact that *every function of a complex variable* w = f(z) *turns out to represent a possible two-dimensional flow pattern of an ideal fluid in irrotational motion*. There are an infinite number of such functions, some simple, some complicated, but they all represent a possible flow of an ideal fluid.[30]

I illustrate this remarkable fact by a particular case. It is a case used by Cowley and Levy to illustrate the inverse method, that is, to show how one can, for whatever reason, begin by looking at a piece of mathematics dealing with complex variables and then realize its significance as a description of a possible flow. The example also shows how every function of a complex variable embodies the formula for *both* the potential lines *and* the streamlines of a possible flow. The potential function ϕ occurs as the real part of the complex function, while the stream function ψ occurs as the imaginary part. The link between the two sets of lines, the streamlines and the lines of equal potential, can be represented by writing $f(z) = \phi + i\psi$.

Cowley and Levy invite their readers to "consider" the function $f(z) = (z + 1/z)$. What is the flow that could be represented by this function? This flow can be found by explicitly writing out the function in terms of the more familiar x and y notation. Recalling that $z = x + iy$ and that $i^2 = -1$, then substitution in the formula for $f(z)$ gives

$$\phi + i\psi = z + \frac{1}{z} = (x+iy) + \frac{1}{(x+iy)}.$$

The numerator and denominator of the last term are multiplied by $(x - iy)$, and then all the terms on the right-hand side can be put over the same denominator $(x^2 + y^2)$. Rearranging and grouping together the real and imaginary parts of the expression gives

$$(x+iy) + \frac{x-iy}{x^2+y^2} = \frac{x(x^2+y^2+1)}{x^2+y^2} + i\frac{y(x^2+y^2-1)}{x^2+y^2}.$$

Examination of the right-hand side of the above equation will reveal that it has the form $f(z) = \phi + i\psi$. It is now easy to read off the formulas for the potential lines and the streamlines of a possible flow.

The potential lines $\phi = c$ are given by the real part of this expression, thus

$$\phi = \frac{x(x^2+y^2+1)}{x^2+y^2}.$$

The streamlines, $\psi = c$, are given by the imaginary part of this expression, thus

$$\psi = \frac{y(x^2 + y^2 - 1)}{x^2 + y^2}.$$

The family of curves ψ = constant gives the equations of the streamlines, and one of these can be interpreted as the outline of the boundaries that constrain the flow, that is, the shape of the body around which, or along which, the fluid is flowing. Consider the streamline represented by $\psi = 0$. This equation calls for either of two conditions to be satisfied. It requires that either $y = 0$ or $(x^2 + y^2 - 1) = 0$. The first of these conditions is satisfied if the x-axis (that is, the line $y = 0$) is a streamline. The other condition is fulfilled if $x^2 + y^2 = 1$. This condition is met by points that fall on the circle of radius 1, whose center is at the origin. Both of these will be streamlines of the flow represented by the function $f(z)$. In other words, *the function can be understood as giving the mathematical description of a flow along the x-axis, which then goes around a circular cylinder of unit radius.* If the formula had been $(z + a/z)$ rather than $(z + 1/z)$, the flow described would be that around a circle of radius a.

What about the velocity of the flow? Recall that the velocity components of a flow are given by differentiating the stream function. Thus

$$u = -\frac{\partial \psi}{\partial y} = -1 + \frac{x^2 - y^2}{(x^2 + y^2)^2} \quad \text{and}$$

$$v = \frac{\partial \psi}{\partial x} = \frac{2xy}{(x^2 + y^2)^2}.$$

To find the velocity at a great distance from the circular cylinder (which has its center at the origin) one must ask what happens when the values of x and y become very large. Because of the squared term in the denominator, the fractional terms in the previous equations will tend to zero. Thus "at an infinite distance," as Cowley and Levy put it, $u = -1$ and $v = 0$. At a large distance from the cylinder, the flow has unit velocity along the x-axis (from right to left) and that is all. On the surface of the cylinder, one point that can be selected for consideration is $x = 1$ and $y = 0$, that is, the very front of the cylinder. Here substitution in the previous formulas gives $u = 0$ and $v = 0$, so this point is called a stagnation point or a stopping point of the flow. The upshot is "that the circular cylinder is stationary and the fluid is streaming past it with unit velocity at infinity" (46).

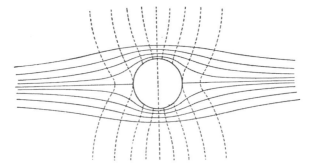

FIGURE 2.4. Flow past a circular cylinder. The continuous lines indicate the streamlines of the continuous flow of an ideal fluid moving horizontally past a circular cylinder. The dotted lines indicate lines of equal potential. From Cowley and Levy 1918, 46.

This worked example illustrates the way that (given sufficient ingenuity) the mathematical behavior of a function and its geometrical representation can be read, retrospectively, as providing a picture of a flow. The diagram given by Cowley and Levy showing the flow generated by $f(x) = z + 1/z$ is reproduced here as figure 2.4. The solid lines represent the streamlines; the dotted lines represent the lines of equal potential. That the two sets of lines form orthogonal sets is evident from the figure. Inspection of the diagram indicates two further points. First, the general rule that streamlines and lines of equal potential are at right angles breaks down at the stagnation points of a flow. Second, when streamlines and lines of equal potential are switched in their roles, it is necessary to adjust the boundary conditions.[31]

The Method of Conformal Transformation

Although many flows were discovered by the indirect method, there are direct methods for describing a flow. How, for example, does the mathematician manage to describe the flow around a straight barrier that is placed facing head-on into a uniform stream of ideal fluid? The flow in question is sketched in figure 2.5, again taken from Cowley and Levy's book. (Because the flow is presumed to be symmetrical around the central streamline of the main flow, only the upper half of the flow need be considered. The central streamline can be treated as if it were a solid boundary.) How can the equations for the streamlines ever be discovered if the mathematician does not have the good fortune to come across a function amenable to after-the-fact interpretation? The answer is by cleverly establishing a relationship between

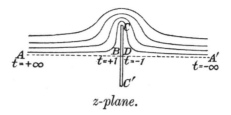

z-plane.

FIGURE 2.5. Ideal fluid flowing irrotationally around a barrier normal to the free stream. From Cowley and Levy 1918, 49.

this complicated flow problem and the simplest of all possible flow problems, namely, the uniform flow along a straight boundary. The method involves *transforming* the straight boundary into the shape desired, for example, the shape of a barrier that is sticking out at right angles into the flow. The process is carried out by means of what is called a conformal transformation.

First, I should explain the word "transformation." Everyone is familiar with the process of redrawing a diagram on a different scale. Suppose a geometrical figure has been drawn on one piece of graph paper, and it is required that the figure be redrawn, to a different scale, on another piece of graph paper. A line three centimeters long in the original is to be, say, six centimeters in the new diagram. A circle of radius four centimeters is to become a circle of radius eight centimeters, etc. The rule, in this case, is to double the length of the straight lines. The original diagram has thus been subject to a very simple, linear "transformation." Other, much more complicated transformations are possible. Not only might a transformation magnify the figure in the original, but it might shift it relative to the origin, or rotate it or even distort it in various ways, turning, say, a circle into an ellipse. This shift will depend on the particular transformation that is being followed, namely, the particular rule that relates the positions of points in the one figure to the points in the other figure. If two figures are related by a transformation, then, if we know one of the figures, along with the rule of transformation, we can construct the other figure. A figure can be subject to more than one transformation so that a figure which results from one transformation can be transformed yet again.

Transformations are important in hydrodynamics for the following reasons. First, the rules governing many transformations can be embodied in mathematical formulas that are functions of a complex variable. These are the conformal transformations. Second, if the flow around one shape is known, and a formula of this kind is available to transform the shape into a new shape, then the flow around the new shape is known. Conformal transformations change the streamlines as well as the boundaries of the figure,

modifying the shape of the flow to fit the new circumstances. Methodologically this is important. It means that, given an appropriate transformation, it
is possible to move from *simple* flows, with *simple* boundaries, to the description of *complicated* flows with *complicated* boundaries. All this can be done
once it has been established that the transformation maps the boundaries
of the two flows on to one another. Cowley and Levy sum up the situation,
tersely, as follows: "It must be noticed that as long as complex functions are
dealt with, the hydrodynamical equations will be satisfied and it will only
be necessary therefore to consider boundaries. If a functional relation existing between two planes is such as to provide a correspondence between the
boundaries in these planes it is the transformation required" (47). The "two
planes" referred to in this quotation are, in effect, just the two pieces of graph
paper I mentioned at the outset. In this case, however, the idea is that one
plane (usually called the w-plane) has the boundaries of a simple flow drawn
on it, while the other plane has the, transformed, boundaries of the more
complicated flow. This is usually called the z-plane and the transformation,
or the sequence of transformations, links the two planes.

The problem is to find the necessary rule, or rules, of transformation.
Fortunately there are general theorems that deal with the subject of transformation which can be put to use. For example, there is a powerful result called
the Schwarz-Christoffel theorem which proved central to classical hydrodynamics and, as we shall see in later chapters, also played an important role in
the history of aerodynamics. The Schwarz-Christoffel theorem is applicable
to the present problem, namely, finding the flow around a barrier across the
flow of the kind shown in figure 2.5. This theorem, used by Cowley and Levy
in their book, transforms the interior of a closed polygon on one plane (the
z-plane) into the upper half of another plane (usually called the t-plane) and
turns the boundary of the polygon into the real axis of the t-plane. If the
t-plane can then be related to the basic, simple flow along the horizontal axis
in the w-plane, then the requisite connections have been made. The simple
flow with its simple boundaries can be turned into the complicated flow. The
bridge is symbolized by $w = f(z)$. Although the details need not be described, I
want to sketch the way the theorem is used. The first step is to explain where,
and why, polygons come into the story.

The polygon is familiar from school geometry and is usually defined as
a many-sided figure whose sides are straight lines. A "closed" polygon obviously has an inside and an outside. The exterior angles must add up to four
right angles. The interior angles add up to $(n - 2)\pi$, where n is the number
of vertices. Thus a rectangle is a simple case of a closed polygon that has just
four vertices and in which each of the four interior angles is also equal to $\pi/2$.

The Schwarz-Christoffel theorem is embodied in the following, daunting, formula:

$$\frac{dz}{dt} = A(t - t_1)^{\frac{\alpha}{\pi} - 1} (t - t_2)^{\frac{\beta}{\pi} - 1} \ldots$$

The letter A represents a constant and α, β, ... are the internal angles of the polygon. The numbers t_1, t_2, ... are real numbers ranging from minus infinity to plus infinity, with one number for each vertex. In order to put the formula to work to transform a given polygon, it is necessary to insert the values for the interior angles of the polygon, α, β, etc., into the formula and to assign the vertices of the polygon to the positions t_1, t_2, etc. on the real axis of the t-plane. (Some of these assignments can be made arbitrarily, while some depend on the shape of the polygon. In a moment I shall show how Cowley and Levy made the assignment.) Having filled in the appropriate values in the formula, we must then integrate it, and the result is a function of a complex variable $z = f(t)$.

Why is this result useful when the aim is to find the flow around a barrier? The answer is that the complicated boundary, represented by the barrier in figure 2.5, can be counted as a closed polygon for the purposes of the theorem, and this fact can be exploited to get the desired flow. Given the picture of a polygon that comes to mind from school geometry, such a designation seems counterintuitive. The streamline along the axis of symmetry combined with the barrier normal to the flow doesn't look like the polygons drawn on a school blackboard. Clearly, the words "polygon" and "closed" have been given a wider meaning. The justification is that the sides of a polygon can be made "infinitely long," and the vertices dispatched to "infinity," provided that the appropriate conventions are still kept in place regarding what counts as the interior and the exterior of the polygon. In this extended sense a polygon can even take on the appearance of, say, a single straight line.[32] Crucially, it can also take on the appearance of the boundary in figure 2.5 that represents a straight barrier jutting out into a fluid flow.

How is the diagram of the barrier-as-polygon connected to the Schwarz-Christoffel transformation formula? Look at Cowley and Levi's figure, that is, my figure 2.5. The "vertices" of the "polygon" are marked A, B, C, D, A'. Inspection of the figure shows that A and A' are both located at "infinity." The points B and D are at the front and back of the base of the barrier, while C is at the top of the barrier. The "internal" angles can also be located. In moving along the boundary the point B is the location of a right-angle turn at the front of the barrier, while at C there is a turn through 180° at the top edge of the barrier, and there is another right-angle turn at D on the rear face of the barrier. These are the angles α, β, etc. to be inserted into the formula. Cowley and Levy's diagram also shows how they have assigned t-values to these vertices. The one assignment not shown in the figure is the point C, the top of the barrier, which is given the value $t = 0$.

Once these particular values have been inserted into the formula it is ready to be integrated. After integration the constant A in the formula, as well as the constants of integration, can be evaluated by using the initial and boundary conditions of the problem. Proceeding in this way gave Cowley and Levy a formula connecting z and t, namely,

$$z = l\sqrt{(t^2 - 1)}.$$

The process is, however, not quite finished. The basic, simple flow itself now needs to be expressed in terms of the t-plane. The t-plane is an intermediary between the z- and w-planes. Only when the t-plane has been linked to the w-plane will the desired connection have been made. The general form of the simple flow on the w-plane and the boundaries on the t-plane suggest that the link will be a simple one having two constants and taking the general form $w = at + b$. Consideration of the velocity of the flow at a great distance from the barrier, and the disposition of the bounding streamlines, allows the constants to be evaluated. The transformation connecting w and t is then given by the formula $w = l\,V\,t$, where V is the free-stream velocity and l is the half-length of the plate.

Combining the two formulas by eliminating t gives the result that has been sought, the complex function expressing the flow around the barrier. The desired formula is

$$f(z) = V\sqrt{z^2 + l^2}.$$

Separating out the imaginary part, ψ, gives an expression for the streamlines of the flow, and from this the velocities and pressures on the boundary can be calculated. The formula for ψ turns out to be a complicated one, but it allows the curves to be drawn by setting ψ = constant. The formula is

$$\psi^4 + V^2(x^2 - y^2 + l^2)\psi^2 - V^2 x^2 y^2 = 0.$$

Now the streamlines of the flow of an ideal fluid around a flat barrier placed head-on to the flow can be calculated and represented with mathematical precision.

The remarkable fact that functions of a complex variable such as $f(z) = (z + 1/z)$ and $f(z) = V\sqrt{z^2 + l^2}$ are all descriptions of irrotational flows has undoubtedly left its mark on the development of classical hydrodynamics.[33]

It also raises a question. Why *should* the functions of a complex variable, containing esoteric mathematical entities such as the square root of negative numbers, yield pictures of fluid flows? Consider the formula for the flow around a circular cylinder. The formula itself, $f(x) = (z + 1/z)$, is not remarkable and is familiar to any student of mathematics (and we meet it again in a later chapter). It is hardly surprising that the formula is to be found in G. H. Hardy's famous, Tripos-oriented textbook *A Course of Pure Mathematics*, first published in 1908. It crops up in the miscellaneous examples at the end of the chapter on complex numbers.[34] But Hardy's student reader was set the purely mathematical task of proving that $(z + 1/z)$ transforms concentric circles into confocal ellipses. There was no mention of streamlines. The formula merely provided the occasion for an exercise in analytical geometry. That is what is puzzling. What has geometry got to do with fluids?

Part of the answer is provided by noticing that the functions that describe the complicated flows do so by virtue of being transformations of the simplest possible flow, namely, the uniform flow of an infinite fluid along a smooth, straight barrier. But that merely pushes the problem back. Why should mathematics furnish a description of even the simplest of fluid flows, and why should that applicability survive the transformations leading to the complicated cases of flows that go around circular cylinders and encounter barriers? Does it all, perhaps, hint at a preestablished harmony between mathematics and nature? Metaphysical responses of this kind have a long history. Famously, Galileo declared that God wrote the Book of Nature and did so in the language of geometry.[35] Such reactions should not be dismissed. They represent an attempt to address a real question, and they are not confined to the past. Even contemporary physicists have been struck by the "unreasonable" effectiveness of mathematics in the natural sciences. The implication is that something beyond reason is at work, something mysterious and even miraculous.[36] In the present case, however, any hint of the noumenal will be quickly dispersed when the *empirical* track record of the theory of ideal fluids is examined. I now turn to this side of the matter.

Paying the Price of Simplification

I have shown how, in order to generate and solve the equations of motion of an ideal fluid, all manner of simplifications had to be introduced. What price, if any, had to be paid for the advantages that the simplifications brought with them? What does it cost, for example, to bring the investigation under the scope of Laplace's equation and confine the theory to motion in which fluid elements do not rotate? Those who developed classical hydrodynamics hoped that the price would not be a large one. The hope was reasonable because water and air were only slightly viscous. Unfortunately, the transition from small viscosity to zero viscosity sometimes had a very large effect on the analysis. In important respects the difference between real fluids and the theoretical behavior of a perfectly inviscid fluid was dramatic. The price of the approximation was high, and it was extracted in a surprising way.

Suppose that you hold a small, rectangular piece of thin cardboard by one of its edges, for example, a picture postcard held by its shorter edge. Move the card rapidly through the air so that the card faces the flow head on (not edgewise). It is easy to *feel* the resistance to the motion, and (within a Newtonian framework) this means feeling the force that the motion through the air exerts on the card. In the same terms, one can also *see* what must be the effect of the force by the way that the card bends. Experts in aerodynamics want to be able to calculate the magnitude and direction of the force that is so evidently present. A good theory would furnish them with an accurate description of the characteristics of the flow and an explanation of how the flow generates the forces. The theory should permit answers to questions such as: Does the air flow smoothly around the card? What are the streamlines like? Does the card leave a turbulent wake in the air? What happens at the edges of the card? How does the force vary with the angle at which the card is held, that is, with the "angle of attack"?

It turns out that if the air were a perfect fluid, there would be *no resultant force at all on the card*. The simplifications led to a mathematically sophisticated analysis but also to a manifestly false prediction. Why is this? The answer can be seen by looking at the form of the flows that are generated under the idealized conditions assumed by the mathematician. The (theoretical) flow of an ideal fluid in irrotational motion over the postcard or lamina normal to the flow looks like the flow sketched in Cowley and Levy's diagram shown in figure 2.5. In the diagram the card is represented as fixed, and the ideal fluid that stands in for the air is shown approaching the card and moving around it. This example differs from the experiment in which the card moves through the air (which is assumed to be at rest), but scientists typically prefer to adopt this convention. The justification is that dynamically the two things are equivalent. As far as the forces are concerned, all that matters is the *relative* motion of the air and the obstacle. Pretending that the card is still and the air moves, rather than the other way round, turns out to be easier because seen from the standpoint of the card the flow is steady. It also makes the diagram fit more closely to experiments that are done in wind tunnels.

From figure 2.5 it can be seen that half of the air (or ideal fluid) that impinges on the front face moves away from B, the front stagnation point, up to the edge C, while the other half (not shown) will move down to the edge C'. The fluid then curls sharply around the corner at each of these edges and approaches the point D (the rear stagnation point) and then continues on its way. The fluid farther from the plate follows a similar path to the fluid near the plate but with less abrupt changes of direction. At the stagnation points the lines representing the flow meet the surface of a body and can be thought of as splitting into two in order to follow the upper and lower contour of the body. At a stagnation point, mathematical consistency is preserved by taking the direction of the flow to be indeterminate and the speed of the flow to be zero.

Inspection of the diagram for the steady flow around the flat plate as shown in figure 2.5 allows the direction and some indication of the speed of the flow to be read off. It can be seen that the flow moves rapidly around the edges of the plate. Inspection also shows that the flow is symmetrical about an axis that lies along the plate as well as being symmetrical about an axis that is normal to (that is, at right angles to) the plate. (A mathematician would spot the symmetry from the equation for the streamline because all the x- and y-terms appear as squares.) These symmetries have important consequences for the pressure that the flow exerts on the plate. According to Bernoulli's law, as the fluid impinges on the plate and is brought to a halt, it exerts its maximum pressure. As it moves along the plate and gathers speed, it exerts a lesser pressure. Because the fluid is perfectly free of viscosity, there will be no tangential traction on the plate and all the forces will be normal to the plate. The pressure on both the front and the back will be high near the stagnation points and low near the edges of the plate. The symmetry of the flow around an axis along the plate means that the pressures exerted on the front of the plate will be of the same magnitude as the pressures exerted on the back of the plate. The pressures will be in opposite directions and will thus cancel out. There will be no resultant force.

The forces on a plate moving relative to a body of ideal fluid are therefore fundamentally different from those on a plate (such as the postcard) moving relative to a mass of real air. Both experiment and everyday experience stand in direct contradiction to the mathematical analysis. Treating a *slightly* viscous fluid (such as air) as if it were a *wholly* inviscid fluid may have seemed a small and reasonable approximation, but the effect is large. Neglecting a small amount of compressibility caused no trouble; neglecting a small amount of viscosity proved vastly more troublesome. The disconcerting conclusion that the resultant force is zero does not just apply to a flat plate running head-on against the flow. Consider again the flow around a circular cylinder. This was Cowley and Levy's other textbook example and was shown

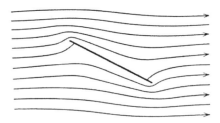

FIGURE 2.6. Continuous flow of an ideal fluid around an inclined plate. From Tietjens 1929, 161. (By permission of Springer Science and Business Media)

in figure 2.4. The closeness of the streamlines indicates that the flow speeds up as it passes the top and bottom of the circular cylinder. Since the fluid is free of any viscosity, all the pressure on the cylinder will be directed toward the center. The symmetry of the flow means that any pressure on the cylinder will be directly counteracted by the pressure at the diametrically opposite point on the cylinder. Again, counter to all experimental evidence from cylinders in the flow of a real fluid, there will be no resultant force on the cylinder. In reality the pressure distribution is not the same on the front and rear faces.[37]

Do these results depend on the obstacle and the flow possessing symmetry? The answer is no. The results apply to objects of all shapes and orientations. Consider the flow around a flat plate that is positioned not normal to the flow but at an oblique angle to the flow. The situation is represented in figure 2.6.

Such a flow introduces certain additional complexities into the analysis, but the outcome is still a zero resultant force. The extra complexity is that the forces at points on immediately opposite sides of the plate are not equal, which can be seen from the way the front and rear stagnation points are not directly opposite one another. The front stagnation point is near to, but beneath, the leading edge, while the rear stagnation point is near to, but above, the trailing edge. As a result the plate is subject to what is called a "couple" that possesses a "turning moment." A "couple" arises from two forces that are equal and opposite but act at different points. Here they exert leverage on the plate, causing it to rotate and to turn so that it lies across the stream. As the plate rotates, the two stagnation points move. The front stagnation point moves away from the leading edge toward the center point of the front of the plate. The rear stagnation point moves away from the trailing edge to the center of the back of the plate. When the plate is lying across the stream, the two stagnation points are directly opposite one another and there is no more leverage they can exert. For the inclined plate, then, there is an effect produced by the forces, but it is still true that there is no *resultant* force. A force at a given point on one surface of the plate will still have a force of equal magnitude and opposite direction at *some* corresponding point on the other face. Overall, the forces will still sum to zero, as they did for the circular cylinder or the plate that was initially positioned normal to the flow.

Similar considerations apply to any complex shape and would therefore also apply to a shape chosen for an aircraft wing. If the air were an ideal fluid, there might be a turning effect exerted on the wing but there would be no resultant aerodynamic force. *There would be neither lift nor drag.* The zero-resultant theorem for plates and cylinders had been established many years before the practicality of mechanical flight had been demonstrated and had long been a source of some embarrassment. Interpreted as a prediction about either air or water, its falsity was evident, but it continued to haunt the theoretical development of aerodynamics. In France this old and well-established result was called d'Alembert's paradox, in Germany it was called Dirichlet's paradox, and in Britain it wasn't called a paradox at all. Remember that, for Cowley and Levy, the mathematics just *defined the nature of the fluid.* The zero-resultant theorem simply establishes that this is how an ideal fluid would behave were there such a thing. Interpreted in this way, the zero-resultant theorem brings out the difference between a real fluid and an ideal fluid, so it can be taken as a powerful demonstration of the unreality of ideal fluids.[38] But if real and ideal fluids are so different, how are theory and experiment ever to be brought into relation to one another? This was a long-standing problem. Failure to resolve it had generated a sharp distinction between hydrodynamics, which was largely a mathematical exercise, and hydraulics, which was largely an empirical practice—hence the two different chapters and the two different authors in the *Encyclopaedia Britannica.*[39] Was aerodynamics to take the path of empirical hydraulics or the path of mathematical hydrodynamics? There were strong social forces pulling in each of these opposing directions.

New Approaches to Ideal Fluid Theory

Something was badly wrong with the picture of air behaving and moving as an ideal fluid. It was mathematically impressive but empirically defective. What exactly was wrong? Was it the assumption of zero viscosity itself that should be dropped or were there perhaps other, unnoticed, assumptions at work in the picture of the flow that might be the cause of the trouble? What about Laplace's equation and the assumption of irrotational motion?

This question was addressed by a number of late nineteenth-century experts whose investigations greatly deepened the understanding of ideal fluid theory. They began to explore some possibilities that previously had been neglected. But why did they not simply abandon ideal-fluid theory as empirically false and turn directly to the analysis of viscous fluids? Attempts were made to do this but with very limited success. The reason was that the mathematics of viscous fluids was so difficult. It was possible to write down the equations of motion of a viscous fluid by taking into account the traction forces along the surface of the fluid element, but it was another matter to solve the equations. The full equations of viscous flow are now called the Navier-Stokes equations, though in Britain they used to be called just the Stokes equations. Of course, they contain a term involving the symbol μ standing for the coefficient of viscosity. If the value of this coefficient is set at zero to symbolize the absence of viscosity, that is, $\mu = 0$, the Navier-Stokes equations turn back into the Euler equations that have been described earlier in this chapter. In a later chapter I look more closely at the status of the Navier-Stokes equations and the different responses to their seemingly intractable nature. For the moment it is only necessary to appreciate the problem they posed. No one could see how to solve and apply the equations except in a few simple cases. Because of their intractability, any attempt to avoid the impasse thrown up by the zero-resultant theorem had to be one that stayed with the Euler equations and thus within the confines of ideal-fluid theory.

The crucial insight that permitted the further development of ideal-fluid theory was provided by Helmholtz and Kirchhoff in Germany and Rayleigh in Britain. These men realized that the solutions to the Euler equations that gave the streamlines around an obstacle were not unique. More than one set of streamlines were possible and consistent with the equations. More than one kind of flow could satisfy the equations and meet the given boundary conditions. What is more, some of these flows could generate a resultant force. There were in fact two very different kinds of flow that might have this desired effect and, in principle, allow the zero-resultant outcome to be evaded. Rayleigh contributed to the study of both. Both approaches involved the limited introduction of fluid elements that possessed rotation and vorticity. The strict condition of irrotational motion was dropped. On one approach this involved the introduction of just one singular point in the flow that rotated and constituted the center of a vortex. On the other approach a sheet or surface of vorticity was postulated. In both cases the remainder of the flow was still irrotational. These two approaches provided, respectively, the basis for the two different theories of lift that I mentioned in the introduction and called the circulatory or vortex theory of lift and the discontinuity theory of lift. Historically, the first of the two approaches to be developed in detail was the one that led to the discontinuity theory. I now introduce the ideas underlying this approach. The other approach and the other theory of lift are introduced in chapter 4.

Surfaces of Discontinuity

Consider the idealized model of the postcard experiment, that is, the stream-lines around a flat plate normal to the flow. The flow could take the form shown in figure 2.7 as well as that already shown in figure 2.5. Instead of curl-ing around the edges of the plate and moving down the back of the plate, the flow of ideal fluid can break away at the edges. Behind the plate the flow is not a mirror image of the flow in front of it but consists of a body of "dead air," or dead fluid, bounded by the moving fluid which has met the plate, moved along the front face of the plate, and separated at the edges. The pressure in the "dead air" will be the same as that at a great distance from the plate and can be equated with the atmospheric pressure. In a real, viscous fluid, the moving fluid and the dead or stationary fluid would interact. There would be a transition layer, with a speed gradient created by the stationary fluid retarding the moving fluid while the moving fluid sought to drag the station-ary fluid along with it. In an ideal fluid there will be no such transition layer because there will be no traction between the two bodies of fluid. The free

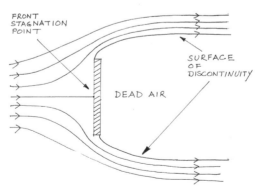

FIGURE 2.7. Discontinuous flow of an ideal fluid around a barrier normal to the free stream. The surfaces of discontinuity or "free streamlines" represent the abrupt change between the moving fluid and the dead fluid behind the barrier.

stream will pass smoothly over the dead fluid so there will be a sudden transition from fluid with zero velocity to fluid with a nonzero velocity. Mathematicians call this sudden transition a discontinuity in the velocity because there are no intermediate values. This term gives rise to the general label for flows of this kind, which are called discontinuous flows. The streamline that marks the mathematically sharp discontinuity between the moving and stationary bodies of fluid is called a free streamline. It is a line of intense vorticity along which the flow possesses rotation in the technical sense defined earlier in the chapter.

This attempt to make mathematical hydrodynamics more realistic was introduced by Helmholtz in 1868 in a paper titled "Über discontinuirliche Flüssigkeits-Bewegungen" (On discontinuous fluid motions).[40] Helmholtz argued that all the flows that had produced d'Alembert's paradox had depended not only on the assumption that the flow was inviscid but also on the assumption that the velocity distribution was continuous. Helmholtz explored flows involving surfaces of separation (*Trennungsfläche*) or (what is mathematically equivalent) sheets of vorticity (*Wirbelfläche*).[41] Of course, said Helmholtz,

> Die Existenz solcher Wirbelfäden ist für eine ideale nicht reibende Flüssigkeit eine mathematische Fiction, welche die Integration erleichtert. (220–21)

> The existence of such a vortex sheet for an ideal inviscid fluid is a mathematical fiction to make the integration [of the equations] easier.

But fiction or no fiction, Helmholtz had raised the hope that the glaringly false consequences of the standard picture of ideal-fluid flow could be avoided.

If a steady, discontinuous flow is to be possible, certain conditions must be satisfied. It must be the case that the static pressure on either side of the free streamline is the same, otherwise the flow pattern would not be in equilibrium and would modify itself. Since the flow at a great distance in front of the plate is assumed to have a constant speed V and to be at atmospheric pressure p_a, while the dead air is also at atmospheric pressure, then the speed of the flow along the free streamlines that bound the dead air must also be V. This conclusion follows from Bernoulli's equation relating speed and pressure. Bernoulli's law also leads to the conclusion that a flow of this kind will generate a greater pressure on the front of the plate than on the back.

Consider the streamline that terminates at the stagnation point at the front of the plate. What is the pressure on the front of the plate at the stagnation point? Call the pressure p_s. Everywhere along the streamline that goes to the stagnation point, the static and dynamic pressure will sum to the same constant value, that of the Bernoulli constant or the total pressure head. The value of the constant, or the total pressure, at a distance from the plate is $H = p_a + \frac{1}{2} \rho V^2$. On the plate, at the stagnation point, the speed is zero. There will be no dynamic pressure but only a static pressure that will equal the total pressure, therefore $p_s = H = p_a + \frac{1}{2} \rho V^2$. The pressure produced by bringing the air to a standstill at the stagnation point thus exceeds the atmospheric pressure p_a by the quantity $\frac{1}{2}\rho V^2$. But the pressure on the back of the plate is also p_a, so at this point there is an excess pressure on the front of the plate.

This argument only applies to the stagnation point, which is the point of maximum pressure on the front of the plate. What happens at other points on the front of the plate as the fluid moves away from the stagnation point and moves toward the edges? The fluid will speed up so its pressure will drop. But the pressure exerted by the moving fluid only drops to atmospheric pressure as it reaches the free stream velocity at the edges. It follows that, 'at all points on the front of the plate, there will be a higher pressure than the atmospheric pressure on the rear of the plate. On this account, therefore, the forces on the plate do *not* cancel out, except at the very edges, and there *is* an overall resultant aerodynamic force on the plate.[42]

Discontinuous flows of this kind thus avoid the paradoxical-seeming zero-resultant outcome found by D'Alembert, but it is still necessary to ask whether the predicted forces are the right size. It is one thing to avoid a blatantly false outcome and another thing to do so by giving the right answer in quantitative terms. The question still remains: Do the forces predicted on the basis of discontinuous flow fully correspond to the observed forces? Quantitative knowledge of the forces on the plate calls for a quantitative knowledge of the speed and pressure of the flow along the front of the plate, not just at the stagnation point and the edges. Until this information could be provided, the picture was merely qualitative. Working independently of one another, Rayleigh and Kirchhoff provided testable answers.[43]

The quantitative analysis of discontinuous flows was not an easy task, but by the use of ingenious transformations, it proved possible to connect the discontinuous flow around a flat plate to the simple, uniform, horizontal flow. There was no guaranteed way to find the required steps leading to the simple flow. It called for a high order of puzzle-solving ingenuity. The character of the thinking required can be glimpsed from the first few steps of the process. Rayleigh and Kirchhoff noticed that in the original flow, the *direction* of the boundary streamline along the plate was known but not the velocity. For the free streamline, the reverse held: the *velocity* was known but not the direction. If the flow could be redrawn on a diagram where one axis was proportional to speed while the other axis was proportional to direction, then both parts of the streamline would be transformed into straight lines. This was a step toward the desired simplicity because the straight lines could be interpreted as "polygons" of the kind to which the Schwarz-Christoffel theorem could be applied. Neither Kirchhoff nor Rayleigh explicitly used the Schwarz-Christoffel theorem but used a number of ad hoc transformations to achieve the same goal.[44] But once the formula describing the flow had been found, pressures and velocities could be calculated and quantitative predictions made.

Rayleigh's achievement was to generalize Kirchhoff's analysis, which dealt with plates that were normal to the flow, and give the analysis required for plates that were oblique to the flow. This classic result in hydrodynamics was published in 1878 and provided the starting point for the work of the Advisory Committee for Aeronautics when its members tried to explain the lift generated by an aircraft wing. The work was officially overseen by Rayleigh himself as president of the ACA. It was monitored on a day-to-day basis by Glazebrook and other mathematical physicists who were closely associated with Rayleigh. In the next chapter I describe this early British work on the lift and drag of a wing, which was based on the idea of discontinuous flow.

Nonmathematical Summary

The main points of this chapter may be summarized by the following ten items. The list begins with a résumé of some of the terminology of the field. This terminology is taken for granted in the subsequent discussion.

1. A flow is called a two-dimensional flow when it can be drawn in cross section and the drawing taken as representative of the flow at any other cross section. Thus if the flow around a barrier, or some other obstacle, is drawn in two dimensions, this ignores the complications introduced into the flow by what happens at the edges not shown in the picture (that is, below the page and above the page). If the drawing shows the cross section of, say, a wing, then the picture does not portray what is happening at the wingtips, that is, the third dimension of the situation. This absence can be justified if the wing is very long and the immediate concern is with the flow at parts of the wing that are distant from the tips. The diagram can then adequately represent the flow around the central sections of the wing. A wing that is long enough to justify this approximation is often called an infinite wing. The word "infinite" is much used in hydrodynamics. References to, for example, "the flow at infinity" usually mean the flow as it is at a great distance from some disturbance so that the effects of the disturbance can be ignored.

2. The main theoretical resource used in early aerodynamics came from

classical hydrodynamics. Hydrodynamics offered a mathematically sophisticated theory of the flow of an "ideal" fluid, that is, a fluid that was incompressible and also completely devoid of viscosity. Of these two idealizations, the most contentious was the neglect of viscosity. The differential equations that govern the flow of an ideal fluid are called the Euler equations. These equations give the speed of the flow at a specified position and time. To make it easier to solve these equations, mathematicians introduced two further idealizations. First, it was often assumed that the flow was steady. This meant that rate of change with time was zero and could be ignored. Second, it was assumed that the fluid elements did not rotate. There was an emphasis on irrotational flows because it simplified the mathematics. Unfortunately the benefit of mathematical simplicity was purchased at the cost of making the flows being analyzed less than realistic as models of real fluid flows.

3. Fluid elements (that is, the small volumes of fluid whose velocities and rotations are under study) are not to be identified with molecules or atoms or material particles, although occasionally such identifications seem to have been made. Fluid elements are mathematical abstractions that enable the methods of the differential and integral calculus to be applied to fluids.

4. One logical consequence that can be derived from the Euler equations is a highly useful result called Bernoulli's law. With the assumption of a steady, irrotational, and incompressible flow, the law takes on a simple form. It states that the pressure and the velocity at a point in the flow are related by a simple law that implies that as the speed increases the pressure will decrease, and as the speed decreases the pressure increases. Speed and pressure trade off against one another. The use of the law makes it important to distinguish between three different meanings that are attached to the word "pressure." There is (i) static pressure, (ii) dynamic pressure, and (iii) total pressure. Total pressure equals the sum of static pressure and dynamic pressure. Static pressure is the pressure on the sides of a pipe or the surface of a wing. Total pressure is the pressure felt when a body of fluid is brought to a standstill. Dynamic pressure is the name given to the quantity $\frac{1}{2}\rho V^2$, where ρ is the density of the fluid and V is the speed of flow. In the simplified conditions dealt with in early aerodynamics, the total pressure can be considered to have a constant value. As speed V increases and hence dynamic pressure increases, then static pressure must go down. Care is needed to ensure its correct application, but Bernoulli's law plays an important role in (i) calculating the forces on an object that is immersed in a flowing fluid, for example, a wing in a stream of air, and (ii) understanding the operation of instruments such as the Pitot probe, which registers total and static pressure and (via Bernoulli's law) permits the computation of velocities.

5. The restriction to irrotational flow permitted the mathematical description of a wide variety of two-dimensional flows such as the flow of a steady stream around a circular cylinder and the flow around a barrier facing head-on into the stream. The streamlines of these flows could be drawn on the basis of the formula (called the stream function) that furnished the mathematical description of the flow. In a steady flow (but not in an unsteady flow) the streamlines give the path taken by the fluid elements. As well as streamlines a flow can be described by what are called lines of equal potential. These are orthogonal to the streamlines except at points called stagnation points, which are points where streamlines come to a halt on the surface of a body. Streamlines and potential lines can be switched in the sense that the potential lines can be interpreted as the streamlines of a new flow. The old streamlines then become the potential lines of the new flow. Just as the streamlines are specified by the stream function so the potential lines are specified by a potential function.

6. The possibility of arriving at a mathematical description of a flow was greatly improved because a large number of familiar mathematical functions (called functions of a complex variable) turned out to be interpretable as possible fluid flows. The geometrical patterns generated by these functions (that is, the lines of the curves plotted on graph paper) could be read as the patterns made by a flowing fluid and the boundaries that constrain them. Exploring a function and then giving it an after-the-fact interpretation in terms of a flow of ideal fluid was called the indirect method of arriving at the equations of the flow. I illustrated this process by means of a function that could be understood as describing the flow of a uniform stream around a circular cylinder. The example, along with the overall presentation of the material in this chapter, was taken from one of the standard textbooks of the World War I period, namely, Cowley and Levy's *Aeronautics in Theory and Experiment* published in 1918.

7. A more direct line of attack was sometimes available to the mathematician in search of a mathematical description of a flow pattern. This method involved constructing a set of equations that related the flow to be understood to a very simple flow that was already understood, for example, the uniform flow along a straight boundary. If the boundaries of the simple flow could be transformed into the boundaries of the more complicated flow, then the methods of transformation would also turn the simple streamlines into the more complicated streamlines of the desired flow. A particular set of transformations called conformal transformations played a central role in this process. Many such transformations had been studied as exercises in pure mathematics and geometry but were found to be important resources

in the study of fluid flow. One such important transformation was called the Schwarz-Christoffel theorem.

8. The main problem with the hydrodynamics of an ideal fluid was that, although it became mathematically sophisticated, it appeared to provide no resources for explaining the resistance that an object experiences when placed in the flow of a real fluid such as water or air. When an ideal fluid flows around, say, a flat plate or a circular cylinder, the flow exerts no resultant force on the object. This is often called d'Alembert's paradox, although whether it is a paradox in the true sense of the word is examined in more detail later. What is beyond dispute is that the result presented a problem for anyone who wanted to understand the air by likening it to an ideal fluid. The use of the theory of ideal fluids led to the false result of zero resistance or zero drag.

9. One possible response to this "paradoxical" result would be to reject ideal fluid theory as useless for the study of real fluids such as air. Why not develop a more realistic hydrodynamic theory devoted to viscous fluids? This project was begun, and the equations of motion of a viscous fluid were formulated. They are now called the Navier-Stokes equations. (The British just called them the Stokes equations.) Frustratingly they could only be solved in a few very simple cases, which gave special significance to the search for new ways to make ideal-fluid theory more realistic. In principle there were two ways to do this—hence the existence of two competing theories of lift. Only one of these ways (called the theory of discontinuous flow) was described in this chapter. The alternative, the vortex theory, is discussed in chapter 4.

10. The theory of discontinuous flow was proposed by Helmholtz and carried forward by Kirchhoff and Rayleigh. Helmholtz argued that the "paradoxical" result of zero resistance or drag arose because an ideal fluid could wrap itself around an object and exert pressure from all sides in a way that canceled out any resultant force. The discontinuous flow approach exploited the possibility that there could be discontinuities in the velocity of different bodies of ideal fluid that were in direct contact with one another. The flow was assumed to break away from the edges of an obstacle and create a wake behind it. The wake would be "dead water" or "dead air," and the main body of ideal fluid would flow past it. (The assumption here is that the body is stationary and the fluid moving. This is the situation of a model airplane in a wind tunnel.) Such a flow pattern in an ideal fluid, with a wake of dead fluid, turned out to be compatible with the Euler equations. Furthermore, it could be established that, given such a discontinuous flow (see fig. 2.7), the pressure on the front face of an object would be greater than the pressure of the dead fluid on the rear. The forces did not cancel out and d'Alembert's paradox was

avoided. If the resultant force proved large enough, here was a theory that could, in principle, explain the lift of a wing as well as the resistance to motion, the "drag." Such was Rayleigh's idea for explaining the lift on an aircraft wing, and it was taken up by the British Advisory Committee for Aeronautics. The results that emerged from the theoretical and experimental study of this model are described in the next chapter.

Early British Work on Lift and Drag:
Rayleigh Flow versus the Aerodynamics of Intuition

To the scientist an aeroplane is merely a complex body moving through a fluid, and until he understands how a simple body moves he has no chance of understanding the fundamental principles of aeronautics.

G. I. TAYLOR, *"Scientific Method in Aeronautics"* (1921)

The research agenda drawn up at the Admiralty and endorsed at the first meeting in the War Office accurately prefigured the approach that was to be adopted by the members of the Advisory Committee in their work on lift and drag. The immediate research aim was to provide a mathematical analysis that would predict the forces exerted on a flat or curved plate immersed at an angle to a flowing fluid. Of course, this was not the ultimate aim. The plate was to function as a simple model of an aircraft wing, and the mathematically idealized fluid, necessary to perform the calculations, was to act as a model of the air. To calculate the forces, researchers needed a precise and quantitative picture of the flow around the wing. What would that flow look like? For the British, the best available guess was provided by Rayleigh's important work on discontinuous flow. Although the work was over thirty years old, and it was obvious to everyone that the analysis was highly idealized, it appeared to the Advisory Committee that here was the rational place to start. Initially, therefore, as far as lift was concerned, all the research effort of the ACA, both theoretical and experimental, went into studying the theory of discontinuity. I now describe this work and then, later in the chapter, contrast it with the ideas about lift put forward by the leading representative of the "practical men." The contrast in style is stark.

Rayleigh's Paper of 1876

In 1876 Rayleigh had published a paper called "On the Resistance of Fluids."[1] It contained one of the most striking results of classical hydrodynamics, which came to be reproduced in all the advanced treatises on the subject. By using conformal transformations Rayleigh had arrived at a formula for the force exerted on an inclined, flat plate subject to a uniform flow of an ideal fluid (see fig. 3.1). The plate is at an angle α to a horizontal flow, and the fluid has a density ρ and a speed V. If the length from the leading to the trailing edge of the plate is l, then the resultant force R was

$$\frac{\pi \sin \alpha}{4 + \pi \sin \alpha} l \rho V^2.$$

The resultant is perpendicular to the plate, that is, inclined backward at an angle α so that the vertical (lift) component would be $R \cos\alpha$ and the horizontal (drag) component would be $R \sin\alpha$. Rayleigh was also able to work out the position of the center of pressure, that is, the precise distance of the resultant force from the leading edge of the plate. The analysis was carried out in two dimensions, that is, the plate was assumed to be very long. The diagram thus represents a cross section in the middle of the plate, and what happens at the ends of the plate is ignored. The dead fluid can be seen to form a "wake" stretching downstream to an indefinite extent.

At the time of its publication Rayleigh had presciently remarked that his result had interest because "it will be of vital importance in the problem of artificial flight" (431). The diagram in figure 3.1 is thus a drawing of a wing. Nearly thirty years before the success of the Wright brothers, Rayleigh had been attuned to the problem of explaining the lift of an aircraft wing and had offered a theoretical analysis, and perhaps even a solution, to the problem of how it generates lift. Rayleigh had the reputation, as a physicist, of being somewhat conservative.[2] His forward-looking orientation to the problem of explaining lift is therefore all the more noteworthy, given that contemporaries such as Lord Kelvin were declaring that they had "not the smallest molecule

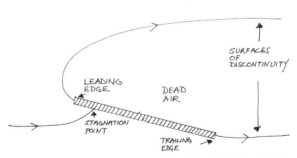

FIGURE 3.1. The discontinuous flow of an ideal fluid around an inclined plate is often called Rayleigh flow. Rayleigh saw this flow as a model of the flow around an aircraft wing and in 1876 calculated the resultant aerodynamic force on the plate.

FIGURE 3.2. John William Strutt, Lord Rayleigh (1842–1919). Rayleigh, who had made classic contributions to fluid dynamics, became the first president of the Advisory Committee for Aeronautics. He held this post from the committee's inception until his death in 1919. From Schuster 1921. (By permission of the Royal Society of London and the Trustees of the National Library of Scotland)

of faith in aerial navigation."[3] In 1909 the task the Advisory Committee set itself was to see if, and how, Rayleigh's early approach could be carried further. Greenhill, the mathematician on the committee, and himself an authority on hydrodynamics, set about consolidating and extending Rayleigh's mathematical analysis. Mallock and his collaborators, meanwhile, surveyed the work of other laboratories and used the resources of the NPL to gather relevant empirical data.

When Rayleigh (see fig. 3.2) first published his formula, there was little experimental data available on air or water resistance against which it could be tested. He had to rely on some old experiments made with an unsatisfactory type of apparatus called a whirling arm, which consisted of a horizontal beam or arm that rotated around a vertical axis. The arm had the test object at one end and a means (such as a spring) for measuring the force on it. The problem with an apparatus of this kind was that the test object was repeatedly exposed to the turbulence caused by its previous orbits. These experiments had, however, shown that resistance depended on the sine of the angle of incidence, and this result accorded with Rayleigh's formula.

The first impression, in 1876, was that theory and experiment agreed "remarkably well" (437). The impression did not last. By 1891, when Rayleigh reviewed Langley's *Experiments with Aeronautics* for *Nature*, he knew that the experimentally determined relation between the angle of incidence and the aerodynamic force on a flat plate diverged from that stated in his formula.[4] If the results are expressed as a graph and the lift force on the plate is plotted against angle of incidence, the curve that derives from Rayleigh's theory is strikingly different from that derived from experimental measurements of these quantities.[5]

Rayleigh had a good idea where the trouble lay. It was a matter of what happened on the rear face of the plate. He cited experiments in which the pressure on the back of a plate had been measured. This was done by using a hollow plate and making a hole in the rear surface. A thin pipe was led from the hole through the hollow plate and connected to a manometer. These measurements showed the presence of a suction effect. Rather than the assumed atmospheric pressure, there was a *lowering* of pressure, which was inconsistent with the model. As Rayleigh conceded, "It will naturally be asked whether any explanation can be offered of the divergence . . . from the theoretical curve. . . . It seems probable that the cause lies in the suction operative, as a result of friction, at the back of the lamina. That the suction is a reality may be proved without much difficulty by using a hollow lamina . . . whose interior is connected with a manometer" (495). Rayleigh's 1876 analysis, of course, was based on ideal fluid theory, and therefore the results of friction and viscosity had been ignored.[6]

Rayleigh had also ignored all the eddies in the flow that would be expected on the basis of simply observing, say, the flow of water past a barrier. Lord Kelvin (Sir William Thomson) objected in the pages of *Nature* that the "dead water" behind a barrier, or the "dead air" on the upper surface of the wing, was far from dead: it was full of turbulence and instability.[7] But if the dead fluid wasn't really dead, then the free streamlines would be unstable and the entire picture of the flow would be compromised. Rayleigh did not deny that there was a problem here but initially sought to play down its significance. Thus, "it was observed by Sir William Thomson at Glasgow, that motions involving a surface of separation are unstable. . . . But it may be doubted whether the calculations of resistance are materially affected by this circumstance, as the pressures experienced must be nearly independent of what happens at some distance in the rear of the obstacle, where the instability would first begin to manifest itself" (437). And there the matter was left.

In the analysis of the lift and drag of a wing, the situation was therefore very different from that which prevailed in the study of stability. The program of research begun in 1909 by the Advisory Committee was, from the outset, plagued with doubts and anomalies. Quantitatively, the predicted lift and drag were not accurately rendered by Rayleigh's model, and qualitatively they did not seem to correspond to what was known about the physical features of the flow. Undeterred, the committee pressed on with their program of research.

Greenhill's Memorandum

Greenhill's task was to carry Rayleigh's mathematical analysis forward. The result was the Advisory Committee's lengthy Reports and Memoranda No. 19, published in 1910. Its full title was "Report on the Theory of a Stream Line Past a Plane Barrier and of the Discontinuity Arising at the Edge, with an Application of the Theory to an Aeroplane."[8] (The word "aeroplane" here means "wing.") Greenhill addressed neither the empirical shortcomings of Rayleigh's model (acknowledged by Rayleigh himself), nor the issue of instability and turbulence in the flow that had been raised by Kelvin (and brushed aside by Rayleigh). He discharged his duty by assembling everything that was known about the mathematics of discontinuous flow. As Greenhill put it, "the object of the present report is to make a collection of all such problems solved so far, and to introduce a further simplification into the treatment" (3). It was not unsolved problems but solved problems and their further refinement that engaged Greenhill's attention. Particular attention was given in the report to the work of two other Cambridge mathematicians, Michell and Love. J. H. Michell was fourth wrangler in 1887. He became a fellow of Trinity in 1890 and a fellow of the Royal Society in 1902. In 1890 Michell had written a seminal paper titled "On the Theory of Free Stream Lines."[9] A. E. H. Love's Cambridge credentials have been mentioned in chapter 1. His paper "On the Theory of Discontinuous Fluid Motions in Two Dimensions" was published in 1891 and provided a development of Michell's work.[10]

Michell and Love introduced two new methods into the repertoire for turning free streamlines into simple straight line flow. One method was to make a transformation by taking logarithms. Such a transformation has the effect of turning the arc of a circle into a straight line. It led to a new diagram of the flow which had the angle of flow as one axis and the logarithm of the reciprocal of the speed as its other axis. In this way the streamlines were turned into a polygon in the extended, mathematical sense of the word. The other contribution was to make explicit use of the Schwarz-Christoffel theorem, which put the process of finding the necessary transformations on a more systematic basis. Compared with Rayleigh's original discussion, the

FIGURE 3.3. Sir Alfred George Greenhill (1847–1927). Fellow of Emmanuel College, Cambridge, and professor of mathematics at the Woolwich Arsenal. Greenhill was one of the founding members of the Advisory Committee for Aeronautics and wrote a detailed report on the theory of discontinuous flow as a basis for aerodynamic theory. Photo by J. W. Hicks, in H.F.B. 1928. (By permission of the Royal Society of London and the Trustees of the National Library of Scotland)

number of transformations had been increased, but the new approach gave the analysis a more routine character, and Greenhill (see fig. 3.3) applied it, indefatigably, to case after case.

There is no doubting the mathematical sophistication of the material that Greenhill gathered together. The report was a virtuoso display of wrangler skills. One could say that it was Tripos aeronautics in full flight—but for one oddity. Where were the airplanes? The cases he collected appear to have little to do with aeronautics. Apart from comparisons with electrical phenomena, drawing heavily on the work of Maxwell and J. J. Thomson, the bulk of the examples treated themes such as the flow of water through orifices, spouts, and mouthpieces. Jets of water impinge on plates, and water flows through channels, around barriers, past piers, and over weirs. Walls, bridges, and pillars feature more prominently than flying machines. The puzzle is not the extent to which the discussion deals with water rather than air; these can properly be dealt with together. The problem is why this particular range of examples has been introduced. Given that Greenhill had little to offer that might directly strengthen the connection between Rayleigh's mathematics and the wings of aircraft, what did he take himself to be doing?

The clue lies in the diagrams. All the cases that Greenhill discussed could be reduced to simple configurations of straight lines. They were all shapes that could be turned into the "polygons" needed for the application of the Schwarz-Christoffel theorem. Interpreting them as "mouthpieces," "reservoirs," "weirs," "piers," and the like was distinctly post hoc. This was especially true in the few cases where an attempt was made to link the diagram to aeronautics. Thus Greenhill gave an analysis, analogous to Rayleigh's, of a flow against an inclined plane, but the main line in the diagram was positioned between two further lines, one above it and one below it. This, said Greenhill, "may be taken to represent a rudder boxed in" between two planes (17). A variant of this figure effectively dispensed with the upper line by locating it at infinity, "so that the analysis will serve for an aeroplane flying horizontally near the ground" (20). The words "may be taken to represent" and "will serve for" reveal the derivative character of the interpretation. The "indirect method" was at work. The examples had been gathered, not because of their relation to wings or aircraft, but because of their relation to a certain, favored mathematical technique. Relevance to what Greenhill called the "analytical method" of the report—which from the outset he identified as the deployment of the Schwarz-Christoffel theorem—was the real principle of selection.[11]

Greenhill's report to the Advisory Committee was modeled on a species of document characteristic of Cambridge mathematics and perhaps unique to it. It had become customary for the examiners of the Mathematical Tripos to publish the questions they had set in the previous year, compiling books of problems along with their approved solutions. This practice contributed to a cumulative archive of mathematical work which progressively deepened and refined the Tripos tradition.[12] The archive was vital for the coaches in honing the skills of the next cohort of would-be wranglers. It enabled them to identify the main theorems that would be tested so that they could teach their students to recognize all the possible applications of the result, however diverse the fields, and disguised in outer form, they might be. Routh had published such a collection in 1860 when he had acted not only as a coach but also as an examiner.[13] The most famous collection was Joseph Wolstenholme's *Mathematical Problems* of 1867.[14] Greenhill himself, after his stint as a Tripos examiner, had published *Solutions of the Cambridge Senate-House Problems and Riders for the Year 1875*.[15] The 1910 report was just such a collection of problems and solutions. One hears the voice of the conscientious coach as Greenhill provided useful hints to his readers to help them avoid errors and traps. "The signs are changed when the area is to the left hand," warned Greenhill, "so it is useful to employ an independent check of the sign" (5). "It simplifies the work to take $i = \infty$" (6). Again, "we introduce an angle φ, not to be confused with φ the velocity function" (10).

Although its contribution to aeronautics was close to zero, Greenhill's R&M 19 soon joined the papers of Michell and Love in the list of canonical sources that were cited in Horace Lamb's *Hydrodynamics*.[16] A. S. Ramsey, of Magdalene College, who had been seventh wrangler in 1889, did not mention Greenhill by name but introduced his extensive discussion of discontinuous flow in his 1913 *Treatise on Hydromechanics* by saying, "such problems have recently acquired a new interest because of their relation to Aerodynamics."[17] Others of a more practical bent were less appreciative. The review in the *Aeronautical Journal* for 1911 was signed "B.G.C."—presumably Bertram G. Cooper, who was to become the editor in 1913. Cooper was exasperated by Greenhill's report: there were 96 large-format ("foolscap") pages of text and 13 sheets of diagrams with, "on the average, about 8 lines of the vernacular to each fsc. page, the rest being mathematical equations."[18]

> It would doubtless be expecting too much of human nature to ask that the mathematician and the practical man should make up their minds to cooperate. Only, however, by a reasonable combination of the methods of both can the best results be obtained. If, therefore, the Advisory Committee were to lay their heads together and produce a volume giving a *quantitative comparison between solutions of problems as calculated mathematically and as obtained by actual experiment*, they would clear the ground enormously, and incidentally would do something towards fulfilling the function which the average man (doubtless from the depths of his ignorance) considers that they exist to perform. The publication of an expensive work, such as this, giving no results or deductions in English, is highly to be regretted. (94)

The anonymous reviewer in *Flight* was equally aghast, calling it the "most extraordinary book yet published relating to the subject of aeronautics."[19] It would be unintelligible to 9,999 out of every 10,000 potential readers. Would some other member of the Advisory Committee, asked the reviewer, please write a nonmathematical report explaining the "practical deductions" to be drawn from Greenhill's work? The reviewers in the technical journals clearly believed that the Advisory Committee was throwing down the gauntlet to the practical men, and their reaction was predictable. It would seem, however, that these robust responses from the nonmathematical reviewers had an effect. Subsequent publications, when not entirely empirical, typically involved

a comparison of theory and experiment. Nothing quite like Greenhill's report
was seen again.[20]

Greenhill's Lectures at Imperial College

Greenhill's contribution was not confined to the daunting R&M 19. As well as
working on the mathematics of gyroscopes and problems of airship stability,
in 1910 and 1911 he gave a series of lectures at the Imperial College of Science
and Technology. The course was published a year later in a book titled *The
Dynamics of Mechanical Flight.*[21] Greenhill explained that the lift of a wing
depended crucially on "the opening out of the stream lines" (40) behind the
wing. This occurrence would be the expected effect of the surfaces of discontinuity enclosing the "dead" air region above and behind the wing. Greenhill
went on to contrast the truth, as he saw it, of the picture of discontinuous
flow with the error of certain popular conceptions about the flow of air over
a wing. (The two different ideas of the flow are represented in figs. 13 and 14
in his book.) The passage in which he contrasts them is a revealing one: "A
popular figure of the stream lines past a cambered wing as here in Fig. 13,
showing no such broadening, would imply at once to our eye an absence of
all thrust and lift; the figure should be more like Fig. 14" (41). The diagrams
to which Greenhill was referring are shown as my figures 3.4 and 3.5 (with
Greenhill's numbering identified in the captions).

   Greenhill's second diagram that was meant to describe the correct flow
indicates turbulence in the "dead" air, though his mathematical analysis does
not make provision for this. The embellishment seems to be a concession to

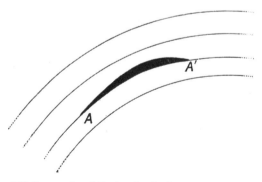

FIGURE 3.4. Greenhill's figure 13. Greenhill referred to this figure as a popular, but erroneous, conception of the flow of air over a wing. He argued that it would generate neither lift nor drag. From Greenhill
1912, 41.

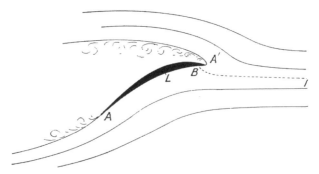

FIGURE 3.5. Greenhill's figure 14. This figure represented Greenhill's conception of the correct picture of the airflow over a wing, which corresponds to discontinuous Rayleigh flow. From Greenhill 1912, 41.

Kelvin. We know why Greenhill believed that there was lift in the case of the discontinuous flow, because of the reasoning set out by Rayleigh, but why was it obvious to Greenhill's eyes that the popular flow picture, my figure 3.4, would be devoid of all thrust and lift? The reasoning may have been that without surfaces of discontinuity, the flow must have the character of the original, continuous flow of an ideal, frictionless fluid—with the "paradoxical" result of zero-resultant force. It looks as if Greenhill took the diagram of smooth, streamlined flow over a wing to imply that the air was being treated as a continuous, ideal fluid in irrotational motion. In other words, it was taken as a flow in which there would be no resultant and where d'Alembert's paradox would be applicable.

If this was the reasoning, then two significant details of Greenhill's first drawing were wrong. The flow is not pictured accurately at the leading or trailing edge. The front stagnation point should be below the leading edge, while the rear stagnation point should be on the upper surface of the wing in front of the trailing edge. Instead the air is shown coming away smoothly from the trailing edge itself. Greenhill would certainly have noticed this error, and he gave the correct form of the diagram for a flat plate on page 47 of his book. He presumably put the inaccuracy down to the approximate character of the "popular" representation, as he had in an earlier criticism of drawings of leading-edge flow (22). In any event, he seems to have taken the popular diagram as an attempt to depict the kind of idealized, continuous perfect fluid flow where, as any mathematician would know, all the forces (except the turning couple) canceled out.

Although Greenhill made more effort in the book than in the ACA report to bring real aircraft into the discussion, it was still full of examples and mathematical technicalities of questionable relevance. Predictably, it did not go down well with the practical men. The anonymous reviewer for *Aero-nautics*, who had apparently attended the lectures, said the book confirmed the earlier impression that the calculations were really aimed at providing a diverting recreation for the mathematical mind. "Practical value they lack wholly; the data on which Sir George Greenhill's mathematical excursions are based are theoretical without fail."[22]

G. H. Bryan's review was very different.[23] Up to the present, said Bryan, there had been a lack of understanding about the role of mathematics in aeronautics. The subject has failed to attract our best mathematicians, while "practical men" make claims "in utter disregard for the fundamental prin-ciples of elementary mathematics and physics" (264). Under these "chaotic conditions" it would be useful to have a work "by so reliable a mathematical authority" as Greenhill (265). Bryan acknowledged the presence of drastic simplifications involved in Greenhill's approach but insisted that, in spite of these shortcomings, "the theory of discontinuous motion affords the best opening to the study of pressures on planes from the mathematical stand-point" (266). He listed the "great mathematicians" who had developed the theory, but noted that it had only been applied to flat plates and not yet to bent or cambered planes (which would make better models of the aerofoils in practical use). Some calculations of this kind, the reader was told, were now under way. Bryan did not once ask if the theory of discontinuous flow gave empirically adequate answers. Rayleigh, of course, knew that, as the theory stood, it did not give the right answers, and so did his experimentally inclined colleague Mallock. I now turn from the mathematical to the experimental study of discontinuous flow to see how matters were carried forward on this front.

The Empirical Study of Rayleigh Flow

Reports and Memoranda No. 16, of September 1909, was a note prepared by Mallock on experiments that involved moving flat and curved plates through a tank of water and measuring the forces on them.[24] The plates were rect-angles whose sides were in the ratios of 2:1, 1:1, and 1:2 and were inclined at various angles between 0° and 90° to the direction of motion. Measurements were taken at various speeds between 150 and 500 feet per minute. The plates were attached to scales in order to read off the forces exerted on them. The resultant force on the plate was broken down into a resistance force (the drag directly opposing the motion) and a lateral force at right angles to the direc-tion of travel (a lift force). A graph of the results showed that the measure-ments "differ greatly from the values calculated on the assumption that the pressure on the rear surface is uniform and equal to that of the fluid at a distance" (40), that is, calculated on the assumption that there was "dead water" behind the plate. Mallock's explanation for the disparity was the same as Rayleigh's.[25] There can, said Mallock, "be no doubt that the distribution of the pressure over the rear surface and its difference from the pressure prevailing in the fluid at a distance, account both for the peculiarities of the resistance and the lateral thrust curves" (40).

There was a negative pressure at the back of the plate, and Mallock suggested that the cause of this suction effect was eddy formation: "The fact of negative pressure being found on the down-stream side of the surface is ultimately connected with the formation of eddies . . . nor will a satisfactory explanation of such features . . . be obtained until the formation of eddies under varying conditions has been investigated" (40). These ideas had already found expression in one of the very first Reports and Memoranda, that of May 19, 1909, titled "Memorandum on General Questions to Be Studied." Much of this publication was devoted to the results of several series of experiments Mallock had performed on plates in currents of air.[26] He described the general form of the curves relating angle of incidence to lift and drag. The lift increases in a roughly linear fashion but only up to a certain, critical angle. After this it declines more or less sharply. In the region of the critical angle, he noted, the airflow had an unsteady, oscillating character accompanied by the formation of eddies.

Mallock went on to make the following significant remark about the difference between perfect fluids and real fluids: "The details of eddy formation are also important. The eddy in a real fluid differs greatly from the ideal vortex ring. The latter is a separate entity which could not be made, and, if existing, could not be destroyed. The eddies in real fluids are composite structures in which layers of originally different velocities are wound up together" (22). Mallock was arguing that the turbulence that characterizes real flows cannot be modeled by the kind of vortices that can exist in a perfect fluid. He was implicitly citing one of the most important results of classical hydrodynamics. Kelvin had provided a rigorous proof of a result that had long been known in one form or another, namely, that vortex or rotational motion in an ideal fluid can be neither created nor destroyed. The governing equations of ideal flow did not allow a vortex to be generated (for example, by the motion of a solid body immersed within the fluid), nor could any existing vortex be damped down or dissipated. Mallock concluded that an adequate understanding of the flow of a real fluid (such as air), around a real object (such as a wing), cannot be achieved using a theory of an ideal, perfectly inviscid, fluid. Like Rayleigh, he now believed it was necessary to take account of viscosity.

A further criticism of Kirchhoff-Rayleigh flow was contained in Reports and Memoranda No. 24, dated April 21, 1910. This was written by T. E. Stanton, the head of the Engineering Department at the NPL, and his assistant Leonard Bairstow.[27] The aim of the research was to discover the relative efficiency of the various rudders and lifting surfaces that were used on airships. Could these be modeled as flat planes set at an angle to the flow, and could Rayleigh's formula be used to predict the forces on them? Recall that Rayleigh had deduced the formula for the distance of the center of pressure of the resultant aerodynamic force from the leading edge of the plate. Stanton and Bairstow carried out measurements in the air channel to test this. They found that "the experimentally determined position of the centre of pressure for the pattern of rudders used are in all cases ahead of the theoretical position for a thin plate" (75). Here was a second blow to the utility of Kirchhoff-Rayleigh flow. First it underestimated the force on wing; now it emerged that it located that force in the wrong place.

The next step was to try to render the flow around wings and plates visible and to capture them in photographs. The photographs were provided for the Advisory Committee in 1912 by C. G. Eden in Reports and Memoranda No. 58.[28] Eden injected a mixture of aniline and toluene into the flow of water in a small water channel, a 3×4-inch cross section, containing a small, 1-inch-wide model wing that could be given various angles of incidence. The flow had a speed of 1 inch per second and could be illuminated by means of an arc lamp. The specific gravity of the injected fluid was adjusted to match that of the water so that it did not alter the character of the flow. The optical properties of the oil were such that at an angle of about $70°$ to the incident beam, the oil drops appeared as bright dots and could be photographed. Commenting on the resulting photographs, Eden described the effect of putting the wing at different angles of incidence, saying, "At small angles up to $9°$ no eddies are formed, but it will be seen that between $9°$ and $10°$ a change takes place, and at $10°$ there is a small 'dead water' region at the back of the plate and the eddying of the flow in the wake is clearly shown" (99). At first it may seem that the photographs were evidence in favor of Kirchhoff-Rayleigh flow. Eden detected eddies that the theory could not explain, but he reported a clearly visible region of "dead water," which is the most characteristic feature of the flow. The real significance of the photographs, however, came out when they were related to model wings as they were being studied in the wind channel.

Bairstow and Melvill Jones produced two papers in 1912 that had a direct bearing on the meaning of the photographs. The first, Reports and Memoranda No. 53, was a general assessment of the properties of aerofoils, "as deduced from the results of various aeronautical laboratories."[29] These in-

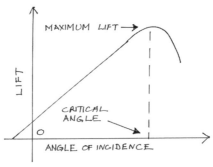

FIGURE 3.6. The typical relation between the lift and the angle of incidence of a wing. The lift increases in a linear way with increased angle of incidence up to a critical angle, at which the lift begins to decline rapidly. At this angle the wing "stalls." Notice that a typical wing still generates some lift for small negative angles of incidence, that is, when sloping down.

cluded Eiffel's near Paris and Prandtl's in Göttingen. These results, said Bairstow and Jones, made it possible to identify the salient features of aircraft wings, both qualitatively and quantitatively. Suppose a wing is aligned with the general direction of airflow. It generates a lift force at right angles to the direction of flow, and as the angle of incidence is slowly increased the lift increases in an approximately linear manner. At a certain point, around 10° to 15°, the lift reaches a maximum and then declines, sometimes very sharply. This was Mallock's "critical" angle. Finally, because the wing produces lift when it is horizontal, the position of zero lift must occur when the wing has a slight negative angle of incidence. The general form of the graph relating lift to incidence, as described by Bairstow and Melvill Jones, is indicated in figure 3.6.

In their next report, "Experiments on Models of Aeroplane Wings," of March 1912, Bairstow and Melvill Jones made two important moves.[30] First, they reinforced the evidence that the factors regulating the amount of lift in a wing are precisely those that are excluded by the discontinuity theory, namely, the shape of the upper surface. Second, they made the crucial link to Eden's photographs. To show the significance of the upper surface, they used a series of aerofoil sections with a flat base but (1) varied the camber of the upper surface and then (2) altered the position along the chord of the maximum ordinate. (The camber was measured by the ratio of the maximum height of the curved upper surface to the chord.) They demonstrated how changes in camber altered the ratio of the lift to the drag. Thus, starting from a nearly flat wing, an increase in camber at first increased and then decreased this ratio, the maximum being for a wing with a curvature of 0.05. The optimum position of the highest point in the upper surface turned out to be one-third of the chord from the leading edge. Variation of the curvature of the lower surface of the wing, by contrast, produced little effect on the lift or drag. Thus the systematic accumulation of inductive evidence about the behavior of wings consolidated Rayleigh's early doubts about his classic results. The most significant aerodynamic processes on a wing had, on his analysis, been located on the wrong surface.

Eden's photographs showed a winglike shape at various angles below and above the critical angle. At small angles there were no eddies, so the flow was smooth and stayed close to the surface of the plate or wing. Around 9°, eddies formed behind the plate and the smooth flow broke down. Here the illuminated oil drops stayed as spots of light and did not show up as lines because they did not move during the exposure time. As Bairstow and Melvill Jones put it, "It seems clear that the alterations in pressure at the critical angle are due to the sudden breakdown in the character of the fluid flow in the neighbourhood of this angle, and in this connection the photographs . . . are of interest. . . . It will be noted that above the critical angle the fluid near the upper surface is practically 'dead'" (10).

The implication was that dead air does indeed form behind a wing, as in Rayleigh flow—but only *after* the wing has passed the critical angle. Bairstow, Melvill Jones, and Eden were all aware that this reasoning had a weakness. The experiment was in water while the conclusion was about air. The conditions of similarity, required for a confident inference from the one phenomenon to the other, were not strictly satisfied.[31] The fact remained that the photographs that looked most like Kirchhoff-Rayleigh flow showed a wing that was over the critical angle.[32] This information pointed to an important but disconcerting conclusion. The mathematical studies devoted to Kirchhoff-Rayleigh flow were not describing at all how a wing can be so effective in producing lift. In as far as this model of flow over a wing approximated to reality, it was actually describing the breakdown of the pattern that was required for the efficient working of a wing. If they described anything, the formulas of Kirchhoff-Rayleigh flow were the mathematics of a stall, when lift fails. They described the situation when an aircraft was about to fall out of the sky.[33]

An Old Anomaly or a New Crisis?

How did the leading British aerodynamicists react? The theory of lift derived from the picture of discontinuous flow could not be right, at least, not as it stood. Was this situation to be treated (in Kuhn's terms) as a "crisis"? Did it call for a radical response involving the use of new models and the creation of a wholly new approach to lift? Or was it no more than an "anomaly"? Was there still the possibility that some refinement of the old, discontinuity approach would eventually allow the problem to be met? The answer is that the problem was seen as both an anomaly and a crisis, but it was seen in these different ways by different groups. The split was roughly along disciplinary lines. For the more purely mathematical contributors, the emerging results represented, or were collectively treated, as an anomaly rather than a crisis; for the experimentalists and physicists, even the mathematical physicists, the findings were more than merely anomalous: they were taken to herald a full-blown crisis.[34]

Greenhill, Bryan, and a number of other research mathematicians continued to work on the problems of discontinuous flow. They aimed to refine the analysis by taking account of the curvature that was characteristic of the cross section of a wing. This involved complicating the Schwarz-Christoffel transformation, which applied to straight-sided figures. In 1914 G. H. Bryan and Robert Jones published a paper called "Discontinuous Fluid Motion Past a Bent Plane, with Special Reference to Aeroplane Problems."[35] They were able to align the analysis with some of the qualitative facts, for example, that a moderate degree of camber can increase lift without increasing drag. In 1915 J. G. Leathem (the sixth wrangler in 1894 and a fellow of St. John's) published "Some Applications of Conformal Transformations to Problems of Hydrodynamics," which was meant to put the introduction of "curve-factors" on a systematic basis.[36] Also in 1915, Hyman Levy published "On the Resistance Experienced by a Body Moving in a Fluid," in which he set out to link a discontinuity analysis to recent work on vortices by von Kármán.[37] In 1916 Greenhill published a substantial appendix to his R&M 19 which was titled "Theory of a Stream Line Past a Curved Wing." Greenhill noted that a curved surface could be approximated by a large number of short, straight surfaces so that the way was open to work with a more realistic model of a wing. He added to his previous discussion by surveying the contributions to discontinuity theory of French and Italian mathematicians such as Brillouin, Villat, Cissotti, and Levi-Civita.[38] The same year, 1916, saw Levy's "Discontinuous Fluid Motion Past a Curved Boundary."[39] The investigation was again justified by its relevance to aerodynamics. The author asserted that "In aeronautics alone recent developments have shown the practical necessity for an effective discussion of the case where the plane is cambered" (285).

What kept the mathematicians at work?[40] Partly, they hoped to bring discontinuity theory into closer contact with experimental results, as indeed they did; it would be wrong, however, to overstate the optimism associated with this project. The main factor seems to have been the lack of any perceived alternative. The question they faced was whether there was any chance of digging beneath the equations of perfect fluid theory and making progress with the full, governing equations of viscous flow. If there was no chance, or very little chance, then it was reasonable to carry on as before. Consider the stance of Cowley and Levy. They concluded that the fatal flaw in the theory of discontinuous flow was that it depended on the assumptions of the theory of a perfect fluid, although, they argued, "it is remarkable . . . that the results obtained for the resistance are comparable at all with those derived from experiment" (65), and they speculated that perfect fluid flows with vortices might be used to simulate the flows that could be photographed in a turbulent, viscous fluid.[41]

While Cowley and Levy spoke of the mathematics of viscous flow as "not yet sufficiently developed" (75)—thus holding out hope—a more pessimistic induction is hinted at by G. H. Bryan's remarks in the *Mathematical Gazette* of 1912.[42] With a nod to Greenhill's article on hydrodynamics in the *Encyclopaedia Britannica*, Bryan said that the subject really consists in the study of certain partial differential equations and not "town water supply, resistance of ships, screw propellers and aeroplanes" (379). There was not much inducement for the mathematician to adapt his work to the needs of engineers. If he were going to do that, he might as well become an engineer "and give up most of his mathematics, relying on the introduction of *constants* or *coefficients* to save him from running his head against insoluble differential equations" (379). As for the hypothetical conditions that make hydrodynamics so unreal, these have "pretty well done their duty when they have been made use of to write down differential equations" (379). So it was not the empirical status of these conditions (that is, their falsity) that counted, but their power to help the mathematician frame tractable equations. The real issue was "remarkably simple": if you give up ideal fluid theory, you get equations for which nobody can find the integrals, "at least, mathematicians have tried over and over in vain to find them" (379). And the same argument applied to the simplifying assumption of steady motion, which involved, and justified, ignoring "for example, eddy formation in the rear of planes" (380). Bryan's position was the same as the one he adopted in his main field of research into stability. There the equations turned out to be accurate, but for Bryan, this was a bonus rather than something that was necessary for justifying the work. Even if inaccurate, he said, the analysis might still furnish a useful basis for the interpretation of experimental data.

Such was the reasoning by which a small number of high-status mathematicians justified their continued elaboration of discontinuity theory and sustained a pessimistic form of "normal science." For Bairstow, and others at the NPL who were more experimentally inclined, the emerging problems indicated the end of the road for discontinuity theory. The theory was artificial and doomed to failure because it was grounded in the unreal conception of a perfect, frictionless fluid. What was needed was a return to the full equations of viscous flow and the attempt to develop new methods of approximation. For the moment, however, Bairstow accepted that the complex flow around a wing was beyond the comprehension of the mathematician.

In a lecture at the Aeronautical Society, on February 12, 1913, Bairstow asked what shape of aerofoil or strut would give the most lift or the least resistance.[43] "A true theory of aerodynamics," he said, "would answer these questions for us completely, but unfortunately for us the answers to such questions are beyond the reach of our present mathematical knowledge." To reinforce the point Bairstow showed his audience a photograph that, he said, "illustrates a motion which has defied the mathematician" (117). The photograph was one of the NPL water-channel pictures showing a wing with the characteristic turbulent wake associated with a stall, that is, a wing exhibiting Kirchhoff-Rayleigh flow. Greenhill was in the audience—but he did not defend his mathematical model of discontinuous flow around a wing, nor did he challenge Bairstow's conclusion. It is difficult to know how to interpret this disregard. Greenhill initiated the discussion that followed the lecture, but only to make jocular comments on the law of mechanical similarity. According to Greenhill's calculations, if angels existed in the form in which they are usually depicted, then they would have to be about the size of a bee.[44] This lack of an explicit response to the shortcomings of the discontinuity theory was wholly characteristic. There had been no discussion of the demise of discontinuity theory at the meetings of the Advisory Committee (or none at which minutes were taken). Greenhill, though in regular attendance, appears to have made few contributions to the business of the committee. Anecdotal evidence reveals that Sir George was actually prone to fall asleep during meetings. On one such occasion Mervin O'Gorman, a talented artist, drew a sketch of the slumbering mathematician and left it on the table. The caption was from the well-known hymn: "There is a green hill far away."[45] But sleep patterns are not really very illuminating. Perhaps the reason for the reticence was simply that Greenhill, and his fellow wranglers on the committee, shared something of Bryan's pessimism. Without an analysis of viscous flow, the choice was between inviscid theory and empiricism—and inviscid theory had failed.

Bairstow's slightly more optimistic view, that progress of some kind was possible with the equations of viscous flow, was shared by Geoffrey Ingram Taylor. Taylor came up to Trinity in 1905, took part I of the Mathematical

Tripos in 1907 and part II of the Natural Sciences Tripos in 1908. He was given a major scholarship at Trinity and in 1910 was elected to a fellowship. After war was declared on August 4, 1914, Taylor hurried to submit his dissertation for the Adams Prize, unsure whether it would be awarded because of the uncertainty of the international situation. He volunteered his services to the military on August 5, hoping to work in meteorology, but was immediately drafted to Farnborough and later co-opted onto the Aerodynamics Sub-Committee of the ACA. Soon after his arrival at Farnborough, Taylor took his first flight. He flew with Edward Busk, on the day before Busk's death. Later, after gaining his "wings" with the Royal Flying Corps, Taylor carried out numerous pieces of research, including measurements of the pressure distribution across the wing of a BE2. His work enabled comparisons to be made between wind-channel results on models and full scale data.[46] In the early months of the war, however, he used every spare moment to bring his thesis to completion. It was titled "Turbulent Motion in Fluids."[47]

The preface and introduction of the thesis were used to set out the current situation in fluid dynamics. Taylor's position contrasted starkly with that of Bryan, who placed the emphasis on getting hold of some differential equations rather than worrying about what had to be assumed or discarded *en route*. In Taylor's view: "In no other branch of applied mathematics is the danger of neglecting the physical basis of the subject greater than it is in hydrodynamics" (5). It has been possible, he said, to get "rigorous mathematical solutions" in hydrodynamics, but they have no relation to what is found experimentally. Unfortunately, mathematicians adhere to these physically unrealistic assumptions simply because it makes the mathematics easier. As an example Taylor cited the theory of discontinuous flow. The theory is based on the assumption that the pressure in the dead-water region is the same as that of the undisturbed flow, when measurements show it is less than this. Referring to measurements made by Melvill Jones at the NPL, Taylor drew an unequivocal conclusion: "This, I think, finally disposes of the discontinuity theory, which . . . must now be placed among the curiosities of mathematics" (4). The rigor of the old methods, argued Taylor, was well worth sacrificing if there was a chance of explaining the turbulent behavior of real fluids—and this is what he set out to do in the remaining sections of the bulky thesis.[48] His aim was to develop a new (statistical) theory of turbulence.

For Taylor, like Bairstow, discontinuity theory and Rayleigh flow were things of the past. The same judgment can be read into what was said, and what was not said, in an important lecture given by Glazebrook in June 1914. His title was "The Development of the Aeroplane," and the aim was to describe the achievements of the National Physical Laboratory. He mentioned the work of Stanton and Bairstow and explained how pressure measurements reveal the important role of the upper surface of the wing (from the outset, the weak point of the discontinuity theory). Glazebrook mentioned no names and did not make it explicit that the one candidate for the explanation of lift that had been taken seriously in British aerodynamics was now being quietly abandoned. All the attention was directed toward the successful work on stability. The occasion of Glazebrook's assessment was the second of the Wright Memorial Lectures. It was an important event so the assessment would have been carefully considered.[49] The unspoken message was that, notwithstanding the opinion of a few of the older mathematicians, the discontinuity theory of lift was dead.

Intuitive and Holistic Aerodynamics

The practical men did not like "scientific" aerodynamics.[50] So what sort of aerodynamics did they like? I begin to answer this question by identifying what might be called their "practical epistemology." Then I look in more detail at the accounts of lift that are to be found in books written for the designers of airplanes and in articles that appeared in the *Aeroplane, Flight,* and *Aeronautics.*

The epistemology of the practical man was intuitive and qualitative. It was formulated in conscious opposition to the pedantic concern with accuracy and irrelevant detail attributed to the despised figure of the mathematician.[51] Reality must be grasped in all its complexity rather than simplified and broken down into imagined elements. In this sense their epistemology was holistic. It was also artistic. A good designer could rely on his eye, his experience, and his judgment. In a literature review in *Aeronautics* the editor said: "I don't deny the infinitely valuable role of pure science, still less that of theory, but science should have some relation to practice, since it is its foster-mother. There is more than one aeroplane designer who knows just enough mathematics to make twice two work out at four, but he will turn out machines equal in performance to the best. We in this country know, as they do in the United States, of eminent designers who *see* a new type of machine rather than *design* it."[52]

Grey made the point more bluntly with no genuflection in the direction of science: "Never mind what the scientists calculate. Trust the man who guesses, and guesses right."[53] The claim was that some designers have a track record of guessing rightly, and these are the people to trust. We may not be able to see how they do it, but we should not let this put us off. Trust rather than understanding lies at the root of things. This was indeed Grey's view:

there were not only unknown factors involved in the design of aircraft but there were actually *unknowable* factors, and this was something the "slide-rule scientists" could not grasp.[54]

The implication was that the reasons behind practical success will remain mysterious. This notion implied a species of intellectual pessimism or even nihilism. Such pessimism was not unusual among practical men and was sometimes echoed by those in the other camp. For example, writing as J. C., a reviewer of G. P. Thomson's *Applied Aerodynamics* recommended the book to practical designers (even though it was the product of Farnborough) and said, "One of the first ideas that arises in the reading is the state of ignorance that still exists in aerodynamics; it is safe to say that we know practically nothing of the reasons for the experimental results that we find. The amazing thing is that we are able to make aeroplanes as well as we can."[55]

At least two of these statements come from spokesmen of the practical men rather than from designers themselves, but they seem to articulate a widely held view. Grey's characteristic denunciations were repeated in a foreword he wrote in 1917 for the book *Aeroplane Design* by F. S. Barnwell, who was the chief designer at the British and Colonial Aeroplane Company. This firm, usually known as the Bristol Company, became famous during the Great War for the Bristol fighter, which was designed by Barnwell.[56] Much harm had been done, said Grey, "both to the development of aeroplanes and to the good repute of genuine aeroplane designers by people who pose as 'aeronautical experts' on the strength of being able to turn out strings of incomprehensible calculations resulting from empirical formulae based on debatable figures acquired from inconclusive experiments carried out by persons of doubtful reliability on instruments of problematic accuracy."[57] If one asks what is left when all the hated calculations, experiments, and instruments have been swept away, the answer is intuition. This was Grey speaking, not Barnwell, so we cannot be sure that Barnwell endorsed it. Authors do not necessarily agree with what others say in the forewords of their books, but it is reasonable to expect general agreement.

W. H. Sayers, a strong critic of the National Physical Laboratory, was involved with the development of seaplanes during World War I. In an article written after the war, in 1922, called "The Arrest of Aerodynamic Development," Sayers described the current conception and form of the airplane.[58] It was, he said, "the hybrid product of two utterly different and independent methods of development." From 1908 to 1914, its evolution was "the result almost entirely of individual adventure." There were, he insisted, no wind-tunnel results worth mentioning, the mathematics of stability had no apparent connection with the facts, and even engineers regarded the airplane as a mechanical curiosity. "Individual designers worked, as artists worked, by a sort of inspiration as to what an aeroplane ought to be like, and built as nearly to their inspiration as the limited means, appliances and increasing knowledge they possessed would allow them" (138). Sayers went on to deplore the degree of standardization that had set in with regard to design. This, he said, gave a spurious sense of understanding and control. In reality we did not know how to predict what would happen outside the limited range with which we had become familiar. Similarly, the laboratory workers had been in error in concentrating on simple bodies, especially "such simple bodies as might be used as components of the standard type of aeroplane" (138). The result, he said, was a bias toward an additive conception of the different aspects of design and a tendency to overlook large, qualitative effects such as the interference of different components.

Like many other practical men, Sayers was skeptical about model work.[59] In his view, aerodynamicists did not yet know what dynamic "similarity" really was, so that inferences from models remained doubtful. Full-scale experimentation was the real basis of knowledge. Grey could be relied upon to give the relatively measured prose of Sayers, his frequent contributor, a more colorful rendering: "I would back any one of a dozen men I know to find out more about streamlines in a month at Brooklands, with the help of a borrowed racing car, a jobbing carpenter, and a spring-balance, than the combined efforts of the National Physical Laboratory, Chalais-Meuden, the Eiffel Tower, the laboratory at Kouchino, and the University of Göttingen have discovered since flying first attracted the attention of that section of humanity which the Americans expressively call 'the high-brow.'"[60]

This cavalier dismissal of all the major aerodynamic laboratories of Europe dramatizes the anti-intellectual strand in the epistemology. Not all of its expressions were so markedly of this character, but there is no denying a tendency in this direction. Nor can one deny a certain justice in the stance. If scientists have a tendency to simplify the complex and decompose it into its elements, where does this leave the designer who has to reassemble the elements in novel ways? Even if simple principles can be discovered, it can still be unclear how these principles interact when they work together. Design is still a matter of judgment about their combination and compromise in their balance.

Grey's dismissive attitude toward Gustav Eiffel's work was not shared by all practical men. The impression created by articles and reviews in the technical journals is that Eiffel was seen as an engineer who could be relied upon to operate in a practical way. If Eiffel's large, empirical monograph, replete with tables of data, graphs and diagrams of airplanes, is laid side by side with

Greenhill's mathematical report, there can be no more striking visual proof of the extremes of style that can be represented in aeronautical work. What is more, Eiffel's work was frequently compared favorably with the experimental work of the NPL. Where the two laboratories diverged, the practical men backed Eiffel.

The reviewer of Eiffel's *La resistance de l'air et l'aviation*, for the *Aero*, in March 1911, was enthusiastic: "One is hardly going too far in describing this book as the most authoritative work on the subject that has yet appeared, and it is especially valuable in as much as the experiments have been evolved with an eye specially inclined toward their value in practical aeronautics . . . while experiments of a more purely academic interest have . . . been relegated to the background." This, the reviewer continued, was strikingly different from the situation that "obtains in more than one experimental laboratory."[61]

Writing in July 1916, the editor of *Aeronautics* invited readers to compare Eiffel, "working almost single handed," with the National Physical Laboratory: "It would not be unjust to say that Eiffel attains practical results, neglecting a slight margin of error, accounting probably 2 per cent. in extreme cases, which for the time being and for practical purposes is inappreciable. On the other hand, the N.P.L., in its beautiful work, seems rather to strive for the meticulous elimination of this negligible margin of error and passes by the major facts." Ask Eiffel for the air resistance of, say, an airship hull and the job is done "in a couple of days," while it would last "heaven knows how many weeks" at the N.P.L.[62]

The report of the Advisory Committee for 1911–12 noted that, between Eiffel's laboratory and the NPL, there were differences of some 15 percent between the values of the lift coefficient for certain wings. The probable reason, it was said, was observational errors. The ACA resolved to investigate the matter and to ensure that a high degree of accuracy was maintained at Teddington. The "Editorial View" in the *Aero* was that to the "lay mind" such differences are "disquieting," and the writer of the editorial chose to read the ACA's response "almost as an acknowledgement of error on the part of Teddington."[63]

Neo-Newtonianism and the "Sweep" of a Wing

G. H. Bryan described the approach to lift adopted by the practical men as neo-Newtonian.[64] The label accurately identified two salient features of their work. First, like everyone else, the practical men operated within the framework of Newton's mechanics. Ultimately the wing must act on a mass of air, accelerating it downward, thus ensuring, in accordance with Newton's third

law of motion, that the wing suffered an equal and opposite reaction. This reaction was the ultimate source of the lift. Second, the practical men adopted
a line of reasoning that was, in some respects, analogous to one that Newton used in the *Principia* when he compared the forces exerted by a flowing
fluid on a sphere and a cylinder "described on equal diameters."[65] Recall that
Newton assumed that his fluid, or "rare medium," consisted of a number
of independent particles which would hit the sphere and cylinder and give
up their momentum. (This was the model I previously likened to a shower
of hailstones). The practical men greatly simplified the analysis by addressing the case of a flat plate exposed to the uniform flow of this rare medium.
While Newton was no doubt conscious of the distinction between his hypothetical fluid and real air, this difference tended to be blurred in some of the
later aerodynamic discussions. By applying the reasoning to a simplified wing
moving in air, the following argument was constructed.

Suppose that a flat plate has area A and is at an angle θ to a uniform,
horizontal flow of a fluid that was, like Newton's, composed of independent
particles. Suppose, further, that the collisions are inelastic so that the particles
simply slide along the plane after impact. Let the particles in the main flow
move at speed V units of distance per second. Then the volume of fluid striking the plate each second is given by multiplying the vertical projection of the
plate ($A\sin\theta$) with the velocity V. The projection $A\sin\theta$ was the "sweep" of
the wing and $\sin\theta$ was the "sweep factor." Now multiply the volume $AV\sin\theta$
by the density ρ (presumed to be the same as that of the air) to give the mass,
and then multiply the mass by the velocity component normal to the plate
($V\sin\theta$) to give the momentum exchanged per second. This is the source of
the pressure P whose vertical component, $P\cos\theta$, is the lift and whose horizontal component, $P\sin\theta$, is the drag. This neo-Newtonian argument gave
the formula for the resultant aerodynamic force P on the plate as

$$P = \rho A V^2 \sin^2\theta.$$

The formula was often called Newton's $\sin^2$ law, although it is not to be found,
in an explicit form, in the text of the *Principia*. There are two reasons for
its absence. First, Newton was dealing with a curved surface not a flat plate,
and second, his reasoning was geometrical in form, so that the trigonometric
terms appear as geometrical ratios.[66] The label is, however, a reasonable one.
All of the subsequent work of the practical men involved versions of, and
variations on, this formula.

For the range of angles relevant to aeronautics, $\sin\theta$ is a small quantity,
so its square is very small indeed. The Newtonian formula condemns any
predicted lift to be small, except where the magnitude of A and V^2 can offset

the smallness of the squared $\sin\theta$ term. On this analysis, lift would demand enormous velocities or unrealistic wing areas. Had the formula been true it would have rendered artificial flight a practical impossibility. It is little wonder that in his 1876 paper Rayleigh had expressed satisfaction that his own formula made pressure proportional to $\sin\theta$ rather than, as in Newton's formula, to $\sin^2\theta$. In following the Newtonian tradition the practical men inherited a serious problem and resorted to a variety of expedients in an attempt to overcome it. A number of examples will show how comprehensively they failed to meet this challenge.

Four Practical Men

In 1907 Herbert Chatley, lecturer in applied mechanics at Portsmouth Technical Institute, published *The Problem of Flight: A Text-Book of Aerial Engineering*. The book went through two further editions, in 1910 and 1921. In the preface Chatley explained that, in terms of mathematics, he followed the practice, "well established in engineering," of omitting factors that appear unimportant: "The formulae are therefore 'engineering formulae' in the strict sense of the word, i.e. they are not the result of a deep mathematical analysis which it is, in the majority of cases, almost impossible to apply."[67]

Chatley's account of lift was eclectic. It involved elements of perfect fluid theory and discontinuity theory along with the Newtonian analysis and its problematic $\sin^2$ formula. To overcome the difficulties he added various empirical corrections. These cut across the deductive links between the formulas in a manner that might have been calculated to offend the sensibilities of a wrangler. He began by mentioning the continuous flow of a perfect fluid over an inclined plane. This, said Chatley, is described in Lamb's *Hydrodynamics*, and its reality has been demonstrated by photographs taken by Prof. Hele-Shaw. (In fact Hele-Shaw's photographs show the behavior of a viscous fluid in slow motion between two glass plates that are very close together. Stokes was able to show that, under these conditions, "creeping motion," as it is called, provides an accurate simulation of the flow of a perfect fluid. The forces at work and the boundary conditions are different, but the photographs show what a perfect fluid flow would look like.)[68] Chatley went on to assert that at greater speeds this flow breaks down and is replaced by one showing surfaces of discontinuity enclosing pockets of turbulence on the rear of the plate. These eddies reduce the pressure. So far, Chatley's qualitative picture is approximately that of Kirchhoff-Rayleigh flow, combined with some of Kelvin's ideas about the turbulence in the dead-water region.

Chatley then introduced some mathematics and deduced Newton's $\sin^2$

formula in the manner just described. What Chatley meant by an engineering formula became clear when he asserted (without giving evidence) that the effect of the eddying on the rear of the plate is to augment the pressure by half as much again. He therefore repeated the above formula, but now multiplied by 3/2, and called it N_{max}. Chatley claimed that this formula agrees "very fairly" with some experimental results of Coulomb, although it was conceded that, for small angles, experimenters disagree greatly. In what is presumably a reference to Rayleigh's paper, he went on: "The latest results are almost unanimous in making the variations of thrust as $\sin\theta$ and not as $\sin^2\theta$. All these following expressions are thus divided by $\sin\theta$" (29). Thus the $\sin^2\theta$ term in the Newtonian formula was simply altered to $\sin\theta$. In the 1921 edition this abrupt step was justified by saying that the original formula is approximately correct for large angles of incidence (from 60° to 90°), but for small angles, "owing to the continuity of the air, $\sin\theta$ must be substituted for $\sin^2\theta$" (31). In the 1910 and 1921 editions the reader was told that the end result is "practically correct" for plane surfaces and, with "slight correction to the coefficients," also applies to curved surfaces.

Algernon Berriman was the chief engineer at the Daimler works in Coventry and the technical editor of *Flight*. In 1911 both *Flight* and *Aeronautics* published accounts of his lectures titled "The Mathematics of the Cambered Plane," and in 1913 he published *Aviation: An Introduction to the Elements of Flight*.[69] This book was based on lectures he had given at the Northampton Polytechnic Institute. Berriman also started from the Newtonian idea of action and reaction. Lift came from the reaction on the wing of the mass of air that was, by some means, forced downward by the wing. Thus, "the wing in flight continually accelerates a mass of air downwards, and *must* derive a lift therefrom."[70] The basic formula is Force = Mass × Acceleration, but how is this formula to be applied? What is the mass of air that is involved? The original Newtonian picture must have underestimated this mass, hence the underestimation of the lift that can be generated.

Berriman assumed that the wing sweeps out, and pushes down, a greater area than is suggested by the simple geometry of an inclined plane. The wing exerts an influence on "all molecules within an indefinite proximity to the plane; in other words a stratum of air of indefinite depth."[71] Instead of stating that a wing of area A engages with a quantity of air determined by the frontal projection $A\sin\theta$, the assumption was made that it sweeps out a larger area, AS, whose sweep factor S is typically much larger than $\sin\theta$. Berriman said that "practical considerations" gave reason to believe that "the effective sweep of a cambered plane may be defined in terms of the chord of the plane" (5). In other words, Berriman put $S = 1$. This equation was based on the experience of the pioneers and experimenters, such as Langley, who found by trial and error that they got the best results with a biplane when they positioned one wing about one chord length above the other.[72] The reasoning was hardly rigorous, but the assumption allowed Berriman to avoid the troublesome sine-squared term.

Although the concept of "sweep" was popular with the practical men, it had few practical advantages. All that could be done was to determine the lift empirically and then deduce that the quantity called sweep must have such and such a numerical value. No one could go in the other direction. There was no way to predict the lift from the sweep. One might argue, inductively, that similar wings will have similar sweeps, but one could also say that similar wings have similar lifts, so in practice nothing is gained by introducing the concept. G. H. Bryan, after expressing irritation with Berriman's casual way with trigonometric formulas,[73] identified the source of the difficulty:

> there is no such thing as "sweep" except in Newton's ideal medium of non-interfering particles satisfying the sine squared law. In a fluid medium the disturbance produced by a moving solid theoretically extends to an infinite distance, gradually decreasing as we go further off. Mr. Berriman's "sweep" is, physically speaking, an impossibility. If, however, "sweep" is defined as the depth of a hypothetical column of air, the change of momentum in which would represent the pressure on the plane, then the introduction of this new quantity is only a useless and unnecessary complication. Instead of facilitating the determination of the unknown data of the problem, it merely replaces one variable which is physically intelligible and capable of experimental determination by another variable satisfying neither of these conditions. (265)

The practical men never found a way to use the idea of sweep so that they could sustain what Kuhn called a "puzzle-solving tradition."[74]

The work reported in Albert Thurston's *Elementary Aeronautics* of 1911 was based on the hope of identifying the significant properties of the flow, such as the sweep of a wing, in an empirical manner.[75] Thurston had worked for Sir Hiram Maxim and then became a lecturer in aeronautics at the East London Technical College. He took numerous photographs of airflows made visible by jets of smoke as they streamed past objects of various shapes. The objects ranged from rectangular blocks to aerofoil shapes, or "aero-curves" as Thurston called them. He concluded that the important qualitative factor in the flow over a wing was that the entry at the leading edge was smooth and avoided the "shock" detectable in the case of a simple, flat plate. The avoidance of shock was possible because of the rounded and slightly dipped front edge characteristic of a wing, a shape whose advantages had been discovered empirically by Horatio Phillips and Otto Lilienthal.[76] The essential thing, according to Thurston, was to maintain a smooth "streamlined" flow. The attempt to impose sudden changes in the velocity of the air merely produces surfaces of discontinuity (20). The photographs showed that with a good, winglike shape at small angles of incidence, "the air divides at the front edge and hugs both sides as it passes along; its resistance to change of motion causing a compression on the lower side of the plane and a rarefaction or suction on the upper side. As the inclination is increased a critical angle appears to be reached, after which the stream line ceases to follow the upper side and forms a surface of discontinuity with corresponding eddies" (21–24).

As with the work of Eden, Bairstow, and Melvill Jones at the NPL, such photographs revealed that the model of discontinuous flow was not going to provide a basis for understanding lift. The photographs did, though, support ideas about the extended sweep of a wing. Thurston avoided using the word "sweep" but referred to the "field" of a wing and, on the basis of his photographs, asserted: "The air affected by an aeroplane [= wing], that is the field of an aeroplane, is greater than the air lying in its path. Thus . . . it will be seen that air, which is considerably above the front edge of the plane, is within the range of the plane, and is deflected downwards" (26–27).

Even with photographs of smoke traces in the flow, it proved impossible to identify the sweep in a quantitative way, and the underlying causes of the qualitative effects visible in the photographs remained obscure. Thurston, however, was convinced that the secret of good design, both for wings and other components, was attention to streamlining, that is, ensuring that the lines of flow in the immediate neighborhood of the body coincide with the surface of the body. Like Chatley, Thurston drew on the work of Hele-Shaw to show how streamline flow works and what it looks like.[77]

Frederick Handley Page also appealed to Hele-Shaw's photographs. Handley Page was one of the celebrated pioneers in the British aviation industry. Despite its early inability to deliver BE2s, his firm was later to achieve fame for its manufacture of large bomber and passenger aircraft. In April 1911 he gave a lecture at the Aeronautical Society titled "The Pressure on Plane and Curved Surfaces Moving through the Air."[78] He began by discussing the result that had caused so much difficulty for the discontinuity theory, namely, that the formation of "dead" air behind the plate happens when it has passed the critical angle. Practical aeronautics by contrast, said Handley Page, deals with small angles of incidence where the flow hugs the back of the plate or aerofoil. There is then a maximum of lift to drift and a minimum of eddy disturbance. Commenting on a Hele-Shaw photograph of the flow of a perfect fluid past an inclined plane, Handley Page said, "The air on meeting the plane divides into two streams . . . the streams meeting again at the back of the plane. At high velocities the eddies and turbulence at the rear of the plane completely obscure this, but up to the critical angle at which the 'live' air stream leaves the plane back, the effect is still the same" (48).

Handley Page, like Chatley and Thurston, took ideal fluid theory to provide an accurate picture of real flow round a wing, even when there are no surfaces of discontinuity or other complications such as vortices in the flow. Hele-Shaw's photographs, however, depicted d'Alembert's "paradox" in action, not a wing delivering lift. The practical men were thus walking into the trap that Greenhill had identified in his lectures at Imperial. They were proposing a picture of the flow which the mathematician would immediately recognize as one that gave neither lift nor drag.

No one in the audience at the Aeronautical Society mentioned this problem in the subsequent discussion. Even if the point had been raised it would have had little impact on Handley Page's eclectic argument because, immediately after this appeal to hydrodynamic theory, the perspective was changed. He adopted the neo-Newtonian approach but suggested refining the idea of sweep by dividing it into two parts. This modification revealed his real interest in the Hele-Shaw pictures. The portrayal of the flow at the leading edge suggested to Handley Page that two different processes were at work. There was the sweep associated with the flow upward from the stagnation point to the leading edge, and the sweep associated with the downward flow toward the rear edge. This complication enabled him to refine the mathematics of the sweep picture, but it did not get round Bryan's objections: it merely doubled the number of unknowns. Handley Page still had to infer the total sweep from the observed lift and had no way to apportion the contributions of the two components of the sweep that he postulated.

Handley Page's lecture was generally well received, though no mathematicians contributed to the discussion—if, indeed, any were present. Cooper, who had been so scathing about Greenhill, congratulated Handley Page on getting a formula that applied to experimental results: "it is not everybody," he added, "who does that." Cooper was either being polite or had failed to see how little had been achieved. He went on to say that he thought Handley Page's analysis applied more to the flat plate, where the leading edge caused a "shock" in the flow, than it did to an aerofoil with its rounded, dipping front edge. Here, claimed Cooper, the stagnation point would be on the very front, not below the leading edge on the underside of the wing, as it was in Hele-Shaw's picture of the plate. In what may have been meant, at least in part, as a response to this point, Handley Page said, "It seems to me that the entering front edge is only a kind of transformer . . . a curved plane is more efficient than a flat one because you have a more efficient transformer" (63). Unfortunately, no attempt was made to explain the metaphor of the "transformer."

A few years after this exchange Handley Page introduced the famous, and commercially lucrative, Handley Page wing. The standard wing was modified by introducing a slot along the leading edge which changed the flow at the leading edge by directing air from underneath the front of the wing onto its upper surface. It was, in effect, a small extra aerofoil that ran along the leading edge of the main wing. The resulting change in the flow significantly increased the lift and delayed the stall.[79] Could this innovation have been a result of the earlier conception of the leading edge as having a capacity to "transform" the flow? When Handley Page described his invention in the *Aeronautical Journal*, he made no mention of the metaphor of the transformer. Indeed, he never gave a clear account of the thought processes behind his invention, so the question must remain unanswered. (He had originally tried making slots that ran from the leading edge to the trailing edge, that is, along the *chord* of the wing rather than along the *span*. This suggests that the process of invention was trial and error, rather than theory-led.) The value of the leading-edge slot is an example of that intriguing phenomenon "simultaneous discovery" and, almost predictably, gave rise to a priority dispute.[80] The slot was developed independently in Germany by G. Lachmann, and for many years Thurston also argued his claim to be recognized as the inventor.[81]

As The Lamps Were Going Out

Another of the practical men—and among the most interesting—was the designer A. R. Low, of Vickers, whose name has already been mentioned. In 1912 he had lectured at University College, London, on aerodynamics and, on March 4, 1914, with O'Gorman in the chair, gave a talk to the Aeronautical Society titled "The Rational Design of Aeroplanes."[82] Low discussed the same range of ideas that I have already identified in the writings of other practical men, but he did so with a lively, critical intelligence and a breadth of knowledge that makes his work stand out. Since he will play a significant role in the postwar story of the reception of Prandtl's work, Low's position in 1914 is worth appreciating.

Low argued that hydrodynamics is useful for providing a sound, general "outlook" on fluid flow but that in order to get "reasonably accurate numerical values we shall see . . . that we are thrown back on experimental methods" (137). He showed his audience the diagrams from Greenhill's report representing discontinuous flow around plates, some of which were normal to the flow and some at an angle to the flow. These were compared with photographs of real flows taken by the Russian expert in fluid dynamics, Dimitri Riabouchinsky. There was a "strong resemblance between the theoretical boundary line between the stream and the back water, and the experimental boundary line between approximately steady flow and the region of marked turbulence" (138). But although Lord Rayleigh was "the first to give a formal value" for the reaction of a fluid on a barrier, his predicted value was only about half of the observed value for the small angles relevant to aerodynamics. An error of 50 percent is "quite intolerable to physicists and engineers" (137).

Low was clear that the idea of sweep was not the way forward. He did not use the word "sweep" but spoke of an "equivalent layer." This approach, he said, introduced a number of variables, and there was no way of apportioning the energy losses between them. There were, said Low, an infinite number of possible ways of assigning the energy losses. Perhaps experimental methods, such as injecting colored dyes into the flow, could shed light on the question, but until then the picture was essentially arbitrary (140). Only the empirical study of lift and resistance was left. Low then turned to a discussion of some experimental graphs and empirical formulas produced by Eiffel. It was known experimentally that, over a wide range, resistance varied as the square of the speed. The desired equations would then have the general form $R = KAV^2$, where R is the force on the wing, V is the velocity relative to the air, A is the area of the wing, and K is a constant that must be empirically determined. Given the number of variables involved, such as incidence, aspect ratio, and camber, Low observed that finding a formula that yielded the correct value would not be easy. A further complication was that the performance of the wing interacts with the flow round the rest of the machine. The point that Low wanted to stress was the "formidable series of special developments of engineering science" that were necessary before designers could be confident that any given set of drawings would turn into an airplane with predictable performance and air-worthy qualities. He concluded: "That nation will take the lead whose scientists and technical engineers, and whose works engineers, and whose pilots best understand each other and work together most cordially" (147). This plea for cooperation and coordination was timely and well meant, but Low must have known that none of these preconditions for taking the lead was satisfied. The divisive bigotry of C. G. Grey and his ilk put an end to any of the requisite cordiality.

The conclusion must be drawn that in 1914, on the eve of the Great War, none of the British workers in the field of aerodynamics, whether they were mathematicians or practical men, had any workable account of how an airplane could get off the ground. As the lamps of Europe were going out, vital parts of the new science of aeronautics were also shrouded in darkness.[83] The mathematicians had a sophisticated theory that addressed the right questions but, being based on the theory of discontinuous flow, gave disconcertingly wrong answers. The high-status and mathematically brilliant experts of the Advisory Committee were reduced to empiricism. The practical men were simply paddling in the shallows and, with the exception of a few, such as A.R. Low, appeared to be oblivious of the fact. Their ideas were vague, confused, and frequently failed to engage with either practical experience or experimental results. The mathematicians (unlike the practical men) could handle many of the problems about stability with confidence and rigor, but on the question of the origin and nature of lift, and the relation of lift to drag, they too had been effectively brought to a standstill.

The demands of the war years that followed seemed to discourage rather than encourage any fundamental reappraisal of the British approach. As far as the fundamental theory of lift and drag were concerned, British experts came out of the war little better than they went into it. In 1919 George Pagett Thomson summed up the situation as it had appeared to the British during those years. The terms of his assessment are sobering: "In spite of the enormous amount of work which has been done in aerodynamics and the allied science of hydrodynamics there is no satisfactory mathematical theory by which the forces on even the simplest bodies can be calculated with accuracy."[84] Thomson's judgment was that for British experts, practice still ran ahead of theory, as it had at the beginning of the war and as it had from the earliest days of aviation. Throughout the war British aircraft could certainly get off the ground. They flew and their wings worked. For this, both the trial-and-error methods of the practical men and the experimental work done by the Advisory Committee must be thanked. But *why* aircraft flew remained a mystery to those of a practical and a theoretical inclination alike. Only the most general principles of mechanics could be invoked by way of explanation, but these only indicated what, in terms of action and reaction, the wing *must* be doing, not *how* and *why* it did it. Probing the more specific workings of the aircraft wing remained in the realm of experiment, a process consisting of case-by-case empirical testing that was guided, or misguided, by intuition.[85]

4

Lanchester's Cyclic Theory of
Lift and Its Early Reception

The commonly distinctive feature of a modern mathematical treatise, in any branch
of physics, is that the investigation of any problem is initially conducted on the widest and most comprehensive basis, equations being first obtained in their most general form. . . . The author has endeavoured to minimise any difficulty on this score by
dealing initially with the simpler cases and afterwards working up to the more general
solutions.

F. W. LANCHESTER, *Aerodynamics, constituting the
First Volume of a Complete Work on Aerial Flight* (1907)[1]

By the beginning of the Great War the British experts on the Advisory
Committee who were responsible for research in aerodynamics had effectively abandoned the discontinuity theory of lift. There was, however, a
known alternative: the circulatory or vortex theory that had been developed
by Frederick Lanchester. It would be reasonable to expect that this theory
would now become an object of some interest even if it had been ignored at
the outset of the committee's work when they had concentrated on Rayleigh's
achievements. But, rather than turning to the circulation theory, the ACA
again treated it as if it were of no merit. Lanchester was a member of the
committee but his ideas were passed over—for a second time. Given that the
circulation theory later came to be accepted as the correct account of lift, this
insistent rejection has long been seen as a puzzle. Why did it happen? In this
chapter I lay the foundations for an explanation of this negative response.
The explanation is developed and tested as the analysis is carried further in
subsequent chapters.

I begin by introducing the basic ideas and technical vocabulary of the circulatory theory.[2] This will give access to the (largely) qualitative version of the
theory developed by Lanchester and lay the basis for my discussion in later
chapters of the quantitative versions that were subsequently developed in
Germany. In this chapter I also have more to say about Lanchester's treatment
at the hands of the so-called practical men. Their opinion of Lanchester was
divided. While some recognized him as one of their own, others saw him as
selling out to the state-funded academic scientists, theorists, and mathemati-

cians who were so reviled by the spokesmen of industry. Lanchester hit back in a characteristically forthright way and became involved in some bruising, but revealing, encounters with the antigovernment press. While Lanchester was defending the Advisory Committee, the Royal Aircraft Factory, and the National Physical Laboratory, the experts within these bodies were articulating a systematic rejection of the circulatory theory of lift. They were hostile or indifferent to Lanchester's approach. Lanchester was thus in the unenviable position of being attacked for belonging to a group that, while not actually excluding him, was certainly marginalizing him. It is vital to understand both the external politics and the internal politics that were woven together in this episode. Both are described in this chapter. The first step, however, must be to understand the conceptual basis of Lanchester's theory.

The Basis of the Circulation Theory

Within the framework of Newton's mechanics, the flow of air around an aircraft wing can only support the weight of the aircraft if the flow generates a force that is equal and opposite to that of the weight. In level flight the upward force, the "lift," must be in equilibrium with the downward force of gravity. Expressed in terms of fluid dynamics, the lift must be the result of air pressure on the wings. There must be an overall pressure imbalance between the upper and lower surfaces of the wing. The pressure of the moving air on the upper surface of the wing pushes downward. This must be surpassed by the pressure on the lower surface of the wing which pushes upward. It is the excess of the upward over the downward pressure that constitutes the lift and is therefore the central fact to be explained. It cannot be assumed that the resultant downward pressures and the resultant upward pressures act through the same point. In general they will not; the pressures on the wing will not only have the capacity to produce a lift, but they will also generate a turning moment that causes the wing to pitch. These pitching moments played a significant role in the analysis of stability carried out by G. H. Bryan. In what follows, however, I am mainly concerned with the resultant lifting effect of the pressures on the wing. I have already introduced Bernoulli's law which implies that, if the air behaves like an ideal fluid, then the faster the air flows over the wing the lower will be the pressure it exerts, and the slower the flow the higher the pressure. If it is also accepted that the airflow around a wing is not discontinuous Rayleigh flow but follows the surfaces of the wing, then the problem of lift is simplified. It reduces to that of explaining why the air immediately below the wing is moving more slowly than the air immediately above the wing.

Here it is necessary to avoid a popular misconception. A cross section of a typical wing has a flat base and a curved upper surface. The airflow divides at the leading edge, and some air takes the upper route over the curved surface while some takes the lower route along the flat and straight surface. Looking at such a shape, one can easily imagine two molecules of air parting company at the leading edge and joining up with one another again at the trailing edge. Like two travelers they wave farewell at the parting of the ways and then shake hands when they meet up later. But the low road is straight while the high road is circuitous, so the traveler who took the high road must have sped along more swiftly in order to meet up with the traveler who took the shorter path. Is this how it is with the air? Equal transit time plus a path difference certainly implies a speed difference, but this is not the secret of the wing.[3] The questionable assumption is that the traveling companions, that is, the two molecules, meet up again. There are decisive reasons why this theory cannot be right. First, the increase in speed necessary to pass over the curved, upper surface of the wing would not generate the observed amount of lift. The path difference is not great enough. Second, the theory would have the consequence that an aircraft could not fly upside down. Once inverted, the curved surface would become the lower surface. The theory would then imply that the aerodynamic force would reinforce gravity rather than counteract it. But aircraft can fly upside down, so the theory cannot be right.[4] This false theory, based on path difference and equal transit time, must not be confused with the circulation theory of lift. The circulation theory offers a very different account of the speed differences above and below a wing, as I shall now explain.

The flow of air over the cross section of a wing is a complicated phenomenon, but, argued the supporters of the circulation theory, it can be thought of as built up out of two, simple flows. These are (1) a steady wind of constant speed and direction, and (2) a swirling vortex that goes round and round a central point. The two components are shown separately in figure 4.1. I discuss each component flow in turn and then explore the flow that arises when they are superimposed on one another. The steady wind arises from supposing a steady, relative motion of the wing and the air. In reality the air is stationary and the wing moves, but, as previously noted, aerodynamic processes are frequently described in terms of the situation in a wind channel where it is the air that moves. Let the steady wind have a constant speed V and move horizontally. At any given point the flow can be represented by a vector, that is, an arrow pointing in the direction of the flow whose length is proportional to the speed. All the vector arrows representing the steady wind are therefore of the same length and can be assumed to lie horizontally, as shown in the figure. The streamlines of the flow are then equally spaced horizontal lines.

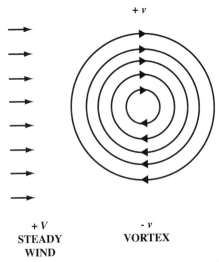

+ *v*

+ V   **- v**
STEADY   **VORTEX**
WIND

FIGURE 4.1. A steady wind and a vortex motion. When combined they produce resultant flow with a speed $V + v$ above the center of the vortex and speed $V - v$ below. These two component flows were central to the circulation theory of lift.

The vortex flow is more complicated, but the early work on aerodynamics was confined to a particularly simple form of vortex. Let the vortex swirl in a clockwise direction around a central point that is assumed to be in a fixed position. Unlike a normal vortex, in water or air, this one is not carried along by the stream. This special sort of vortex came to be called a "bound" vortex. All the streamlines in the vortex flow have the form of concentric circles. The fluid elements at any given distance from the center of the vortex are assumed to move with the same, constant speed around one of these circles. The elements do not get drawn into the center of the vortex. This is expressed by saying that they have a constant "tangential" velocity and no "radial" velocity. Just as the velocity of the fluid elements in the steady wind can be represented by vector arrows, the same can be done for the fluid elements in the vortex. In this case the arrow is a line whose length is proportional to the speed but whose direction always lies along a tangent to the circular streamline. The direction of the tangent varies as the fluid element proceeds around the streamline, although its length stays the same. Figure 4.1 shows a vortex with a clockwise rotation and gives the arrows of speed and direction at two important positions. At what may be called the six o'clock position the arrows are horizontal and facing into the steady wind, while at the twelve o'clock position they are horizontal but point in the same direction as the uniform wind.

Typically one further important assumption was made about the structure of the simple vortex. It was specified that the fluid elements that circle around the vortex near the center move along their assigned path with greater speed than do fluid elements circling at a greater distance from the center. The speed drops off uniformly with distance from the center. The greater the radius of the streamline, the smaller the tangential velocity. The assumed relation can be expressed more precisely by saying that, for the kind of simple vortex under consideration, the speed of the flow (v) at any given point is "inversely proportional" to the radius (r) of the circular streamline that runs through that point. In mathematical terms the formula relating speed and radius is then $v = k/r$, where k is the constant of proportionality.

Now imagine that the constant wind and the vortex are superimposed. The two flows, which have hitherto been treated as separate cases, are now combined. What is the result? In reality, the mixing together of two flows, whether in water or air, is accompanied by all manner of eddying and turbulence produced by viscosity and other physical features of the fluid. In the analysis developed for aerodynamic purposes, all of these complications were put aside and an extremely simple process of combination was assumed to provide an adequate description. Because the two flows that are combined are steady, the new flow will also be steady and all that was necessary to describe it was a process called vector addition (see fig. 4.2).

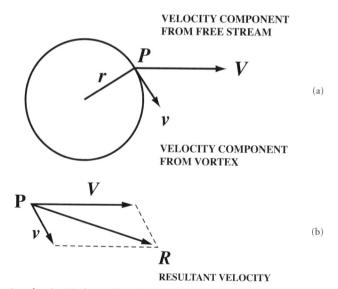

FIGURE 4.2. At each point P in the combined flow of a steady wind and a vortex, two components are combined and determine the velocity of the resultant flow (*a*). This is done through vector addition, as shown in (*b*), which involves completing the parallelogram of velocities to give the speed and direction of the resultant velocity.

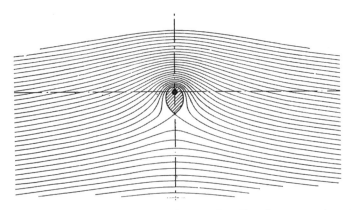

FIGURE 4.3. The flow that results from the combination of a uniform free stream and a vortex. The streamlines above the center of the vortex are closer together than below it, showing that above the vortex the speed is higher, and the pressure lower, than below the vortex. There will be a resultant force directed upward, that is, a lift force. From Lanchester 1907, 164.

At any given point P in the new flow, there are two vector arrows to take into consideration. One, provided by the steady wind, is horizontal; the other, provided by the vortex, is at an angle determined by the position of P relative to the center of the vortex. P is located at some radial distance from the center of the vortex (this determines the speed), while the direction of the vortex component is determined by the direction of the tangent of the streamline that passes through P. A typical case is shown in figure 4.2a. The procedure needed to combine the effects of the two flows is shown in figure 4.2b. The resultant velocity is given by a geometrical construction called "completing the parallelogram," whose intuitive meaning can be read directly off the diagram. Completing the parallelogram gives the speed and direction of the new flow at that point. A picture of the combined flow can be built up by the carrying out of this process at a large number of points. It will have the general appearance of figure 4.3, which is taken from Lanchester's *Aerodynamics.*

The diagram shows that the streamlines are closer together above the center of the vortex than below, and this difference indicates a speed difference. The speed of flow above the vortex is greater than the speed below. How does this arise? To explain this occurrence it is sufficient to focus on two particularly important positions in the vortex, namely, the twelve o'clock and six o'clock positions, which are directly above and directly below the center of the vortex. Here the vector addition effectively reduces to simple arithmetical addition because there is no angle between the contributions of the two flows. At both points, but only at these points, the effect of the vortex is exactly aligned with the horizontal wind. At a point positioned some given distance r directly above the center of the vortex, the speeds of the two flows are going to add together to produce a flow with the speed $V + v$. At a diametrically opposite point, a distance r directly below the center, the two flows will oppose one another to produce a reduced speed $V - v$. Elsewhere in the flow, at points not directly above or below the vortex, the contribution of the vortex component will augment the upper half of the flow, and diminish the lower half of the flow by less than v, but the general effect will still be present. Hence the spacing of the streamlines visible in Lanchester's diagram.

The crucial step is the next one. The supporters of the circulatory theory supposed that, as it moves through the air, an aircraft wing (viewed in cross section) somehow generates a vortex effect around itself. There is, they argued, a vortex "bound" to the position of the wing. The effect of the wing is to be represented by a vortex, even though the wing profile has an elongated shape, while the vortex is circular in form and centered on a geometrical point. Why a wing has this effect on the air and why it can be represented in this way were problems for the supporters of the theory, but they proceeded on the assumption that this was the case. They accepted that near the wing the flow could not look exactly like a combination of a steady wind and a vortex but that the picture became more accurate at a distance. Following the reasoning set out here, and applying it to the case of the wing, they argued that if the flow around the wing consisted (approximately) of a uniform flow combined with a vortex, then some of the air at a given distance above the wing would reach a maximum speed of $V + v$ and some of the air at the same distance below it would drop to the speed $V - v$. Here was an explanation of the required speed differential in the flow over the wing, which in turn accounted for the pressure differential, and thus for the lift. Or, to be more precise, here was an explanation of lift *if* the assumption is granted that the wing generates a vortex. But should this point be granted? The question epitomizes all the subsequent arguments over the circulatory theory.

Does the circulation theory imply that, during normal flight, molecules of air make a journey around the chord of the wing? No, this is not what its supporters were saying. Such a picture may be conjured up by abbreviated formulations, such as "lift is created by the circulation of air around a wing," but these words depend on a technical meaning of the word "circulation" and do not mean what they may seem to mean. It is true that in an isolated vortex, such as a whirlwind, the air does indeed make a circular journey around the center of the vortex, but the theory does not require this to happen in the case of a wing delivering lift. The actual flow involves fluid elements curving up slightly to meet the leading edge of the wing. They then travel along the chord of the wing and leave with a slight downward inclination of the streamlines at the trailing edge. The claim is merely that during normal flight, the vortex exists as a *component* of this overall flow pattern.

In 1903 the Cambridge logician Bertrand Russell argued that "the component of any . . . vector sum, is not part of the resultant, which alone could be supposed to exist."[5] Russell (who was seventh wrangler in 1893) did not have aerodynamics in mind but was writing about the nature of mathematical concepts in general. His position suggests that only the resultant flow of air over a wing really exists, whereas the uniform flow and the vortex, being mere components, do not really exist. Such a conclusion does not do justice to what the supporters of the circulatory theory were saying. The component flows were meant to describe *real tendencies* existing in the resultant flow. These tendencies can be "supposed to exist" even when not manifesting themselves in isolation from other tendencies. This realistic way of speaking seems more natural than Russell's formulation and better covers the range of empirical possibilities that would have been evident to those working in aerodynamics. First, the realistic idiom implies that if one contrived to bring a moving (and lift-generating) wing to a sudden halt in midair, then the circulating tendency would have nothing to modify it and *would* reveal itself in its full form.[6] In these circumstances there would be air swirling around the wing. Second, as a general fact about vortex flow, if a *very strong* vortex is combined with a uniform wind, *some* of the air close to the center of the vortex *actually will* go around in a closed loop. (An examination of Lanchester's diagram in fig. 4.3 shows that it represents a flow of this kind.) These considerations suggest that Russell was wrong and that the components of the vector addition can be as real as the resultant. Whether these real tendencies display themselves as independent phenomena is merely a matter of how strong they are relative to the other components.

The Mathematical Definition of Circulation

The behavior of a simple vortex, of the kind just described, can be used to introduce the mathematical definition of the quantity called "the circulation" of a flow.[7] So far I have used the concept of circulation in an intuitive way, but for the purposes of aerodynamics it was important to deploy a precise definition. To follow the history of the dispute over the circulatory theory of lift, readers must grasp the general features of the more precise, mathematical concept. I introduce them in two stages. First I confine the discussion to a simple, isolated vortex of the kind already introduced and then I move to a more general definition that can be used to detect and measure circulation in complicated flows where the vortex is just one component. The ideas discussed in this section are now standard, textbook material, but that was not always the case. They found no place in Cowley and Levy's textbook of 1918. These authors were still under the influence of the discontinuity theory of lift and did not take the circulation theory seriously.

Consider again the family of simple vortices where the streamlines are concentric circles of radius r and the uniform speed of flow along the streamlines (the tangential velocity v) diminishes in proportion to the distance r of the streamline from the center of the vortex. This model has already been expressed by the formula $v = k/r$, where k is the constant of proportionality. Three features of this family of vortices are immediately evident. First, the product of v and r, that is, the speed along a streamline multiplied by the radius of the streamline, always has the same value, namely, k. Second, as the radius r gets very small, the velocity v gets very large. Mathematically, as $r \to 0$, $v \to \infty$. Third, as r gets very large, the velocity of the flow gets very small. In the limit, as $r \to \infty$, then $v \to 0$. The flow is effectively stationary at great distances from the center of the vortex. All vortices of this simple kind therefore cover the entire range of velocities from infinite velocity at the center to zero velocity at the distant periphery. The situation at the very center of the vortex, at $r = 0$, where the flow rotates at infinite speed, is obviously physically unrealizable. It indicates the abstract character of the vortex model. The problematic point at $r = 0$ is called a "singular point" of the formula.

Intuitively some vortices are said to go faster than others or to be stronger than others. How are these distinctions to be expressed given that the simple k/r formula covers the full range of velocities from zero to infinity? The answer is that at any given distance from the center, the speed of flow of some vortices will be greater than others at that distance. In order to make the distinctions that are desired, the speed of flow must always be related to the distance from the center. This requirement must be built into any definition of the strength of a vortex. Such a relation is embodied in the constant k that has already been introduced in the formula relating tangential speed v and radius r. The constant k characterizes a vortex and distinguishes the faster from the slower vortices. For any given distance r from the center, the bigger the value of k the faster the vortex goes.

Suppose now that the word "circulation" is introduced with the aim of describing a vortex. The aim is to discriminate between them so that a fast, strong vortex has a "greater circulation" than a slower, weaker vortex. One plausible candidate for the definition of the concept would be to equate circulation with the value of k. Historically the chosen definition was close to this but not identical. The measure of circulation that was chosen made it proportional to k but not equal to k. It was decided that the circulation was to be $2\pi k$. Stated in words, the circulation around a simple vortex was defined as the product of the tangential velocity v along some streamline and the circumference of the streamline, which is $2\pi r$. This (provisional) definition refers to the speed along a contour multiplied by the length of the contour. Using the symbol Γ (gamma) for the circulation, the circulation around a simple vortex is given by the formula $\Gamma = v\,2\pi r$ or by the formula $\Gamma = 2\pi k$. The value of the circulation around a given vortex has the property that it comes out the same whatever streamline is used to arrive at the value.

The next step is to generalize the definition so that it is not tied to flows with circular streamlines. The move toward generality exploited the fact, noted earlier, that in a simple vortex it does not matter which circular streamline is used to arrive at the value of the circulation. The circulation could be computed in part by using the speed v_1 along an arc of the circle of radius r_1 and in part by using the speed v_2 along an arc of the streamline of radius r_2 (see fig. 4.4a). The jump from one arc to the other could be taken along an appropriate radius. The jump would not contribute to, or alter, the circulation because all the speed in the vortex is tangential and the speed along any radius is zero. As long as the two circular arcs were combined so as to make a closed circuit around the vortex, the value of the circulation would be no different from the value calculated using a single circular streamline as the relevant contour. And what holds for jumping between two different circular streamlines holds for jumping between three or four or any number. This too makes no difference to the value attributed to the circulation. In the limit, the jumps can be imagined to be so numerous, and the streamlines can be imagined to be so close, that the contour along which the circulation is computed could weave any path through the field of flow (see fig. 4.4b). The upshot is that the contour can be arbitrary. All that matters is that it forms a closed loop around the center of the vortex.

In the original definition of circulation around a simple vortex, where the contours were concentric circles, the speed used in the calculation was the tangential speed along the circumference. When the contour is arbitrary, the speed used in the computation must be the speed along this arbitrary contour. Stated generally, at any point P on the contour, the flow can be assumed to have a speed q and a direction that makes an angle θ to the contour (see fig. 4.5). To compute the circulation, the component of speed along the contour, namely, $q\cos\theta$, must be used. The contribution to the circulation at that point is $d\Gamma = q\cos\theta\,ds$. Summing or integrating this quantity around the closed contour gives the circulation around the contour. This leads to the general definition of the circulation Γ around a closed contour C as

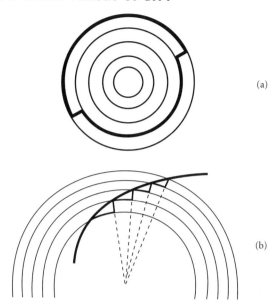

(a)

(b)

FIGURE 4.4. In (a) the product of the speed and circumference is the same for all streamlines of the vortex. The value of the circulation can therefore be computed using a contour that jumps from one streamline to another. Part (b) shows how this fact can be exploited and any contour used to compute the circulation. An arbitrary contour can be thought of as a limiting case involving an infinite number of jumps. The strength of the circulation is thus contour-independent provided that the contour encloses the vortex.

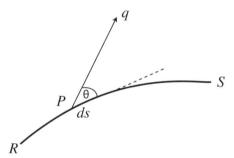

FIGURE 4.5. The speed of the flow at P is q. The component of q along the curve RS is q cosθ. The "flow" along RS is defined as ∫q cosθ ds where the integration is taken along the contour RS. If the contour is extended to form a closed loop, the flow counts as the "circulation" around the closed contour.

$$\Gamma = \int_c q \cos\theta \, ds$$

The previous, and provisional, definition of the circulation around a simple vortex can now be seen as a special case of this more general definition. In the special case, C is a circle of radius r. The speed q is equal to the constant tangential velocity v and always lies along the contour so that $\cos\theta = \cos 0 = 1$.

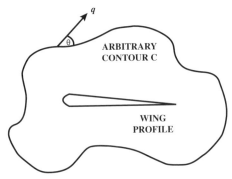

FIGURE 4.6. The circulation around an aerofoil is defined as the value of the integral $\int q\cos\theta \, ds$ where the integration is taken around any contour enclosing the aerofoil. The integration adds all the infinitesimal components of speed along the tangent.

These constant quantities can be taken outside the integral sign, leaving a line integral which equals the circumference $2\pi r$. Thus, as before, the circulation around the vortex is $\Gamma = v2\pi r$.

The general definition identifies a circulation in a flow even when the vortex merely acts as one component. If there are no vortices in the flow, the circulation is zero. The important point for aerodynamics, and the circulation theory of lift, is that the measure of circulation should be *independent of the contour* so the selection of the contour can be arbitrary, as in figure 4.6. All contours that form a closed loop containing an aerofoil, and the vortex that it is assumed to generate, should yield the same measure for the circulation around it. A large circle could be drawn around a two-dimensional aerofoil, or a rectangle could be used, and the numerical value of the integral along these contours should be the same.

Where the integral produces different values for different contours, the concept of "the circulation of the flow" is not well defined. This can happen, for example, when the fluid is viscous. Contour-independence will prove to be of great significance later in the story, when I describe how, in the 1920s, the personnel of the National Physical Laboratory attempted to conduct experiments to establish, once and for all, whether the circulation theory gave the right account of the lift of a wing moving through air.[8]

Circulation without Rotation

In chapter 2 I introduced the idea of describing a flow by means of a stream function. What is the stream function for a vortex? From the definition of a vortex, the flow is in concentric circles of radius r so there is no velocity component u' along the radius, but there is a tangential velocity component v' along the circumference of the circles. It will be recalled that the speed of flow is given by the rate of change of the stream function ψ. In polar coordinates the two relevant equations relating v' and u' to the stream function are

$$\frac{1}{r}\frac{\partial \psi}{\partial \theta} = u' = 0 \text{ and}$$

$$-\frac{\partial \psi}{\partial r} = v' = \frac{\Gamma}{2\pi r}.$$

Integration of the second of these equations answers the question posed earlier and gives the stream function for the vortex as

$$\psi = -\frac{\Gamma}{2\pi}\log r.$$

Both Lanchester and his critics made repeated references to flows with circulation but that were also described as "irrotational." The flow around a simple circular vortex of the kind I have been describing is an example of circulation without rotation. It may seem puzzling to say that the air in such a vortex is not rotating. The problem lies with a divergence between technical and common usage. The difficulty is resolved, at least in part, by distinguishing between the overall path of the fluid element, that is, its circular orbit around the vortex, and its behavior while going round that path. The distinction is sometimes illustrated in textbooks by reference to a Ferris wheel, where the chairs are carried around but always hang vertically and so maintain the same orientation, that is, the chairs "circulate" but do not "rotate." Circulation is also compatible with an absence of rotation when a vortex exists as one component in a flow. Thus the combination of a uniform wind and a vortex motion, of the kind assumed in the circulatory theory of lift, can also be conceived as an irrotational motion with circulation.

One final feature of vorticity and circulation should be mentioned before introducing Lanchester's own use of these ideas. In 1869 Kelvin established a remarkable theorem about circulation in an ideal fluid. He proved that it does not and cannot change with time.[9] In an ideal fluid obeying the Euler equations the value of the circulation cannot change as a result of any processes such as the movement of a body through the fluid. Circulation can exist, but once in existence it can be neither augmented nor diminished. Thus if the circulation starts as zero, it stays zero. If the amount of circulation is symbolized by Γ, then Kelvin's theorem is expressed by saying that the rate of change of Γ with time is zero. In mathematical symbols,

$$\frac{d\Gamma}{dt} = 0.$$

The proof of Kelvin's theorem depends on the condition that the contour of integration used to establish the circulation is one that moves along with the fluid. The contour must be made up of the same elements of fluid at all times. Such a contour is sometimes called a material loop and stands in contrast to the kind of contour mentioned previously, which is a purely geometrical entity and can be selected on grounds of mathematical or experimental convenience. This restriction of the theorem to the circulation around material loops has consequences that are both important and subtle. It made the precise implications of the theorem difficult to decide. Later we see that Kelvin's theorem was interpreted in different ways by different groups of experts. The divergence of opinion had a significant impact on the discussion of how (or whether) a wing can generate lift by generating circulation. The generation of circulation by the movement of a wing through the air would involve a variation of circulation with time. Is it ruled out by Kelvin's theorem? If it is, then what does this mean for the circulation theory of lift? I shall come back to these questions, but now, having laid the foundations, I turn to Lanchester's pioneering statement of the circulation theory.

Lanchester's Treatise

Frederick William Lanchester (fig. 4.7) was born in 1868, the son of an architect.[10] He was educated at the Royal College of Science and Finsbury Technical College. He began work in 1889 with a company making gas engines, and in 1895 he began to develop his own motorcar. Until 1919 he was managing director and then the consultant engineer of the Lanchester Motor Company Ltd. He was responsible for some important patents for devices that successfully reduced the vibration that plagued early engines. During this time Lanchester had also been working on the problems of artificial flight and experimenting with model gliders. In 1907 he assembled the ideas about lift that he had been developing since the mid-1890s and published them in the form of a bulky volume called *Aerodynamics*, the first of a two-volume *Treatise on Aerial Flight*.[11] This work is now recognized as the locus classicus of the circulation theory of lift, though it does not read like a modern textbook on aerodynamics. The circulation theory is only one strand in the argument that had evolved over some dozen years and had been changed to bring it more into line with the concepts used in, for example, Lamb's *Hydrodynamics*.[12] The precise character of the changes and the form in which the theory was

FIGURE 4.7. Frederick William Lanchester (1868–1946). Lanchester published a treatise on aerodynamics in 1907 in which he presented the circulatory theory of lift. He was a founding member of the
Advisory Committee for Aeronautics. His ideas were quickly welcomed in Göttingen and his work translated into German, but the ACA did not take his ideas seriously until after the Great War. (By permission
of the Royal Aeronautical Society Library)

first conceived are not known, though they may, to some extent, be guessed
from the variations in the uneven text.

Aerodynamics shows the traces of at least five interwoven lines of argument: (1) an evolutionary perspective, (2) the concept of a wing being carried
on a wavelike airflow, (3) a quasi-Newtonian idea of the "sweep" of a wing,
(4) examples of the theory of discontinuous flow, and (5) versions of the
theory of circulation or the vortex theory. Although Lanchester devoted all
of chapter 3 of his book to an exposition of basic hydrodynamic ideas, the
assimilation was incomplete. He did not avail himself of the mathematical
formula expressing the circulation as an integral, though he did accept the
ideas behind it. Furthermore, his use of the word "circulation" was not consistent. It was often used informally to refer to fluid that was displaced by a
body and pushed from the front to the rear.[13] Tracking the word "circulation" in Lanchester's text does not necessarily reveal those places where the
circulation theory of lift was being developed. Terminologically, Lanchester
preferred to speak of "cyclic flow."

The theoretical centerpiece of the 1907 book was chapter 4, called "Wing Form and Motion in the Periptery." The word "periptery" was coined by Lanchester to refer to the characteristic form of airflow in the vicinity of a lifting surface. The chapter began with an evolutionary argument and a criticism of an existing theory of lift. In order to perform its biological function, argued Lanchester, the wing of a bird must have evolved into a shape that conforms to the pattern of airflow necessary to provide lift. It should therefore be possible to read off this pattern of flow from the shape of the wing. All such naturally occurring wings show a similar "design and construction" that involves an arched profile and a slight downward inclination of the front edge. From this Lanchester inferred that the air must be moving *upward* as it approaches the leading edge of the wing and *downward* as it leaves the trailing edge.

The advantages of the dipping front edge was first recognized by Horatio Phillips, who made it the subject of two patents in 1884 and 1891, but, said Lanchester, Phillips gave an incorrect account of it. According to Phillips the air impinged on the sloping, upper surface of the leading edge and was deflected upward, off the surface of the wing, leaving a partial vacuum on the upper surface. Lanchester rejected this in favor of an explanation based on principles drawn from Newton's mechanics. The central point, he said, was the exchange of momentum. The air, which was rising at the front of the wing, had to have the vertical component of its motion reduced to zero. The air then had to be given a downward direction, and thus supplied with another vertical component of motion, but this time in the opposite direction. It was important, said Lanchester, that during this process the flow of air remained conformable to the shape of the wing and that no surfaces of discontinuity were created.

These ideas were developed by means of a thought experiment involving the fall of a flat plate. The plate was to fall so that it presented its full surface-area to the direction of its descent. During the fall the air would be pushed around the edges from the lower to the upper surface. "There is *at first* a circulation of air round the edge of the plane from the under to the upper surface, forming a kind of *vortex fringe*" (145). (Notice that here the word "circulation" refers to the air being literally displaced from the front to the back of the object.) Lanchester then supposed the falling plate was given a horizontal velocity. This, he said, made the case equivalent to an inclined plane moving horizontally. If, following Newton, air is treated as if it is made up of independent particles, the analysis gives the wrong answers. Lanchester concluded that it was necessary to take account of the continuous nature of the fluid medium but knew of no way to do this for the flow under consideration. This led him to introduce a (more or less) arbitrary "sweep" factor to define the amount of air that was involved in the momentum exchange. It was Lanchester's references to "sweep," rather than circulation, that were picked up by the practical men.

Lanchester then explored a number of different approaches. The reader was told that "the peripteroid system may be regarded as a fixed wave" (156), though this idea was nowhere adequately explained. It seems to have been part of an early version of the theory. A few pages later Lanchester explained that because the disturbances in the neighborhood of the aerofoil possess angular momentum, it can be inferred that the flow comprises a cyclic motion. Lanchester went on: "The problem, then, from the hydrodynamic standpoint, resolves itself into the study of cyclic motion superposed on a translation" (162). He then used the formulas of mathematical hydrodynamics to plot the streamlines and potential lines for the flow around a circular cylinder on the assumption that the flow contained a circulation or cyclic component. Graphical methods were used to establish that the imbalance of pressures furnished a lift force. The flow depicted in the plot did not look like a picture of the flow round a wing, because there was a circular cylinder, rather than a winglike profile, at the center of the action, but, said Lanchester, we may look upon this figure "as representing in section a theoretical wing-form, or aerofoil, appropriate to an inviscid fluid" (163). He justified this statement by observing that from "the hydrodynamic standpoint," that is, with a perfect fluid, the shape of the aerofoil section is irrelevant.

Lanchester then moved from perfect fluids to real fluids and from the infinite wing to the wing of finite length where the behavior of the air at the tips had to be considered. Finite wings could be understood by supposing that the cyclic flow extends beyond the wingtips in the form of two vortices issuing from the ends of the wing and trailing behind it. The trailing vortices could be assumed to extend back to the point on the ground from which the aircraft took off. Such a picture meets the requirement, first identified by Helmholtz, that a vortex can only end on a surface of the fluid. Lanchester acknowledged that, because of Kelvin's theorem, the creation of such vortices in a perfect fluid presented a problem. He argued that viscosity had to be invoked to start the process, but inviscid theory could be used for the subsequent description. He also noted that the two trailing vortices would interact with one another. As Lanchester put it, "We have seen . . . that the lateral terminations of the aerofoil give rise to vortex cylinders. . . . Such a supposition presents no difficulties in a viscous fluid. . . . Now we know that two parallel vortices, such as we have here, possessed of opposite rotation . . . will *precess* downwards as fast as they are formed" (173). Lanchester then referred his readers to the diagram that is reproduced here as figure 4.8.

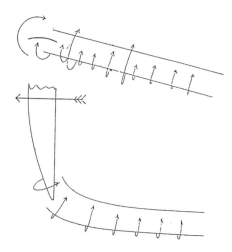

FIGURE 4.8. Lanchester's pictures of trailing vortices. From Lanchester 1907, 172.

Lanchester now had a qualitative account of lift that fulfilled the following conditions: (1) it was based on "cyclic" flow, that is, a flow around a vortex with circulation; (2) it was applicable to a finite wing; (3) it identified the role of trailing vortices; and (4) it made appeal to the viscosity of air as well as to conceptions derived from ideal fluid theory.[14]

Anyone for Tennis?

Lanchester offered no quantitative expression for the dependence of lift on circulation. He presented many of his main results in a verbal or pictorial form. A modern textbook of aerodynamics would give a mathematical formula to express the central ideas of Lanchester's theory of lift. The formula would be

Lift = $\rho \Gamma V$,

where ρ is the density of the air, Γ is circulation (that is, the value of the integral described earlier), and V is the speed of the wing relative to the undisturbed air.

It is a matter of some historical interest that this important formula had been published in 1877 by Lord Rayleigh. Rayleigh did not, however, offer it as a law that might govern the lift on an aircraft wing. He reserved that role for the very different formula he had published in the previous year describing the pressure on a flat plate created by a discontinuous flow. How and why did Rayleigh arrive at the $\rho \Gamma V$ formula? I want to look into this matter and then come back to Lanchester.

The short paper in which the formula appeared was titled "On the Irregular Flight of a Tennis Ball."[15] In it Rayleigh posed the question of why a spinning ball sometimes veers to one side in flight: "It is well known to tennis players that a rapidly rotating ball in moving through the air will often deviate considerably from the vertical plane. There is no difficulty in so projecting a ball against a vertical wall that after rebounding obliquely it shall come back in the air and strike the same wall again" (344).

In posing his question Rayleigh would have been thinking of the old game of royal or real tennis, which was played indoors and involved a net strung between two walls. In this form of tennis the players are permitted to use the walls and hence produce the effect that Rayleigh mentioned.[16] Real tennis courts only existed in a few country houses and ancient universities, but there were (and still are) courts in Cambridge. It was Rayleigh's chief recreation as an undergraduate.[17] Rayleigh rejected the idea that the trajectory of the ball deviated from a straight line because the ball "rolled" on a cushion of condensed air that had formed in front of it. That, he noted, would produce a deflection in the wrong direction. The correct answer, he said, had been given in a qualitative form by the Berlin scientist Gustav Magnus in 1852. Magnus had been interested in artillery shells rather than the more gentlemanly tennis ball.[18] The present paper, said Rayleigh, was designed to supplement Magnus' answer with a mathematical formulation.

To simplify the problem Rayleigh assumed that the ball was stationary and the air flowed uniformly past it. If the ball is not spinning, all the forces, frictional and otherwise, will be in the direction of the air stream. To produce a deviation in the trajectory there must be a lateral force. This force depends on the rotation of the ball, which, by friction between the ball and the air, produces what Rayleigh called "a sort of whirlpool of rotating air" (344). If the whirlpool is combined with the uniform stream, the velocities oppose one another on one side of the ball and augment one another on the other side. This velocity difference produces a pressure difference and hence a side force. This time the force is in the correct direction. "The only weak place in the argument," said Rayleigh, "is in the last step, in which it is assumed that the pressure is greatest on the side where the velocity is least. The law that a diminished pressure accompanies an increased velocity is only generally true on the assumption that the fluid is frictionless and unacted upon by external forces; whereas, in the present case, friction is the immediate cause of the whirlpool motion" (345). The law relating diminished pressure with increased velocity was Bernoulli's law, and Rayleigh was rightly sensitive to the preconditions of its legitimate use. Despite the questionable character of this step in the argument Rayleigh continued to develop the analysis in terms of a circular cylinder "round which a perfect fluid circulates without molecular rotation." What was meant by "molecular" here? Rayleigh was explicitly working with an ideal or perfect fluid, so the "molecules" would actually be fluid elements. Rayleigh would not have been confusing them with the molecules of the chemist. He was simply postulating an irrotational flow of an incompressible, inviscid fluid with circulation.

Rayleigh proceeded to write down the stream function for the flow he was postulating. He was able to do this by adding together the stream functions for the two components of the flow, namely, the stream functions for a uniform flow round a cylinder and for a vortex round a point. The formula he used was, in his notation,

$$\psi = \alpha\left(1 - \frac{a^2}{r^2}\right)r\sin\theta + \beta\log r,$$

where r and θ are the polar coordinates of any point in the fluid and a is the radius of the cylinder. The two symbols α and β are constants proportional to the velocity of the free stream and the circulation, respectively. Examination of Rayleigh's formula shows that it can be related to the examples I have discussed in previous sections. The first part is the stream function that was discussed in chapter 2 for the flow around a circular cylinder. The second part, involving the log term, was introduced earlier in this chapter as the stream function of a vortex.

Rayleigh used this formula to arrive at an expression for the velocity and hence the pressure at any point on the surface of the cylinder. The lateral component was integrated around the circular cylinder to give the resultant lateral force. The expression that Rayleigh arrived at was

$$\int_0^{2\pi} (p - p_0)a\sin\theta d\theta = -2\pi\alpha\beta,$$

where p is the pressure at the surface of the cylinder and p_0 is the pressure at infinity. The other symbols have the meanings already assigned. The conclusion indicated by the final term of this equation was that the lateral force causing the irregular flight of the tennis ball was "proportional both to the velocity of the motion of circulation, and also to the velocity with which the cylinder moves" (346). This is essentially the result on which Lanchester's analysis depended, and the right-hand side of the above equation is equivalent to the law relating lift to density, circulation, and velocity. The minus sign on the right-hand side of the above equation indicates direction, the circulation is $\Gamma = 2\pi\beta$, while the free-stream velocity is $V = \alpha$. Rayleigh took the density of the air to be of unit value, which explains why the term ρ did not appear in his final formula.

Having arrived at the lateral force on the tennis ball, Rayleigh asserted without argument (but citing the authority of Kelvin) that, under the simplified conditions of the example, the trajectory of the tennis ball would be along the arc of a circle, the motion having the same direction as the circulation.[19] Rayleigh's analysis was taken up and extended by Greenhill in a paper published in 1880 in which the equation of the trajectory was deduced in an explicit way.[20] Further, whereas Rayleigh had ignored external forces such as gravity, Greenhill reintroduced this complication and showed that its effect was to give the trajectory the form not of a circle but of a trochoid.[21]

How seriously did Rayleigh take his analysis of the tennis ball and, by implication, the artillery shell? His aim in the paper, he explained, had been to "solve a problem which has sufficient relation to practice to be of interest, while its mathematical conditions are simple enough to allow of an exact solution" (345). These words reveal the element of judgment that entered into the exercise. Relevance was being traded with simplicity. A balance was being struck, but Rayleigh evinced some doubts about the terms of the exchange. His final words in the paper referred back to what he had called the weak links in the argument. The behavior of a real fluid would give rise to a wake behind the ball, and this aspect of the flow finds no recognition in the analysis. Rayleigh thus ended his paper with a warning: "It must not be forgotten that the motion of an actual fluid would differ materially from that supposed in the preceding calculation in consequence of the unwillingness of stream-lines to close in at the stern of an obstacle, but this circumstance would have more bearing on the force in the direction of motion than on the lateral component" (346). The informed reader of Rayleigh's paper would know that there was a mathematical approach that could represent this "unwillingness" of the streamlines to close in behind an obstacle. Such an analysis was provided by the theory of discontinuous flow and free streamlines. Perhaps this would do more justice to the complex reality of the case. Explicitly, then, the reader was given a circulation analysis but, implicitly, was being directed toward the theory of discontinuous flows.[22]

Lanchester also had an account of the swerving tennis ball in his book on aerodynamics. His account did not invoke the circulation theory but appealed to the discontinuity theory, that is, to the picture of Kirchhoff-Rayleigh

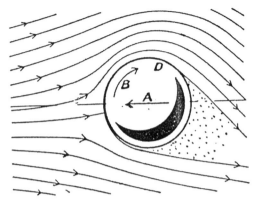

FIGURE 4.9. Lanchester's account of why a spinning ball generates lift. Notice the presence of the dead air behind the ball and hence the appeal to the discontinuity theory. From Lanchester 1907, 43.

flow. Lanchester's discussion of the spinning ball was given in a section of his book titled "Examples Illustrating Effects of Discontinuous Motion." The diagram he used to illustrate his analysis is given here in figure 4.9. The figure shows a ball moving through a viscous fluid, such as air, where the direction of motion is shown by the arrow *A*, and the rotation by arrow *B*. In this case the ball will veer upward. Like Rayleigh, Lanchester drew attention to the increase in the speed of the air on one side (here the top) and its diminution on the other side (the bottom). Where Rayleigh merely hinted, Lanchester was explicit that the flow round the ball creates dead fluid behind it. He argued for this concept on experimental grounds. The effect of the spin was then said to modify the shape of the dead air, as is shown in the figure. Lanchester claimed that at the top of the dead-air region, the spin helps to expel the dead air. At the bottom, the spin helps to draw in the dead air and spread it along the lower surface of the ball. This action was said to have the effect of giving the ball an upward thrust.

In retrospect, familiarity with the circulation theory of lift makes it easy to assimilate the case of the spinning tennis ball and the case of the wing and to see the possibilities inherent in treating both of them as involving circulation. It is evident that for Rayleigh and Lanchester the situation was not clear cut and both inclined toward an eclectic account. In broad terms it can be said that Rayleigh used the discontinuity theory to explain wings but the circulation theory to explain tennis balls. Lanchester did the opposite and used discontinuity theory to explain the swerve of tennis balls and circulation theory to explain the lift of wings.

Professor or Practical Man?

Not long after its publication, the first volume of Lanchester's *Treatise* became the topic of a seminar discussion in Göttingen, and Lanchester was approached for permission to produce a German translation. He received a letter to that effect from Prof. Carl David Runge. At first Lanchester had no idea who Runge was, but inquiries soon revealed that he was an eminent applied mathematician. He had done work on methods of numerical approximation used to solve differential equations as well as having discovered empirical formulas to predict spectral lines.[23] Runge had been brought from the *technische Hochschule* in Hanover to the university in Göttingen at the behest of Felix Klein. During the session of 1907–8 Runge had joined Prandtl in giving a seminar on hydro- and aeromechanics. They discussed Lanchester's book, and Runge penned his letter to Lanchester in March 1908. In September, Lanchester visited Runge in Göttingen to discuss the translation, and they became lifelong friends. Runge was impressed by Lanchester's vigor and humor, which were very different from the demeanor of the academic Englishmen he had met before. Lanchester was a guest of Runge and his wife and amused their young sons by flying model gliders. Lanchester recalled that one evening Prandtl and von Parseval (the airship designer)[24] "blew in" for a beer. A convivial time was had, though communication was limited by Lanchester's lack of German and Prandtl's lack of English. Runge, by contrast, spoke English with a perfection that amazed Lanchester.[25] The German translation of the *Treatise* was a collaborative effort by Runge and his wife. The first volume was published in 1909.[26]

Lanchester's appointment to the Advisory Committee for Aeronautics in 1909 is not difficult to understand in general terms—though exactly how it came about remains unknown. He had published a substantial work on aeronautics and was an engineer with a proven record of innovation. His knowledge of engine design was considerable, and the difficulty of producing light, powerful, and reliable engines remained a major stumbling block to progress in aviation. Lanchester grumbled retrospectively that Glazebrook thought that this was his only area of expertise: "he always treated me as if I was appointed purely for my knowledge of the internal combustion engine." In the hope of being taken seriously on a broader front, Lanchester had distributed six copies of his treatise *Aerial Flight* to his fellow committee members. Rayleigh responded with an appreciative letter and gave Lanchester four volumes of his collected papers in return. Of the rest, said Lanchester, only one acknowledged the gift, while the others "ignored it entirely."[27] Despite these slights Lanchester worked hard, both for the main committee and for the two subcommittees on which he served: the Engine Sub-Committee and the Aerodynamics Sub-Committee. His biographer described how Lanchester would attend some forty to fifty meetings a year as well as taking on the responsibility of writing a significant number of papers published as Reports and Memoranda.[28]

While Lanchester's appointment to the ACA is intelligible, his position on the committee was anomalous. He was a representative of the world of practical engineering on an administrative and scientific body dominated by influential Cambridge academics. He was not the only engineer or the only nonacademic or, indeed, the only non-Cambridge man on the committee, but (the military representatives aside) the others were more closely integrated into academic life. Joseph Petavel was a professor of engineering at Manchester and Horace Darwin, of the Cambridge Scientific Instrument Company, had close connections with the university. This left O'Gorman and Lanchester as the only authentic examples of what the British called "practical men," but this did not spare them from the hostility that was directed against the "theoreticians" of the Advisory Committee.

Lanchester's treatise was published before his appointment to the hated committee, so how was it received in the aeronautical and engineering press? The first volume was reviewed in *Aeronautics* in February 1908. The review was unsigned but was probably written by the editor, J. H. Ledeboer. The reviewer recommended Lanchester's book for thorough study by all those who take an interest in aeronautical problems, "since it includes a mass of valuable theory." The aircraft designer could gain "many a hint" from the book. The reviewer did not, however, identify the cyclic theory of lift among the "mass."[29] The anonymous reviewer in the *Aeronautical Journal* likewise greeted Lanchester's first volume as a valuable, but difficult, contribution to aerodynamical theory and also failed to convey to the reader the cyclic character of the flow postulated in Lanchester's book.[30]

The same pattern was repeated in the anonymous reviews published in the *Engineer, Engineering,* and the *Times* engineering supplement.[31] Florid congratulations were offered by the reviewer, but there was a signal failure to convey the actual content of the theory of lift. The words "circulation" or "cyclic" did not feature in any of the discussions. There were also warning signs suggesting that the reviewers might have opposed the theory had they grasped it. Thus, the reviewer in the *Times* spoke of "doubtful hydrodynamic analogies" (6), while the reviewer in the *Engineer* noted that "Evidently the author has a deep admiration for the work of Horace Lamb" (617) and went on to say that Lanchester "somewhat naively" described how he had added the hydrodynamic interpretation to his original ideas in order to give it a more secure basis. The naivety, presumably, lay in Lanchester's failure to realize that, in certain quarters, invoking Lamb would reduce rather than increase credibility. The reviewer in *Engineering* also took issue with Lanchester's diagrams: "The diagram given on page 145 seems to indicate that particles originally in contact with the lower surface of the plane finally reach the upper by describing circular paths round the edge. The true condition of affairs is, we believe, much more nearly represented by a diagram given later in the volume—viz. on page 266" (461). The complaint is revealing. As Lanchester had explained on the page before the diagram favored by the reviewer, at that point in the discussion "the flow is of the Rayleigh-Kirchhoff type" (265). The reviewer was responding to, and endorsing, a diagram that shows a discontinuous flow—precisely the picture that Lanchester was challenging.

Most of the "practical men" who read Lanchester initially extracted from him not the idea of circulation but the neo-Newtonian idea of "sweep." The realization that circulation played a central role in Lanchester's thinking came later. For example, Arthur W. Judge in his eclectic *Properties of Aerofoils and Aerodynamic Bodies* followed Lanchester in giving an account of both circulation and sweep, including a reproduction of some of the diagrams showing trailing vortices.[32] Judge's book was published in 1917, rather than being an immediate reaction, so there was more time for Lanchester to have disseminated his ideas.[33] Nevertheless, Judge did little to integrate the circulatory theory with empirical data, despite his references to some of the Göttingen publications.[34]

The second volume of Lanchester's treatise was reviewed in *Aeronautics* a year after the first volume, and this time the hand of the editor was signaled by the initials J. H. L.[35] He expressed "whole-hearted admiration" for the work but soon took the discussion into the vexed domain of the relation between theory and practice. "The present work once again raises the oft-debated question of the advantages in aeronautical research of the theoretical or mathematical and practical methods" (6). In the reviewer's opinion, "victory" rested with the practical methods. Admittedly, he went on, people such as Bryan, Rayleigh, "and, we may now add, Lanchester" show the path along which research is to proceed, but the fact remained that "the most successful machine of all," that of the Wright brothers, was built without the aid of mathematical formulas. This, then, was, "the main reproach we could make against Mr. Lanchester's work: it is far too theoretical" (6).

The designer J. D. North had apparently arrived at a similar conclusion. He later recalled that he had not found Lanchester's book as useful as the more empirical, data-rich volumes of Eiffel.[36] This opinion was not universal, but clearly many people saw Lanchester in this way, that is, as a representative of an overly theoretical approach. In February 1914, Herbert Chatley wrote an article in *Aeronautics* designed to head off the reading of Lanchester's book as just another example of mathematical "x-chasing."[37] Chatley acknowledged that there was controversy in the technical press about the value of applying mathematics to aeronautical problems. One dimension of the problem, he said, was generational. Older engineers had been trained practically and resented the attempts of younger colleagues "to dominate engineering practice with mathematical theorems." Chatley wanted to do justice to both sides and to get the competing parties to see that the problem did not have a black or white answer. The issue, he said, was always one of degree.

Chatley illustrated his conclusion by an example: "There are three books published in England by men with big reputations dealing with the theory of aviation, viz. Lanchester's 'Aerial Flight,' Bryan's 'Stability in Aviation,' and Greenhill's 'Dynamics of Mechanical Flight.' To those unfamiliar with mathematics these three books are classed together and, I imagine, are regarded by the majority of practical aeronautical engineers as beyond the limits of practical application" (46). It was a mistake, he said, to treat these books as three of a kind. Lanchester's book was not obscure and was of the greatest use to designers; Bryan's book was, indeed, mathematically much more demanding, but it was genuinely engaged with real problems; Greenhill's book, however, was quite different. It was even more mathematically difficult than Bryan's but "very little of it has at present any application to practical problems." As well as making discriminations of this kind, said Chatley, both sides, the practical men and the mathematicians, must recognize the weakness in their own positions. The former must give up reliance on "mechanical instinct," while the latter must "assimilate more complex conditions" rather than "elaborate hypotheses on too simple foundations" (46).

Chatley was seeking a compromise, but this desire was wholly foreign to C. G. Grey, editor of the *Aeroplane*. On September 12, 1917, Grey described Lanchester as "several government officials rolled into one, and a mathematician, and an advisory expert, and several other equally exalted things at the same time."[38] Grey took pleasure in baiting Lanchester, who held no academic post, by calling him "Professor."[39] While granting that Lanchester's books had their "lucid moments," Grey claimed that, overall, they were "utterly ununderstandable . . . to the ordinary man, and even to some trained mathematicians." He called on Lanchester to mend his ways: "No! Mr. Lanchester. Forget your professorship for a while. Be a practical engineer" (158). In an article headed "On a Professor," Grey described Lanchester's main publications in unflattering terms: "He has written two ponderous tomes, entitled

'Aerodynamics' and 'Aerodonetics' respectively, which profess to be mathematical expositions of the theory of flying. On these volumes his reputation in aeronautics was founded. Not being a mathematician myself I can only accept the statement of others better informed, who say that these expositions are not mathematics in the accepted sense of the term, but some science of symbols invented by Professor Lanchester."[40]

What had called forth this invective? The answer is that Lanchester had ventured into what an editorial in *Flight* had called "the cesspool of politics."[41] He had been goaded into responding to articles that Grey had published in the *Sunday Times* and the *Observer* on July 29, 1917. Lanchester's reply was called "A Campaign of Slander." It appeared on August 15, 1917, in the short-lived journal *Flying*.[42] He said Grey was indulging in "a venomous and unrestrained attack on the Air Board and upon the two branches of the Air Service."[43] These attacks repeated earlier accusations made by Pemberton Billing and Joynson-Hicks in Parliament—accusations, said Lanchester, that had been thoroughly investigated and found to be groundless. Lanchester was referring to the "murder" charge leveled at the government and to the Bailhache inquiry whose final report had been published on November 17, 1916. "No documents," said Lanchester, "could constitute a worse indictment of the self-appointed critics, or a greater vindication of the ability, character and honour of those attacked, unless possibly that document be the record of the proceedings themselves. . . . It is probable that in our time no public enquiry has been called upon to investigate charges so recklessly made and based on so slender a foundation" (51).

Now, a year later, Grey was repeating the old complaints. Notice, said Lanchester, how those making the attacks were always "wily enough to avoid putting them in a form which would permit of their being challenged in the Courts" (51). They always expressed themselves in a vague and general way. They talked of "appalling waste," "official designers," a "little clique," etc. It was a case, said Lanchester, of simply throwing mud in the hope that something would stick. The ground of the attack, as before, was the allegation of official incompetence and favoritism. Even the old BE type, an aircraft currently being replaced, was still under attack. "Its present survival in active service," said Lanchester, "is talked about as an offence attributable to some interested and evil-disposed persons in authority. The whole tenor of these long abusive articles is based upon the *existence* of machines, any machines whatever, of official design, when (it is alleged) far better machines of proprietary design are available" (51).

This was the nerve of the issue: the existence of official designs. The question with which we are faced, said Lanchester, "is why and to whom is the said design an offence?" (52). Had these machines failed to perform their required duties? Were there any better designs available when the BE2C was put into manufacture? The answer to both questions, said Lanchester, "is clearly in the negative" (52). The BE type had been the backbone of the reconnaissance service for three years of war. It had to bear the blame for its shortcomings as a fighter and for being not quite the equal in speed and climb to the Fokker, but, insisted Lanchester, it had acquitted itself "in a manner which is little short of astonishing" (52).

Lanchester then addressed Grey's complaints about engine shortages and the mobilization of manufacturing capacity in order to improve aircraft supply. In Lanchester's opinion, none of Grey's dramatic remedies could stand up to scrutiny. Grey had no understanding of real conditions, and his ideas about economics were naïve. Grey was responding to an inevitable feature of large-scale production but was treating it as if it were a defect that could be swept away. Thus, said Lanchester,

> He condemns offhand everything which is not the best. If a new machine is produced to-day which is better in its performance . . . on trial than something in service, then the latter is at once condemned, and the people who tolerate its use are obvious fools or incompetent rogues! It is spoken of as murder still to employ the B.E.2c for reconnaissance, or an 80-h.p. scout as a fighter. Mr. Grey has evidently had no experience of quantity production, otherwise he would know that *it is always the case, so long as there is progress, that the latest and the best designs will not be available in sufficient quantity.* It cannot in the nature of things be otherwise. (53)

Grey may have been the mouthpiece of the aircraft manufacturers, but in Lanchester's opinion he betrayed no understanding of the system of manufacture and development.

Lanchester returned to these themes later in 1917, in the December issue of *Flying*. The title of Lanchester's article was "The Foundation Stones,"[44] and he set out to explain the scientific and research basis on which airplane design depended. The airplane as we know it today, argued Lanchester, doesn't merely depend on the experience of the draughtsman, the intuition of the designer, or the cunning of the craftsman but "is the ultimate outcome of scientific research" (354). Theory, both mathematical and otherwise, played a role, while engineering and physical research, carried out in the workshop, the laboratory, and with full-scale aircraft, consolidated the results already explored theoretically. Lanchester's aim, however, was not simply to draw attention to the diversity of the scientific foundations of aeronautics. The different contributions had to be properly coordinated: "In order to secure continued progress it is vitally necessary that forces of many different kinds should act in concert, and nothing but evil can result from the worker in any particular field trying to encroach on other fields" (354). Lanchester listed the various contributors and delimited the proper scope of their activities. The mathematician may be useful but is "utterly incompetent" when it comes to "specifying or designing an aeroplane" (345). The mathematician's field is important but narrow. Unfortunately, many mathematicians are men of "childlike simplicity" and fail to understand that there is anything beyond their symbols. Physicists also have a vital role but "only one degree less narrow" than that of the mathematician. "For the physicist to invade territory outside his own ken would certainly not lead to satisfactory results" (354). The engineer is responsible for design and construction. "His job is frequently that of a buffer." He has to listen to both the mathematician and the physicist but he must also be responsive to the requirements of the user, and pay attention to economic considerations and, in times of war, to shortages of materials and the problems of substituting one material for another. The pilot also had a legitimate claim to be heard, as did the military and naval authorities who ordered the aircraft and were responsible for its deployment.

Lanchester had no illusions about a system of this kind. As a realist he knew there would always be clashes of interest. They were inevitable and could not be wished away, but they were not always bad in their effect. In a tilt at the press campaigns he declared: "Now there are many who think that friction in the management of our Air Services as between one department and another, or between different departments and manufacturers, or between the military authorities at the Front and those at home, etc., is essentially a sign of bad management and muddle. No such idea is justified" (354).

Theoretically, if a single "autocrat" were in control, friction might be avoided, but even then the autocrat would have to defer to those with specialized knowledge. In reality the answer lay in the proper division of labor and clear, long-term policies. "The greater part of the difficulties that have arisen in the past have been due to two causes, one a failure to define properly the spheres or fields of activity of the different contributory factors; the other an absence of foresight and far-sighted policy, both in detail and in the gross" (354).

As an example of a failure of foresight, Lanchester pointed to the attacks on mathematicians and physicists and their condemnation as impractical. At times these specialists may have made themselves too prominent, but their contributions are indispensable. The best way to achieve the desired harmony between the different factions "can only come from each man putting up as stout a case as he can for his particular views," while being prepared to listen to the views of others. Such negotiations have something of the character of a game of chess. We should not forget that when a mistake is made, its consequences "may often be so remote as to defy immediate analysis," or that the winning move "may have every appearance at the time it is made of being a blunder" (355).

In Lanchester's opinion, "indiscriminate Press criticism and stump oratory" only served to distract from the serious business of the war (355). Journalists who thought they were omniscient, he went on, merely traded on the ignorance of their fellows. The proper way to proceed was shown by the late E. T. Busk, who acted as a "bridge between the scientific man, the engineer and the pilot" (356). In the period leading up to the war, opinion was divided on the value of the inherent stability of an aircraft. Physicists and mathematicians, said Lanchester, strongly supported stability, while many pilots had been against it (as had the Wright brothers). The military authorities didn't know what to do. Busk resolved this dispute and "proved the value of inherent stability" (356). Lanchester accepted that since Busk's work, in 1914, knowledge of how to balance the relative virtues of stability and instability had been deepened. Despite these subsequent advances, the "BE2C machine was the immediate outcome of his work, and it is worthy of note that every Zeppelin brought down in night flying in this country was brought down by a machine of this type. For a long time it was the only machine which could safely be sent up or flown in the dark. Beyond this, for the first twelve months of the war, and more (up to the spring of 1916), nearly two thirds of the total enemy aeroplanes brought down on the Flanders front were brought down by the BE2C" (356).

Lanchester was the only member of the Advisory Committee for Aeronautics to engage in public with the likes of Grey and Pemberton Billing. He was appalled at their activities and had the civic courage to say so. He was also disturbed by the mounting evidence that Billing was conducting his campaign with the help of inside information provided by a member of the Advisory Committee itself. That member, Lanchester concluded, was the naval representative Murray Seuter. Lanchester complained in letters to A. J. Balfour and G. A. Steel at the Admiralty that Seuter was a slacker who did not pull his weight and abused his position of trust. Lanchester's suspicions had been aroused when he had shown Seuter an article he intended to publish in one journal but had then decided to publish elsewhere. Within an hour of showing the document to Seuter, one of Billing's supporters had rung up the original (that is, the wrong) publisher to raise objections. "I think," said Lanchester pointedly, "Commodore Seuter should be given a change of air."[45]

Although some practical men treated Lanchester as if he were a mere theoretician, in his own understanding he was very much the engineer. He had a sharp sense of the proper relationship of the engineer to the mathematician and physicist and of the differences in their perspective. I shall now look at how this sense of difference was reciprocated by mathematicians and physicists themselves. I review, in roughly chronological order, a sequence of critical responses to the circulatory theory of lift, as they were advanced by the high-status, mathematically sophisticated British experts in aerodynamics, and I identify all of the main objections. Running against the flow of these objections were some experimental results that might have worked in Lanchester's favor, but, strangely, these had little impact. Why this was so is part of the problem to be addressed.

Anonymity and Connectivity

Lanchester's *Aerodynamics* was reviewed anonymously in *Nature* on August 18, 1908.[46] The overall judgment was ungenerous and negative. No reader was likely to come away with the idea that the book contained striking insights into the nature of flight but instead that Lanchester was proposing a theory that was neither original nor successful. The theory was, perhaps, the product of a lively mind, but not a mind whose powers could be relied upon. The parts of the book that contained Lanchester's most characteristic opinions were described by the reviewer as "the more shaky theoretical chapters" (338).

There was qualified praise for some of the more empirical sections, which described Lanchester's experiments on viscosity and skin friction. The glider experiments, conceded the reviewer, gave results that were "remarkably consistent." Lanchester's account of the "chief methods and results of hydrodynamics," which lay at the basis of his theory of lift, were described as "on the whole very clearly written," but the reader was warned that Lanchester was "not, however, content to follow orthodox theory." It was in chapter 4 of the book, noted the reviewer, that Lanchester "leaves behind the solid ground of orthodox theory" and "attempts to work out the motion of a curved lamina," that is, a winglike surface (338). Furthermore, Lanchester's originality was challenged: "It seems to us that the author is wrong in claiming to be the first to give a theory of the motion of curved surfaces, and [in claiming] that Lilienthal had only practical acquaintance with the curved form, for Lilienthal clearly realised that the effect of curvature was to diminish eddy motion and to give an increased upward pressure due to the centrifugal force of the air. The theory has been worked out mathematically by Kutta, and his results are in fair agreement with Lilienthal's experiments" (338).

The reviewer then turned to Lanchester's own explanation of how a curved plate generates lift. It was introduced and dismissed in one sentence: "The author of the present volume attempts to work out the problem by applying the theory of cyclic motion to the motion of a surface in two dimensions, but it is difficult to see how this can have any application to the case of a lamina moving in free air" (338). Before looking into this expression of doubt I must address two preliminary points that concern the reviewer's mention of Kutta. First, it looks as if the reviewer did not appreciate that Kutta had put forward a cyclic theory. Second, Kutta's main contributions were published in 1910 and 1911, two or three years *after* the review. So what was the reviewer's source?[47]

Other than personal contact, there were two possible sources of information. One was a brief account of his work that Kutta himself published in 1902 in the *Illustrirte Aëronautische Mittheilungen*.[48] He gave his main results in the form of a complicated and opaque formula (not the simple product of density, circulation, and free-stream velocity). Kutta said that to reach the given formula he had used conformal transformations, but the assumptions behind his analysis were not explained. The other source was a footnote reference to this article by Sebastian Finsterwalder, Kutta's research supervisor at the *technische Hochschule* in Munich.[49] Finsterwalder had contributed the article on aerodynamics to Felix Klein's multivolume *Encyklopädie der Mathematischen Wissenschaften*. The relevant volume had been published before the *Nature* review appeared. The cyclic character of Kutta's theory was not apparent in the 1902 paper, though its relation to Lilienthal's work was explicit.[50] The same holds true of the Finsterwalder reference: there was no mention of the role of circulation. If these were the sources used, it could account for the misleading way in which Kutta was invoked in the review.

Why did the reviewer find it "difficult to see" how an account of a two-dimensional, cyclic motion could have any application to the motion of a lamina in free air? The reasons behind the difficulty were not explained, so it is necessary to make a conjecture about the argument that was probably in the reviewer's mind. The worry was about the move from two dimensions to three dimensions. Why should there be a problem about generalizing an account of cyclic or vortex motion in this way? The answer lies in the properties of the space around the wing that mathematicians call "connectedness"—a topological theme with which all Cambridge-trained mathematicians would be familiar.[51]

Connectedness refers to the conditions under which a contour in the form of a closed loop can be shrunk into a point or stretched and distorted so that it coincides with another closed loop. A "simply-connected" space is one in which every closed loop can be changed into any other closed loop without going outside the space. A "multiply-connected" space is one that is divided by barriers so that it ceases to be true that any two arbitrary loops can be made to coincide. Now a loop enclosing the infinitely long wing cannot be unhooked from it. It can be transformed into any other loop that is itself already around the wing, but it cannot be transformed into a loop that does not go around the wing. The space around an infinite wing is thus "doubly connected," while the space around a finite wing is "simply connected."

The move from a two-dimensional analysis to a three-dimensional analysis thus involves a move from a multiply connected space to a simply connected space. But why should this matter? A mathematically sophisticated reviewer will have known that, in a simply connected region, the only possible form of irrotational motion is acyclic.[52] In an acyclic motion there is no circulation and hence no lift. The reviewer seems to have assumed that Lanchester was exploiting a special, topological feature of two-dimensional flow but was then illegitimately applying the analysis to the three-dimensional case.[53] This assumption may explain why it was "difficult to see" how a theory of cyclic motion in a surface of two dimensions could have any application to a lamina moving in free air, that is, in three dimensions.

Was Lanchester's work really vitiated by these considerations? The answer is no. If this was the reviewer's argument, it was wrong. Lanchester had attended with some care to issues of connectivity. He stated explicitly that "we are consequently confined, in an inviscid atmosphere, strictly to the case where the aerofoil is of infinite extent, for a cyclic motion is only possible in a multiply connected region" (162).

How did Lanchester, having formulated the topological problem for himself, get round it? He needed some way to render the space of the three-dimensional case multiply connected. Lanchester did this by appeal to the trailing vortices issuing from the wingtips and reaching back to the ground. This method divided the space in such a way as to destroy its simple connectivity.[54] In figure 81 of his book (175), Lanchester gave a clear diagram of the vortices reaching back from the wingtips to the ground. But if Lanchester had anticipated and solved this problem, there was still another issue left. If circulation now makes mathematical sense, there is still the physical problem of how it gets started. Lanchester conceded that, as long as the atmosphere was viewed as an inviscid fluid, his vortices could be neither created nor destroyed. Such a system, he said, "in a fluid that is truly inviscid would be uncreatable and indestructible" (174). His response was to appeal to the viscosity of real fluids: "In dealing with a real fluid the problem becomes modified; we are no longer under the same rigid conditions as to the connectivity of the region" (175). Lanchester's remarks were perceptive, but the problem of the creation and destruction of vortices, and thus the problem of how circulation could arise, would continue to haunt the theory.

G. H. Bryan Reviews Joukowsky

Who was Lanchester's anonymous reviewer?[55] The most likely candidate was
G. H. Bryan. There are three reasons for drawing this conclusion. First, Bryan
had been involved with Klein's mathematical encyclopedia (contributing the
article on thermodynamics) and so was in a position to have come across
the mention of Kutta in that work. Second, Bryan was the usual reviewer on
aeronautical topics used by *Nature* and was later to review the second volume
of Lanchester's treatise.[56] Third, there is a piece of internal evidence. The 1908
review of Lanchester broached one of Bryan's pet themes: the dependence of
aerodynamics on hydrodynamics and the more fundamental status of hydrodynamics compared to that of the new, would-be discipline. As far as mathematical theory was concerned, said the reviewer, "aerodynamics as applied
to problems of flight does not differ from hydrodynamics" (337). This denial
of the independent status of aerodynamics was taken up again a few years
later in a review of Joukowsky's work that appeared in *Nature*, under Bryan's
name, on February 15, 1917.[57]

Joukowsky had already published a German-language account of the circulation theory in the *Zeitschrift für Flugtechnik* for 1910 and 1912.[58] In 1916 a
book-length exposition of Joukowsky's seminal work, based on his lectures,
appeared in French under the title *Aérodynamique*, and it was this that Bryan
reviewed.[59] The subject matter of Joukowsky's book, insisted Bryan, was not
of a sufficiently distinct character to form the nucleus of a new science—
aerodynamics. It was "hydrodynamics pure and unadulterated" (465). Bryan
also pointed out that there were two ways of "reconciling the existence of
a pressure on a moving lamina with the properties of a perfect fluid." One
was by assuming a circulation, and this, he said, appeared to be the basis of
Joukowsky's work. The other, "which has now been greatly elaborated in this
country," was the theory of discontinuous motion. "Of this theory," sniffed
Bryan, "Prof. Joukowski's treatment is practically *nil*" (465).

Bryan did not explain why Joukowsky *should* have discussed the discontinuity theory. Though Bryan still adhered to it, most British experts had
abandoned it, so some justification for the reproach would have been appropriate. Nor did Bryan say what might be wrong with the circulation theory.
The absence of any detailed engagement with the theory suggests that it was
simply considered to be a nonstarter and that the reviewer believed he could
count on his readers' agreement in this matter. But if this part of the argument was implicit, other parts were explicit. Bryan insisted at some length
that Joukowsky's book was of an elementary nature from which little was to
be learned—except, that is, by a certain class of engineer. "According to the

usual conventions in this country," said Bryan, "practical and experimental considerations regarding the motion of fluids are classified under the designation of *hydraulics*" (465). He went on to insist that both hydraulics and hydrodynamics should form the basis of a good engineering education:

> It is very important that engineering students who are proposing to take up aeronautical work should be equipped with a knowledge of the necessary hydrodynamics and hydraulics, and Prof. Joukowski's lectures were probably admirably adapted to the students in his classes. But the book goes only a very little way towards covering the subject-matter contained in the English treatises on hydrodynamics of more than thirty years ago, with their chapters on sources, doublets, and images, motion in rotating cylinders in the form of lemniscates and cardioids, motions of a solid in a liquid, tides and waves, and detailed treatment of discontinuous motion in two dimensions. (465)

The subjects mentioned by Bryan look suspiciously like the syllabus of an aspiring wrangler. This suspicion is confirmed when Bryan goes on to recommend that any "advanced student" revisit the standard, English treatises for "a thorough grounding in hydrodynamics" rather than rely on the "more superficial and fragmentary treatment of the same subject" offered by Joukowsky. Both the tone and the content of Bryan's review suggest that Joukowsky's book was not taken seriously in its own terms but was being judged as a Tripos textbook in hydrodynamics—and found wanting. It might do for the engineering students in Joukowsky's technical-college classes, but it would not get anyone through their Senate House examinations.

A Firm Basis in Physics

Further objections to the circulation theory came from G. I. Taylor, one of Cambridge's most brilliant young applied mathematicians. I have already mentioned his Adams Prize essay of 1914.[60] In that work Taylor did not confine himself to rejecting discontinuity theory; he also rejected the circulatory account of lift. Critical of the unreality of the textbook hydrodynamics that Bryan so admired, he argued that "the important thing in the earliest stages of a new theory in applied mathematics is to establish a firm basis in physics" (preface, 5). After describing the central idea of Rayleigh-Kirchhoff flow and pointing out its empirical shortcomings, Taylor turned briefly to Lanchester's theory. This too was faulted because of its lack of a firm basis in physics. Taylor's dismissal of Lanchester was swift: "Besides these [discontinuity] theories of the resistance of solids moving through fluids, Mr Lanchester has proposed the theory that a solid moving through a fluid is surrounded by an irrotational motion with circulation. This theory, as far as I can see, has nothing to recommend it, beyond the mere fact that it does give an expression for the reaction between the fluid and the solid" (4–5).

All that was granted to the theory, in its two-dimensional form, was that it had the (minimal) virtue of avoiding d'Alembert's paradox. It permitted the researcher to deduce "an expression" for the resultant force on the body, but that is all. The formula, however, was not, in Taylor's opinion, grounded in a real physical process. The theory provided no understanding of the mechanism by which the circulation round the body could be created. The problem came from Kelvin's proof that circulation can neither be created nor destroyed. If Lanchester's theory was an exercise in perfect fluid theory, then the premises of the theory precluded the creation of the very circulation on which it depended. Setting a material body in motion in a stationary fluid would not create such a flow. An aircraft, starting from rest on the ground in still air, and moving with increasing speed along the runway, would never generate the lift necessary to get into the air (not, at least, if the air was modeled as an ideal fluid). This consequence put Lanchester's theory in no less an embarrassing position than discontinuity theory. As far as it described any reality, discontinuity theory was a picture of a stalled wing, that is, of an aircraft dropping out of the sky. If Taylor was right, Lanchester's theory was equally hopeless because it would leave the aircraft stranded on the ground and incapable of flight.

Taylor thought Lanchester's theory was, if anything, worse than the version of perfect fluid theory that generates d'Alembert's paradox, that is, the version in which the perfect fluid has neither discontinuities nor circulation. Referring to this version as the "ordinary" hydrodynamics of an irrotational fluid, Taylor said that it, at least, gave a rigorous picture of the flow that *would* arise *if* an object were moved in these hypothetical circumstances, though, of course, this picture bore "no relation whatever" to reality. "The advantages of the ordinary irrotational theory is that it does, at least, represent the motion that would ensue if the solid were moved from rest in an otherwise motionless perfect fluid, and if there were perfect slipping at the surface. By taking irrotational circulation round the solid, Mr Lanchester loses the possibility of generating the motion from a state of rest by a movement of the solid" (5). Taylor drew the conclusion that "in searching for an explanation of the forces which act on solids moving through fluids, it is useless to confine one's attention to irrotational motion" (5).

The correct strategy, Taylor argued, is to address flows where the fluid elements possess rotation as a result of viscosity and friction (6). In this way turbulence and eddying might be brought into the picture so that a physically realistic fluid dynamics could emerge. Taylor was aware that the direct deduction of turbulent and eddying flow, starting from the full Stokes equations of viscous flow, presented insuperable obstacles. Progress would be impossible "if one were to adhere strictly to the equations of motion, without any other assumptions" (11). He therefore proposed to begin by a "guess at some result which I think would probably come out as an intermediate step in the complete solution of the problem" (11). On the basis of this guess he would deduce consequences that could be tested by experiment, and if "the observations fit in with the calculation I then go back to the assumptions and try to deduce it from the equations of motion" (11–12).

Taylor's reaction to Lanchester depended on his assimilating Lanchester's analysis to the classical framework of perfect fluid theory, that is, to the equations of Euler and Laplace's equation. The brevity of the argument attests to the taken-for-granted character of this assimilation. It must have seemed obvious that this is what Lanchester was presupposing. There was no hesitation or qualification, nor any suggestion that alternative readings were available. Admittedly, due to the sudden onset of war, Taylor did not have Lanchester's book in front of him.[61] He was recalling the essential point of the theory, and this involved the irrotational flow of a perfect fluid with a circulation. As such, the theory fell under the scope of Kelvin's theorem and hence could never cast light on the creation of the circulation.

Lanchester was aware of the theorem (which he called Lagrange's theorem) that rotation and circulation within a continuous body of ideal fluid can be neither created nor destroyed. He even expressed the point with a striking analogy. Once created, he said, a vortex of perfect fluid, unlike a real vortex, would "pervade the world for all time like a disembodied spirit" (175). He knew this meant that an infinite (that is, two-dimensional) wing starting from rest and moving within an initially stationary ideal fluid cannot then generate a circulation. He was prepared to face the consequences. "It is, of course, conceivable," he said, "that flight in an inviscid fluid is theoretically impossible" (172). As an engineer working with real fluids, such as air and water, he hardly expected mathematical idealizations to be accurate. The important thing was to learn what one could from the idealized case but not to be imposed on by it. As he remarked ruefully, "The inviscid fluid of Eulerian theory is a very peculiar substance on which to employ non-mathematical reasoning" (118). Discussing the "two parallel cylindrical vortices" that trail behind the tips of a finite wing, he accepted that the mechanics of their creation would not be illuminated by standard hydrodynamic theory: "for such vortex motion would involve rotation, and could not be generated in a perfect fluid without involving a violation of Lagrange's theorem. . . . In an actual fluid this objection has but little weight, owing to the influence of viscosity, and it is worthy of note that the somewhat inexact method of reasoning adopted in the foregoing demonstration seems to be peculiarly adapted, qualitatively speaking, for exploring the behaviour of real fluids, though rarely capable of giving quantitative results" (158). For Lanchester, the mathematical apparatus of classical hydrodynamics played a subsidiary and illustrative role. It was merely a way of representing some of the salient features of the flow. Nothing of this complex, if informal, dialectic linking ideal and real fluids found any recognition in Taylor's characterization.

Taylor's response to Lanchester remained unpublished, but it tells us something about the assumptions of some of Lanchester's readers. If Taylor read the work in this way, then presumably others will have read it in a similar way. The case is different with the next objection. It was not made in private but was very public and was acted out before a large audience at one of the major professional institutions in London.

A Public Confrontation

In March 1915, Lanchester gave an exposition of his theory at the Institution of Automobile Engineers in London.[62] In the audience of over 150 members and guests was a fellow member of the Advisory Committee, Mervin O'Gorman, as well as Leonard Bairstow of the National Physical Laboratory. Lanchester devoted the first part of the lecture to the theory of lift or "sustentation."[63]

The presentation started from the observed differences in pressure between the upper and lower surfaces of an aircraft wing. For maximum efficiency, argued Lanchester, the flow of air over the wing must conform closely to the surface of the wing. Conformability, rather than the separation characteristic of Kirchhoff-Rayleigh flow, was the central assumption. At the tip of the wing, however, complications enter into the story. The higher pressures on the lower surface cause the air to move around the tip from the lower to the upper surface. When combined with the motion of translation of the wing through the air, the circulating motion at the tips has two consequences. First, it gives the flow over the top of the wing an inwardly directed component, toward the center line, but an outwardly directed component on the lower surface. Second, at the tips themselves, the circulation is swept backward to form two trailing vortices coming away from the ends of the wings. To complete the dynamical system, argued Lanchester, the two trailing vortices must be joined, along the length of the wing, by a vortex that has the wing itself as its solid core. The vortex provided the circulatory component of the flow around the wing and accounts for the velocity difference between the flow over the upper and lower surfaces. This in turn accounts for the pressure difference, and hence the lift.[64]

Lanchester combined his exposition with some methodological observations. He began by distinguishing the theoretical approach to aerodynamics from the purely empirical approach and noted that the two methods can, to a great extent, be followed independent of one another. Nevertheless, he insisted that engineering needed theory and that experiment without theory was "inefficient." When variables were effectively independent, simple empirical methods of keeping everything constant except one variable might suffice; when variables were dependent on one another, this method obscured the crucial connections. At the conclusion of his lecture he returned to these methodological points, saying, "It has not been found possible in the present paper to do more than give an outline of the theory of sustentation, with sufficient examples and references to practice and experiment to illustrate the importance of the theoretical aspect of the subject as bearing on the experimental treatment; the latter has hitherto been dealt with almost without considerations of theory, and has degenerated into empiricism pure and simple" (207). Although Lanchester was making a general claim about the guiding role of theory, there can be little doubt that he had the neglect of his own theory in mind. This was certainly how he was understood by some of his audience.

Lanchester's lecture impressed at least some of the practical men, and it was greeted by an enthusiastic editorial in *Flight*.[65] The immediate reception by the audience was, however, mixed. Mervin O'Gorman began the discussion after the lecture by congratulating Lanchester on his freshness of outlook and went on to offer empirical support for Lanchester's theory. Experiments had been done on full-sized wings at the Royal Aircraft Factory that demonstrated the predicted inward and outward flow on the respective upper and lower wing surfaces.

> We fastened pieces of tape at one end of the upper surfaces of the leading edge of the tips of an aeroplane wing, and arranged a camera, worked by a Bowden wire, to photograph them in flight; they were not put there for the purpose indicated by the author, but we got exactly what he says we should get, and I am glad to confirm him so far. (228)

Leonard Bairstow (fig. 4.10) then rose and adopted a different tone. He announced to the audience that he was not convinced by Lanchester's ideas.

> I quite agree with Mr. O'Gorman that the paper is extremely interesting, but I also find it extremely controversial, and I disagree with his final conclusions. (229)

By "final conclusions" Bairstow was referring to Lanchester's suggestion that aerodynamics had degenerated into pure empiricism. Bairstow took it personally:

> Many references have been made in the paper to experimental work at the National Physical Laboratory, which work is generally under my charge, and the author has done his best to put the N.P.L. on its defence for not making practical application of his theory. (229)

Given that much of Bairstow's work had been on stability, and had been guided by the theory developed by G. H. Bryan, it is easy to understand why the general criticism might have struck Bairstow as unjust. The work on stability was certainly not mere empiricism. But Lanchester was talking about lift. Here the charge of empiricism was more plausible. For example, Joseph Petavel, a fellow member of the Advisory Committee and the future director of the National Physical Laboratory, had given the Howard Lectures in March and April of 1913 at the Royal Society of Arts. He had devoted them to aeronautics, but his treatment had been purely empirical.[66] He simply presented his audience with a stream of graphs and empirical coefficients. There was no mention of either the discontinuity theory or the theory of circulation. And had not Bairstow himself admitted the resort to empiricism when he had addressed the Aeronautical Society that same year?[67]

This was true, but all that Bairstow needed to claim to rationalize his position was that Lanchester's theory was not acceptable because it was a bad theory. He was saying, in effect, show me an adequate theory and I shall use it to guide my experiments, but as yet no such theory is on offer. Bairstow's objection was that Lanchester's theory covered some, but not all, of the facts that were of interest to the aeronautical engineer. Bairstow had come prepared to prove his point: "I will not pretend to follow the analytical steps between the author's statements of the vortex theory and his applications, but I will deal with two experiments made at the N.P.L." (229). With this heavy hint that Lanchester's position lacked logical clarity, Bairstow proceeded to show the audience two photographs. They depicted a square, flat plate set at an angle of 40° to a stream of water. The water was injected with ink to make the flow visible. Both photographs were taken from above, the first being at a slow speed of flow, the second at a faster speed. Referring to the first picture, Bairstow conceded that it looked to him like the flow that Lanchester had described and as it had been presented in a line drawing (called figure 6) in Lanchester's talk. Two trailing vortices could be seen coming from the sides of the plate (which Bairstow described as a low-aspect-ratio wing). The higher speed flow, however, presented a very different appearance. If one

FIGURE 4.10. Leonard Bairstow (1880–1963). Bairstow was the principle of the Aerodynamics Division at the National Physical Laboratory, where he did extensive testing and development of G. H. Bryan's work on stability. Bairstow was skeptical of the circulatory theory of lift and of any approach that ignored the viscosity of air. As a young man he had a reputation for intellectual pugnacity. (By permission of the Royal Society of London)

photograph fitted the theory, the other certainly didn't. Introducing the first photograph Bairstow said: "The resemblance of this photograph to Fig.6 of the paper is very marked, and up to this point I am thoroughly in accord with the author as to the probable, and in fact almost certain, existence of the type of flow postulated in the early part of the paper" (230).

Moving on to the second picture with the more rapid flow, he added: "The type of flow is now very different from that to which the author's theory applies. The fluid round the model aerofoil leaves it periodically in spinning loops. The spiral showing the spin inside the arch of one of the loops is very distinct" (230). He conceded that Lanchester's theory might fit "the very best aerofoil that can be designed at its very best angle of incidence" (230), but the theory said nothing about the full range of significant flow patterns. The word "stall" was not used, but Bairstow's argument was that Lanchester could not explain what happens when a wing stalls: "There appear, then, to be exceptions to the author's theory, or rather, there are cases of fluid motion of interest to aeronautical engineers which do not satisfy the conditions that the surface shall be conformable to the streams" (230).

Lanchester gave a robust reply. First, he put Bairstow in his place by reminding him of their relative positions in the hierarchy of command. While Bairstow was in charge of much of the experimental work on aerodynamics at the NPL, he, Lanchester, was on the Advisory Committee for Aeronautics, which controlled that work. Would he, Lanchester, be denigrating the very institution for which he had responsibility?

> Mr Bairstow has suggested that my paper is in some degree an attack on the National Physical Laboratory, or at least he states that I have done my best to put the Laboratory on its defence. I will say at the outset that the National Physical Laboratory is an institution for which I have the greatest possible respect, and I am happy to count amongst my friends members of the Laboratory staff, whose work and whose capacity are too well known to be injured by friendly criticism. Beyond this, any criticism which is to be incidentally inferred as implied by my remarks is not only criticism of our own National Laboratory, but equally of every aerodynamic laboratory with whose records I happen to be acquainted. Finally, on this point, any destructive or detrimental criticism of the work being done in the aeronautical department of the N.P.L. must reflect adversely on myself, since I am a member of the Committee whose duty it is to direct or control the particular work in question. (241)

Having sorted out the status question, Lanchester turned to Bairstow's photographs and the accusation that the circulation theory would only apply to a good aerofoil at the best angle of incidence. Is this really a fault asked Lanchester?

> Put bluntly, my answer to this is that it is equivalent or analogous to saying that the theory of low speed ship resistance as based on streamline form, and skin friction, is invalid because it does not apply to a rectangular vessel such as a packing-case, and is only true if applied to the very best design of hull with the finest possible lines. (242)

If the theory applied to a few important facts that was triumph enough. All Bairstow's photographs, Lanchester went on, dealt with flows outside the scope of his theory.

> I consider it quite preposterous to suggest that my theory should be tested by its applicability to the case of a square plane at 40 degrees angle as to test the theory of streamline ships' forms by tank experiments on a coffin or a cask of beer. (243)

Bairstow claimed that theories of wide scope served the interests of aeronautical engineers, but Lanchester argued that they cut across, rather than expressed, the engineer's pragmatic standards. Most practical solutions, said Lanchester, were narrow in scope. No one would expect to compute the "resistance of a ship in sidelong or diagonal motion through the water" by the same methods and equations "as those applicable in the ordinary way" (251).

The Balanced-Flap Anomaly

As well as the empirical support briefly mentioned by O'Gorman, some further evidence favorable to Lanchester emerged in the following year. It came from Bairstow's own laboratory and arose from the attempt to clarify some disconcerting experimental results about control surfaces. With the construction of ever-larger aircraft, the forces that pilots had to exert on the controls became correspondingly greater. To overcome this problem the controls were "balanced," that is, part of the area of the control surface was positioned in front of the hinge around which it turned so that some of the aerodynamic forces worked with, rather than against, the pilot. An experimental study of balanced controls was carried out at the National Physical Laboratory in early 1916 by John Robert Pannell and Norman Robert Campbell. Pannell, who had been on the staff since 1906, was the senior assistant in the Aerodynamics Department. He was a familiar figure who bicycled to work every morning, arriving on wet days with an umbrella held aloft in one hand while steering with the other. His main concern was with tests on full-size airship, and he was to die in 1921 when the R38 met with disaster on its trial flight.[68] The balancing experiments, however, were conducted using the "flaps" on an aircraft wing. "Flap" was the old name for the lateral control surfaces, today called ailerons, which allow the pilot to bank and roll the aircraft.

Pannell's co-worker in these experiments was a Cambridge experimental physicist who had played a prominent role in early debates about relativity theory.[69] Campbell, who had been seconded to the NPL for war work, was in the process of writing a book on scientific method, *Physics: The Elements*, which was published after the war and was to prove an influential work in the philosophy of science. Campbell argued for the importance of models in scientific inference and theory construction.[70] On this occasion the models that interested Campbell were not models of the atom or the electromagnetic field but scale models of the wings of a 110-foot-span biplane that was under construction at the Royal Aircraft Factory.[71]

The "flaps" of the projected aircraft ran along the rear edge of the outer portion of the wings but also included the tips of the wings themselves. When the part of the control surface that was on the trailing edge was lowered, then the part of the wingtip that was connected to it, and that was in front of the axel on which it pivoted, went up. The whole tip of the wing was thus part of the flap. This construction was meant to give the desired balance. Pannell and

Campbell wanted to find the proportion of the area that should be in front of the axel. The model wings were placed vertically in the 4 × 4-foot tunnel at a wind speed of 40 feet per second. Different proportions of fore and aft area were tested with the wings set at 0°, +4°, and +12° to the wind and with the flaps (and tips) put at a variety of angles relative to the main wings.

It proved impossible to find a fully satisfactory balance. Frustratingly, there was no ratio of the areas, fore and aft of the pivot, that fully balanced over the desired range of angles. Also, when looked at in detail, the results had some odd features. On occasion, where the experimenters had expected to be able to detect forces at work on the wingtip, there weren't any: "In particular it was found that when the main planes of a biplane were inclined at +12° to the wind, there was no moment on the portion of the wing flap forward of the hinge, if this flap was inclined at an angle of −5° to the wind."[72] This result suggested that the flow of air near the wingtip was itself at a negative angle to the undisturbed flow. Given the conventions for designating angles positive or negative, this meant that the air near the wingtip was moving upward relative to the wing. The air was going round the tips from what, during normal flight, would count as the lower to the upper surface.

In order to shed light on this, Pannell and Campbell conducted a further, qualitative investigation of the flow near the tips.[73] Using a direction and velocity meter to plot the velocity components of the moving air, they found what they called "a very simple and obvious explanation" for their "remarkable results." It became clear that air was indeed flowing round the wingtips from the lower to the upper surface. This was associated with a movement of air along the span of the wing, that is, not just from the leading to the trailing edge of the wing but lengthwise along it. There was a component of outward movement, toward the tips, on the lower surface and an inward movement, away from the tips, on the upper surface. Pannell and Campbell argued that

> The presence of this flow round the wing tips affords, in outline at least, an explanation of the result on the balancing of wing flaps. . . . For, if there is a marked flow near the wing tip directed from the lower to the upper surface, a plane parallel to this flow will experience no wind force. Now it was precisely when the balancing flap was inclined at a negative angle to the wind, so that its plane lay along a flow having a component from the under to the upper side of the plane, that the experiments indicated that there was no force on it. (141)

The qualitative study also addressed the component of flow along the chord of the wing, that is, in the direction of flight rather than around the tips or along the span. Measurements were taken with the direction and velocity meter to build up a picture of the disturbed flow. Next, the steady, undisturbed flow was subtracted from it. If the resultant flow was made up of two parts, the free stream plus a circulation, this subtraction would expose the circulatory component. This is exactly what it did. In the experimenter's words, it emerged that "the component of the wind disturbance which is parallel to the direction of flight is in the direction of flight almost everywhere below the wing, and in the opposite direction everywhere above the wing. There is therefore some indication of the cyclical motion of the air round the wing in the vertical plane of flight which has been assumed by Mr. Lanchester in his discussion of the theory of the aerofoil" (142). Although the full path of the circulation had not been traced, those parts of it above and below the wing had been factored out and, as it were, exposed to view. Here again was evidence for the reality of the circulatory component that was central to Lanchester's theory.

A Theory with Zero Probability

Pannell and Campbell's work received no mention when, in February 1918, their colleagues at the National Physical Laboratory, W. L. Cowley and Hyman Levy, published their authoritative *Aeronautics in Theory and Experiment*.[74] (This book is the one I used in chap. 2 to introduce some of the basic ideas of classical hydrodynamics.) Cowley and Levy's chapter titled "The Mathematical Theory of Fluid Motion" was, for the most part, devoted to the discontinuity theory. The weaknesses of the approach were candidly acknowledged, but it was still deemed "remarkable" that its predictions agreed as well as they did with the experimental results. By contrast, their treatment of Lanchester's theory was brief. It consisted of just two paragraphs. Cowley and Levy clearly believed it was beset by a fundamental hopelessness. They began by noting that, apart from the discontinuity theory, "the only other serious attempt to originate a reaction between a body and a perfect fluid is that strongly advocated by Mr F. W. Lanchester, among others, in which he supposes a cyclic motion, about the aerofoil say, superposed on the ordinary steady streaming" (65). Note the word "suppose." Lanchester did not *explain* the cyclic motion; he merely *supposed* that it was present. Cowley and Levy then showed how, granted the supposition, there would be a resultant pressure. This would have been the moment to mention that Pannell and Campbell had actually detected the presumed circulation. The authors did not take this opportunity. Instead, the exposition was followed by two terse sentences giving objections to the cyclic theory: "It can be shown that this force is proportional to the intensity of the cyclic motion and is thus apparently quite arbitrary. The method moreover gives rise to a lift on the aerofoil and no drag" (66). Both points are correct. First, consider the no-drag problem. The force generated by combining a circulation and a uniform horizontal flow is directed vertically upward. There is no horizontal component, so there will be a lift without a drag. But no wing can move through the air without experiencing some drag force, however small. On factual grounds, therefore, this theoretical analysis is doomed from the outset to give empirically false results.

Second, why was the amount of circulation said to be arbitrary? Here Cowley and Levy were pointing to a structural feature of the mathematics. Their point can be illustrated by going back to Rayleigh's analysis of the tennis ball. Rayleigh's formula for the stream function of the flow had two parts to it. Call them ψ_1 and ψ_2. The first part dealt with the uniform flow of speed V, which went toward and over the ball, while the second dealt with the circulating flow of strength Γ, which went around it. Rayleigh's formula thus looked like this:

$$\psi = \psi_1 + \psi_2.$$

The two parts of the formula are independent of one another. The speed of flow in the first part can be varied without altering the strength of the circulation, and the strength of the circulation can be altered without affecting the speed of the oncoming flow. There is no grand, overarching stream function ψ^* from which these two features of the flow can be deduced. They are not values to be deduced but merely parameters to be specified. And what was true of the analysis that Rayleigh gave of the flow with circulation around a tennis ball applied to the flow with circulation around a wing. This is what Cowley and Levy meant by "arbitrary." Applied to Lanchester's theory it meant that there was nothing in the theory that actually predicted the amount of circulation, and therefore nothing that predicted the amount of lift.

Isn't the intensity of the circulation determined by the shape of the aerofoil, for example, its curvature or thickness? Plausible though this is, Cowley and Levy could point out that nothing in the mathematics indicated any such connection. The theory simply implied that lift was proportional to the density, the speed of the free stream, and the circulation. These were the only variables, and shape does not feature in the list.[75] The formula applies to any shape of cylinder and not just to those whose cross section looks like that of a typical wing. Lanchester, as we have seen, was aware of this. He expressed the point succinctly in his *Aerodynamics* when he said that all bodies, of whatever shape, must count as being "streamlined" in a perfect fluid (22). Lanchester then drew the obvious, but striking, conclusion: "From the hydrodynamic standpoint irregularity of contour is no detriment, as obstructing neither the cyclic motion nor that of translation. The consequence is that peripteroid motion [that is, motion of a kind that generates lift] is theoretically possible in the case of a cylinder of infinite extent, no matter what its cross-section. This conclusion applies naturally only in the case of the inviscid fluid" (163).

The circulation theory, when it is based on the behavior of a perfect fluid in irrotational motion, thus has some disconcerting features, both factually and formally. First, the circulation is independent of the shape of the wing and, second, it cannot be created by the movement of the wing through the air. Both problems derived from the same source and appeared to be ineradicably connected with the mathematics of a perfect fluid. This was the root of all the trouble. Both circulation theory and discontinuity theory were doomed because they were built on the same unreal foundation. Until a way was found to overcome the limitations of classical hydrodynamics, progress in this branch of aerodynamics would be impossible. As Cowley and Levy put it, "the failure of the various treatments of the problem of the motion of a body and the forces experienced, to approximate to that of practice is evidently due to the supposition that the fluid dealt with is perfect" (66). The need was for a general theory of viscous flow around a wing. As yet, said Cowley and Levy, no such theory existed, but if and when it did, it would "clarify at one stroke the whole problem of aerodynamics" (75).

Why did Cowley and Levy make no mention of Pannell and Campbell's results, given that the results came from their own laboratory? Could the explanation be that their book was written under wartime restrictions? Certainly it would not have been prudent in such circumstances to advertise problems with a prototype aircraft, but such restrictions hardly explain the negative assessment of Lanchester. Even if the details of the evidence could not be given, it is difficult to see why the mere existence of experimental support could not have been admitted. This neglect suggests that the evidence was deemed inadmissible on scientific grounds rather than for reasons of security.

What scientific grounds could ever justify the neglect of evidence? The answer calls for a brief look at the principles of scientific inference. Philosophers sometimes analyze science in terms of what they call Bayesian confirmation theory.[76] The analysis depends on a mathematical theorem associated with the name of Thomas Bayes. The idea is that new experimental evidence that confirms the predictions of a theory increases the scientist's assessment of the probability that the theory is true. The size of the increase is given by a simple formula derived from the calculus of probabilities. The degree of belief in a theory h, given the new evidence e, depends on the initial or a priori probability of the evidence $p(e)$ and the initial or a priori probability of the theory $p(h)$. If the theory h entails, that is, predicts, the evidence, so that $h \rightarrow e$, then the posterior probability, that is, the probability of h given e, written $p(h/e)$, is given by Bayes' theorem as

$$p(h/e) = \frac{p(h)}{p(e)}.$$

The initial probability of the theory is divided by the initial probability of the evidence to give the new, increased, probability that the theory is true. If the predicted evidence is itself surprising and improbable, then the value of $p(e)$ will be smaller than if the prediction is less surprising. A successful but surprising prediction will thus increase the posterior probability of the theory to a greater extent than a less surprising prediction.

Suppose that Lanchester's theory is symbolized by h. Pannell and Campbell knew about the theory and did not dismiss it, but they had no independent grounds for expecting the flow around the wingtips. Accordingly, they called that piece of evidence "remarkable." They also knew that Lanchester had predicted it. If they were behaving like Bayesians, the probability of Lanchester's theory $p(h)$ would have been enhanced by the evidence e so that $p(h/e) > p(h)$. Their subjective degree of belief would have been increased.[77] What about Cowley and Levy? They too must have known about the result, so why were they unmoved by it? Their response makes sense in terms of Bayes' theorem provided that one, simple, further condition is satisfied. The a priori probability they accorded to the theory must have been zero. For Cowley and Levy, $p(h) = 0$. This has the result that $p(h/e) = 0$, whatever the value of $p(e)$. The a posteriori probability will always be zero if the a priori probability is zero. Mathematically this follows because any number multiplied by zero again yields zero. Psychologically, it means that if a scientist starts with a zero degree of belief in a theory, then the subsequent course of belief will be wholly unresponsive to new evidence in its favor.

Scientists do not behave exactly like Bayesian calculating machines, but the model dramatizes the logic of the situation. The association between Lanchester's circulation theory and perfect fluid theory was sufficient, in the minds of some scientists, to render his account of lift irredeemably false. It represented an *essential* failure, and the failure was fatal. As Cowley and Levy put it: "The absence of reaction between body and fluid is extremely unfortunate for it implies an essential failure in the application of results obtained for a perfect fluid to a real case. Mathematical physicists have striven for years to introduce some new assumption into the nature of the flow that will avoid this fatal result, but it is clear that no matter how ingenious the suggestions may be, they must of necessity be artificial since they attempt to simulate the action of viscosity without actually assuming its existence" (53). The argument was that perfect fluids are mathematical fictions. A theory built upon such a foundation cannot possibly offer a true account of the world. It followed that Lanchester could not possibly be right.

Two Traditions:
Mathematical Physics and Technical Mechanics

For the engineer and the physicist are acquainted with exactly the same facts, but the manner in which they approach their subjects is quite different.

PHILIPP FRANK, *Relativity: A Richer Truth* (1951)[1]

That it is Applied Physics is to me the most inspiring definition of engineering; and if this be true for engineering in general, as I think it is, especially true is it of aeronautics.

H. E. WIMPERIS, *"The Relationship of Physics to Aeronautical Research"* (1926)[2]

The circulation theory of lift was developed by Lanchester, who was an engineer. The reasons advanced against it were proposed by men such as G. I. Taylor who were not engineers but who worked in the British, and particularly the Cambridge, tradition of mathematical physics. This is a clue that needs to be followed up. If the objections were the expressions of a disciplinary standpoint, located at a specific time and place, then perhaps the resistance to the circulatory theory would be explicable as a clash of cultures, institutions, and practices. Such an explanation would not imply any devaluation of the reasons that were advanced against the circulatory theory. It would not be premised on the assumption that these reasons were not the real reasons for the resistance. On the contrary, the intention would be to take the objections against the theory in full seriousness and to probe further into them. To do this it is necessary to understand the sources of their credibility and why the reasons were deployed in precisely the way that they were. I shall now begin that process. By the end of the chapter I shall be in a position to outline a theory that could explain the negative character of the British response to Lanchester's theory.

Scope and Rigor

Consider the scope of Lanchester's theory. His narrow focus on small angles of incidence was not shared by critics. Bairstow invoked standards of assessment appropriate to a wholly general theory of fluid resistance. Lanchester found this preposterous. He said it was like ignoring useful knowledge about how water flowed round a ship in normal motion because it did not also explain the flow when it moved broadside. But it is not difficult to see how for the critics, if not for Lanchester, genuine knowledge of the one case also meant having knowledge of the other. Lanchester's commonsense plea for theories of limited scope was at odds with the forms of generality routinely exhibited in classical hydrodynamics. This can be seen from the textbook treatment of the flow around an elliptical cylinder moving through a fluid. The elongated, elliptical cylinder bears a certain visual likeness to the plan of a boat, which was the case cited by Lanchester. Lanchester's critics could point out that the mathematics of the flow does *not* single out, as being of special significance, any particular angle of inclination of the major axis of the ellipse to the direction of motion. It makes no difference to the mathematics whether the ellipse moves like a ship going forward or like a ship moving broadside, that is, awkwardly and inappropriately. Both motions are but special cases of the same general formula. Mathematically they merely depend on whether the real or the imaginary part of the complex potential is set to zero. This fact would have been familiar to any Cambridge student of hydrodynamics, or to anyone, such as Bairstow, Cowley, or Levy, schooled in a similar tradition. It was clearly not a significant reference point for Lanchester.[3]

The issue of scope also arose in another way. We have seen that circulation theory explained lift but not drag. The critics had rejected discontinuity theory because it could not yield accurate predictions of resistance, so on grounds of consistency circulation theory should be, and was, treated likewise. The false prediction of zero drag was not lost on Lanchester, but it did not worry him in the way it did his critics. Unfortunately, Lanchester did not articulate a clear rationale for his stance, so the critics may have been tempted to see it as indicating a certain laxity on his part, compared to their own greater concern with truth and rigor.

There is some evidence that Cambridge physicists involved in aerodynamics were prone to misperceive the difference between their mental habits and those of engineers as the difference between rigor and sloppiness. Reflecting on his work as a physicist at Farnborough during the Great War, George Paget Thomson, the son of J. J. Thomson, drew attention to this cultural divide. Scientific work during wartime, said Thomson, "might properly be described as engineering."[4] He recalled how difficult it was for physicists to adopt the requisite point of view. As the author of *Applied Aerodynamics*, which had been well received by the "practical men," Thomson could not be accused of lack of sympathy with engineers. But even he was inclined to exemplify, rather than bridge, the disciplinary divide he described. Thomson spoke of the need for engineers to make up their minds on the basis of "insufficient evidence" and of the need to "compromise between conflicting requirements." He concluded: "What is perhaps harder for the scientist to realize is the doctrine of 'good enough.' The better is the enemy of the good" (3).

Could it be that Lanchester, as an engineer, was prepared to accept the circulatory theory and perfect fluid theory because they were "good enough" for him even though they were not "good enough" for a physicist? The implication is that Lanchester, unlike his critics, was content with "insufficient evidence." But there is another explanation of why a supporter of the circulation theory might find the lift-but-no-drag result an acceptable one. Rather than expressing a compromised standard of empirical accuracy, the response might simply embody a different standard and one that is not necessarily lower. The lift-without-drag result might be taken to be a true and accurate assertion about an "ideal wing," that is, the sort of wing *at which an engineer might aim*. The result should perhaps be seen not as a false statement, but as an engineering ideal. This was not a defense explicitly offered by Lanchester, but as we shall see, it was how Ludwig Prandtl, a fellow pioneer of the circulation theory, expressed the matter.

Consider now the objection that the circulation is "arbitrary." Both Kutta and Joukowsky were aware of the mathematical rationale behind this objection, namely, that the theory contained no way of deducing the amount of circulation around a wing. Nevertheless they responded in a very different way to the British critics. They stipulated that the circulation be of precisely the amount necessary to ensure that the flow comes away smoothly from the trailing edge of a wing. The rear stagnation point must be on the trailing edge so that the flow does not have to wrap itself around a sharp corner. For a given angle of incidence, and a wing with a sharp trailing edge, this stipulation provides an unambiguous specification of the amount of circulation and is often called the Kutta condition. It derives its significance, and its nonarbitrary nature, from the empirical fact that the flow of air over a (nonstalling) wing in a steady state settles down so that there is indeed an approximately smooth flow at the trailing edge. The Kutta condition tells the theorist what value of the circulation to assume, and thus what value of lift is predicted when this value is substituted into the formula $\kappa \rho U$, the lift equation.

Were Cowley and Levy, who made the complaint about arbitrariness, unaware of this solution to the problem? The answer is that they were fully aware of the Kutta condition. In 1916 Levy had explicitly mentioned it in correspondence with Lanchester.[5] In Levy's view, however, Kutta's proposal did not remove the arbitrary character of the amount of circulation. The argument presented to Lanchester was that, in reality, the trailing edge of a wing is not mathematically sharp but rounded. It therefore provides no mathematically unambiguous location for the rear stagnation point. The point on the curve that is selected for this role will itself be arbitrary. The amount of circulation needed to bring the stagnation point to this location will, therefore, also be arbitrary. The only thing that would remove this feature of the theory would be some means of deducing the circulation from first principles, given relevant data about the wing, for example, its shape and angle of incidence. No such method was known. For Cowley and Levy the word "arbitrary" clearly meant "not deducible from the basic equations of fluid dynamics." They operated with a mathematical criterion and were looking for a mathematical solution to the problem, not an empirical one.

A Precedent

Lanchester's critics insisted on reading his work as an exercise in inviscid fluid theory. Bryan, Taylor, Bairstow, Cowley, and Levy all made this move. This should be puzzling. It must have been evident to any reader of Lanchester's book that he was not simply thinking in terms of perfect fluids. He was constantly moving back and forth between viscous and inviscid approaches, trying, as it were, to negotiate some rapprochement between them. How could this have been overlooked? One answer is that it was not overlooked at all. Perhaps it was perceived clearly but seen as a weakness in the text. The inclusion of both viscous and inviscid strands in the argument may have seemed like mere ambiguity. Reading Lanchester as a purely inviscid theorist may have been a way to repair the ambiguity. Wouldn't this be a natural thing for mathematically sophisticated readers to do?

There may be some truth in this suggestion, but it cannot be the whole story. It does not explain why the ambiguity was resolved by turning Lanchester into an exponent of the inviscid approach rather than an exponent of a viscous approach. Either would have resolved the ambiguity, so why did the inviscid reading prevail? Here is a possible answer to that question. There was a precedent for the preferred assimilation, namely, the received understanding of Rayleigh's paper on the irregular flight of the tennis ball. Rayleigh had done exactly what Lanchester had done, that is, work with an informal mixture of ideal-fluid theory and viscous considerations. The mathematics of Rayleigh's tennis ball paper dealt with an inviscid fluid, but the circulation described by this mathematics could not have arisen from any processes conceptualized within it. Friction was needed to account for the circulation. The fluid was therefore taken to be viscous at one point in the account and inviscid at another point, thus rendering the argument logically inconsistent. How did readers respond to this oscillation between viscous and inviscid fluids? In Cambridge the tennis ball paper was absorbed into the literature on inviscid theory. The ambiguity was resolved by playing down the appeal to viscosity and treating the analysis as if it began at the point where a circulation could be taken as a given. This approach had the virtue of focusing on the part of the work that was easiest to develop mathematically, and indeed it may explain why the assimilation went this way rather than the other. It was in these terms that Rayleigh's tennis ball result found its way into the Cambridge examination papers.[6]

A Tripos Question

The examination for part II, schedule B, of the Mathematical Tripos of 1910 was held at the Senate House and began at nine o'clock on Thursday, June 2. Question 8 of part C of the paper consisted of a typical, but daunting, combination of book work and problem solving.[7] The question read as follows:

> C8. Prove that in irrotationally moving liquid in a doubly connected region the circulation is the same for all reconcilable circuits and constant for all time.
>
> A long elliptic cylinder is moving parallel to the major axis of its cross section with uniform velocity U through frictionless liquid of density ρ which is circulating irrotationally around the cylinder. Prove that a constraining force $\kappa \rho U$ per unit length of the cylinder must be applied at right angles to the direction of motion, where κ is the circulation round the cylinder.

To a modern reader, versed in aerodynamics, the expression $\kappa \rho U$ would be identified as the fundamental law relating the circulation, density, and velocity to the lift on a wing. In modern aerodynamics it is called the Kutta-Joukowsky relation. If interpreted aerodynamically, the "elliptic cylinder" would be a mathematically simplified substitute for the cross section of a wing, and the "constraining force" would be the weight supported by the lift. It is doubtful, however, whether any of the Tripos candidates of 1910 would have thought in this way. The year 1910 was when Kutta and Joukowsky, independently, published their results in German journals, and it is unlikely that the news had reached Cambridge. Admittedly "aeroplanes," that is, aircraft wings, had been the subject of Tripos questions in the past, but the reference had been to Rayleigh's paper on an inclined plate in a discontinuous flow, not to his tennis ball paper.[8] Thus question C8 was unlikely to have evoked aeronautical associations. If candidates attributed any technological significance to the formula $\kappa \rho U$, it would have referred to ballistics not aeronautics.

Perhaps some of the candidates had read Rayleigh's tennis ball paper and

Greenhill's extension of the analysis. More probably, they were calling up in their memories the relevant pages of Lamb's *Hydrodynamics* and relying on the hours of coaching and drill to ensure that their recall was accurate. A well-prepared candidate would have remembered Lamb's treatment of the irrotational flow of a perfect fluid around a circular cylinder. In article 69 of both the 1895 and 1906 editions, Lamb laid out his version of the analysis originally developed in the papers of Rayleigh and Greenhill. Like these writers, Lamb was mainly concerned with trajectories, but his analysis would have given the candidates both the general idea behind the question and the derivation of a formula which included an expression identical to that of the constraining force mentioned in question C8.

After explaining how the circulation augments the speed of flow on one side of the cylinder and diminishes, or even reverses, it on the other, Lamb had gone on to calculate the forces. Rayleigh had approached the problem using the stream function, while Lamb used the velocity potential. To begin, Lamb wrote down the velocity potential φ for the flow, assuming the cylinder moving at any angle. He then differentiated this expression with respect to time to give $\partial\varphi/\partial t$. Next he derived a term for q, the velocity of the flow. These results were substituted in a general form of Bernoulli's equation to get a value for the pressure, and the pressure was then integrated round the surface of the cylinder to yield the resultant force. Lamb's treatment was more general than Rayleigh's, although there was no mention, even informally, of the role of friction. The two components of the force on the circular cylinder came out as follows, first in the direction of motion:

$$-M'\frac{dU}{dt}$$

and then at right angles to the motion:

$$\kappa\rho U - M'U\frac{d\chi}{dt}$$

where κ is the circulation, ρ is the density of the fluid, U is the relative velocity of the fluid and cylinder, $M' = \pi\rho a^2$ represents a mass of fluid equivalent in volume to the cylinder with radius a, while χ is the angle that the direction of motion of the cylinder makes with the x-axis. The term $\kappa\rho U$ can be seen on the left-hand side of the second formula. Where the conditions of steady motion were specified as they were in question C8, the derivatives dU/dt and $d\chi/dt$, giving the rate of change with time, will be zero. The only remaining force on the cylinder will then be $\kappa\rho U$ at right angles to the motion.

Recollection of this result would have helped the candidates, but it would not have given them all they needed. The examiners of the 1910 paper had added a further complication to ensure that the mere reproduction of textbook material would not suffice. Lamb's derivation referred to a *circular* cylinder, but the examiners had specified an *elliptical* cylinder. This made the question more difficult and required the candidates to demonstrate a facility with elliptic coordinates and elliptic transformations. If they could make the necessary transformation, they would then be in a position, for the rest of the deduction, to follow the pattern of the simpler case given by Lamb for the circular cylinder. Candidates would then have found that all the extra complexity actually produced terms that cancelled out, or went to zero in the course of the integration, thus leaving them, in the case of steady motion, with the same resultant force of $\kappa\rho U$.

Senate House records for 1910 show that A. S. Ramsey of Magdalene and A. E. H. Love of St. John's were the two examiners who would have had responsibility for the hydrodynamics questions. The other examiners, A. Berry of King's and G. H. Hardy of Trinity, would have dealt with the more "pure" topics.[9] Three years later Ramsey wrote a textbook on hydrodynamics in which circulating flow around an elliptical cylinder featured prominently.[10] After discussing the case of the circular cylinder, and working through some of the intermediate steps in reasoning that Lamb had omitted, Ramsey showed the reader exactly how to address the problem of the ellipse. The move to elliptical coordinates was explained along with advice about the types of function that would satisfy Laplace's equation and hence describe a possible flow. Ramsey also included question C8 from the 1910 Tripos paper in the exercises at the end of his chapter, which was called "Special Problems of Irrotational Motion in Two Dimensions" (119).

The title of Ramsey's chapter conveys the point that I want to make. It shows the assimilation of Rayleigh's tennis ball paper to the theory of perfect fluid flow in two dimensions. Ramsey did not wholly bypass the role of friction. At the end of his discussion of the circular cylinder case he said: "The transverse force depending on circulation constitutes the mathematical explanation of the swerve of a ball in golf, tennis, cricket or baseball, the circulation of the air being due through friction to the spin of the ball" (101). Friction was therefore mentioned, but the student was told that the *mathematical* explanation was to be found in *inviscid* theory. Ramsey, like Rayleigh, knew that this "mathematical explanation" could not furnish an account of how a spinning ball *created* the circulation. The point was implicit in the Tripos question which had two parts. The first part asked the candidates to prove that, under the conditions of the question, the circulation is constant for all time, that is, Kelvin's theorem. The question then called on the candidates to generalize Rayleigh's tennis ball result within this taken-for-granted inviscid framework.

It is now easy to see how G. I. Taylor could decide that Lanchester's work was unacceptable. Taylor would have found himself confronted with something very familiar and having little relevance for his research on eddies and turbulence. He would have known all about theories of circulation based on irrotational, perfect fluids and would have known how little they had to say about physical reality. He would certainly have been familiar with the mathematical expression $\kappa \rho U$. It was the sort of thing that Tripos students were expected to deduce as an exercise in mathematical manipulation. Everyone in Cambridge knew that the cyclic approach gave an expression for a force but took away the possibility of generating that force. No wonder Taylor dismissed the theory of circulation as readily as he dismissed the theory of discontinuity. Lanchester's theory was not a new discovery; it was the stuff of old examination questions.

Lamb's *Hydrodynamics*

Sir Horace Lamb's famous textbook started life in 1879 as *A Treatise on the Mathematical Theory of the Motion of Fluids*, which was based on the lectures Lamb had given as a fellow of Trinity.[11] Lamb (fig. 5.1), who had been taught by Stokes and Maxwell, left his Cambridge fellowship in order to marry. He took a chair at Adelaide and then taught for many years at the Victoria University of Manchester. He returned to Trinity in 1920 as an honorary fellow. During this time the small *Treatise* was renamed *Hydrodynamics* and grew into the imposing volume known to generations of students in applied mathematics. It went through a total of six editions between 1879 and 1932. The gap between the third and fourth editions, that is, from 1906 to 1916, covered the pioneering phase of aerodynamic theory and the emergence of the circulation theory of lift. Most of this aerodynamic work was too late for inclusion in the 1906 version but found a response in the updated 1916 volume. Here, for the first time, one finds the names of Kutta, Joukowsky, Prandtl, Föppl, von Kármán, and Lanchester.

The structure of the 1916 edition was very close to that of the previous editions. The new work on aerodynamics was not allowed to upset the pre-existing framework of hydrodynamic theory.[12] Consider the relation between the accounts of viscous and inviscid flow. The book was some 700 pages long, and the discussion of viscosity began on page 556 with chapter 11. Viscosity, said Lamb, is a phenomenon "exhibited more or less by all real fluids, but

FIGURE 5.1. Horace Lamb (1849–1934). Lamb was a pupil of Stokes' at Cambridge and the author of *Hydrodynamics*, the leading British treatise on fluid dynamics. (By permission of the Royal Society of London)

which we have hitherto neglected" (556). Up to that point all the analysis had concerned an ideal fluid. The Euler equations had been sufficient to solve the problems discussed in the previous chapters, but it was now necessary to confront the formidable Stokes equations of viscous flow. The equations, which Stokes had arrived at in 1845, were duly derived and set out on page 573. I give a form of the equations below. I have simplified them and changed Lamb's notation slightly so that they can be compared more easily with the Euler equations for ideal fluids as I gave them in chapter 2. Like the Euler equations, they are partial differential equations that relate together the components of fluid velocity u and v and pressure p, but this time allowance has been made for viscosity represented by μ. Stripping away the negligible effect of external forces such as gravity and treating the fluid as incompressible, and the flow as both two dimensional and steady, the Stokes equations for a viscous fluid can be written as follows:

$$-\frac{\partial p}{\partial x} = \rho\left(u\frac{\partial u}{\partial x} + v\frac{\partial u}{\partial y}\right) - \mu\nabla^2 u \text{ and}$$

$$-\frac{\partial p}{\partial y} = \rho\left(u\frac{\partial v}{\partial x} + v\frac{\partial v}{\partial y}\right) - \mu\nabla^2 v,$$

where

$$\nabla^2 = \frac{\partial}{\partial x^2} + \frac{\partial}{\partial y^2}.$$

If the coefficient of viscosity, μ, is put equal to zero, the equations lose the terms on the extreme right and they assume the simpler form of the Euler equations for an ideal fluid. This does not mean that any solution to the Euler equations is also a solution to the Stokes equations. The presence or absence of the viscous terms alters the character of the equations. The Euler equations do not have to satisfy all the boundary conditions of the more complicated equations. A solution to the equations of viscous flow must satisfy the condition that the fluid adheres to any solid boundary and thus has zero velocity along the boundary as well as zero velocity normal to it. An ideal fluid is not required to adhere to a solid boundary but can slide along it with perfect smoothness.

After the derivation of the Stokes equations, Lamb considered a number of applications, for example, the flow of a viscous fluid between two flat plates that are very close together, and the motion of a sphere falling through a very viscous fluid. The former case approximates the study of lubrication. It also provided the occasion for Lamb to discuss the intriguing photographs taken by Hele-Shaw. These rendered with great accuracy the appearance of the flow of an *inviscid* fluid. By introducing a (very thin) cylinder between the plates, and forcing the fluid to flow around it, the flow could be studied even in cases that defied direct mathematical analysis. Lamb recapitulated the mathematical explanation of these photographs first given by Stokes. Stokes had been able to demonstrate why a viscous flow could, under the circumstances of creeping flow, simulate the behavior of an inviscid flow.[13]

The case of the sphere falling through a viscous fluid was important for the study of meteorological phenomena that involved droplets of water in

the atmosphere.[14] Because the drops fell very slowly, it was possible to simplify the equations and arrive at a law giving their speed. In 1910 this law had
been applied to the oil droplets in Robert Millikan's famous experiment to
measure the unit of electric charge. The oil drops in the apparatus obeyed
Stokes' law, as it came to be called, which gave their speed of fall in terms
of their radius, relative density, and the coefficient of viscosity of the fluid.[15]
Lamb showed how this law was derived from the basic equations. Important
as these and similar results were, the introduction of viscous forces into the
analysis had to be accompanied by a corresponding limitation in the role
played by the inertial forces. The motions under study had to be very slow
or the dimensions very small. Without this restriction the Stokes equations
were, in general, intractable.

The 1916, fourth edition of *Hydrodynamics* contained a new section, "Resistance of Fluids," which Lamb added to the end of the chapter on viscosity.
It was here that he addressed the aeronautical work. The location and the
name of the new section suggest that Lamb saw the problem of lift as falling
under the rubric of viscous flow. The position and title carried the message
that lift was not to be analyzed on the basis of perfect fluid theory. Resistance,
said Lamb, "is important in relation to many practical questions" (664). He
mentioned the propulsion of ships, the flight of projectiles, and wind forces
on, for example, buildings, and added that although resistance "has recently
been studied with renewed energy, owing to its bearing on the problem of
artificial flight, our knowledge of it is still mainly empirical" (664).

Lamb then discussed Kirchhoff-Rayleigh flow and drew attention to its
empirical failings, particularly the failure to account for the suction effect on
the upper surface of a wing. The reader was referred to publications by Stanton and Eiffel for information on "the experimental side." After this came
an account of the circulation theory. It was introduced as an explanation of
how a body may be supported against gravity that had been "put forward
from a somewhat different point of view," that is, somewhat different from
the theory of discontinuous flow (666). A footnote then made reference to
Lanchester's *Aerodynamics*, but no attempt was made to expound Lanchester
in his own terms. Instead the reader was briskly referred to the earlier section in *Hydrodynamics*, article 69, which dealt with Rayleigh's tennis ball.
Lanchester's theory, said Lamb,

> is based on the result of Art. 69, where it was shown that a circular cylinder
> will describe a trochoid path, the motion being mainly horizontal, if the sur-
> rounding fluid is frictionless, and its motion irrotational, provided there is a
> circulation (κ), in the proper sense, about it. In particular the path may be a

horizontal straight line, the lifting force (which is to counteract gravity) being then

$$Y = \kappa \rho U$$

per unit length, where U is the horizontal velocity. (666)

Lamb went on to show that the formula held for a cylinder of any cross section, not just for the case of a winglike section. Here Kutta and Joukowsky were mentioned and references given. Lamb then cited the result for an elliptic cylinder (that is, the 1910 Tripos question) and went on to say that Kutta "had treated the case of a lamina whose section is an arc of a circle." Lamb's summary is instructive: "He [Kutta] assumes the circulation to be so adjusted in relation to the velocity of translation that the infinite value of the fluid velocity which would otherwise occur at the *following* edge is avoided, whilst an infinity remains of course at the *leading* edge. It is supposed that in this way an approximation to actual conditions is obtained, the 'circulation' representing the effect of the vortices which are produced behind the lamina in real fluids; and a good agreement with experiment is claimed" (667).

All the standard British objections found an expression in this compressed passage. Kutta's theory was not offered as a description of reality. There was merely the supposition that there was some "approximation to actual conditions." In a real fluid there are vortices in the flow behind a lamina, but these received no recognition in Kutta's analysis. The circulation (in quotation marks) merely *represents* the effects of a certain phenomena but (the wording implies) does not correspond to its real nature.

It would be difficult to devise a description of Kutta's work that was as brief and as accurate as Lamb's but that contained more qualifications and implied question marks. Perhaps this was to be expected, given that the discussion of Kutta's inviscid analysis was located in a chapter devoted to viscous processes. The message was that Kutta, Joukowsky, and Lanchester were trying to represent essentially viscous processes by an inviscid theory. Lamb was highlighting the artificiality of their analysis and pointing to the need to ground it in Stokes' equations of viscous flow.

The Real and the Ideal

Two characteristics have now been identified in the British response to the circulation theory of lift. First, there was a desire for theories of wide scope that embrace complex viscous phenomena beyond the reach of the theory. Second, there was a tendency to read Lanchester as contributing to an inviscid theory and therefore as committed to a simplified and unreal representation of fluid flow. Both of these indicate the importance that British experts attached to the distinction between real fluids and ideal fluids. Taylor insisted that fluid mechanics should have a firm basis in physics and dismissed the idealizations of classical hydrodynamics. Cowley and Levy described inviscid theory as fatally flawed and spoke of the need for a theory of viscous flow that would solve the problems of aerodynamics at a stroke. Bairstow agreed that it was fundamentally impossible to represent real fluids in terms of ideal fluids and duly turned to the study of viscous flow. What Bairstow had asserted with characteristic acerbity, Lamb had hinted at with characteristic restraint. The different objections and formulations all point to one conclusion. The distinction between viscous and inviscid fluids is to be seen as the axis around which British thinking revolved.[16]

It is important not to view this distinction as self-evident or something that was understood in the same way by all competent operators in the field of fluid dynamics. In reality it was treated differently in different institutional settings. How then should the distinction between viscous and inviscid fluids be understood? Formally, it centers on whether μ, the symbol for viscosity in the Stokes equations, is to have a value of zero or of nonzero. Was $\mu = 0$, or $\mu \neq 0$? Logically it must be one or the other and it can't be both. Empirically, whether Stokes' equations turn out to be true, and Euler's false, (or vice versa), is something to be settled by reference to experiment. But these truisms do not tell us how to interpret the difference between putting $\mu = 0$ or $\mu \neq 0$; nor do they indicate what physical meaning is to be given to the mathematical limit when $\mu \to 0$. They do not tell us whether the distinctions involved are qualitative or quantitative or whether the boundaries under discussion are strong or weak or for what purposes they might be important or unimportant. This is the point. The conceptual boundary between viscous and inviscid fluids is more than merely formal. Rehearsing the elementary mathematical properties of the distinction does not tell us what *methodological* implications are attached to it by the scientists concerned. I shall now illustrate the broader, methodological significance of the distinction by reference to Lamb's own discussion of viscosity.

Lamb began his account of aerodynamics, in the 1916 edition, by pointing out that the analysis of Kirchhoff-Rayleigh flow was the first attempt, "on exact theoretical lines," to overcome the result that a perfect fluid exerts no resultant force on a body. He added: "The absence of resistance, properly so called, in such cases is often referred to by continental writers as the 'paradox of d'Alembert'" (664). Why did Lamb think that "absence of resistance" was the more proper description? What was wrong with talking about a "paradox"? The reasoning behind Lamb's remark went back to the first edition of his book, where he had originally addressed the well-known discrepancies between the empirical facts of hydraulics and the mathematical deductions of hydrodynamic theory. He traced the problem back to "the unreality of one or more of the fundamental assumptions" of the theory (244). The empirically false conclusion about resistance came from an empirically false premise, namely, the inviscid character of the postulated fluid. However, d'Alembert's reasoning was sound, and the logic of the situation was clear. An inviscid fluid is correctly characterized by the absence of resistance. This is how ideal fluids behave or would behave. It is a simple fact about them, and there is nothing paradoxical about it.

A paradox is more than a falsehood, even a blatant falsehood. A paradox must involve a seeming contradiction. Suppose that experiments on a fluid F showed that it exerts a resultant force on a submerged body, while a mathematical analysis of F entails a zero resultant. Suppose, further, that the experiments on F seemed wholly reliable and the mathematical analysis of F seemed wholly correct. That would be paradoxical. Contradictory specifications of F have been generated from sources that seem undeniable. This is not the case if the experiments refer to a real fluid F_R, and the mathematics refers to an ideal fluid F_I. There is now no single point of reference as there was with the "paradoxical" fluid F. Two conditions are thus required to make d'Alembert's result a genuine paradox: (1) there must be two plausible specifications that exclude one another, and (2) the two specifications must be applied to one and the same fluid.

Lamb avoided paradox by treating the two specifications as referring to different things. He drew a boundary between the referent of the experiment and the referent of the theory and thus rejected condition (2). In eschewing the word "paradox," Lamb's language was meant to carry a methodological message. It was a way of saying that viscous fluids were one thing and perfect fluids were another and never should the two be confused. This was an admirably straightforward position, but was it the only tenable position? To address this question I consider a line of reasoning advanced by Ludwig Prandtl and Georg Fuhrmann in Göttingen. It will become clear that these experts did not distinguish between ideal and real fluids in precisely the same way as their British counterparts did.

Prandtl and Fuhrmann on Airship Resistance

Georg Fuhrmann had been Prandtl's pupil at the *technische Hochschule* in Hanover, where his ability in the course on technical mechanics caught Prandtl's attention. After completing his training as an engineer in 1907, Fuhrmann

joined Prandtl at Göttingen and played a significant part in setting up the
wind tunnel. It was Fuhrmann in 1910 who wrote the review of the German
translation of Lanchester for the *Zeitschrift für Flugtechnik*.[17] His warm recommendation stood in marked contrast to the coldness of the *Nature* review. In 1911 Fuhrmann carried out important theoretical and experimental
research on the resistance of model airships. He was to die in action in the
first few weeks of the war in 1914.[18]

The experiments were designed to compare the predictions of ideal-fluid
theory with wind-tunnel measurements.[19] To make his theoretical predictions Fuhrmann had used a standard technique from classical hydrodynamics in which complex flows were built up from simpler flows, for example,
from an array of sources and sinks, each of which can be represented by a
simple velocity potential. He assumed a theoretical distribution of sources
and sinks in a uniform flow of perfect fluid and arranged them in such a way
that they gave rise to airship-like configurations of streamlines. The basis of
Fuhrmann's procedure can be conveyed intuitively by examining figures 5.2
and 5.3. If a source is combined with a uniform flow, then the fluid from the
source pushes the free stream aside as shown in figure 5.2. At the same time
the streamlines radiating out from the source are distorted and bent back.
The streamlines of the new flow coincide with the streamlines of the flow
around a long, blunt-nosed body. Selecting an appropriate streamline of the
new flow and imagining that it is suddenly solidified gives the surface of the
body.

If now a line of sinks is introduced directly downstream of the source and
spread along the axis of the body, the overall flow is modified once again,
as in figure 5.3. The fluid injected into the flow at one point by the source is
now drawn out of the flow at other points by the sinks. If the intake of the

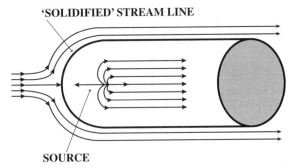

'SOLIDIFIED' STREAM LINE

SOURCE

FIGURE 5.2. A single source in a uniform flow of ideal fluid creates a flow pattern similar to that around
a blunt-nosed object of infinite length. The surface of the object consists in an appropriate streamline of
the flow that is imagined to be solidified.

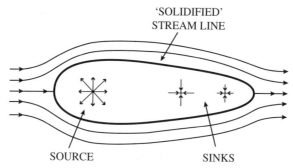

FIGURE 5.3. Fuhrmann and Prandtl used a system of sources and sinks placed in a uniform flow of ideal fluid to simulate the flow around a closed solid. An appropriate distribution of sources and sinks produces streamlines similar to those over an airship.

sinks equals the output of the source, the streamlines of the combined flow can close up again. While the obstacle represented in figure 5.2 was infinitely long, the obstacle in figure 5.3 is of finite length and resembles the outline of the hull of an airship. Fuhrmann used Bernoulli's law to calculate the pressures at various points on its surface. These surface pressures, in accord with d'Alembert's paradox, summed to zero, but it was the distribution of pressure that was the focus of interest.

By assuming different distributions of sources and sinks, Fuhrmann could produce theoretical configurations representing different shapes of airship, for example, some with blunter noses or longer tails than others. He worked out the streamlines for six different shapes. The next step was to construct a set of hollow model airships, made out of metal, which accurately conformed to these theoretically generated shapes. Fuhrmann placed the models in the Göttingen wind tunnel and measured the pressures at a number of points on their surface. He did this by means of small holes in the surface that were connected to a manometer. Next, after removing the holes and piping, the models were suspended from wires and attached to scales so that wind-tunnel measurements could be made to find their total drag.

Careful corrections had to be made to allow for both the resistance and the stretching of the supports. Fuhrmann conceived of the drag on the airship as divided into two parts: the pressure drag and the friction drag. The pressure drag was the result of pressures normal to the surface; the friction drag was tangential. Normal pressures could be generated by a perfect fluid, but it takes a viscous fluid to create a tangential traction. Fuhrmann reached three important conclusions. First, the graph of the observed pressure distribution of the air flow was very close to that predicted from ideal-fluid theory except at the very tail of the airship models. The only exception was a blunt-nosed model, where there was deviation from the predicted pressure at the nose as well as the tail. Second, models with a rounded nose, slender body, and long tapered tail had astonishingly low resistance, for example, at 10 meters per second they had less than one-twentieth of the resistance of a sphere of the same volume. Third, nearly all of this small, residual drag could be accounted for by the frictional drag of the air on the surface. Even given the slight deviation at the tail and, of course, the effect of the air in immediate contact with the surface of the airship, the air behaved like an ideal fluid.

In the period immediately after the Great War, Prandtl wrote an account of the Göttingen airship work for the American National Advisory Committee for Aeronautics. It appeared in 1923 in English as the NACA Report No. 116.[20] The first part of the report included a survey of ideal-fluid theory, and the second part began with an account of the resistance measurements on model airships carried out by Fuhrmann. Prandtl declared that the agreement between theory and experiment in Fuhrmann's work had given them "the stimulus to seek further relations between theoretical hydrodynamics and practical aeronautics" (174). "Theoretical hydrodynamics," here, meant perfect fluid theory. Even more striking was how Prandtl described the character of the agreement that had so encouraged them. "The theoretical theorem that in the ideal fluid the resistance is zero," he said, "receives in this a brilliant confirmation by experiment" (174).

Prandtl's British counterparts such as Bairstow, Cowley, Lamb, Levy, and Taylor did not speak in this way. For the British, a "theoretical theorem" would be the result of deduction from the premises of the theory and would be something to be judged by logical, not experimental, criteria. It described an ideal fluid not a real fluid. Even if ideal-fluid theory could, on occasion, generate an empirically correct answer, this would only be because false premises can sometimes produce true conclusions. Properly speaking, experiment could never provide a "brilliant confirmation" of what was essentially a mathematical theorem, and certainly not of a theorem that referred to an acknowledged mathematical fiction. The first would be wholly unnecessary and the second wholly impossible.

Prandtl's enthusiastic formulation was slightly qualified when Fuhrmann's experiments were discussed in his Göttingen lectures, which were published a few years later.[21] In 1931 Prandtl described Fuhrmann's result as follows:

Diese Tatsache kann man bis zu einem gewissen Grade als einen experimentellen Nachweis ansehen für den Satz der klassischen Hydrodynamik, daß in einer reibungslosen Flüssigkeit der Widerstand eines bewegten (hier allerdings stromlinienförmigen!) Körpers Null ist. (153)

Up to a certain degree one can regard this fact as an experimental proof of the classical hydrodynamic theorem that the resistance of a moving body (at least, a streamlined one!) in a frictionless fluid is zero.

The brilliant confirmation had become a proof up to a "certain degree," but when Prandtl came to spell out the basis of this more qualified judgment, it is clear that this did not bring him nearer to the British position. In his lectures, Prandtl dealt with Fuhrmann's work in a section devoted to bodies of small resistance. He introduced the section by identifying the area in which inviscid theory has a legitimate application to the real world. It was, he said, an area of great technological significance and included airships, aircraft wings, and propellers.

Während die klassische Hydrodynamik der reibungslosen Flüssigkeit durchweg in allen denjenigen Fällen versagt, in denen es sich um Strömungsvorgänge mit beträchtlichem Widerstand handelt, lässt sie sich mit Vorteil anwenden bei Flüssigkeitsbewegungen mit geringem Widerstand. In den meisten praktischen Fällen—so besonders in der Flugtechnik und im Luftschiffbau—handelt es sich aber darum, den meist schädlichen Widerstand auf ein Mindestmaß zu bringen, so daß gerade hier ein grosses Anwendungsgebiet der Methoden der Hydrodynamik reibungsloser Flüssigkeiten vorliegt. Auf diesen Umstand ist es zurückzuführen, dass die Flugtechnik und Luftschiffahrt so ausserordentlich durch die neueren Untersuchungen der Luftbewegungen (Luft aufgefasst als reibungslose Flüssigkeit) gefördert wurde—wir erinnern nur an die Ausbildung der günstigsten Luftschiffform, an die Tragflügel- und Propellertheorie—und dass umgekehrt die praktischen Probleme der Flugtechnik der Theorie eine grosse Anzahl dankbarer Fragestellungen gegeben haben. (150–51)

While the classical hydrodynamics of a frictionless fluid always fails in those cases where the flow must cope with a considerable resistance, it can be applied with advantage to fluid motion with small resistance. In most practical cases—particularly in aviation and the construction of airships—it is a matter of bringing the most damaging forms of resistance down to a minimum. It is precisely here that there lies a large field for the application of the methods of the hydrodynamics of frictionless fluids. It is for this reason that aviation and airship travel received such benefit from the new investigations into the flow of air (where the air was conceived as a frictionless fluid)—one calls to mind the development of the most satisfactory shapes for airships, and wing and propeller theory. Conversely, the practical problems of aviation have presented to the theory a large number of fruitful questions.

This passage gives Prandtl's argument for conceiving air as a frictionless fluid. Lamb kept the two things separate, putting one in a box marked "real" and the other in a box marked "ideal." Prandtl put them both in the same box.

Giving theory and experiment the same referent is necessary for turning d'Alembert's result from a mere theorem into a genuine paradox. This was why Lamb adopted the strategy of assigning them separate referents. In giving them the same referent, did Prandtl intend to generate or embrace a paradox? Or, if this was not his intention, was it the unwitting consequence of his position? The answer is neither. Prandtl's stance was not paradoxical. He avoided paradox, but he did so by rejecting precondition (1) rather than, as Lamb did, precondition (2). To create a paradox it is necessary that two sources of information about a common object contradict one another. Prandtl said they did not contradict one another. The theory predicted zero resistance—and this was (very nearly) what was found by experiment.

The situation here was not, as G. P. Thomson might suspect, a case of engineers working in the realm of "good enough." It was the opposite. Prandtl and Fuhrmann found they could use the theory of ideal fluids to design airships that were very close approximations to the zero resistance entailed by the theory of perfect fluids. It helped them to identify the places where smooth flow was breaking down so that they could reduce it further. Their efforts were informed by an ideal they were striving to attain. The ideal was not kept distinct from practice, or set in opposition to it, but was integral to it and gave practice its direction and purpose. Max Munk, a distinguished pupil of Prandtl, looking back over some seventy years, recalled the Prandtl and Fuhrmann experiments and clearly thought that their methodological significance had not been properly appreciated.[22] Munk said: "The wind tunnel was asked whether the actual pressure distribution was sufficiently equal to the one computed for a perfect fluid. It was asked whether the study of the motion of a perfect fluid was helpful for practical aerodynamics. The wind tunnel answered with a loud Yes. This was a very great achievement of Prandtl, one for which he did not get enough credit" (1). Munk did not indicate who had been reluctant to give due credit, but it is clear that, had he wished to do so, he could have pointed to the British stance and the methodological assumptions behind it.

The Status of Stokes' Equations

All real fluids are viscous, but not all viscous fluids are real. A mathematician may construct a model of a fluid which makes provision for viscosity, but it remains an open question whether any real fluid satisfies the specifications of the model. Lamb was very clear on this matter. He raised it in connection with the derivation of Stokes' equations. All such derivations must start from assumptions, and these typically involve simplifications. A few years before Stokes' work, Navier in France had arrived at these same equations and so they are often known as the Navier-Stokes equations.[23] Navier, however, worked from assumptions about the supposed forces operating between the particles that made up fluids. Stokes is generally considered to have improved on this account by finding a way to avoid speculating about the ultimate particles of a fluid. He treated a fluid as a continuum and confined himself to considering the tangential stresses and shear forces on the sides of a fluid element. This modification avoided Navier's assumptions but inevitably introduced others. What laws were obeyed by the stresses and forces? Stokes made the assumption that there was a linear relation between the shear force and the rate of shear. Lamb was careful to point this out to the reader.[24] The assumption of linearity, he said, was exactly that—an assumption. He hastened to add that the assumption was plausible and the success of the equations, where they had been tested empirically, gave every reason to believe it was correct.

> It will be noticed that the hypothesis made above that the stresses . . . are *linear* functions of the rate of stress . . . is of a purely tentative character, and that although there is considerable *a priori* probability that it will represent the facts accurately in the case of infinitely small motions, we have so far no assurance that it will hold generally. It was however pointed out by Reynolds that the equations based on this hypothesis have been put to a very severe test in the experiments of Poiseuille and others. . . . Considering the very wide range of values over which these experiments extend, we can hardly hesitate to accept the equations in question as a complete statement of the laws of viscosity. (571)

Assumptions had been made, but it turned out that the assumptions were correct. The Stokes equations were not approximations in competition with other approximations. Evidence, said Lamb, shows that the equations are to be accepted as a "complete statement" of the laws of the real-world phenomenon of viscosity. In a word, the Stokes equations were true.

Given the immense authority behind this judgment, it can be difficult to realize that it was not necessitated by the facts. Lamb could have drawn a different conclusion. He was making a methodological choice and did not have to choose as he did. Others adopted a different stance toward Stokes' equations and the experimental evidence that Lamb cited.[25] I illustrate this point by reference to the work of the applied mathematician Richard von Mises (fig. 5.4). As well as his broad literary and philosophical interests, von Mises made important contributions to aerodynamic theory by generalizing the mathematical technique for creating aerofoil shapes by conformal

FIGURE 5.4. Richard von Mises (1883–1953). A leading applied mathematician who worked extensively in fluid dynamics and aerodynamics, von Mises adopted an empiricist or "positivist" stance toward the equations of fluid dynamics and treated both the Euler and the Stokes equations as abstractions.

transformations.[26] He also corresponded extensively with Prandtl about fluid dynamics and contributed to boundary-layer theory.[27] Von Mises had lectured on aerodynamics to military aircrew in Berlin as early as 1913 and had himself learned to fly at Adlershof. Before the war von Mises held a chair at the University of Strassburg, where he published *Elemente der technischen Hydromechanik* (The elements of technical hydrodynamics).[28] On the title page von Mises was styled as a *Maschinenbau-Ingenieur,* or "mechanical engineer." During the Great War he returned to Vienna, served as a pilot and an instructor, and then worked on the design of a giant aircraft for which he had provided the wing profile.[29] Toward the end of the war he published his military lectures in the form of a textbook, *Fluglehre.*[30] The little book was warmly welcomed by Prandtl because it was written by someone who could handle both the scientific and the technical sides of the aeronautics.[31]

In 1909, in an article on the problems of technical hydromechanics, von Mises had made a proposal that was designed to rationalize the relation between perfect fluid theory and the theory of viscous flow.[32] He called it the "hydraulic hypothesis" and claimed that it was implicit in many of the practical applications of hydrodynamics, even if it was not usually made explicit. Rather than emphasizing the fundamental difference between viscous and inviscid theory (for example, by saying that one referred to something real while the other referred to something unreal), the hydraulic hypothesis embodied the view that they were intimately connected. Von Mises still used the hypothesis many years later in his advanced textbook on aerodynamics, the *Theory of Flight*, first published in English in 1945.[33]

According to von Mises, so-called ideal fluids represent a process of averaging out the statistical fluctuations always present within real fluids. The implied relation between the ideal fluid and real fluid may be illustrated by an analogy. An element of an ideal fluid stands to the elements of a real fluid, that is, the molecules, in roughly the way that, say, the average taxpayer stands to the array of real taxpayers. The behavior of an element of perfect fluid mathematically encodes real information about a specified collection of real things, without itself constituting a further item in the collection. The only fluids are real fluids, just as the only taxpayers are real taxpayers. The concept of an ideal fluid is an instrument by which we talk about, reason about, and refer to real fluids. Indirectly, equations that involve ideal fluids have a real reference, just as statistical data about taxpayers have a real reference. The Euler equations capture the mean values of a statistically fluctuating reality.[34] In one formulation von Mises put it like this:

> the flow around an aerofoil in a wind tunnel is doubtless a turbulent flow of a viscous fluid. But if the small oscillations are disregarded, the remaining steady velocity values agree very well with those computed from the theory of perfect fluids. . . . The hydraulic hypothesis does not contend that the viscosity effects are negligible. On the contrary . . . the viscosity is responsible for the continual fluctuations or for the turbulent character of the motion. It is left undecided whether the instantaneous (fluctuating) velocities of the real fluid follow the Navier-Stokes equations or not. The hydraulic hypothesis states only that the mean velocity values satisfy, to a certain extent, the perfect-fluid equations. (84–85)

This wording comes from *Theory of Flight* and therefore dates from 1945, but it is entirely consistent with the original formulation of the hypothesis.

What, on this view, is the relation between Euler's equations and Stokes' equations? Von Mises' answer is interesting. He insists that *both* are idealizations. In neither case do their concepts have objects that are to be simply or directly identified with real fluids. Both have an indirect relationship. "It should be kept in mind that the 'viscous fluid' as well as the 'perfect fluid' are idealizations. In introducing the viscous fluid the presence of shearing stresses is admitted, and thus a broader hypothesis is used, which can be expected to give a better approximation to reality. However, we are not entitled to call 'real fluid' what is still only an idealization" (76–77). The term "real fluid," said von Mises, should only be used, "when reference is made to observed facts" (77). Real fluids are encountered in experiments and practical engineering. They always stand in contrast to the equations of the mathematician—a point that holds whether the equations describe a perfect or a viscous fluid. Both are idealizations, approximations, and constructions, and neither can be identified with reality.[35] This was a very different position from that adopted by Lamb and his colleagues. Although Lamb acknowledged the idealization that entered into the construction of Stokes' equations, he concluded that experiment had confirmed their truth. Such confirmation lifted the equations out of the realm of conjecture and put the stamp of reality on them. The idea that, formally, Stokes' equations stood in the same relationship to real fluids as the equations of Euler would have blurred the fundamental distinction that Lamb and his British colleagues wanted to make.

For certain purposes, some idealizations may be better than others. Von Mises acknowledged that, by taking into account the shearing stresses in the fluid, Stokes had offered a "broader hypothesis" and a "better approximation" than that provided by a perfect fluid. This point must be handled with care. It is surely correct but it does not follow that Stokes' equations will always give a more accurate answer than that given by the Euler equations. It does not follow that a viscous fluid idealization will always outperform an inviscid idealization. Calling one a "better approximation" than the other may create a certain presumption to that effect, but, given that they are both idealizations, this should not be taken for granted. Von Mises' own discussion of Poiseuille's results provides a salutary reminder.

Poiseuille's experiments concerned the uniform flow of a viscous fluid down a straight tube of circular cross section. The fluid will travel more quickly along the middle of the tube than it will closer to the perimeter, and it will be stationary on the walls of the tube itself. Of course, the fluid will have an average velocity, and this will depend on the pressure gradient. The velocity vector will always be parallel to the axis, so the flow is "laminar." In these simple conditions it can be deduced from Stokes' equations that the velocity will be distributed over the diameter of the tube in the form of a parabola. The shape of the parabola is determined by the result that the maximum velocity, on the axis, turns out to be exactly twice that of the average velocity. Poiseuille established these facts experimentally, and it was Stokes' ability to deduce them theoretically that Lamb cited as the grounds for the truth of his equations. But the deductions only hold good if the velocity of the flow is below a certain critical speed. Von Mises reported that, for air in a one-inch pipe, the critical speed is a little below 4 feet per second. Above that speed the analysis fails because the flow ceases to be laminar and becomes turbulent.

In turbulent flow the velocity distribution in the cross section of the pipe alters markedly. Instead of the parabolic distribution, a much flatter distribution prevails where the maximum is only a few percent higher than the average. Did this directly contradict Stokes' equations? It remained unclear whether this behavior contradicted them or not. The relation to the equations could not be determined, and no one could predict the pattern of turbulent flow from them. But while the equations of viscous flow were no help, it was evident that the flat distribution looked strikingly similar to that predicted on the assumption of an inviscid fluid. A frictionless fluid would not adhere to the sides of the pipe, so the fluid there would not be retarded relative to that near the center. There would be no parabolic distribution of velocities but a uniform march forward on a straight front. And this is very nearly what happens in turbulent viscous flow. As von Mises put it in the *Theory of Flight*: "This uniform velocity distribution of the perfect fluid flow agrees much better with observations under turbulent conditions than the velocity distribution of a laminar viscous flow" (83). The perfect fluid provided a better approximation to the complicated case of turbulent flow than did the equations of viscous laminar flow.

Nor did this superiority hold just for the case of a fluid in a pipe. Von Mises argued that it applied to other practically interesting flows such as those through curved channels and those with varying cross sections, and, as we have seen, to the flow in wind tunnels. "If the small fluctuations are disregarded and attention is given only to the average values at each point, there appears a marked resemblance to the irrotational flow pattern of a perfect fluid. The mean values of the velocity are distributed very much like the instantaneous velocities in a perfect fluid" (84).

The hydraulic hypothesis was a particular expression of a more general view that von Mises adopted toward the state of mechanics in the early decades of the twentieth century. His understanding of both the Euler equations and the Stokes equations brought them into line with his views on probability theory and his understanding of the modern scientific picture of the world—"das naturwissenschaftliche Weltbild der Gegenwart." He was impressed by current developments in quantum theory and understood them to mean that behind the differential equations of classical mechanics there lay a reality governed by statistical rather than causal laws.[36]

Whatever one makes of the hydraulic hypothesis and the ultimate indeterminism of physical laws, the essential point that von Mises was making about the Stokes equations still holds good. Even if von Mises' statistical interpretation of the Euler equations were to be rejected, his claim that both the Euler and Stokes equations were idealizations would not be directly threatened and could be defended on independent grounds. The point was forcefully made by the American mathematician Garrett Birkhoff. In his book *Hydrodynamics: A Study in Logic, Fact and Similitude*,[37] Birkhoff lists molecular dissociation and ionization at hypersonic speeds, chemical kinetics, and sound attenuation as some of the physical effects not covered by the equations. He also noted that "the first supersonic wind-tunnels were plagued by condensation shocks due to water vapor in the air—another 'hidden variable' ignored by the metaphysics of Navier and Stokes" (31). Impressive though they are, the Stokes equations are not a complete statement of the laws of viscosity. They should be seen, as von Mises saw them, as idealizations covering a very incomplete range of phenomena in a very partial manner.

I now put these divergent responses to perfect fluids, d'Alembert's paradox, and Stokes' equations into context. I identify two divergent traditions of mathematical work, one British, the other German. The tradition with the strong boundary between ideal and real fluids might be called "Cambridge-style mathematical physics." The other, with the weaker boundary between the real and the ideal, is the tradition of technical mechanics as it was developed in the German system of technical colleges. I start with a characterization of the Cambridge approach and, again, take Horace Lamb as my reference point.

The Cambridge School

At 9.30 a.m., on August 24, 1912, Lamb took the chair of section III of the Fifth International Congress of Mathematicians that was being held in Cambridge.[38] Section III was devoted to mechanics, physical mathematics, and astronomy. Lamb wanted to say a few words before getting down to business. He noted that, in spite of the subdivision of the field, the scope of the section was still a wide one. He then went on to offer a classification of the different styles of work that were to be represented. He also identified the predominant style of what he called the "Cambridge school" within this typology. His words are revealing.

It has been said that there are two distinct classes of applied mathematicians; viz. those whose interest lies mainly in the purely mathematical aspect of the problems suggested by experience, and those to whom on the other hand analysis is only a means to an end, the interpretation and coordination of the phenomena of the world. May I suggest that there is at least one other and an intermediate class, of which the Cambridge school has furnished many examples, who find a kind of aesthetic interest in the reciprocal play of theory and experience, who delight to see the results of analysis verified in the flash of ripples over a pool, as well as in the stately evolutions of the planetary bodies, and who find a satisfaction, again, in the continual improvement and refinement of the analytical methods which physical problems have suggested and evoke? All these classes are represented in force here today; and we trust that by mutual intercourse, and by the discussions in this section, this Congress may contribute something to the advancement of that Science of Mechanics, in its widest sense, which we all have at heart. (1:51)

The tone may have been lofty but Lamb had a purpose. He was making a plea for the representatives of the different tendencies in the discipline to communicate and cooperate. The remarks suggest a background anxiety that there might be problems on this score, and Lamb may have known about the convoluted and acrimonious arguments between pure and applied mathematicians that had been taking place in Germany. We must also remember that Lamb was addressing a gathering of men of powerful intellect, many with significant achievements behind them and reputations to make or break. Larmor, Levi-Civita, Darwin, Moulton, and Abraham were all in the audience, while the Göttingen laboratory was represented by the presence of two of Prandtl's colleagues and former assistants, Theodore von Kármán and Ludwig Föppl. All of these men played an active part in the session that followed. Given his unifying purpose, Lamb could not have risked caricaturing the different classes of mathematician.

The care with which Lamb would have chosen his words lends a particular interest to his description of the Cambridge school. Lamb saw the characteristic concern of its practitioners as lying between pure mathematics, on the one hand and, on the other, a purely instrumental view of mathematics, one in which its role was simply the interpretation and coordination of data. The point on which he placed the emphasis was that mathematical results should be verified by the interplay of theory and experience. Lamb obviously saw this process as more than mere success in the ordering of data. Truth and correspondence with reality were the central aims. He described this concern as "aesthetic"—a word chosen, surely, to portray an intellectual involvement that was dignified rather than merely useful. The emphasis on truth was certainly consistent with what Lamb had said elsewhere, for example, in the discussion of Stokes' equations in his *Hydrodynamics*. Applied mathematics, as practiced at Cambridge, was to be justified by its capacity to portray the nature of physical reality, not by its employment of useful fictions.

Other Cambridge luminaries expressed themselves somewhat differently but conveyed a similar orientation. At a different session of the same conference the Cambridge mathematical physicist Joseph Larmor voiced sentiments that reinforced Lamb's message. Larmor asserted that the role of the mathematician and the physicist were essentially identical.[39] A. E. H. Love had spoken out in support of Larmor at the conference.[40] Love was reiterating a position already developed in his authoritative *Treatise on the Mathematical Theory of Elasticity*.[41] This volume contained a historical introduction in which tendencies and distinctions similar to those identified by Lamb were rehearsed and evaluated. Love declared, as one of his aims, that he wanted to make his book useful to engineers and this had led him "to undertake some rather laborious arithmetical computations" (v). But he also wanted to "emphasise the bearing of the theory on general questions of Natural Philosophy" (v), and it was clear that this was where his heart lay. His historical comments were judicious, but he went out of his way to emphasize the non-utilitarian origins of the subject matter he was about to expound. Thus,

> The history of the mathematical theory of Elasticity shows clearly that the development of the theory has not been guided exclusively by considerations of its utility for technical Mechanics. Most of the men by whose researches it has been founded and shaped have been more interested in Natural Philosophy than in material progress, in trying to understand the world than in trying to make it more comfortable. From this attitude of mind it may possibly have resulted that the theory has contributed less to the material advance of mankind than it might otherwise have done. Be this as it may, the intellectual gain which has accrued from the work of these men must be estimated very highly. (30)

Technical mechanics is to be distinguished from natural philosophy, and he, Love, was doing a species of natural philosophy. Any resulting failure to contribute to material progress did not seem to distress him unduly. He was more interested in the link with fundamental physics and in recounting the detailed discussions that had taken place over the number and meaning of the elastic constants. These had thrown light on "the nature of molecules and the mode of their interaction" (30). The wave theory of optics and the theory of the ether had benefited from advances in the theory of elasticity, as had, even, certain branches of pure mathematics. Though Love and Lamb expressed themselves differently, we see a similar distancing of applied mathematics from issues of utility and an affirmation of the fundamental character of the relation between mathematics and physical reality. G. I. Taylor's demand, made a few years later in his Adams Prize essay, that applied mathematics should have a firm basis in physics was the expression of a stance already endorsed by figures of authority on the Cambridge scene and already characteristic of the Cambridge school.[42]

The demand for a firm basis in physics had not always characterized what had passed as "mathematical physics" or "mixed mathematics" at Cambridge. Mathematicians of earlier generations had often been happy to see mathematics arise from physical problems but had then developed it independently of experimental data or with only a loose or analogical link to physical reality. An example of this earlier phase, which was still evident as late as the 1870s, was James Challis' *Essay on the Mathematical Principles of Physics* in which he offered a speculative, hydrodynamic cosmology.[43] The closer connection between mathematics and real physics that Lamb and, later, Taylor were taking for granted had originally been forged in the work of Stokes, Thomson, and Maxwell, who were critical of the earlier style.[44] Lamb, however, still felt the need to express himself carefully when he said that the Cambridge school provided "many examples" of the intermediate path between an overly abstract and an overly utilitarian approach. He thus acknowledged a continuing diversity in Cambridge work. This should come as no surprise since traditions, even vigorous traditions, will always encompass a range of positions as they change and develop. Rayleigh, like Lamb, spoke of "the Cambridge school," and he too noted a certain inner complexity and development. In connection with Routh's textbook on dynamics, Rayleigh took the view that the earlier editions had been overly abstract, whereas later editions evinced a closer engagement with genuine scientific problems.[45] In other words Routh had shifted toward the position that Lamb, like Rayleigh himself, saw as the strong point of the Cambridge school.[46]

Föppl's *Vorlesungen*

The influential vision of the turn-of-the-century Cambridge school of mathematical physics, as Lamb, Love, and others presented it, stood in contrast to the German idea of technical mechanics. This body of work came out of the great system of German technical colleges or *technische Hochschulen*, such as that at Charlottenburg, or in Munich where Prandtl had studied, or Hanover where both Prandtl and Runge had taught before their call to Göttingen. A representative example of this style of work is provided by August Föppl's

influential lectures on technical mechanics. His multivolume and vastly popular *Vorlesungen über technische Mechanik* was published in many editions
around the turn of the century. Föppl originally worked in industry and had
spent a number of years teaching in a trade school. He later rose to become
the professor of theoretical mechanics at the Munich *Hochschule* and the director of their materials laboratory.[47] A versatile mathematician, Föppl had
written the first book introducing Maxwell's work on electromagnetism into
Germany. The book was later revised and coauthored with the experimental
physicist Max Abraham, and it is known that one student who was influenced
by it was the young Einstein.[48] Föppl had been Prandtl's teacher at Munich
and had supervised his doctoral research on the buckling of loaded beams.[49]
In 1909 Prandtl married Föppl's eldest daughter, Gertrud. In Prandtl's biography, written by his own daughter, there is evidence of a certain tension
between Prandtl and his father-in-law, occasioned by the older man's au-
thoritarian attitudes, but there was no lack of scientific respect.[50] Ludwig
Föppl, who, along with Abraham and von Kármán, had been in the audience
at Cambridge when Lamb spoke, was one of August Föppl's two sons. The
other son, Otto Föppl, worked with Prandtl on wind-tunnel experiments in
Göttingen. Some of Otto Föppl's work is discussed in a later chapter; for the
moment, however, the concern is with August Föppl (fig. 5.5).

What did the many readers of Föppl's published lectures on *technische
Mechanik* learn about the status of their field as they imbibed its carefully
graded and expertly presented content? First, they learned that mathematics
was a means to an end, rather than an end in itself. In the introduction to the
first volume, Föppl wrote that mechanics makes extensive use of mathematics,
but as an auxiliary. Mathematical techniques, he said, were simply the clothing
in which the body of knowledge was garbed. The point was reiterated at the
beginning of the more mathematically demanding third volume, but this time
with more stress on just how important, on occasion, these aids could be:

> Analytische Entwicklungen betrachte ich immer nur als ein Mittel zur Er-
> kenntnis des inneren Zusammenhangs der Thatsachen. Wer auf sie verzichten
> wollte, würde das schärfste und zuverlässigste Werkzeug zur Verarbeitung der
> Beobachtungsthatsachen aus der Hand geben. (1900a, viii)

> I only consider analytical processes as a means for understanding the intimate
> interconnections of the facts. Those who want to renounce them are letting
> go of the sharpest and most dependable tools for working with the facts of
> observation.

Mathematics provided what Föppl called *Hilfsmittel* and *Werkzeuge*, "aids"
and "instruments." Föppl's language is important here. There had been in-

FIGURE 5.5. August Föppl (1854–1924) was a versatile mathematician and a professor at the *technische Hochschule* in Munich. He was the author of an extremely influential textbook on technical mechanics that was based on his Munich lectures. He was also the father-in-law of Ludwig Prandtl. Photograph from Bäseler et al. 1924.

tense, not to say wearisome, debate in German academic circles over whether mathematics was to be seen as a *Hilfswissenschaft* or as a *Grundwissenschaft* with respect to technology.[51] Was it an auxiliary to, or a foundation of, technology? The debate was really a coded argument over the status of mathematicians in the technical college system and their role in the education of engineers. Föppl was signaling that mathematicians had to earn their living by making themselves useful to engineers. The function of mathematics was to further technology and engineering.

The first chapter of volume 1 of the *Vorlesungen* was devoted to the origin and goals of mechanics. Föppl acknowledged that mechanics was part of physics and, like all the sciences, was grounded in experience. To grasp experience, he argued, it was always necessary to work with simplified, easily imagined "pictures" (*Bilder*) of reality. The ideas of a point particle and a rigid body were two such pictures. Both were valuable and had their appropriate range of application, but they must not be mistaken for physical realities.[52] Föppl also drew a distinction between the *Naturforscher* and the *Techniker*—the natural scientist and the engineer. His book was for the latter, not the former, and dealt with a mode of knowledge having special characteristics that differentiated it from natural science in general.

> Bei der technischen Mechanik tritt als bestimmender Beweggrund für ihre Fassung zu der Absicht einer Erforschung der Wirklichkeit . . . noch die andere Absicht, ihre Lehren nutzbringend in der Technik zu verwerthen. (11)
>
> In the case of technical mechanics there is a definite motive for its approach over and above the intention to investigate reality . . . and that further intention is that its theories be usefully applicable in technology.

Föppl was just the sort of utilitarian, applied mathematician from whom Lamb and Love had distanced themselves. Indeed, Lamb's typology might have been expressly contrived to ensure that the Cambridge school did not get caught in the cross-fire between the champions of mathematics as *Hilfswissenschaft* and as *Grundwissenschaft*. Be this as it may, Föppl certainly didn't have the Cambridge tone. His technical mechanics was not natural philosophy. Furthermore, Föppl differentiated technical mechanics from mechanics in general because there are many cases when the general doctrines of mechanics do not, or do not yet, provide rigorous answers to the questions that have to be confronted by the engineer. Natural scientists and engineers, he said, stand in a wholly different relationship to these cases: "solchen Fällen steht aber der Naturforscher anders gegenüber als der Techniker" (11). The engineer *must* produce an answer and *must* forge concepts to deal with the problem. The natural scientist can wait for inspiration or more information; the engineer cannot:

> Der Techniker dagegen steht unter dem Zwange der Nothwendigkeit; er muss ohne Zögern handeln, wenn ihm irgend eine Erscheinung hemmend oder fördernd in den Weg tritt, und er muss sich daher unbedingt auf irgend eine Art, so gut es eben gehen will, eine theoretische Auffassung davon zurechtlegen. (11–12)
>
> The engineer, by contrast, is subject to the force of necessity. He must, without delay, deal with the matter when some phenomenon interferes and interposes itself in his path. He must, in some way or other, arrive at a theoretical understanding of it as best he can.

The demands of this enforced creativity may generate concepts that do not meet the logical demands of existing mechanics. Here, said Föppl, was the deep reason for separating out technical mechanics as a special branch of knowledge—"diese Absonderung der technischen Mechanik als eines besonderen Zweiges der Wissenschaft" (11). Its practitioners must have the freedom to develop concepts of their own, and these might be distinct from those acceptable in the more reflective and leisurely branches of knowledge. For example, the application of hydrodynamic theory to turbines developed by Prášil and H. Lorenz, depended on certain artifices or tricks (*Kunstgriffe*) involving the idea of "forced accelerations."[53] The approach had been controversial, but Föppl defended it. He went on to say that in such cases subsequent developments in science might permit a reconciliation. The anomalous concepts, special to engineering and technical mechanics, might be absorbed back into the main body of knowledge. But this was an open question, something for the unspecified future rather than the urgent present. His main concern was to emphasize the restless force running through the scientific life of modern technology, which was, he said, like the life force in a tree that continually generated new branches.

In the peroration rounding off the introductory chapter, however, Föppl suddenly changed the metaphor. The life force of a tree was replaced by another kind of force. Knowledge was power, said Föppl, the power of a modern technological state. The bucolic image was replaced by a military one. Those who first possessed the right theory ("die richtige Theorie") might be able to intervene in nature at will. In this way, said Föppl, science, put at the disposal of humanity and its peoples, is the most powerful of all weapons— "und darum ist die Wissenschaft die gewaltigste Waffe, die Menschen und Völkern zu Gebote steht" (12).

What did Föppl understand by the "right" theory? His normative standards were predictably active and pragmatic. Like many in his position in the *technische Hochschulen*, Föppl was fighting a war on two fronts. On one side were "humanistic" critics. These were usually outside the technical college system, or only passing through it on their way to posts in universities. In as far as they wanted mathematics at all, they wanted it "pure." On the other side were critics, often from within the technical colleges, who placed all the emphasis on practicality and were suspicious of any form of higher mathematics.[54] These were the counterpart of the "practical men" in Britain who were so hostile to the work of the Advisory Committee for Aeronautics. In responding to the local, German, variant, Föppl dismissed such people as mere *Praktiker*.[55] He had no time for them or their slogans about the conflict between theory and practice—"dem Gegensatze zwischen Theorie und Praxis" (3:vii.). For Föppl there was no such conflict:

> Diese Behauptung lasse ich aber auf dem Gebiete der technischen Mechanik durchaus nicht gelten; hier kann nur von einem Gegensatze zwischen falscher oder unvollständiger Theorie und der richtigen Theorie die Rede sein. Die richtige Theorie ist immer in Übereinstimmung mit der Praxis. (3:vii)

> I do not admit this claim as having any validity in the realm of technical mechanics. Here one can only speak of the conflict between false or incomplete theories and the right theory. The right theory is always in agreement with practice.

Did Föppl mean that a theory was practical *because* it was right, or that it *counted* as right because it worked in practice? Taken in isolation, his wording was ambiguous. If, however, we recall Föppl's insistence on the overpowering, practical necessities that dominate the life of the engineer—the "Zwang der Nothwendigkeit"—then the formal ambiguity can be resolved. In Föppl's world, practice was the effective criterion not of an abstract and future truth but of acceptability and viability for the pressing moment. In the colleges of Cambridge, if Lamb is to be believed, the pursuit of truth had an aesthetic character. In the colleges of technology a theory was counted as right if, and only if, it worked.

Others in the field of applied mathematics in Germany may have made the case in different words, but Föppl's general orientation toward engineering represented a widely held view. For example, in 1921 Richard von Mises started a new journal for applied mathematics—the *Zeitschrift für angewandte Mathematik und Mechanik*. Prandtl and von Mises had been in correspondence after the war on new institutional arrangements for encouraging applied mathematics. Prandtl mentioned that he and von Kármán had been discussing the founding of a society to promote technical mechanics, "eine Vereinigung für technische Mechanik," and these exchanges were part of the process that culminated in the journal.[56] On the first page of the new publication von Mises set out his conception of the task and goals of the discipline and the role of the journal. He would have been conscious of stepping into a long-standing discussion about the role of mathematics in the German academic world but he had no desire to equivocate. He insisted that the core of the journal would be devoted to mechanics whose cultivation, he said, today lies almost exclusively in the hands of engineers, "deren Pflege heute fast ausschließlich in den Händen der Ingenieure ruht."[57]

Mathematics, said von Mises, covered a wide spectrum of activities so that the partition between pure and applied mathematics was a relative one, located differently by different practitioners. Each would count what was (so to speak) on their "left" as pure and what was on their "right" as applied. But there was not only this dimension to consider: the very content of mathematics itself changed with time as new areas (for example, the concept of probability) were brought within the scope of quantitative analysis. We must, said von Mises, accept this "two-fold relativity" in the identity of applied mathematics. To overcome the definitional problems this created, he concluded that a practical, rather than a theoretical, specification of the field was called for. Applied mathematics and mechanics were to be defined as what was done, at that time, by scientifically oriented engineers. Thus,

> Angesichts dieses Tatbestandes zweifacher Relativität der Begriffsabgrenzung müssen wir nun eine praktische Erklärung dafür suchen, was wir hier im Folgenden unter "Angewandter Mathematik" verstehen wollen. Es ist selbstverständlich, daß wir uns auf den Boden der Gegenwart stellen, und es sei hinzugefügt: auf den Standpunkt des wissenschaftlich arbeitenden Ingenieurs. (3)

> Given the facts of this twofold relativity of the conceptual boundary, we must now seek for a practical explanation of what, in the following, we want to understand by "applied mathematics." It will be obvious that we take our stand on the basis of the present and, let it be added, on the standpoint of the scientific work of the engineer.

Clearly the two mathematical traditions that I have delineated had different orientations: one more toward physics; the other more toward engineering. Obviously, Cambridge mathematical physics and German technical mechanics still had much in common. There were many respects in which they overlapped, and it was possible for results to be passed from the practitioners of one to those of the other. Prandtl's early papers on elasticity and Föppl's volume of the *Vorlesungen* devoted to the strength of materials were mentioned in Love's treatise, while, in return, Föppl advised his more advanced readers to consult Love's work. Representatives of the two traditions attended the same conferences, even if this caused Lamb a touch of anxiety. Klein admired Cambridge pedagogy and tried, though without much success, to introduce it in Germany.[58] Lamb's *Hydrodynamics* was translated into German in 1907, again at Klein's prompting, though later von Mises added a lengthy supplement to the book designed to build a bridge to the more technical concerns and less formal orientation ("weniger formalen Richtung") of German readers.[59] Although G. H. Bryan evinced disdain for the intellectual level of the engineers in Joukowsky's classes, he could write a respectful review of Föppl's *Vorlesungen* in *Nature* saying, "Prof. Föppl's treatises on technical mechanics are of a far more advanced character than the mechanics taught commonly to technical students in this country."[60] In his own way Bryan wanted to further the cause of applied mathematics in this country and was ready to hold up German efforts when it was expedient to do so. To this extent the acknowledgment of communality between British and German mathematical cultures was real enough.

Given this mixture of divergent tendencies and common ground, it is not surprising that the members of the respective traditions did not themselves always have a clear awareness of the relations between them. This was epitomized in an exchange of letters that took place between G. I. Taylor and Ludwig Prandtl a number of years after the events described here. By the time the letters were written, in the 1930s, the magnitude of Prandtl's contributions had become known and widely admired. Taylor had written to say that he thought Prandtl deserved the Nobel Prize in physics. Prandtl's response, in a letter of November 30, 1935, was not only becomingly modest but was also culturally revealing.[61] He said that what he had done would not count as physics in Germany. Rather, it was a contribution to *Mechanik*.

> Nach der in Deutschland üblichen Einteilung der Wissenschaften wenigstens wird die Mechanik heutzutage nicht mehr als ein Teil der Physik betrachtet, sondern steht als selbständiges Gebiet zwischen der Mathematik und den Ingenieurwissenschaften.

> At least according to the division of the sciences that is usual in Germany today, mechanics is no longer considered to be part of physics. Rather, it stands as an independent area between mathematics and the engineering sciences.

While Taylor now assimilated Prandtl's work to physics, the Germans saw it as something distinct from physics and as standing between mathematics and engineering.[62] The different stance toward engineering and its demands perhaps sheds light on why the circulation theory was actively resisted in Britain but accepted and developed in Germany. Before taking this argument further, however, I look at what Lanchester himself said to explain the rough ride given to his work. Lanchester's account will give me an opportunity to look at another variable whose explanatory potential needs to be assessed, namely, the personalities of the main actors.

Personalities

Lanchester came to loath Bairstow and what he called "the Cambridge School"—a group to which he had no hesitation in assigning Bairstow, despite the latter's London provenance.[63] Unlike the positive comments he made about the National Physical Laboratory in 1915, in later years Lanchester expressed resentment at the lack of support he had received from that quarter and identified the majority of those working there as effectively belonging to the "Cambridge School." In a memorandum written in 1936, in which he sought recognition from the Air Ministry for his contribution, Lanchester expressed himself with some bitterness: "The trouble is, or arose from the fact, that with the exception of Lord Rayleigh, the N.P.L. did not take my work seriously. . . . They fell into the error, and for this Leonard Bairstow was mainly to blame, of casting doubt on my work, I believe because my methods did not appeal to them in view of their training. They mostly belonged to the Cambridge School, whereas I was the product of the Royal College of Science (then the Normal School of Science)" (19–20).[64] He recalled that, on more than one occasion, Bairstow had asserted, during meetings of the Advisory Committee for Aeronautics, that "we do not believe in your theories" (20). In an earlier letter of 1931 to Capt. J. L. Pritchard, the secretary of the Royal Aeronautical Society, Lanchester referred to "that man Bairstow who would have nothing of the vortex or cyclic theory and took every occasion when I was a member of the Advisory Committee to laugh and jeer at it."[65]

The minutes of the Advisory Committee do not contain any specific record of episodes of this kind.[66] Whether those writing the minutes drew a veil over such exchanges or whether Lanchester's memory was at fault is impossible to determine. Nevertheless there is no reason to doubt the essential accuracy of Lanchester's account, and the minutes contain clear evidence of Bairstow's opposition. There is also ample corroboration in the public realm. As J. L. Nayler, the secretary of the committee, put it, in his early years Bairstow was "a dominant and almost pugilistic character."[67] In another letter to Pritchard, Lanchester left no doubt as to where he placed the blame for the opposition to his work. "The whole thing," he asserted, "originated with Bairstow backed up by Glazebrook."[68]

The personalized focus of this explanation has been taken up by others. This was the line taken by J. A. D. Ackroyd in his Lanchester Lecture of 1992. After giving an authoritative account of Lanchester's contributions to aerodynamics, Ackroyd posed the question of why there was so little interest in the circulatory theory. "Perhaps part of the problem," he suggested, "lay in the personalities involved."[69] Ackroyd, however, did not place all the emphasis on Bairstow's personality but noted the role of Lanchester's own strong personality and his inclination to be critical of Cambridge and London graduates and the work of the NPL. Perhaps, Ackroyd concluded, there was a mutual antipathy between the persons involved. In developing this argument, Ackroyd cited and endorsed the psychologically oriented explanation that had been advanced some years previously by the eminent applied mathematician Sir Graham Sutton FRS. Sutton pointed to what he called Lanchester's "isolation" and put this down to Lanchester having been one of the great "individualists" of science. "Throughout his life he remained an individualist, perhaps the last and possibly the greatest lone worker that aerodynamics will ever see."[70]

The clash of personalities must be part of the story, but can this really be the explanation of the opposition to the cyclic theory? I do not believe that it can. Consider the role of Bairstow's personality. In the survey that I gave of the reasons advanced against Lanchester, it is clear that Bairstow's arguments were aligned with those offered by others, such as Taylor, Cowley, Levy, and Lamb. Later I shall add more names to this list. I have seen no evidence that suggests they shared Bairstow's main personality characteristic, that is, his aggressiveness. They had their own, quite different, personalities. Levy, for example, always said Cambridge was an unattractive place where the mathematical traditions were too "pure" for his tastes. With his Jewish and Scottish working-class background, he said he did not feel socially or politically comfortable in Cambridge and declined the chance to do postgraduate work there. Levy's class consciousness and bitterness at the blighted lives he had witnessed in the slums never left him.[71] After graduating from Edinburgh, however, Levy used his scholarship funds to visit Göttingen (where he met von Kármán) and then took himself to Oxford to work with the Cambridge-trained Love. The relation between Levy's personal feelings and this career trajectory is not easy to fathom,[72] but perhaps we do not need to understand such matters. What can be said about all these diverse and complex personalities is that they all took a similar stance on the central, technical problems that were in question. *They shared professional opinions and judgments, not individual personality traits.* The explanation in terms of personality, therefore, breaks down. The candidate cause (personality) varies, but the effect (resistance to Lanchester's ideas) stays the same. This means that we must look elsewhere for the real cause.

What, in any case, would be the basis of an account that rested on an appeal to personality? No one believes that certain psychological types are selectively attracted to this, that, or the other preferred pattern of fluid flow, whether viscid or inviscid. Those who invoke "personality" generally do so in order to explain the disruption of a process of rational assessment that (it is assumed) would otherwise have proceeded in a different way. It is offered as a way of explaining why things went wrong. It is meant to explain why a theory was rejected when it should have been accepted, and the answer is found in individual psychological traits. But given that the assessment of Lanchester actually rested on the appeal to shared standards, common to a group of otherwise diverse individuals, this explanatory approach bypasses the most salient feature of the episode. Its outstanding characteristic was its *systematic and shared* nature. It had the character of a concerted action by a group.

A further point needs to be stressed. An examination of the technical arguments that were used against Lanchester suggests that the response to his work was not a disruption in the rational working of science but a routine example of it. It was orderly, consistent, and reasoned and drew upon a refined body of received opinion and technique. It is true that some of the complexity was factored out of Lanchester's text, but that again was a consistent and shared feature of the response, not an individual variable. Personality played its part, but only by giving a different tone, and a different degree of intensity, to the expression of a central core of repeated, and overlapping, argumentation. The common content of the arguments derived not from individual psychology but from participation in a shared scientific culture.

Lanchester himself hinted at an explanation of this kind. As well as his explicit and angry psychological account, focused on his irritation with Bairstow, there was also an implicit, more sociological dimension to his account of the resistance to his theory. This aspect surfaced in his reference to the "Cambridge School" and the common background of training of the scientists at the NPL. We should also recall his 1917 discussion of the organizational characteristics of well-conducted aeronautical research. This, too, can be read for its bearing on the resistance encountered by Lanchester's work. His central preoccupation was that the different parties to the process of research should confine themselves to their proper spheres of competence. No good would come, he argued, of mathematicians and physicists encroaching on territory outside the (narrow) limits of their expertise. What could have been in Lanchester's mind? What examples of invasive physicists might he have cited? The public confrontation with Bairstow, two years previously, when they clashed over the proper scope of a theory of lift could not have been far beneath the surface. Was it necessary to find a universal law of nature, as Bairstow wanted, or would a specialized, practically oriented approach suffice, as Lanchester believed? Whether or not this was the example in Lanchester's mind, it illustrates the general problem to which he was referring, namely, the problem of the division of labor.

The division of labor generates a diversity of specialized perspectives and localized forms of knowledge. Professional subgroups and disciplinary divisions such as those between mathematical physics and technical mechanics are instances of this general phenomenon. What happens when the product of one of these subgroups and perspectives is assessed from the standpoint of another, different subgroup with a different perspective? We have here all the preconditions for a small-scale culture clash. Has the knowledge claim been properly understood, or has it been misinterpreted? Is a contribution to one project being assessed (deliberately or unwittingly) by criteria more appropriate to another project? If I am right, this is exactly what happened when Lanchester's work was assessed so negatively by the "Cambridge School," and it was this problem (although it was not the only problem) that Lanchester was addressing when he discussed the proper organization of aerodynamic research.

An Explanation

Understood in terms of schools, traditions, and disciplines, the question of accounting for the response to Lanchester's work takes on a sociological rather than a psychological form. The psychological machinery of individual cognition must underpin all sociological processes, but that does not mean that individualistic answers can be given to sociological questions. The explanation that is needed must involve the interaction of scientific subcultures and institutions and thus go beyond the personalities of those involved. As long as we atomize the process into a sequence of individual responses, determined by the idiosyncrasies of personality, we shall miss the significance of what happened to Lanchester. The central point that I have sought to demonstrate is that in Britain his work was actively assessed by members of a confident scientific subculture—a subculture from which Lanchester himself was excluded. Its members were steeped in the achievements and exemplars of the Tripos tradition and the research of those who carried the tradition forward. The tradition and the line of research growing out of it were diffused through authoritative textbooks such as Lamb's *Hydrodynamics*. This book and the growing body of work documented in successive editions was a reference point not only for those who had themselves sat the Senate House examinations, but also for those in Britain who trained at universities elsewhere.

As a by-product of this collective assimilation, Lanchester's work was treated selectively. It was reinterpreted and restructured along lines that were familiar to the group responding to it. Such transformations are routine during processes of cultural assimilation. Social psychologists are familiar with the process and have given it a name. They call it conventionalization.[73] In this chapter I have documented the conventionalization of Lanchester's work to the norms and practices of Cambridge-style mathematical physics. The process of assimilation governed the way Lanchester's work was understood. For a number of crucial years it was the precondition for the assessment and the rejection of Lanchester's ideas.

The assimilation and conventionalization had the effect of simplifying the overall structure of Lanchester's argument, introducing into it an exclusive emphasis on the behavior of ideal fluids that was not present in the original text. In itself this simplification could be represented either as a distortion or as an improvement in the formulation of the theory. It certainly increased the precision with which the central ideas of cyclic flow and lift were spelled out. In this respect there can be no doubt that the "mathematicians" saw more deeply into Lanchester's work than did the general run of "practical men." The mathematicians were doing for Lanchester what Maxwell had done previously for Faraday. Qualitative ideas were cast into a mathematical form. This need have had no detrimental effect on the appreciation of Lanchester's achievement. The mathematical reformulation could have been the starting point for work that carried the theory forward—as in Germany. In Britain it had the opposite effect and justified the rejection of Lanchester's work.

The reason was that the local, scientific culture into which Lanchester's work was assimilated had effectively abandoned ideal-fluid theory as a research topic and a research tool. The Euler equations of classical hydrodynamics described territory that had already been conquered, exploited, and left behind by the moving front of fundamental research. British mathematical physicists were confident that ideal-fluid theory dealt only with a mathematical fiction and not with a physical reality. It was material for examination boards, not research committees. Interest had now moved to Stokes' equations and the real behavior of viscous and turbulent fluids. By contrast, other experts from a different tradition, who were responsive to different imperatives, could respond very differently to the simplified, mathematical version of Lanchester's cyclic theory. This is why it was the German engineers schooled in *technische Mechanik* who carried this approach forward. They too were professionally interested in viscosity and turbulence, but their background assumptions and engineering orientation encouraged them, and permitted them, to frame and partition the problems of aerodynamics and fluid dynamics in a different way to their British counterparts. The consequences of this orientation toward engineering is the subject of the next two chapters. These chapters contain a detailed discussion of the German work on aerodynamics and provide a further opportunity to see the tradition of technical mechanics in action and to explore its institutional context more deeply. They consolidate the picture that is beginning to emerge.

6

Technische Mechanik in Action:
Kutta's Arc and the Joukowsky Wing

Die Strömungs- und Druckerscheinungen, wie sie in bewegten Flüssigkeiten, insbesondere auch der Luft, an den dareinversenkten Körpern beobachtet werden, haben schon seit längerer Zeit der hydrodynamischen Theorie einen viel bearbeiteten, nicht ganz einfachen Gegenstand geboten. Seit Otto Lilienthals Errungenschaften, und der neueren Entwicklung und Lösung des Flugproblems haben diese Fragen auch große praktische Bedeutung erlangt.

W. M. KUTTA, *"Über eine mit den Grundlagen des Flugproblems in Beziehung stehende zweidimensionale Strömung"* (1910)[1]

The flow and pressure phenomena, as they can be observed on bodies immersed in a moving fluid, particularly the air, have long provided for hydrodynamic theory a much worked on, but far from simple, object of study. Since Otto Lilienthal's achievements and recent developments in solving the problem of flight, these questions have acquired great practical significance.

In the next two chapters I show *technische Mechanik* in action by giving an overview of the early German (or German-language) development of the circulatory theory. In this chapter I deal with the "infinite wing" paradigm, that is, with an analysis deliberately confined to a two-dimensional cross section of the flow in which the wingtips are ignored. I then devote the next chapter to the more realistic theory dealing with a wing of finite span and the three-dimensional flow around it. It was Wilhelm Kutta in Munich who triggered the striking progress in the field of two-dimensional flow that was made in Germany before and during the Great War. His work is my starting point. Where Rayleigh used a simple, flat plane as a model of a wing, Kutta used a shallow, circular arc. Both men treated the air as an inviscid fluid, but where Rayleigh postulated a flow with surfaces of discontinuity, Kutta postulated an irrotational flow with circulation. Joukowsky, a Russian who published in German, then showed how to simplify and generalize Kutta's reasoning. A variety of other workers in Göttingen, Aachen, and Berlin, starting from Kutta's and Joukowsky's publications, carried the experimental and theoretical analysis yet further. Appreciating why these developments constitute an exercise in technical mechanics, rather than mathematical physics, requires engaging with the details of the scientific reasoning. As a first step I place Kutta and his achievement in their institutional setting.

A Private Man in a Public Context

Wilhelm Martin Kutta (fig. 6.1) was born in Pitschen in Upper Silesia in 1867. He lost both parents at an early age and was brought up in the household of an uncle in Breslau. After attending the university in Breslau from 1885 to 1889, he went to the University of Munich, where he studied from 1891 to 1894. Kutta went on to achieve a lasting place in the history of applied

FIGURE 6.1. Martin Wilhelm Kutta (1867–1941). In 1910 and 1911 Kutta published and extended an analysis of the flow of air around the wing of Lilienthal's glider that he had worked out in 1902 in a dissertation at the *technische Hochschule* in Munich. Kutta assumed that the flow contained a circulation and showed how to link the flow around the wing to the simpler and already solved problem of the flow around a circular cylinder. He was then able to make a plausible prediction of the lift of the wing. After these pioneering papers, Kutta published nothing more. (By permission of the Universitätsarchiv Stuttgart)

mathematics for two reasons. First, in his doctoral work of 1900, he developed a numerical method for solving ordinary differential equations. This has become known as the Kutta-Runge method and is to be found in all textbooks on the subject.[2] Second, he produced a pioneering paper on aerodynamics which appeared in 1910,[3] with further developments published in 1911. These papers were based on methods he had developed in his *Habilitationschrift* of 1902, which he wrote at the *technische Hochschule* in Munich.[4] (This institution is often referred to by its initials as the THM and, for brevity, I follow this practice.) Unfortunately no copies of the *Habilitationschrift* appear to have survived.[5] From the brief summary that was published in 1902, however, it seems to have been the first, mathematical analysis of lift that was based on the circulation theory.[6]

Kutta was a conscientious teacher who, over the years, introduced hundreds of engineering students to the methods of applied mathematics. His mathematical knowledge was said to be of enormous scope and his help was frequently requested by colleagues. He had a deep knowledge of history, literature, and music, a command of languages, including Arabic, and was widely traveled. He never married, however, and was something of a recluse. A colleague of long-standing, Friedrich Pfeiffer, who had been a student under Kutta at the THM, wrote an obituary for Kutta after the Second World War.[7] In the article, Pfeiffer recalls that Kutta would typically sit alone in the most remote corner of the Mathematical Institute at Munich. After Kutta's retirement, said Pfeiffer, he and other colleagues would sometimes encounter Kutta, though this happened infrequently. The lack of contact was put down to Kutta's reticence. When they did meet, Pfeiffer was unhappy with what he found. In later years, he said, Kutta obviously lacked a loving and caring hand ("wie sehr ihm eine liebende und sorgende Hand fehlte"). He went on:

> Oft habe ich Kuttas Leben reich und beneidenswert gefunden wegen seiner Aufgeschlossenheit für so viele Seiten menschlichen Geisteslebens, oft aber fand ich es auch arm und bedauernswert in seiner Einsamkeit und Zurückgezogenheit. (56)

> I have often found Kutta's life rich and enviable because of his openness to so many aspects of human cultural life, but I have also often found it poor and rather sad in its solitariness and seclusion.

How were things really, asked Pfeiffer, and did not know the answer. But if Kutta's inner life was closed to his colleagues, and must be closed to us, his work is open to inspection. Seclusion notwithstanding, he published work that bore the stamp of a time and a place. It was the product of a specific, professional milieu.

Kutta's career as an academic began in 1894 when he became a teaching assistant in higher mathematics at the THM. Like all the *technische Hochschulen*, the THM had experienced long-standing tensions over the role to be played by mathematics in the training of engineers. How much mathematics should be on the syllabus? What sort of mathematics should be offered, at what level, and who should teach it? These tensions have now been the subject of close, historical study, and thanks to this work there is much about the overall structure of the situation, as well as the particular circumstances in Munich, that can be sketched with some confidence. It is thus possible to form a picture of the context in which Kutta came to do his work on aerodynamics.

Three points must stand out in any general overview. First, the *technische Hochschulen* (or THs) tended to recruit their mathematics teachers from the universities and, when they were good, lose them again to the universities. This mixture of policy and necessity carried with it certain problems. From the mid-1850s, university mathematics in Germany had been increasingly dominated by a concern with rigor and so-called pure mathematics.[8] Although the THs provided jobs for mathematicians, those who took the jobs often had their eyes focused on matters that fell outside the concerns of the THs. Their teaching, like their research, was abstract and lacked relevance to engineering. Justifiably, this caused resentment among the engineers, with the result that mathematics appointments often turned into a struggle between different factions in the TH.[9]

Second, and predictably, engineers were not a homogeneous group. Some engineers wanted to use mathematics as the model on which to construct a "science" of engineering and the nature of machines. The aim was to create a body of knowledge that was general, abstract, and deductive. This movement, which was designed to improve the status of engineering, was associated particularly with the names of Franz Reuleaux and Franz Grashof and achieved considerable influence during the 1870s and 1880s.[10] These tendencies in the direction of purity and rigor by one part of the profession provoked an angry reaction in the 1890s from some other parts of the profession. The reaction took the form of an antimathematical movement (Anti-mathematische Bewegung) led by Alois Riedler at the TH in Charlottenburg. Riedler presented the issue as one of the very survival of Germany in a world where technological effort must go hand-in-hand with commercial activity and efficient social organization. In this struggle for existence ("Kampf ums Dasein") there was no place for the speculations of the unproductive classes, whether they be literary or mathematical. The practical men who backed Riedler (the *Praktikerfraktion*) argued that mathematical teaching should be cut down to what was, in their opinion, immediately useful.[11]

Third, and finally, in 1899, in a measure backed by Kaiser Wilhelm II, the THs were finally granted the right to issue doctoral degrees, hitherto the prerogative of the universities. As a consequence the status, influence, and size of these technical institutions increased steadily in the years leading up to the First World War. The engineering profession was, in many ways, still a divided and fractious body, but in the course of the expansion, the anti-mathematical movement lost much of its force. The alliance of industry and sophisticated science became increasingly acknowledged as an economic and military necessity. The emergence of aviation and the rapid uptake of this subject in the THs helped to consolidate the position of the applied math-ematician and swing the pendulum back to a less hostile stance toward math-ematically formulated theory.[12]

In his important study of engineers in German society, *New Profession, Old Order*, Kees Gispen quotes, and expresses agreement with, "a certain Friedrich Bendemann," writing in 1907, who commented on this swing back and forth between theory and practice and declared that it was time to redress the present imbalance and reintroduce more theoretical training.[13] Though Gispen does not mention it, the Herr Bendemann in question, who had re-ceived his doctorate from the TH in Charlottenburg, was a significant force in the aeronautical world. He was a specialist in aircraft engines and propel-lers. In 1912 he was to become the director of the Deutsche Versuchsanstalt für Luftfahrt at Adlershof outside Berlin.[14] Bendemann's 1907 comments were a direct, and face-to-face, riposte to Riedler. They suggest the growing con-fidence of the aeronautical community in the THs in the face of old schisms and old campaigns.[15] Those involved with airships and airplanes were begin-ning to think of aeronautics as a natural home for what von Parseval called the "gebildete Ingenieure," that is, the educated or cultivated engineer whose thinking, by definition, combined both theory and practice.[16]

Kutta's career thus began amid some of the more acrimonious attacks on mathematicians, but he was fortunate to be sheltered from the worst ex-cesses of the *Theorie-Praxis-Streit* by the special situation in Munich.[17] The mathematicians at the THM had long made efforts (though with varying de-grees of determination and success) to accommodate the needs of engineers. They had cultivated a geometrical, visual, and concrete mode of teaching. The trend had started when Felix Klein held a chair at the THM and was continued by his successor Walther von Dyck, who was appointed in 1884 at the age of twenty-seven.[18] Von Dyck had been Klein's pupil and remained a friend and confidant. It has been said that von Dyck played an analogous role in South Germany to Klein's role in North Germany.[19] Von Dyck wanted the THM to be an institution of high scientific merit as well as being technologically oriented. He was able to call upon the support of mathematically sophisticated members of the more technical departments at Munich, such as August Föppl, who likewise had no time for the simple *Praktikers*.

Kutta was von Dyck's teaching assistant and frequently took on his classes when von Dyck became involved, as he increasingly did, with running the THM. Kutta also worked with Sebastian Finsterwalder (1862–1951), who held a mathematics chair at the THM. Finsterwalder was significantly more oriented to applied work than von Dyck and has been called "der Prototyp des 'Technik-Mathematikers'"—the prototype of the technologically oriented mathematician.[20] As early as 1893 Finsterwalder was giving lecture courses on the application of differential equations to the problems of technology. He was also an aeronautical enthusiast and a member of the local ballooning club.[21] It was Finsterwalder who suggested that the topic of Kutta's *Habilitationschrift* should be the mathematical analysis of the flow of air over an aircraft wing. This may be guessed from Kutta's thanks to Finsterwalder, but the colleague who wrote Kutta's obituary endorsed the point.[22] He said that the stimulus for the chosen topic would, in any case, be clear:

> das ist aber für denjenigen auch klar, der die Jahre kurz nach 1900 im Mathematischen Intitut der T.H. München miterlebte. Von Finsterwalders regem Interesse an den aerodynamischen Grundlagen der damals in den ersten Anfängen stehenden Luftfahrt wurden auch die jüngeren Kräfte am Institut angesteckt. Ich denke noch daran, mit welchem Interesse Photographien der ersten Flüge—heute würde man bescheidener sagen: Sprünge—die Farmen mit seinem Aeroplan bei Paris ausführte, studiert und ausgemessen wurden, Photographien, die Finsterwalder mitbrachte: es wird so 1906 oder 1907 gewesen sein. (50)

> quite clear to anyone who had been at the Mathematical Institute at the TH Munich in the years after 1900. Finsterwalder's avid interest in the aerodynamic basis of the first beginnings of aviation at that time also infected the younger people at the institute. I think of the interest with which the photographs of the first flights—today one would more modestly say jumps— were studied and measured. These photographs of Farman with his airplane in Paris, which Finsterwalder brought back with him, would have been in 1906 or 1907.

Finsterwalder's suggestion to Kutta must have been made some four or five years before the episode with the photographs recalled by Pfeiffer, and thus before the first powered flights had been made. At this earlier date Finsterwalder would have been preparing his chapter on aerodynamics for Felix Klein's encyclopedia of the mathematical sciences.[23] The aeronautical adventures that were attracting attention at that time were the experiments with hang gliders of the kind pioneered by the engineer Otto Lilienthal. Lilienthal had been killed in a flying accident in 1896 but had left a legacy of both enthusiasm and information. The information was in his book *Der Vogelflug als Grundlage der Fliegerkunst* published in 1889.[24] Kutta was explicit about the connection between his work and Lilienthal's machines in both the 1902 account and the 1910 paper.[25] The link is clearly evident in the circular arc that Kutta took as his representation of a wing profile. This was not only a mathematical simplification; it also corresponded to the profile used by Lilienthal.[26]

After his successful *Habilitationschrift* Kutta continued as teaching assistant in the TH Munich until 1907. He then became an extra-ordinary professor (that is, an associate professor) in the same institution. In 1909 he moved on to become an extra-ordinary professor at the University of Jena, and in 1910 was appointed as an ordinary professor (a full professor) at the TH in Aachen. Finally in 1911, the year of his second paper on the circulation theory, he settled down as an ordinary professor at the TH in Stuttgart, where he stayed until his retirement. After his two papers on aerodynamics, in 1910 and 1911, he published nothing more, although he did not retire until 1935 and lived until 1944.

Kutta's 1910 Paper

Unlike the 1902 report of his work, Kutta's 1910 paper was long and detailed. It was called "Über eine mit den Grundlagen des Flugproblems in Beziehung stehende zweidimensionale Strömung" (On a two-dimensional flow relevant to the fundamental problem of flight).[27] All the basic reasoning in the paper and much of the detailed working were made explicit. It covered some fifty pages of the proceedings of the Royal Bavarian Academy of Science and involved mathematical techniques that ranged from the general and abstract to the concrete and numerical.[28] Using Kutta's own headings, I shall go over the seven sections of the paper to convey the structure of the argument.

SECTION I. INTRODUCTION

Kutta began with a nod not only to Lilienthal but to more recent developments in aviation. These, he said, gave a great practical significance to the old, but difficult hydrodynamic problem of calculating the forces on a body immersed in a moving fluid such as air. Calculating the lift forces on a wing was particularly important. It should be possible to do this because the relevant flow can be understood ("aufgefaßt werden kann") as the superimposition of a circulation and a steady stream. For an explanation of the basic ideas of the circulatory theory, Kutta directed the reader to the first volume of Lanchester's *Aerial Flight* and a 1909 article by Finsterwalder. Finsterwalder's article was titled "Die Aerodynamik als Grundlage der Luftschiffahrt" (Hydrodynamics as the basis of airship flight), but it also dealt with the basics of the circulation theory of lift.[29] Kutta's choice of words, in saying that the flow *can* be so understood, simply indicates that *if* the flow is interpreted in this way, *then* the lift force becomes intelligible. There is, however, no reason to suppose that he doubted the reality of circulation. He was probably stepping carefully because of the highly artificial character of the concepts he was applying to the problem: friction was being ignored, the flow was to be treated as two dimensional, and the fluid was taken to be free of vorticity (that is, the flow was irrotational).

Kutta then made three observations about his own earlier work. First, he said that in 1902 he had discovered a general theorem about lift which was rediscovered by Joukowsky in 1906. He thus made a priority claim for the result that lift is proportional to circulation and is given by the product of density, velocity, and circulation.[30] Second, he said that the 1902 predictions about lift were supported by Lilienthal's data but acknowledged that the discussion had been confined to wings at zero angle of incidence. This restriction would not apply to the more general analysis he was about to offer. Third, in his earlier account, the magnitude of the circulation could be fixed by specifying that the flow was to be smooth at *both* the leading and trailing edge. This was possible because of the symmetry of the arc-like wing at zero angle of incidence. In the more general treatment, with an arc whose chord was at an angle to the flow, adjusting the circulation could only make the flow smooth at *one* edge, for example, the trailing edge. At the leading edge the fluid would divide, generally at a point on the lower surface, and some of it would be forced to flow around the leading edge. Because Kutta represented the wing by a geometrically thin line, this meant the fluid at the leading edge would achieve infinite speeds and pose a significant problem for the analysis.

Kutta indicated to the reader that the lift force on the wing would have to be broken down into two parts. Leaving the explanation until later, he stated that one part of the lift would be produced by pressure on the surface of the arc, while the other part could be represented as a tangential, suction force at the leading edge. He also signaled his intention to make his abstract, geometrical model of the wing more realistic by studying the effects of rounding off the leading edge.

SECTION II. GENERAL APPROACH

In section II Kutta sketched his mathematical method and introduced some of the basic formulas. His aim was to use a number of conformal transformations. The strategy was to exploit the known flow around a circular cylinder by transforming the cylinder into the arc representing the cross section of Lilienthal's wing. The streamlines around the cylinder would be transformed into the streamlines around the wing. What is more, any circulation that is ascribed to the flow around the cylinder would be transformed into a circulation around the wing. The two steps in the procedure are therefore (1) describing the flow around a circular cylinder and (2) finding the transformation to turn a circular cylinder into a shallow circular arc.

Here is the formula that Kutta gave to describe the most general flow round a circular cylinder, where the circle is inscribed on the ζ- (zeta-) plane:

$$W = c_1\left(\zeta + \frac{1}{\zeta}\right) - i.c_2\left(\zeta - \frac{1}{\zeta}\right) + i.c.\log\zeta.$$

Each of the three parts on the right-hand side of the formula characterized one aspect of the flow around the cylinder. The term c_1 specified the component of flow along the horizontal axis of the ζ-plane, while c_2 gave the component of flow in the direction of the vertical axis. The term c toward the end of the formula (without a subscript) was the constant associated with the circulatory component of the flow, that is, the component of flow in concentric circles around the cylinder. Kutta did not use exactly the same formula as Rayleigh. Whereas Rayleigh worked with the stream function for the flow around a circular cylinder, Kutta worked with the more general complex-variable formula which captured both the streamlines and the potential lines.

The problem was to find the right transformation to apply to the circle. How could the circle be turned into a shape resembling Lilienthal's wing profile? There are no rules for finding the right transformation, and Kutta did not spot any direct way to do this. He therefore had to proceed in a piecemeal way. He constructed the required transformation by combining known formulas that could act as intermediate steps and whose combination had the desired outcome. He built a mathematical bridge between the simple case of the circle and the difficult case of a winglike shape, but he did so by starting at both ends and meeting in the middle. In one direction he went from a circle represented on the ζ-plane (which he called the "transcendental plane") to another shape on the t-plane. In the other direction he went from a simplified geometrical description of Lilienthal's wing, on the z-plane (sometimes, today, called the "physical plane"), to another shape on an intermediate plane called the z'-plane, and finally from the z'-plane to the t-plane. The two procedures therefore mapped their respective starting points onto the same shape in the same plane, the t-plane. This was where they met. Kutta then had the required connection between the circle and the wing.

The insight that allowed Kutta to construct this transformational bridge was that he knew a transformation that would turn an arc of a circle into a straight line and another that would turn the exterior of a circle into the top half of an entire plane. As we saw in chapter 2, a straight line counts as a polygon, so once Kutta had a straight line he could use the Schwarz-Christoffel theorem to link the flow to the simple and basic case of the flow along a straight boundary.

SECTION III. THE CIRCULAR CURVED SURFACE

Kutta now carried out the procedures for which he had prepared the ground. He began on the z-plane and specified the detailed geometry of the wing. It was to be an arc of a circle of radius r subtending an angle of 2α. This gave the coordinates of the endpoints A (the leading edge) and B (the trailing edge). The straight-line distance between A and B was the "chord," and the highest point of the arc was to be 1/12 of the chord. Kutta chose to place this highest point at the origin of the coordinate system. He then began the process of transformation. First he used a transformation in which every point was replaced by its reciprocal. Points on the z-plane were linked to those on the z'-plane by the formula $z' = 1/z$. This had the effect of turning the finite, circular arc into what appeared to be two straight lines. One of them ran parallel to the positive part of the x-axis while the other ran parallel to the negative part of the x-axis. Both were at the same height above the axis. They started at equal distances from the y-axis (that is, there is a gap in the middle), and the lines went off to infinity in opposite directions.

It would have helped the reader of Kutta's paper if, at this point, he had provided a diagram. Given the pedagogic values of the *technische Hochschulen*, he would surely have drawn pictures of such transformations on the blackboard when he presented them in lectures. Most mathematicians reading such a paper would sketch the appropriate figures, at least until the transformation had become routine for them. To help us follow Kutta's argument, I exploit an example of this practice. Sometime in the 1920s a young Cambridge mathematics graduate named Muriel Barker had occasion to work through Kutta's article. She carefully wrote out the reasoning, some-

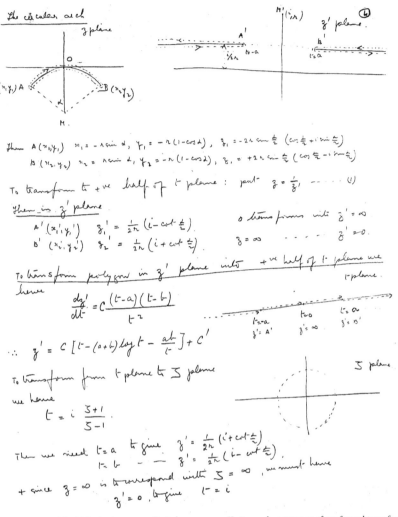

FIGURE 6.2. Muriel Barker's notes on Kutta's 1910 paper. Kutta used a sequence of conformal transformations to map the flow around a circular cylinder onto the flow around a circular arc representing the wing of Lilienthal's glider. (By permission of Dr. Audrey Glauert)

times filling in the steps needed to get from one line to another. She also sketched the conformal transformations. These handwritten notes have survived, and one page from them, containing the sketches, is reproduced here as figure 6.2. Muriel Barker will appear again, later in the story, when the reasons for her interest become apparent. For the moment her notes can help us follow Kutta's thought processes.

On the top left of the page of the notes is a figure labeled z-plane. It is a drawing of the Kutta-Lilienthal wing with the leading edge labeled A and the trailing edge labeled B. The effect of the transformation $z' = 1/z$ is shown next to it in the diagram, on the top right of the notes, labeled z′-plane. Notice how the arc has become two straight lines and the leading and trailing edges A and B of the wing have become the endpoints A′ and B′ of the lines. Following his overall plan, Kutta next mapped these lines onto the t-plane where it would eventually link up with the transformed circle. This he did by using the Schwarz-Christoffel transformation. I have described how this transformation played an important role in the mathematical development of the theory of discontinuous flow. It was central to Greenhill's massive report on this theory for the Advisory Committee for Aeronautics. Kutta used the transformation in a different way and in the service of the circulation theory. He needed it to construct the central arch of his mathematical bridge. The formula of the transformation can be seen about halfway down the page of notes in figure 6.2. It takes the form

$$\frac{dz'}{dt} = C\frac{(t-a)(t-b)}{t^2}.$$

The letter C is a constant, and a and b correspond to the endpoints of the wing. Immediately to the right of the formula is a sketch of the result of the transformation produced by applying this formula. The lines on the z′-plane have become the axis of the t-plane. The new line is shown as dotted in the figure, and the points corresponding to A′ and B′ have been marked in. All that was needed now was to work from the other end in order to map the circle onto the t-plane. The inferential bridge would then have been constructed according to plan. The circle in the ζ-plane is drawn on the bottom right-hand corner of the notes. The formula

$$t = i\frac{\zeta+1}{\zeta-1}$$

is the transformation linking ζ and t. This can be seen in the notes standing to the left of the drawing of the circle. Kutta's aim might be described as getting from the figure at the bottom right to the figure at the top left of the page, but because he could see no way of doing this directly, he made the transition indirectly, by means of the other figures.

Coming back from the Barker notes to the original paper, we see that Kutta was now in a position to evaluate the constants in his formula in terms of the assumed velocity and direction of the free stream relative to the wing. He could also arrive at a value for the circulation on the assumption that the trailing edge is a stagnation point, that is, that the flow does not have to curl around the rear edge. This gave him the following expression for the all-important circulation, which, in the notation used by Kutta, is $2\pi c$. The formula came out as

$$Circulation = 4\pi Vr\sin\frac{\alpha}{2}\sin\left(\frac{\alpha}{2}+\beta\right),$$

where V is the velocity, α the half angle of the arc that constitutes the wing, and β the angle of incidence of the wing to the free stream. The circulation is thus calculable from known or knowable quantities.

SECTION IV. THE LIFT ON THE CURVED SURFACE

Having arrived at the value of the circulation, Kutta immediately multiplied the circulation by the density of the air and the speed of flight to give the lift. He did not state the relevant formula, $L = \rho\,V\Gamma$ (where Γ = circulation), but he used it implicitly. Here, then, was the lift on a Lilienthal-type wing specified in terms of known quantities: ρ, the density of the air; V, the speed of flight; r, the radius of the circular arc of the wing; b, the length of the wing; α, which was half the angle subtended by the arc; and β, the angle of incidence. The formula was

$$Lift = 4\pi\rho V^{2}rb\sin\frac{\alpha}{2}\sin\left(\frac{\alpha}{2}+\beta\right).$$

Kutta did not simply take the general lift formula $\rho\,V\Gamma$ for granted. He announced that he was going to offer a general proof based on energy considerations, which he proceeded to do. The proof not only gave the magnitude of the resultant aerodynamic force as the product of density, velocity, and circulation, but it also carried the implication that the force must be *at right angles to the direction assumed by the free stream at large distances from the wing.*[31] In other words: there was no drag. Given that Kutta was treating the air as an ideal fluid in irrotational motion, this result was a necessary consequence of his premises.

Kutta now had to confront a logical problem. If the fluid is perfect it will slide effortlessly over any material surface. This means that it can only exert a force normally to the surface. Consider a flat plate in a steady flow of ideal fluid and add a circulation around the plate. Suppose that the flow at a distance from the plate is horizontal and that the plate has an angle of attack β to this flow. If the forces on the plate are normal to the plate, then won't the resultant R be normal to it? It will be tilted back at an angle β to the vertical (see fig. 6.3). The resultant R will then have a drag component of $R\sin\beta$. This contradicts the result of the general lift theorem, which Kutta had just proved, where the resultant is vertical, that is, normal to the flow but not

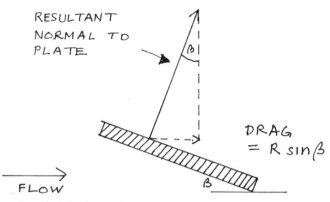

FIGURE 6.3. A paradox? The pressure of an ideal fluid must be normal to the plate. This appears to create a drag and contradict the Kutta-Joukowsky law of lift, according to which the resultant aerodynamic force must be normal to the flow, not the plate.

normal to the plate, so that the drag component is 0. Kutta was primarily considering an arc, not a flat plate, but the same result holds even though the geometry is more complicated. Much of the rest of his paper was spent exploring this apparent paradox.

Kutta said there was no contradiction because the force resulting from the normal pressures was not the only force at work. There must also be another force that operates on the very tip of the plate, hence his remark in the introduction when he said that the lift had two components. Kutta thus identified a suction force that was tangential to the surface at the leading edge. When this force is combined with the normal pressure forces, the resultant is vertical. The forward component of the suction counterbalances the backward component of the pressure forces to produce the zero-drag outcome. Again, the situation can be seen more simply with a flat plate. The tangential suction and the normal pressure forces on the plate are shown in figure 6.4. Introducing the leading-edge suction restores consistency with the results of the kinetic energy proof that Kutta had provided for the law of lift.[32]

Kutta did not treat the leading-edge suction as a mere device to avoid a problem. He proceeded to investigate the flow field near the leading edge by introducing various approximations and assumptions about the shape of the streamlines. An idealized fluid flowing around an idealized, sharp edge would have an infinite speed. This would produce an infinitely large suction force concentrated on an infinitely small area, which suggests that the mathematics would assume the indeterminate form ∞/0. By reasoning that the approximate shape of the streamline would be that of a parabola, Kutta used the results he had already established to argue that the actual force would

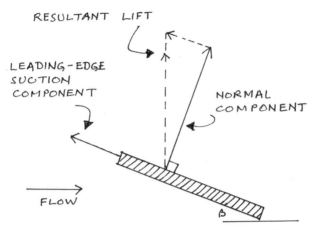

FIGURE 6.4. The "paradox" resolved. There must be another force at the leading edge. The normal pressure on the plate plus the suction force at leading edge give a resultant normal to the flow (but not normal to the plate).

converge to a determinate and finite value. He deduced that this value was exactly that which was required to turn the backward-leaning pressure resultant into a vertical lift and to give it the magnitude predicted by the general lift theorem.

SECTION V. A NUMERICAL EXAMPLE: ROUNDING THE LEADING EDGE

Kutta did not see leading-edge suction as calling for a purely mathematical investigation into a singular point in the equations of flow. He took it as an indicator of a practical problem. It pointed to the presence of high speeds that were physically real and which would result in the breakdown of the flow at the leading edge and the onset of turbulence. It pointed to the presence of vortices and other physical complexities. He therefore looked for an engineering solution by rounding off the leading edge, that is, by the provision of a thickening of the wing at the front which could then be shaped so as to prevent the breakdown of the flow. Here Kutta's mathematics began to make direct contact with the practicalities of building an aircraft wing.

First, he made a rough numerical estimate of the lift of Lilienthal's wing by inserting its specifications into his equations. The wing was an arc subtending an angle of $2\alpha = 37°50'$ and having a curvature giving a maximum height of 1/12 of the chord length. The angle of attack β was set at $\beta = \alpha/2 = 9°27'$. Using the equations he had arrived at by his sequence of transformations, Kutta laboriously calculated the speed of the air and hence the pressure at

22 different points on the upper and lower surface of the wing, beginning at the leading edge and working back to the trailing edge. He assumed a flying speed of 10 m/sec, an atmospheric pressure of 760 mm of mercury, and an air temperature of 10° C. The results were drawn up in tabular form, and Kutta estimated that the lift would be about 10 kg per square meter of wing. This was the sort of result that would enable a wing of practical size to support a significant weight.

Kutta then addressed the practicalities of rounding the leading edge. How much should the front of the wing be thickened? What was the best geometry? Here Kutta felt the force of a dilemma. The suction force, which comes from the high speed of the flow, is important because it cuts down drag. The rounding therefore needs to be slight in order to keep up the speed so that the suction force is maintained. But it must not be too slight, otherwise, in reality, the flow will break down entirely, friction forces will take over, and the suction effect will be destroyed.

> Der Erfolg ist somit, extrem gesprochen, daß im Falle der fehlenden oder gar zu kleinen Abrundung die auf die Schale ausgeübte Auftriebskraft durch den Druckauftrieb allein gegeben ist, da die Saugwirkung fortfällt. (37)

> The result, simply expressed, is that in the absence of a rounding [of the leading edge] or with too small a rounding, the lift force operating on the curved surface will be provided by the pressure forces alone, because the suction effect will have been removed.

He wanted a shape that would change the flow pattern as little as possible from those indicated by the mathematical analysis. For Kutta, the perfect, irrotational fluid flow represented an ideal that should be approached as closely as possible. Thus,

> Wollen wir nur die negativen, physikalisch unzulässigen Drucke vermeiden, und gleichzeitig das Strömungsbild, und damit die Druckverhältnisse in der Nähe des gefährlichen Punktes A möglichst wenig ändern, damit die früheren Formeln noch verwendbar bleiben, so werden wir die Kante parabolisch, nämlich so wie die Strömungslinien bei A verlaufen, abrunden. (31)

> If we want to avoid the physically impermissible pressures near the critical point A [the leading edge], while at the same time altering the flow and pressure relations as little as possible, so that the previous formulas remain applicable, then we should make the edge parabolic, that is, like the streamlines at A.

Using the analysis carried out in the previous section of the paper, and applying it to various forms of leading edge shape (see fig. 6.5), Kutta reached the following conclusions. Given the 2 m chord of Lilienthal's wing, the leading

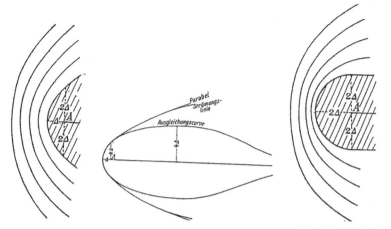

FIGURE 6.5. Kutta's analysis of thickening at leading edge designed to avoid infinite speeds arising from the flow of an ideal fluid around a sharp edge. From Kutta 1910, 32. (By permission of the Bayerische Akademie der Wissenschaften)

edge should have a thickness of 12 cm. The greatest thickness should be about 12–16 cm from the leading edge itself and should merge with the arc of the wing some 40–50 cm from the edge, that is, at about one-fourth of the chord (35). This, he concluded, would produce the optimum leading-edge suction. The surface friction and the vortices coming from the wingtips would still provide some drag, but these effects fell outside the scope of the assumptions he was making.

SECTION VI. FLAT PLATES AND CURVED SURFACES AT VARIOUS ANGLES OF INCIDENCE

Now it was time to compare the theoretical predictions with the results of experiment. Kutta did not perform experiments himself but used existing data. The aim was to see if the predicted relation between lift and angle of incidence was correct. The formula for the circulation shows that the circulation increases with increasing angle of incidence, so lift should likewise grow. Kutta worked out two sets of testable results, one for a flat plate, the other for the curved wing. The two predictions are closely related because the flat plate is just the limiting case of the curved plate. The limiting process greatly simplified the formula and permitted a rough comparison with flat-plate data already published by Duchemin and Langley. The experimental lift was about two-thirds of that predicted by the theory. Kutta declared this "nicht ganz schlecht" (52), which might be rendered as "not too bad." In his discussion of the flat plate, Kutta also established that the center of pressure will be at a point one-quarter the width of the chord from the leading edge and that, unlike for the curved plate, the position stays the same even though the angle of incidence changes.

Coming now to the curved wing, Kutta used Lilienthal's own data, which were generated by experiments on a small model of an arc-shaped wing with a sharp leading edge. Because of the sharp leading edge, Kutta thought that friction effects would be dominant so that the leading-edge suction would be damped down or removed. He therefore compared Lilienthal's measurements with predictions drawn from two different parts of the theoretical analysis. In one case he computed the lift from the general formula showing that the lift = density × velocity × circulation. In the other he used the pressure lift alone (that is, the theoretical lift minus the leading-edge suction). This latter case created a drag because the resultant was tilted backward. On the basis of certain assumptions about the test conditions, Kutta made his predictions for lift and drag for nine different angles of incidence from −9°, through 0°, to +15°. Overall he found that the theoretical predictions of lift were consistently 10−20 percent higher than those arrived at by observation. Kutta concluded:

> Aus der Tabelle scheint also hervorzugehen, daß für die untersuchte gewölbte Fläche und für Luftstoßwinkel unter 15° die beobachtete Hubkraft 80−90 Prozent der errechneten ausmacht—was mit dem Umstande, daß die theoretischen Vereinfachungen sicher auf zu große Zahlen führen mußten, in Übereinstimmung steht. Auch für den Stirnwiderstand ergeben sich einigermaßen brauchbare Zahlen. (54)

> It follows from the table that, for the curved surfaces that were studied, the observed lift force was 80−90 percent of the calculated value for angles of incidence below 15°—which constitutes agreement given that the theoretical simplifications were bound to lead to numbers that were too high. Even the values for the frontal resistance are reasonably useful.

In other words, the theory fitted the data tolerably well given the approximations that had been made.

SECTION VII. CONCLUDING REMARKS

Kutta ended with some ideas about extending his mathematical methods to a variety of different wing profiles with rounded leading edges and with flaps attached to the trailing edge. He mentioned the need to develop a more general form of the Schwarz-Christoffel theorem and indicated the demanding amount of computational effort that would be involved, but Kutta did not feel that the limits of his approach had been reached and hinted at their further application to biplanes.

Kutta's next paper, in 1911, utilized the same mathematical techniques as those adopted in 1910 but it dealt with more complicated cases.[33] The analysis was generalized in two ways. First, Kutta showed how to apply his conformal transformations to an aerofoil whose cross section was composed of not one but two circular arcs in the form of a crescent or sickle shape. Such a sickle-shaped profile was used in the successful Antoinette monoplane, and von Mises has suggested that its practical use was prompted by Kutta's analysis.[34] Second, Kutta generalized the approach in order to describe the flow over a number of wings that could be arranged to make a biplane or a triplane or even a multiwing arrangement in the form of a "Venetian blind." Again, all of these forms had actually been used, or experimented with, by those who tried to build flying machines.

Having made these two outstanding contributions to aerodynamics, Kutta fell silent. He never published anything again. The reason for the silence is unknown.

Locating Kutta

Although Kutta's 1902 thesis appears to have been lost, the historian Ulf Hashagen has discovered the examiner's report in the archives of the THM. Hashagen draws attention to the revealing way in which Kutta's work was described.[35] Finsterwalder, who acted as both adviser and examiner, said that Kutta's thesis was solidly constructed, industrious, and skillful. The calculations presented many difficulties, and these required a detailed knowledge of the theory of functions. But, he went on,

> Erfreulich ist aber auch, daß die Aufgabe noch ein besonderes praktisches Interesse besitzt—wie denn die von Kutta ins Auge gefaßte künftige Lehrtätigkeit gerade auf die Anwendungen der Mathematik sich beziehen soll. Dabei zeigt die gegenwärtige Arbeit, wie auch die früheren, recht guten von ihm verfaßten Abhandlungen und Kuttas ganzer Studiengang, daß er die "Anwendungen der Mathematik" in dem modernen Sinne kennt, welcher sich mit wirklich aktuellen Fragen der Physik und Mechanik beschäftigt, statt—wie dies früher üblich war—nur eine physikalische Einkleidung *rein* mathematischer Untersuchung unter "Angewandter Mathematik" zu verstehen. (257)

> It was also good that the task had a specific practical interest—so it will be just the thing for Kutta to use in the future teaching that he intends to do on the application of mathematics. Like the very good earlier papers that he has authored, and indeed like his whole course of study, this work shows that he knows what it is to "apply mathematics," in the modern sense of those words.

This means getting involved with real questions in physics and mechanics rather than—as used to be the case earlier—dressing up *pure* mathematical investigations in physical clothing and calling it "applied mathematics."

Some mathematicians had been less than genuine in their attempt to accommodate, or be seen to accommodate, the demands for relevance coming from the engineers in the *technische Hochschulen*. Finsterwalder wanted a real engagement. For Finsterwalder, applied mathematics in the "modern sense" would truly embody the ideals of a scientifically oriented technology, and Kutta's aerodynamic work was held up as an exemplary case of the study of mechanics pursued in this spirit.[36]

Finsterwalder's description of the modern spirit sometimes took on a detectably marshal air, as did that of his respected Munich colleague August Föppl.[37] The period around 1910 was one of diplomatic crises and international tension. It was also a time of intense public interest in aviation. There was the triumphant, but troubled, development of Germany's giant Zeppelin airships and Blériot's dramatic flight across the English Channel.[38] The military significance of these events would not have been lost on the attentive Finsterwalder. In the 1909 article on the scientific basis of aeronautics, cited by Kutta, Finsterwalder had already spoken ominously of "the demands of the time" ("Forderungen der Zeit") and the "honorable contest of nations" ("edlen Wettstreit der Nationen"; 32). These were euphemisms for war. Finsterwalder had also noted, pointedly, that in aeronautical matters, unlike nautical ones, all nations could participate equally ("alle Staaten gleichmäßig beteiligt sind"; 31). Given Germany's well-publicized naval arms race with Britain, this comparison could not have been a casual one. It carried the suggestion that what was proving expensive and difficult for Germany in the maritime sphere might be achieved more easily in the sphere of aeronautics because all nations had the same starting point. No wonder the worldly Finsterwalder was now encouraging Kutta to take up aerodynamics once again and prepare an extended version of his old research for publication.

Kutta's work has achieved a classic status, and he has been accorded eponymous honor. The law relating lift and circulation bears his name, and the condition of smooth flow at the trailing edge is frequently referred to as the Kutta condition.[39] His identification of a wing with a geometrical arc can be seen as a precursor to what is called the theory of thin aerofoils.[40] But Kutta's mathematical techniques have not entered the textbook tradition. His work is never described in his own terms; it is always reworked by means of later techniques. Finsterwalder did, however, encourage Wilhelm Deimler, an assistant in Munich, to calculate and publish the precise pattern of stream-lines around Kutta's arc-like wing.[41] Apart from this, and one unpublished doctoral dissertation, there appears to have been little by way of follow up.[42]

Looking at Kutta in retrospect, we may also remark that the empirical support he claimed is open to question. The quality of the data he used was not good. Kutta was aware of some of these problems, though not all of them. One obvious problem was that Kutta's theory was two-dimensional, whereas Lilienthal's data were three-dimensional. Lilienthal's experiments were also conducted in a natural wind and therefore depended on the calibration of the anemometer. Unfortunately, Lilienthal was depending on inaccurate data from other experimenters to enable him to read the wind speed. Finally, when Kutta applied his own theory to a flat plate, he assumed a smooth flow in which the air stayed close to the surface, but the flat-plate data from Lang-ley and others did not satisfy this condition.

These shortcomings have been identified in *Early Developments of Modern Aerodynamics* by Ackroyd, Axcell, and Ruban. The book provides a transla-tion of Kutta's papers of 1902 and 1910 and a valuable commentary from the standpoint of today's aerodynamics.[43] The commentary contains a brief, and negative, evaluation of Kutta's discussion of the rounding off of the leading edge. Kutta is said to have embarked on this project "rather fruitlessly, as subsequent events were to prove" (185). Looking back with the benefit of hindsight this is fair comment. For the purpose of locating Kutta historically, however, it is important not to pass over the clue that this expenditure of ef-fort can give us, however misguided it now seems. In this part of his paper Kutta was engaging with a question that confronted those who were design-ing and building wings. What sort of leading edge should they give the wing? Fruitful or fruitless, Kutta's attempt to grapple with this question provides evidence of the engineering orientation that Finsterwalder wanted. Although today's reader may be tempted to hurry past these sections of Kutta's paper, for Finsterwalder they were evidence of the intimate relation between Kutta's mathematics and technology.[44]

The point deserves emphasis. If we are to understand Kutta historically, we must keep in mind the following characteristics of his work: (*a*) its focus on a specific, technical artifact, namely, the wing of Lilienthal's glider; (*b*) his concern with issues of optimization and trade-off, for example, with regard to the amount of rounding off of the leading edge; (*c*) his willingness to use highly artificial theoretical tools such as perfect fluid theory; (*d*) his aware-ness of the limitations of these conceptual tools but a willingness to postpone asking and answering certain questions, for example, about the origin of the circulation; and (*e*) a determination to bring the theory, however unrealistic, into direct contact with data at the numerical level. This combination of mathematical methods with an engineering orientation placed Kutta's work in the category that Finsterwalder called "modern" applied mathematics. Also, when dealing with turbulence and other complications in the flow, Kutta, in his 1910 paper, used a version of what von Mises called the hydraulic hypothesis. Kutta went on:

> Dennoch halte ich es für möglich, daß diese komplizierten Erscheinungen sich über das hier beschriebene Strömungsbild nur superponieren, und die durchschnittlichen Druck-und Geschwindigkeitsverteilung—besonders die erstere—der geschilderten nahe steht. (51)

> I therefore think it is possible that these complicated phenomena are merely superimposed on the flow formations described here, and the average pressure and speed distributions, especially the former, are close to those that have been presented.

Pioneering work also raises questions about its origins. Consider the breakthrough Kutta made in his 1902 *Habilitationschrift*. His appeal to the circulation theory of lift was made some years before the publication of Lanchester's work. How then did Kutta come to make the link between circulation and lift? From where did he get his ideas? Of course, Finsterwalder may have pointed out the role of circulation when he suggested the research topic to Kutta, but this conjecture only postpones the problem. How did Finsterwalder get the idea?

The proper response is to see that there is something wrong with the question. Neither Kutta nor Finsterwalder *needed* to think up the idea. It was common knowledge that a force can be generated by combining a uniform, rectilinear flow with a circulating flow. This had long been known in Cambridge and would have been equally well known in Munich. It was not the *availability* of the idea that constituted a problem; it was the willingness to *use* it. The thing that needs explaining is not how the idea was generated but why some people saw it as useful while others saw only trouble and futility. It is the mobilization of action that constitutes the real puzzle, not the origin of ideas.

Kutta clearly felt free to use the simple model of circulation in a perfect, irrotational fluid. He did not feel compelled to stop in his tracks because he could not explain how the circulation might arise. He knew that in using the circulation model he had to make physically false assumptions about viscosity. He might have been, but was not, deterred from going ahead on this basis. He might have concluded that he should have been devoting his efforts exclusively to a theory of viscous fluids, but he did not. This is the point. Kutta was prepared to go ahead, while others, such as G. I. Taylor in Cambridge, saw no point in taking this route. This was no personal idiosyncrasy on Kutta's part, any more than the opposite response was a personal idiosyncrasy on Taylor's part. Kutta was simply going with the flow or, at least, with the local flow. He was in a context where this course of action made sense and, if attended by a measure of success, would produce rewards rather than puzzlement.

Recall that August Föppl spoke of the necessity for the engineer to get answers, even if it meant using ideas that could not be justified within the broader framework of established, physical knowledge. One such scientifically unjustifiable idea was that of the perfect, inviscid fluid. Here we have an explanation of the prima facie oddity that it was engineers who displayed the greater tolerance of ideal fluids, while the physicists proved intolerant of them. The engineer's commitment to practicality does not mean that fictional concepts have to be shunned. Rather, it is the physicist's commitment to truth that makes such fictions so unpalatable. In this respect, the engineer's sense of necessity, and the shared awareness of this within the profession and its supporting institutions, generated a certain freedom of action. It meant that a pragmatic move or an expedient step did not meet with immediate criticism, provided its engineering utility could be demonstrated. On the contrary, if such a move contributed to the engineer's project, it would be met with encouragement. Encouragement was precisely what Finsterwalder gave to Kutta. Kutta's willingness to venture where others would not venture becomes intelligible when set in the institutional context of *technische Mechanik*.

Joukowsky's Transformation

In 1910 the Society of German Aeronautical Engineers began to publish the *Zeitschrift für Flugtechnik und Motorluftschiffahrt*—the journal for aeronautics and motorized airship transport.[45] The *ZFM*, as it was called, rapidly became the leading scientific publication in the field. There was no precise British equivalent. The *ZFM* was more technical than, say, the *Aeroplane* or *Flight* and yet more accessible, and certainly more diverse, than the Reports and Memoranda of the Advisory Committee for Aeronautics. As well as scientific reports it contained general survey articles on the state of aviation, accounts of the latest exhibitions and meetings, and reviews of recent publications. There was, however, no close reporting of political controversies of the kind that was conspicuous in the British aeronautical press. Perhaps the nearest British publication was the *Journal of the Aeronautical Society*, but unlike the *ZFM*, this was not a routine vehicle for publishing research results.[46] The lack of any British equivalent hints at the different ways aeronautical knowledge was integrated into the institutions of the two countries. Those who wrote for the *ZFM* communicated across boundaries between theorists and practitioners that seemed more difficult to overcome in Britain, while their silence in the domain of politics shows that there were other boundaries that remained higher in Germany than in Britain.

The advisory board of the *ZFM* was impressive. The journal was edited by the Berlin engineer Ansbert Vorreiter, and the scientific side was under the guidance of Ludwig Prandtl in Göttingen. Alongside Prandtl the board contained Carl Runge, also of Göttingen, along with Finsterwalder of the TH in Munich, Reissner of the TH in Aachen, and von Parseval and Bendemann from the TH in Charlottenburg. The masthead of the journal also carried the name of Dr. N. Joukowsky. His affiliation was given as the University of Moscow and the *technische Hochschule* of Moscow.

In the issue of November 26, 1910, the *ZFM* published the first part of a two-part article by Joukowsky titled "Über die Konturen der Tragflächen der Drachenflieger" (On the shape of the wings of aircraft).[47] The article has come to occupy a notable position in the history of aerodynamics. It is cited as the source of an important methodological shift in the mathematics of lift. The shift consisted in replacing Kutta's complicated conformal transformations with a single, simple transformation, now called the Joukowsky transformation. Not only was it simpler, but it produced a more realistic aerofoil shape. Kutta's method had merely produced a geometrical arc. The arc was an adequate model of Lilienthal's wing, but it did not capture the increasing use of wings with a rounded, rather than sharp, leading edge as well as a slender tailing edge. Kutta had established the logic of the process by which knowledge of the flow around a circular cylinder could be turned into knowledge about the flow around a wing. The next step was to refine and improve this method of analysis. It is in this connection that Joukowsky's paper has, rightly, achieved the status of a classic.

A reader who is aware of its reputation, but who confronts Joukowsky's paper with fresh eyes, might feel puzzled. Where is the bold simplifying stroke? The inner coherence of the mathematics of the infinite wing, so evident in the textbooks that emerged a few years later, is not to be seen. The argument of the paper lacked clarity, and Joukowsky cited formulas without proof and used them without adequate explanation. There was also an edgy concern with issues of priority, particularly Russian priority, and some distracting typographical errors. The formula in the theory of complex variables that is now called the Joukowsky transformation was not actually stated in the paper, although some of its immediate consequences were given a limited application. But any inclination toward disappointment should be resisted. The smoothness of the later analysis is indeed absent from the paper, but that is because the later analysis was the work of others who learnt from Joukowsky and carried his ideas further. It was a collective, not an individual, accomplishment.

What was Joukowsky's own contribution? I answer this question by giving an analysis of the argument of the 1910 paper. Joukowsky began by stating, without proof, two formulas for the lift, P, of an aerofoil that takes the form of a circular arc. The first was for an arc at zero angle of incidence; the second was for an arc at the arbitrary angle of incidence β. The formulas were

$$P = 4\pi a \sin^2 \frac{\alpha}{2} \rho V^2$$

and

$$P = 4\pi a \sin \frac{\alpha}{2} \sin\left(\frac{\alpha}{2} + \beta\right) \rho V^2,$$

where V is the free stream velocity, ρ the density, a the radius of the circular arc, and α is specified as half the angle subtended by the arc at the center of the circle. Clearly, the two expressions become the same when $\beta = 0$. Kutta published both of these formulas, the first, simpler one in 1902 and the second, more general one in 1910, the same year as Joukowsky's paper.[48] Joukowsky, however, said that his colleague Sergei Tschapligin had by this time already discovered the second formula.[49]

Next came a discussion of the general lift theorem $L = \rho V \Gamma$. Here a full proof was provided. Joukowsky approached the problem in terms of the flow of momentum across a control surface. His proof was of a type that has now become standard in modern textbooks. Again, Joukowsky raised issues of priority. He allowed that Kutta discovered this theorem in his unpublished thesis of 1902 but pointed out that he, Joukowsky, in 1906, was the first to publish it.[50] He also noted that Finsterwalder had accepted this priority claim.[51] Joukowsky granted that Lanchester had been the first to explain the relation between two-dimensional and three-dimensional flow by introducing the trailing vortices. At this point the main business of the paper was announced. In studying the problems of Kutta flow, said Joukowsky, he had found contours of a winglike form ("von flügelartiger Form") that did not, like Kutta's arc-wing, give rise to infinite velocities at the leading edge. The aim of the paper was to show how to construct these contours and to test their properties empirically:

> Die Beschreibung der Konstruktion dieser Konturen und die experimentelle Untersuchung der ihnen entsprechenden Widerstandskräfte der Flüssigkeit stellt den Inhalt dieser Arbeit dar. (283)

The content of the work can be represented as the description of the construction of these contours and the experimental study of the corresponding resistance forces of the fluid.

Joukowsky set out, step by step, a geometrical procedure for transforming a circle into the first of his two contours. Whereas Kutta had employed functions of a complex variable, Joukowsky took his readers back to the geometry lessons of the classroom. The procedure involved drawing circles and tangents, labeling significant points and angles in the figure, carrying out some careful measurements on the diagram, and then adding construction lines. To start the process, said Joukowsky, it is necessary to draw a circle whose center is labeled O and whose radius is a. Some arbitrary point E is then chosen which lies outside the circle, and from E two tangents are drawn. The angle enclosed by the tangents at E is called 2α. It is then required to draw a second, larger circle whose radius is called b. The larger circle does not share the same center O as the smaller circle. Rather, its position is determined by the requirement that it encloses the smaller circle but touches it so that it shares one of the tangents. It is this larger circle that is to be transformed into the aerofoil.

The next step was the addition of construction lines. These are needed to connect any specified point M on the larger circle to a corresponding point M' which will lie on the aerofoil. Joukowsky specified which lengths and angles to measure and explained how to use the results to arrive at the position of M'. By selecting, say, ten or twenty representative points around the circle, and following the instructions, the result is ten or twenty points that form an aerofoil shape. The more points that are transformed, the more accurately the outline of the wing emerges.

Joukowsky's own finished diagram is reproduced here as figure 6.6. It looks complicated, but it is not difficult to identify the two main circles and the tangents, meeting at E, which were needed to start the construction. The resulting aerofoil shape can be discerned draped over the top of the diagram with its sharp tail at point C, on the left-hand side, and its rounded nose at M' on the right-hand side. The aerofoil that Joukowsky chose to construct for purposes of illustration has a marked camber and is very thick. This makes it look unrealistic, but such a degree of curvature and thickness is not intrinsic to the method. Joukowsky explained that the shape of the wing is determined by the three parameters, a, b, and α. As the circles are made larger or smaller and the point E is moved closer to, or farther from, the circles, so the shape of the wing is modified, and it can be made more rounded or more slender. In the limit, as $b \to a$, and the larger circle comes ever closer to the smaller

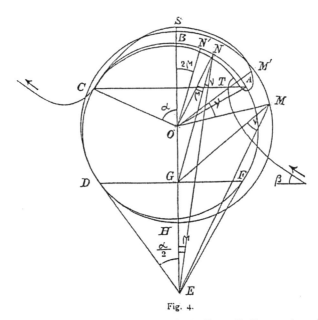

Fig. 4.

FIGURE 6.6. Joukowsky's geometrical construction of a winglike profile. The strongly cambered profile stretches across the top of the figure, having its trailing edge near the letter *C*, on the left, and its leading edge near the letter *M'*, on the right. From Joukowsky 1910, 283. (By permission of Oldenbourg Wissenschaftsverlag GmbH München)

circle, the profile of the wing becomes so thin that it turns into an arc. In fact, it turns into Kutta's arc.

Joukowsky then showed how to construct his second contour. He gave another set of instructions, this time involving trigonometry as well as geometry. Again the process started from two circles, one of radius *a*, and one of radius *b*, with *b* > *a*. The circles have their centers on the x-axis, and so their point of contact must also lie on the x-axis. Using the center of the smaller circle as the origin *O*, each point *M* on the larger circle can be specified by measuring the length *r* of the line joining *O* to *M* and the angle θ between the line *OM* and the x-axis. Joukowsky gave the rules for transforming a point *M* into the corresponding point *M'* on the contour that is to be constructed. The rules gave the x- and y-coordinates of *M'* in terms of the values of *r* and θ that specified *M*. Thus,

$$x = \frac{1}{2}\left(r + \frac{a^2}{r}\right)\cos\theta \text{ and}$$

$$y = \frac{1}{2}\left(r - \frac{a^2}{r}\right)\sin\theta.$$

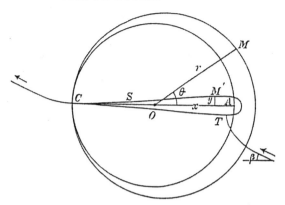

FIGURE 6.7. Joukowsky's second construction gave a strut-like or rudder-like shape. From Joukowsky 1910, 283. (By permission of Oldenbourg Wissenschaftsverlag GmbH München)

Figure 6.7, taken from Joukowsky's paper, shows that the larger circle b is transformed into a streamlined, rudderlike shape lying symmetrically along the x-axis. The thickness of the rudder depends on the relative size of the circles. As $b \to a$, the rudder gets thinner and eventually turns into a straight line of length $2a$ lying along the x-axis. The arc and the line that constitute a sort of skeleton for the thicker shapes were referred to by Joukowsky as the "bases" of his contours.

The second, empirical, installment of Joukowsky's paper was published in 1912, two years after the theoretical part. He focused attention on two, aero-dynamically important properties of his theoretical contours that could be made accessible to empirical testing. The two characteristics were (1) the angle of zero lift, that is, the small (and often negative) angle of incidence at which the wing first begins to produce lift, and (2) the slope of the graph when the coefficient of lift was plotted against the angle of incidence. Both of these angles could be deduced from the basic principles of the circulation theory. Their analysis proceeds in a similar way for all aerofoil shapes derived from a conformal transformation of a circle.[52] This approach enabled Joukowsky to derive his predictions using Kutta's lift formulas and then make experimental comparisons between models of Kutta's arc-like wing and his own wings and rudders. The predictions applied (approximately) to all the profiles.

To address the angle of zero lift, consider again Kutta's formula for the lift P on a circular arc at an angle β to a flow of velocity V. Kutta found that

$$P = 4\pi a \sin\frac{\alpha}{2}\sin\left(\frac{\alpha}{2}+\beta\right)\rho V^2,$$

where a is the radius of the circular arc of the wing and α is half of the angle subtended by the arc. Assuming that the velocity V is not zero, then, if the lift is to be zero, the term $\sin (\alpha/2 + \beta)$ must equal zero. In other words, β must equal $-\alpha/2$. So $-\alpha/2$ is the angle of zero lift, and it is determined by the geometry of the wing. When Lilienthal selected a wing based on a circular arc, and decided that it should subtend an angle of 2α at the center of the circle, he was implicitly fixing the value of the angle of zero lift. More precisely, he was fixing the angle of zero lift, provided all the assumptions of the theoretical analysis held true. It is striking that such a significant parameter should emerge so readily from the theory, and it was a consequence of the analysis that could be easily tested.

The other angle that interested Joukowsky was the slope of the lift-incidence curve. Joukowsky simplified Kutta's formula by supposing that the arc of his wing could be treated as equivalent to two straight lines, one connecting the trailing edge to the highest point of the arc, the other connecting the leading edge to the highest point. The length of the two lines was designated f, and elementary trigonometry showed that $f = 4a \sin \alpha/2$. Substituting this in Kutta's formula for the lift P gave

$$P = \pi \rho f \sin \left(\frac{\alpha}{2} + \beta \right) V^2.$$

Joukowsky then made two further changes to the formula. First, he replaced the lift by a coefficient of lift called K_y. This was done by dividing both sides of the above equation by V^2 and f. Second, for small angles, the sine of an angle equals the angle itself (measured in radians). The equation then becomes

$$K_y = \pi \rho \left(\frac{\alpha}{2} + \beta \right).$$

Joukowsky noted that the angle $(\alpha/2 + \beta)$ represented the angle of incidence as measured from the line of zero lift. If the approximations are reasonable, and if the theory was on the right lines, this formula showed that a graph of the lift coefficient against angle of incidence should have the slope $\pi \rho$. Joukowsky gave the slope the label K. So here was a second testable prediction. He worked out that for a temperature of 20° and an atmospheric pressure of 760 mm, the slope of the graph should be $K = 0.39$.

Joukowsky had built a wind tunnel in the TH in Moscow. The tunnel had a rectangular, working section of 150 × 30 cm and could achieve wind speeds of up to 22 m/sec. The wing sections under test were suspended vertically, with their ends close to the top and bottom of the tunnel, so that they approximated an infinite wing. The sections were rigidly fastened to a framework, and the forces were measured by the weights that were needed to counterbalance them and keep the framework in equilibrium. The wing and rudder contours to be tested had been constructed so that they accorded with the outcome of the geometrical transformations described in the earlier part of the paper. The wing form had been constructed geometrically using a small circle with radius $a = 750$ mm and with the larger circle of radius $b = 762.5$ mm and an angle $\alpha = 20°$. This gave a much thinner and flatter section than the heavily cambered one shown in the diagram in the first installment of the paper. The more slender of the two rudder shapes was generated from two circles $a = 250$ mm and $b = 260$ mm, whereas the fatter model was based on two circles $a = 250$ mm and $b = 270$ mm.

Joukowsky's graphs of his experimental measurements revealed the familiar pattern when lift and drag coefficients are plotted against the angle of incidence. The lift increased in a roughly linear fashion with angle of incidence up to about $\beta = 15°$, while the drag stayed low until about the same point and then increased rapidly. Joukowsky immediately noted that his coefficients of lift and drag had higher values than those reported by Eiffel for comparable shapes. This sort of discrepancy between the wind tunnels in different national laboratories was to plague experimental work for many years. In this case Joukowsky suggested that the Moscow experiment approximated more closely the infinite wing assumed in the theoretical calculations. The important question, though, was whether his experimental graphs corroborated the theoretical predictions.

Joukowsky found that the angle of zero lift for his theoretically derived wing profile fitted more closely to the predicted value than did the Kutta-like arc that Joukowsky called its "basis" or skeleton. But even the model wings that were meant to conform to the Joukowsky profile did not achieve quite the predicted degree of lift. The wing ceased to give lift at $-6°$, and the circular arc that was its basis at around $-4°$ compared with a theoretical value of $(\alpha/2) = -10°$. Some of his computed values of the slope K, however, were very close to the predicted value where $K = \rho\pi = 0.39$. Thus he reported that $K = 0.38$ for the arc, $K = 0.37$ for the wing, but only $K = 0.30$ for the rudder.

The wind tunnel at the Moscow TH was soon to figure again in the pages of the *Zeitschrift für Flugtechnik*. In June 1912, Joukowsky's assistant G. S. Loukianoff published graphs showing the lift, drag, and center-of-pressure characteristics of the wing contours of seven types of aircraft that were currently flying with success: the Bréguet, Antoinette, Wright, Blériot, Farman, Henriot, and Nieuport machines.[53] As von Mises observed, these early Moscow experiments gave a slope for the lift-incidence curve that closely corresponded to the theoretical value, though later experimenters found a slightly smaller value. In general, said von Mises, two-dimensional wing theory overestimates the slope by about 10 percent and underestimates the angle of zero lift by one or two degrees.[54] But it was the theoretical achievement, rather than the experimental work, that proved most significant. Joukowsky's aerofoils, the J-wings, as they were sometimes called, aroused an immediate and positive response in Germany. The interest in the theory was not abstract, aesthetic, or otherworldly. Joukowsky's theoretical profiles became the focal point for a series of developments that brought the mathematical analysis of lift into intimate contact with both physical reality and engineering practice.

Blumenthal Brings Unity

Otto Blumenthal had been Hilbert's first doctoral student at Göttingen and continued to help Hilbert edit the distinguished journal *Mathematische Annalen*.[55] In the winter semester of 1911–12 Blumenthal, now at the TH in Aachen, gave a course of lectures on the hydrodynamic basis of flight. He described, mathematically, the irrotational flow of an ideal fluid over a range of different Joukowsky profiles. Along with his colleagues at Aachen, Karl Toepfer and Erich Trefftz, he drew up diagrams of the precise shape of the profiles. The result of the joint work was published in two papers in the *ZFM* for 1913. The main paper, by Blumenthal, was titled "Über die Druckverteilung längs Joukowskischer Tragflächen" (On the pressure distribution along Joukowsky wings).[56] It was followed by a short note by Trefftz giving a simplified geometrical method for drawing Joukowsky profiles and a graphical technique for rapidly computing the predicted air velocities, and hence pressures, on the surface of the wing.[57]

Blumenthal began by drawing attention to a unifying principle that had not emerged in Joukowsky's original paper. Joukowsky had used two geometrical constructions. The first, which was the more complicated, generated the wing profile, while the second, which was simpler, generated the symmetrical rudder. Blumenthal pointed out that only the second of the two constructions need be used. What is more, the process could be represented by a simple mathematical formula. This formula was the version of the Joukowsky transformation that was to achieve such fame.[58] The formula can yield wing shapes and curved, Kutta-like arcs as well as rudder shapes and flat plates. Only one transformation, not two, was needed. It was all a matter of the position of the circle on the coordinate system of the plane that was to be transformed. The totality of Joukowsky contours, said Blumenthal, could be generated by the set of all circles that can be drawn on the $\zeta = \xi + i\eta$ plane that pass through $\xi = -l/2$, provided they either pass through, or contain, the point $\xi = +l/2$. All that is required is that the circles are then subject to the transformation:

$$z = \zeta + \frac{l^2}{4\zeta}.$$

Those circles that pass through both $\xi = -l/2$ and $\xi = +l/2$ will have their centers on the η-axis and will generate arcs similar to Kutta's wing. The one circle in this family that has its center precisely at the origin, and hence has the line from $\xi = -l/2$ to $\xi = +l/2$ as its diameter, will be transformed into the straight line that is the limiting case of the arc. Wing shapes will be generated by all of the (off-center) circles whose circumference passes through $\xi = -l/2$ but contains $\xi = +l/2$, that is, which are sufficiently large that the circumference goes around the point $\xi = +l/2$. The sharp trailing edge of the wing will be the transformation of the point $\xi = -l/2$, and the curved leading edge will go round the transformation of the point $\xi = +l/2$. As a point moves around the circumference of such an off-center circle, the transformation will trace out the curve of an aerofoil shape with a rounded nose and an elongated tail.[59] These, said Blumenthal, are "the Joukowsky figures in the proper sense" ("die Joukowskischen Figuren im eigentlichen Sinne"; 125).

It was Blumenthal who provided the unity lacking in Joukowsky's original paper but which, today, is so often taken for granted. But Blumenthal's aim was not merely to achieve a formal unity. He was bringing the generation of Joukowsky figures under intuitive control in order to facilitate their practical use. He isolated the features of the construction process that had an aero-dynamically significant effect on the overall geometry of the wing. Where Joukowsky had merely said that the geometrical construction of the wing depended on an angle and two lengths, Blumenthal identified the results of the choices that are to be made.

Blumenthal referred his readers to the diagram reproduced here as figure 6.8. The circle in the figure has center M and passes through the point H, which is at a distance $l/2$ from the origin O. (Notice that $l/2$ featured in the formula that Blumenthal chose to specify the transformation.) The off-center circle in the diagram is to be transformed by means of the Joukowsky formula and turned into a wing profile. The point H (sometimes called the "pole" of the transformation) is to be transformed into the all-important trailing edge. The radial line from M to H cuts the vertical axis at a point labeled M'. Then, explained Blumenthal, the distance OM' (labeled $f/2$) controls the *height of curvature* of the wing, while the distance $M'M$ (labeled δ) controls the *thickness* of the wing. In general, if the center of the circle to be transformed is on the positive vertical axis, the result is one of Kutta's arcs; if the center is on the positive horizontal axis, the result is a symmetrical rudderlike figure; if it is somewhere in between (as in fig. 6.8), the result will be a curved profile of the characteristic Joukowsky type. How curved and how rounded will depend on the factors that Blumenthal had just identified.

Blumenthal gave four examples of Joukowsky profiles to show the effects of modifying these parameters, that is, the effect of moving the center of the circle while ensuring that its circumference still passed through H. Thus the curvature parameter (expressed as the ratio f/l) was given the value of 0, 1/10, 1/5, and 1/5 (again), while the thickness parameter (expressed as the ratio δ/l) was set at 1/10, 1/20, 1/20 (again), and 1/50. The effect of these choices was clearly visible as the Joukowsky profiles that he illustrated went from a symmetrical shape to a markedly curved shape and from fat to thin.

From the velocity q of the flow (provided by Trefftz's speedy method of graphical calculation), the pressure on the surface of the aerofoil could be

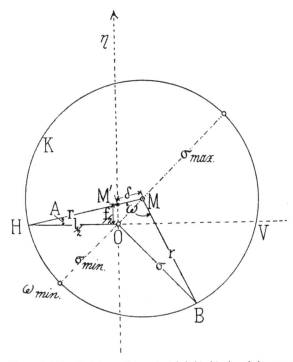

FIGURE 6.8. Blumenthal identified the unifying principle behind Joukowsky's separate treatments of the arc, the symmetrical rudder shape and the curved winglike shape. All of the shapes came from the same transformation formula applied to a circle that passed through a fixed point H on the y-axis. The shape produced depended on the position of M, the center of the circle. From Blumenthal 1913, 126. (By permission of Oldenbourg Wissenschaftsverlag GmbH München)

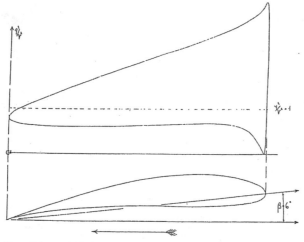

FIGURE 6.9. One of Blumenthal's theoretically predicted pressure distributions along the upper and lower surface of a Joukowsky profile. The part of the graph above the dotted line shows the underpressure (the suction effect) on the upper surface of the wing. The lower graph shows the overpressure on the lower surface of the wing. From Blumenthal 1913, 128. (By permission of Oldenbourg Wissenschaftsverlag GmbH München)

computed, and this led to the most striking feature of Blumenthal's paper. Each of the four Joukowsky profiles that he had constructed was accompanied by a graph showing the theoretical pressure distribution on the upper and lower surfaces. (In all cases Blumenthal assumed that the profile was at an angle of incidence of 6°.) One of his profiles and its accompanying pressure graph is shown in figure 6.9. Summing the areas enclosed by the graphs gave a quantity proportional to the resultant lift. Blumenthal discussed each aerofoil in turn, pointing out the significance of the predicted pressure distribution and its dependence on the parameters of the profile. Some features were common and stood out very clearly in the graphs, for example, the greater contribution of the suction effect on the upper surface of the wing compared to the pressure effect on the lower surface. Others were special to one shape, for example, the presence of a small suction effect even on the lower surface of the symmetrical (rudderlike) aerofoil and the very high speeds at the leading edge of the thinnest profile.

Betz on Pressure Distributions

Knowing the predicted pressure distribution along a specified aerofoil opens up the possibility of subjecting the circulation theory of lift to a demanding empirical test. Does the predicted distribution correspond to the real

FIGURE 6.10. Ludwig Prandtl (*left*) and Albert Betz (*right*) standing in front of one of the Göttingen wind tunnels. In the early years of aerodynamics, Betz was second only to Prandtl in the scope of his theoretical and experimental investigations. (By permission of Zentrales Archiv, Deutsche Zentrum für Luft und Raumfahrt)

distribution in as far as it can be measured? In 1915, two years after Blumenthal's theoretical analysis, a detailed experimental study was published in the *Zeitschrift für Flugtechnik* that was designed to answer this question. The paper was by Albert Betz, Prandtl's close collaborator in Göttingen (fig. 6.10). It was called "Untersuchung einer Schukowskyschen Tragfläche" (An investigation of a Joukowsky wing).[60]

Betz used one of Blumenthal's profiles and worked with a model wing that had a span of 50 cm and a chord of 20 cm. It had a curvature (f/l) of 1/10 and a thickness ration (δ/l) of 1/20 and so corresponded exactly to the second of the four profiles described by Blumenthal (that is, the one shown in fig. 6.9). Betz's aim was to use wind-tunnel data to test Blumenthal's predicted lift and pressure distribution.

The model wing-section was manufactured from metal plate in the form of an airtight, hollow body and made to conform as precisely as possible to the theoretical, Joukowsky profile. Following Fuhrmann's work on model airships, the wing was fitted with bore holes and the hollow interior was connected by a thin pipe leading from the wingtip to a manometer. This enabled pressure measurements to be taken at a number of points on the surface along the chord of the wing. Measurements were taken with one hole at a time exposed while the other holes were smoothly plugged. The line of bore holes was not positioned at the center of the span but was displaced a few centimeters to one side. This was to avoid interference to the flow of air over the holes from the strut that had to be attached to the wing in order to hold it rigidly in place in the wind tunnel. As well as the pressure, Betz also needed to know the overall lift and drag of the wing. For this the wing had to be suspended on a balance so that force measurements could be made. During this phase of the experiment, the pipes leading to the manometer were disconnected and all the bore holes plugged.

It was necessary to make sure that the experimental arrangement provided a good approximation to the infinite wing presupposed in the mathematics. Joukowsky had simply made his wing section run from the top to the bottom of the shallow Moscow wind tunnel. This is the basis of all attempts to realize a two-dimensional flow, but Betz put a lot of effort into refining the technique. His aim was to make the test section as free as possible from disturbing effects produced by the walls of the tunnel and the join between the walls and the ends of the wing. An elaborate system of auxiliary sidewalls, gaskets, and seals was designed and tested to ensure a uniform flow across the experimental cross section of the wing. Once he had an acceptable approximation to two-dimensional flow, Betz's apparatus gave him two sets of data: (1) direct measurements of lift and drag and (2i) pressure measurements distributed over the surface of the wing.

The direct measurements of lift and drag showed the familiar pattern, which partially conformed to, and partially violated, theoretical expectations. The observed lift increased in the predicted, linear fashion with angle of incidence, but only from about −9° up to about +10°, at which point the wing stalled. Even over this range the predicted lift was significantly higher than the observed lift. In fact, the observed lift was only about 75 percent of the predicted value. And, of course, there was an observed drag when, theoretically, it should have been zero. There was also another general feature of the flow that Betz observed. Theoretically each angle of incidence should correspond to one, and only one, value of the lift. Betz found that if he took a sequence of readings in which the angle of incidence was increased in a stepwise fashion, and another sequence in which it was decreased, a given angle might correspond to one value of the lift in one sequence and another value in the other. There were two values of the lift corresponding to each of these angles, not one. This effect was particularly noticeable above the stalling angle. Thus, at +15°, the coefficient of lift had the values of 0.68 *and* 0.55. Such a phenomenon fell wholly outside the scope of the theory.

The most important results, however, were those relating to the distribution of pressure. Here Betz's graphs of the manometer readings showed a definite similarity to the theoretical graphs prepared by Blumenthal and his colleagues at Aachen. Betz's results are shown in figure 6.11. Note that

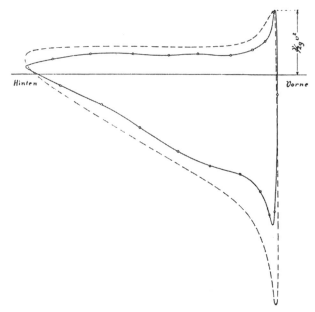

FIGURE 6.11. Betz's pressure graphs for a Joukowsky aerofoil at 6°. Theoretical predictions are indicated by the broken line and experimental results by the solid line. There is a similarity between prediction and observation, but Betz wanted to improve the fit. From Betz 1915, 176. (By permission of Oldenbourg Wissenschaftsverlag GmbH München)

Betz used a convention different than Blumenthal's when drawing his diagrams, and the data for the upper side of the wing are now placed below the base line. Significantly, the general *shape* of the graph derived from theory and that of the graph derived from experiment were the same. Betz, however, pressed the comparison into greater detail. He was interested in getting information about the residual *deviations* between theory and reality, "die Abweichungen der Theorie von der Wirklichkeit" (173). This was the stated purpose of the experiment. He therefore drew attention to where the empirical distribution differed from the theoretical distribution. The *areas* under the empirical graphs were clearly not the same as those under the theoretical graphs. These areas were proportional to the lift, and the theoretical area was significantly greater than the observed area. This was consistent with the fact, mentioned earlier, that the directly measured lift was less than the theoretically predicted lift. What was the cause of the difference, and what should be done in response to it?

Like Fuhrmann, Betz located the source of the difference between the theoretical and empirical flow in the tail region. Theoretically, if the Kutta condition is satisfied and the wing profile is that of a pure Joukowsky contour, and if the air acts like an ideal fluid, then the flow along the upper surface will meet the flow along the lower surface in a smooth way at the trailing edge. In reality, however, the air did not behave in this way at the trailing edge, so Betz made a conjecture. It was a characteristically Göttingen conjecture (see fig. 6.12).

Betz suggested that, although the flow of air along the lower surface runs smoothly along the common tangent, the flow along the upper surface does not. Rather, it detaches from the upper surface before reaching the trailing edge, and this leaves a gap between the two flows. The intervening space between the flows, said Betz, constitutes a turbulent wake filled with "Kármán vortices" (177).

Betz argued that the effect of this separation is twofold. First, it disrupts the pressure relations in the vicinity of the trailing edge and disturbs the equilibrium between the forward-pointing and backward-pointing components of the pressure distribution. Since it is this equilibrium that generates the zero drag of an (ideally) efficient aerofoil, the disturbance must be a contributory cause of the observed drag. Second, the vortices in the wake draw off energy, and this has the effect of lowering the circulation around the aerofoil and hence diminishes the lift. Betz conceded that it was difficult to analyze these processes rigorously but suggested a simple (and intriguing) way to model the situation using the resources of inviscid theory. He proposed rejecting the Kutta condition and relocating the stagnation point. Whereas Kutta had used the position of the stagnation point to fix the amount of circulation, Betz reversed the process. He used the amount of circulation to fix the stagnation point.

Betz began with the value for the lift at a given angle of attack that he had found in his experiment. He then inserted the value into the basic Kutta-

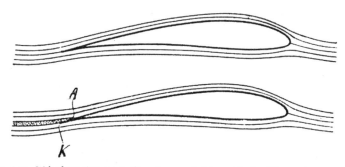

FIGURE 6.12. Wake formation near trailing edge was cited to explain the difference between predicted and observed results. To correct for this error in prediction, Betz abandoned the Kutta condition that the rear stagnation point should be at the trailing edge. From Betz 1915, 177. (By permission of Oldenbourg Wissenschaftsverlag GmbH München)

Joukowsky theorem $L = \rho V \Gamma$. Because he knew the density and velocity of air in the wind tunnel, the formula allowed him to deduce the value of the circulation. He then used this empirical value of the circulation to tell him where the rear stagnation point must be relocated according to the theory of inviscid flow. (Betz was working with a Joukowsky profile, so this point could be calculated by transforming the flow around a circular cylinder.) Given that the empirical value of the circulation is lower than the theoretical value, the stagnation point is on the top surface of the aerofoil, rather than precisely at the trailing edge. It was then possible to replot the theoretical speed and pressure distributions and compare them afresh with the empirical curves. The question was whether the revised location of the stagnation point brought the empirical and theoretical graphs onto closer accord. The method automatically equalizes the areas under the empirical and theoretical curves but the question remains: do the distributions agree? Betz's revised graph, superimposed on the empirical curve, is shown in figure 6.13. The improvement is clear. The graphs are now almost identical. Betz pronounced the agreement to be "extraordinarily good." As he put it,

> Sehen wir von der nächsten Umgebung der Hinterkante ab, für die ja Voraussetzungen der theoretischen Ströming vollständig andere sind wie in Wirklichkeit, so müssen wir die Übereinstimmung der beiden Kurven außerordentlich gut bezeichnen. (177)

> If we disregard the vicinity of the rear edge, where indeed the presuppositions of the theoretical flow wholly differ from reality, then we would have to characterize the agreement of the two curves as extraordinarily good.

Betz had thus brought theory and experiment into closer accord. His procedure involved interweaving them in an interesting way, and the methodological principles implicit in the process are worth looking at with some care.

The Kutta condition had become established as one of the central assumptions of the circulation theory. Could Betz really afford to modify its role in this way? The overriding advantage of the Kutta condition was that it avoided the need for the ideal fluid to move around the trailing edge at an infinite, and physically impossible, speed. How did Betz justify the reintroduction of infinite speeds given that the need to avoid them was routinely cited whenever the Kutta condition was invoked? The point was addressed explicitly in the paper. The argument was as follows: Since perfect fluid theory is known to be unrealistic from the outset, said Betz, one more piece of unreality hardly matters. The entire approach is an artifice, so this should not be disturbing. The infinite speeds are just the way that the complicated, physical

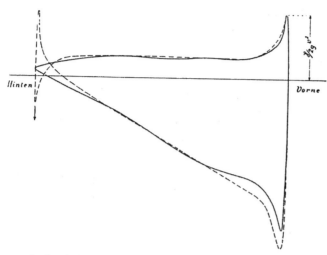

FIGURE 6.13. Betz's revised pressure graphs after modification of the Kutta condition. The observed and predicted curves are now closer. From Betz 1915, 178. (By permission of Oldenbourg Wissenschafts-verlag GmbH München)

processes at the trailing edge receive some manner of recognition within the terms of the theory. Their presence in the analysis simply indicates that further assumptions need to be introduced to mediate between reality and the idealized picture. Outside the wake, the flow can be reasonably modeled by ideal-fluid theory, and the presence of the wake can be taken into account by lowering the value of the circulation. This approach can be seen as the first step toward a better account of the phenomenon. As Betz put it,

> Daß dabei ein Umströmen der scharfen Hinterkante stattfinden müßte, was praktisch unmöglich ist, braucht uns nicht zu stören, da ja die Strömungen ... bis zu dem gemeinsamen Ablösungspunkt nur ein theoretischer Ersatz sind für die in Wirklichkeit vorhandenene Wirbelbewegung. (177)

> That this would have to result in a physically impossible flow round the sharp trailing edge need not disturb us. This is because the flow . . . up to the common point of separation is only a theoretical substitute for the vortex motions that exist in reality.

The assumptions behind this talk of a "theoretical substitute" ("theoretischer Ersatz") can be clarified by noting what Betz had said about d'Alembert's paradox. Notoriously, the theory developed by Kutta and Joukowsky predicted that a wing will have zero resistance. Betz, however, defended the use of an inviscid theory as an approximation, "even though it does not permit any statements to be made about the resistance" ("trotzdem sie über den Widerstand nichts auszusagen vermag"; 173). What did Betz mean by this? Surely, the theory *does* permit a statement to be made about resistance. Indeed, it *requires* that a statement should be made: namely, the false statement that the resistance is zero. Betz knew this, so what he must have meant was that the theory does not permit any *useful* statement to be made. The theory doesn't shed any light on the resistance. His question was: To what practical purpose can the theory be put? The theory was being viewed as a tool rather than a body of propositions. Perfect fluid theory is a useful tool for certain purposes but not for others. Betz was telling his readers that questions about the utility of the theory, rather than its literal truth, should be uppermost in their minds. That is why they should not be unduly disturbed by theoretical deductions that entail infinite speeds.

Generalizing the Mathematics

Not only is a perfect fluid a theoretical substitute for a real fluid, but the geometry of the Joukowsky profile is a theoretical substitute for a real aerofoil section. The mathematics of the Joukowsky transformation of a circle always gives a profile with some highly unrealistic properties. At the trailing edge, the tangents to the upper and lower surfaces of the wing coincide with one another. The trailing edge is like an infinitely thin blade. No engineer would design such a wing, and no workshop could produce one. At most they could produce an approximation of the kind that the Göttingen workshops must have produced for Betz. This raised a question: Could the mathematical advantages of the Joukowsky transformation be retained while avoiding the unrealizable features of the profile? Could a transformation be found that turned a circle into a winglike profile whose trailing edge met at some specified, nonzero, angle? The answer to these questions is yes.

Once again it was members of the *technische Hochschule* at Aachen who provided the answers. In 1918 Theodore von Kármán and Erich Trefftz showed that this job could be done by a transformation from a circle in the ζ-plane to a wing profile in the z-plane that took the form

$$\frac{z-kl}{z+kl} = \frac{(\zeta-l)^k}{(\zeta+l)^k},$$

where k is a constant less than 2 and l is the length that featured in the previous discussions by Blumenthal. Whereas the Joukowsky transformation effectively draws a wing profile that has a circular arc as "skeleton," the

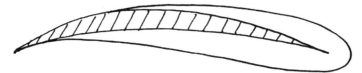

FIGURE 6.14. Kármán-Trefftz profile. Unlike the Joukowsky profile, in the Kármán-Trefftz profile the upper and lower surfaces meet at a nonzero angle at the trailing edge. The "skeleton" of the aerofoil is not the arc of a circle but a crescent. This family of profiles avoids the unreality of the Joukowsky aerofoils, which have an infinitely thin trailing edge.

Kármán-Trefftz transformation produces an aerofoil that has a crescent or sickle shape, made up of two circular arcs, as its "skeleton."[61] Just as the trailing edge of the Joukowsky profile shades into, and becomes, the single arc of its skeleton, the Kutta arc, so the trailing edge of the new profile combines with, and becomes, the endpoint of the crescent (see fig. 6.14).

Von Mises pointed out that the Kármán-Trefftz formula is a close relative of the Joukowsky formula.[62] Starting with a Joukowsky transformation in the form

$$z = \zeta + \frac{l^2}{\zeta},$$

he showed the link in three simple steps. First, subtract the quantity $2l$ from both sides. Second, write down the Joukowsky formula again and, this time, add $2l$ to both sides. Third, form the quotient of these two expressions. The result is another version of the Joukowsky transformation that looks like this:

$$\frac{z-2l}{z+2l} = \frac{(\zeta-l)^2}{(\zeta+l)^2}.$$

The Joukowsky transformation, with its knife-blade trailing edge, is thus a special case of the Kármán-Trefftz transformation, that is, the case where the exponent is $k = 2$. Replacing the exponent 2 by a value of k where $k < 2$, gives the formula for a transformation that generates an aerofoil with a more realistic trailing edge. As k gets smaller, the angle at the trailing edge gets larger.

Von Kármán and Trefftz ended their paper by posing the following question: Given some arbitrary, but plausible, aerofoil shape, is it possible to discover a transformation that will relate it to a circle and thus allow the flow to be predicted? It is one thing to be given, or to discover, a transformation that will go from a circle to an aerofoil-like shape, but starting with an aerofoil and trying to find the transformation is quite a different matter. This is the question that an aircraft designer would pose. What will be the properties of the wing if it is built like this rather than like that?

Von Kármán and Trefftz argued that if a conformal transformation is applied in reverse to some given profile, it may not turn it back into a circle but will turn it into a shape that is not greatly removed from a circle. They then offered a transformation that would, to an adequate degree of accuracy, turn this near circle into a better circle. They thus began to address the way in which ideal-fluid theory could be applied not just to a few favored "theoretical" aerofoils, but to any shape that might come from the drawing board of a designer—shapes that would be strongly influenced by the contingencies of the construction process.

The Kármán-Trefftz transformation showed how to avoid the unrealistic cusp at the trailing edge of the Joukowsky profile, but it did this at the price of a certain complexity. Betz argued that the extraordinary simplicity of the original Joukowsky transformation was worth preserving. The Kármán-Trefftz transformation, he said, was difficult to use in practice. He then exhibited a much simpler way to achieve a finite angle at the trailing edge by a modification of the original graphical method used by Blumenthal and Trefftz. The modification produced a profile with a slightly rounded rear edge, and this again raised the problem of the position of the rear stagnation point. How was the circulation to be determined? Betz declared that from a practical point of view this indeterminacy was of no great significance because the real circulation was always smaller than the theoretical prediction. In reality, even the usual Joukowsky profiles do not unambiguously determine the circulation: "also auch bei gewöhnlichen Schukowsky Profilen nicht eindeutig bestimmt ist."[63] Betz suggested that some point on the rounded edge could be designated to play the role of the sharp edge of the original profile when calculating the circulation.

The cusp on the trailing edge of the Joukowsky profile was not the only problem. There were other respects in which this family of aerofoil shapes differed from those which experience and practice were beginning to favor. Typically, Joukowsky profiles were too rounded and bulky at the front and too thin at the back, even when the zero angle of the trailing edge was avoided. Also, the maximum camber lies near the center of the chord rather than, as was preferred in practice, in the first third of the chord. How were these problems addressed? In a series of articles in the *Zeitschrift für Flugtechnik*, beginning in 1917, Richard von Mises suggested a generalization of the Joukowsky transformation that could yield aerofoils that met almost any specifications of their geometrical properties. Such aerofoils could be designed in a way that

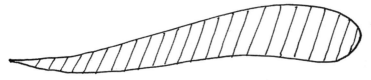

FIGURE 6.15. A von Mises profile has a zero pitching moment for all angles of attack.

avoided the faults identified in the original Joukowsky profiles.[64] Von Mises explored transformations of the following kind:

$$z = \zeta + \frac{a_1}{\zeta} + \frac{a_2}{\zeta^2} + \ldots \frac{a_n}{\zeta^n}.$$

Any aerofoil could be described given a sufficient number of terms in this sequence.[65] The Joukowsky transformation was a special case of the formula for which $n = 1$. Von Mises also wanted to show how the parameters that governed the conformal transformation of the circle were related to the aero-dynamic characteristics of the resulting wing. For example, he showed how to construct a profile in which the resultant aerodynamic force always acts through the same point of the wing, a point that came to be called the aero-dynamic center of the wing. The result was that the pitching moment of the wing was zero for all moderate angles of attack (that is, for the straight-line part of the curve relating lift to angle of attack). This was a property of poten-tial importance for the stability and handling properties of an aircraft. The general shape of a von Mises profile is shown in figure 6.15. Like the Kármán-Trefftz profiles, it avoids the cusp at the trailing edge, but in addition it is characterized by a shallow S-shape with a slight upturn at the rear edge.

Theory and Practice

The way Kutta's creative achievement was reconfigured in terms of the Jou-kowsky transformation, and then subsumed under a sequence of ever more general results, is striking. But generality alone was certainly not the driving force of the development that I have described. The goals that were being pur-sued were not abstract ones. Kutta, Joukowsky, Deimler, Blumenthal, Trefftz, Betz, von Kármán, and von Mises were confronting mathematical puzzles, but their puzzle solving operated within a set of identifiable parameters, and those parameters were set by the practicalities of aeronautics. These men were all aiming to make their mathematical tools work for them so that the ideas involved could be brought into closer contact with the problems faced by engineers who designed wings and built aircraft. Their tools were abstract ones (ideal-fluid theory, conformal mapping, geometry and mechanics), but they were harnessed to engineering goals and exploited or modified accordingly.

The stance the German, or German-language, experts took toward their mathematical apparatus was neither that of the pure mathematician nor that of the physicist. Neither rigor nor purity were central concerns, nor was it their primary goal to test the physical truth of their assumptions. They tested their conclusions for utility rather than their assumptions for truth. Expediency was a prominent characteristic of their mathematical and experimental activity. When Betz looked for deviations between theory and experiment, he was tracking the *scope* of his approach, not trying to expose its falsity (which he took for granted). While no one directly asserted the literal truth of ideal-fluid theory (though Prandtl came close), no one evinced much anxiety about its evident falsity either. Not a single author, in any of the papers described here, even mentioned the problem of how a circulation might arise in an ideal fluid. It was an issue of which they were aware, but it was not a stumbling block.

The particular blend of mathematics and engineering that was visible in Kutta's 1910 paper was sustained throughout all the subsequent developments that have been examined in this chapter. The most vital ingredient in the blend was the orientation toward specific artifacts and the engineering problems associated with them. There is no evidence throughout the developments I have described that practitioners felt the need to make a choice between mathematics and their practical concerns. On the contrary: the former was seen as a vehicle for expressing the latter. Those working in aerodynamics were confident in their ability to combine mathematics and practicality. The continuity and homogeneity of their work suggest an increasingly secure disciplinary identity. Workers in aerodynamics were beginning to form an intellectual community, and they had an institutional basis. Finsterwalder called their discipline "modern" applied mathematics. I have followed August Föppl and brought it under the rubric of *technische Mechanik*.

The particular form of the unity of theory and practice embodied in *technische Mechanik* was eloquently affirmed in a lecture given in 1914 by Arthur Pröll of the TH in Danzig.[66] Speaking at a meeting of the recently formed Wissenschaftliche Gesellschaft für Flugtechnik, Pröll chose as his topic "Luftfahrt und Mechanik" (Aeronautics and mechanics). Pröll surveyed a wide range of topics, including stability and the strength of materials, but he began with the work on lift that had started with Kutta. He described the basic ideas of the circulation theory and reproduced the flow diagrams worked out by Deimler. For Pröll this was a clear illustration of how a "good" theory can work hand in hand with practical concerns ("wie eine 'gute' Theorie mit der Praxis derart Hand in Hand arbeiten kann"). Responding to the rhetoric of the antimathematical movement, he went on:

> Der Kampf ums Dasein mit den Erfordernissen des praktischen Lebens legt auch der wissenschaftlichen Spekulation gewisse Fesseln an und zwingt sie, Überflüssiges oder Unsicheres über Bord zu werfen. Das ist eine erste gute Frucht der gegenseitigen Verständigung von Theorie und Praxis, und eine solche finden wir auch hier bei der Aerodynamik vor. (95)

> The struggle for existence and the demands of practical life impose certain constraints on scientific speculation and force us to throw overboard what is superfluous or insecure. This is the first fruit of the mutual understanding of theory and practice and it is what we actually find here in aerodynamics.

Pröll was not simply reporting a sequence of results in his field. He was making the case for a certain style of work and the methodology that it involved. He was celebrating the utility of technical mechanics in the face of familiar criticisms and characterizing that utility by using the slogan of the unity of theory and practice. He was saying what that unity meant for the practitioners of technical mechanics.[67] This was not lost on his audience, and not all of them accepted his understanding of that unity. Not everyone with an interest in aeronautics was a specialist in technical mechanics, and for them Pröll's claims were not necessarily congenial ones.

On member of the audience was Prof. Friedrich Ahlborn, whose interest in hydrodynamics was empirical not mathematical. Ahlborn was a specialist in, and a pioneer of, the photography of fluid flows.[68] For Ahlborn the mathematics of ideal fluids was just the plaything of theorists who did not realize that experiment alone would yield understanding. In the discussion following Pröll's lecture, Ahlborn was the first on his feet in order to explain these facts to the assembled company. The work Pröll had just described, he said, was mere theory and could be ignored. Ahlborn's remarks about the Prandtl-Fuhrmann work on airships were scathing. As for the new Joukowsky aerofoils, Ahlborn warned aeronautical engineers that they should not assume that they will make good wings. Only experiment could establish that.[69] Pröll, he implied, had ignored experiment. Prandtl, who was also in the audience, sprang to Pröll's defense. The lecture, he insisted, had not been one-sided. Pröll's theme was the unity of theory and practice in aeronautics and that, surely, implied the unity of theory and experiment. If Ahlborn was not convinced, he, Prandtl, was.

To those who were outside the culture of technical mechanics, the work done by the insiders could seem of little value. This did not just apply to those, like Ahlborn, with no mathematical aptitude. It also applied to those whose mathematical expertise was beyond question, for example, to Cambridge-trained mathematical physicists. As G. H. Bryan had made clear in his review of Joukowsky's book, the methods that had proven so fertile in the hands of Blumenthal at Aachen, or Betz at Göttingen, were of no interest to him. They seemed too elementary to be of any value, and they appeared to have nothing to teach a good Tripos man. British experts complained that the Kutta condition was arbitrary and, in any case, could not be applied to a rounded edge. Betz, by contrast, felt free to experiment with different positions of the stagnation point and to explore the flow over a rounded and realistic trailing edge. The mathematically precise position of the stagnation point, he argued, was not of great practical significance. The British, unlike their German counterparts, were greatly exercised by the problem of how a circulation could ever arise in an ideal fluid. But where the German group, in one institutional setting, had surged forward and constructed a cumulative, puzzle-solving, and practically oriented tradition, the British mathematicians, in a different institutional setting, turned their backs on the opportunity, and they felt entirely justified in doing so.

7

The Finite Wing:
Ludwig Prandtl and the Göttingen School

Der alte Göttinger Professor Dirichlet würde sich wohl gefreut haben, wenn er dieses
Resultat hätte sehen können; glaubte man doch gerade seine Potential-Theorie durch
die einfache Tatsache, das ein Widerstand existiert, ad absurdum geführt zu haben.
J. ACKERET, *Das Rotorschiff und seine physikalischen Grundlagen* (1925)[1]

The old Göttingen professor Dirichlet would have been so happy if he could have seen
these results. People just believed that his theory of potential had been reduced to absurdity by the simple fact that there was resistance to motion.

The theory of lift may be divided into two parts: (1) the theory of the wing
profile, that is, the wing sections of the kind studied by Kutta and Joukowsky,
and (2) the theory of the planform of the wing. The planform is the shape of
the wing when seen from above. Wings can be given very different planforms.
The designer may chose a simple, rectangular shape or give the wing a more
aesthetically pleasing curved leading or trailing edge. The wingtips may be
rounded or square, and, most important of all, the wing may be made long
and narrow (high aspect ratio) or short and stubby (low aspect ratio). It was
known experimentally that some features of the aerodynamic performance
of a wing depended on the profile, whereas others (such as the slope of the
curve relating lift to angle of attack) depended on the planform and, particularly, the aspect ratio. Some of the features that depend on the profile were
discussed in the last chapter, for example, the angle of attack at zero lift, the
distribution of pressure along the chord, and the experimentally determined,
but theoretically obscure, point of maximum lift. The minimum drag as well
as the pitching moment were also found to depend on the profile. Now the
discussion turns to the distribution of the lift along the span of the wing and
the properties that a wing possesses in virtue of its finite length and the flow
around the wingtips. Bringing order and understanding to these phenomena
(and predicting unsuspected effects and relationships involving the aspect
ratio of a wing) was the outstanding achievement of Ludwig Prandtl and his
co-workers at the University of Göttingen.[2] Before looking into the technical
details of this achievement, I discuss the intellectual background of the work
and its institutional context.[3]

Prandtl and the Boundary Layer

If Prandtl had never turned his attention to wing theory he would still have occupied a significant position in the history of fluid dynamics. In 1904, at the International Congress of Mathematicians, held that year in Heidelberg, Prandtl had delivered a brief paper called "Über Flüssigkeitsbewegung bei sehr kleiner Reibung" (On fluid motion in fluids with very small friction).[4] In this paper he introduced the now famous concept of the boundary layer. At the time, the full significance of the work escaped most of the audience, though not Felix Klein.[5] Much later the Heidelberg paper came to be seen as one of the most important contributions to science that was made during the twentieth century.[6] It has been likened in its impact to Einstein's 1905 paper on the theory of relativity.[7] The significance of Prandtl's work was that it provided a bridge—a long-sought-for bridge—that connected the behavior of real, viscous fluids and the unreal, inviscid fluid of previous mathematical theory. There had always been a gap between the Stokes equations, which appeared to be true but unsolvable, and the Euler equations, which were known to be solvable but untrue. This logical gap had profound methodological consequences. It attenuated the link between the mathematical hydrodynamics of the lecture theater and the engineering hydraulics of the workshop. It undermined hope in the unity of theory and practice. Prandtl's boundary-layer theory restored that hope. Figure 7.1 shows Prandtl at work on his boundary-layer research.

The theory of the boundary layer can be broken down into four parts: (1) an underlying physical model, (2) an implied technology of control, (3) a mathematical formulation of the model and the technology, and (4) a heuristic resource. I briefly describe each of these dimensions of the theory.

The physical model expressed the idea that, in a fluid of small viscosity, the effects of viscosity arise in, and are often confined to, a thin layer that is in contact with a solid boundary. In the vicinity of the boundary, the fluid layer possesses a sharp velocity gradient. On the actual surface of the body along which the fluid is moving (for example, a wing or the walls of a channel), the fluid is stationary. A short distance away it achieves the velocity of the free stream. The velocity gradient in the *Übergangsschicht*, or transition layer as Prandtl called it, is shown diagrammatically in figure 7.2 (taken from the 1904 paper). As long as the fluid within the layer has the kinetic energy to overcome any adverse pressure gradient, then the boundary layer will conform to the surface along which it is flowing. If it meets too great a pressure, then a backflow will set in and the flow will separate from the surface. This process is shown in Prandtl's diagram. The intense vorticity of the fluid in

FIGURE 7.1. Ludwig Prandtl (1875–1953). Prandtl is shown ca. 1904 at the *technische Hochschule* in Hanover demonstrating his hand-driven water channel used to take flow pictures of boundary-layer phenomena.

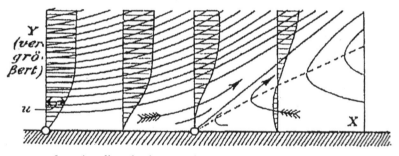

FIGURE 7.2. Separation of boundary layer according to Prandtl. From Prandtl 1904, 487. (By permission of Herr Helmut Vogel)

the boundary layer will then diffuse into the surrounding flow and alter its general character.

The boundary-layer theory thus encompassed the phenomenon of flow separation, which had intrigued Prandtl from his early days as an engineer in industry when he had worked on suction machinery.[8] For Prandtl, as an engineer, the question was how to stop separation and improve the efficiency of the suction effect. A significant part of the 1904 paper implicitly bore upon this engineering problem because it was devoted to the question of boundary-layer control. Prandtl reasoned that if the boundary layer could be removed, then it could not detach itself and modify the rest of the flow. He therefore constructed an apparatus to explore this effect. It consisted of a hollow cylinder with a slit along one side. The cylinder was inserted in a flow of water and, by means of a suction pump, some of the fluid from the boundary layer was drawn through the slit. The result was that on the side of the cylinder with the slit, the remaining flow stayed close to the surface of the cylinder. As predicted, it did not detach itself and cause vorticity and turbulence in the surrounding fluid. Prandtl presented his Heidelberg audience with photographs of this process to show them the difference made by the intervention.[9]

Prandtl was able to express the ideas underlying this process in a mathematical form. He gave the equations of motion for the fluid elements in the boundary layer. He did so by reflecting on the orders of magnitude of the forces and accelerations of the flow in the boundary layer as the viscosity approached zero.[10] This line of thought told him which quantities could be ignored in the original Stokes equations governing viscous fluids. It led to a simplification of the equations that did not involve wholly ignoring either the viscous forces or the inertial forces. It proved possible to keep them both in play. Prandtl thus managed to simplify the Stokes equations without simplifying them too much. Consider the two-dimensional flow of an incompressible fluid in a boundary layer that flows horizontally, that is, along the x-axis. After his simplification Prandtl was left with two equations that described the flow of fluid in the boundary layer by specifying the respective velocity components, u and v, in the x and y directions. If ρ is the density, p the pressure, and μ the viscosity, then Prandtl was able to write

$$-\frac{\partial p}{\partial x} = \rho\left(u\frac{\partial u}{\partial x} + v\frac{\partial u}{\partial y}\right) - \mu\frac{\partial^2 u}{\partial y^2}$$

and

$$\frac{\partial p}{\partial y} = 0.$$

On the basis of these two equations Prandtl worked out an approximate, but reasonable, value for the drag on a horizontal plate acting as the solid boundary along which the fluid was flowing. He was also able to arrive at an expression giving the thickness of the boundary layer and show that the thickness approached zero as the viscosity approached zero. In 1908, in a Ph.D. thesis supervised by Prandtl, Blasius fully solved the boundary-layer equations for the case of the flat plate and improved on the original estimate of the drag.[11] Other Göttingen doctoral students—Boltze, Hiemenz, and Toepfer—refined Blasius' procedure and extended the analysis to circular cylinders and bodies of rotation.[12] Although work on the boundary layer began slowly and, for a decade, was confined to Göttingen and the circle around Prandtl, the theory gradually became the focus of extensive empirical and theoretical research in Europe and America. The idea of the boundary layer eventually found application in every branch of technology where fluid dynamics plays a role.[13]

Given this idea's wide applicability, it is worth noting some of the logical characteristics of Prandtl's equations and reflecting on their methodological status. I have written the equations in a way that brings out their similarities and differences with the Euler equations and the Stokes equations. It is easy to see that the first equation is more complicated than the corresponding Euler equation but simpler than the corresponding Stokes equation. But notice in particular the second, and shorter, of the above equations. It indicates that, given the approximations that are in play, there is a zero rate of change of pressure perpendicular to the plate. The pressure is constant along the y-axis as it cuts through the boundary layer. Clearly, Prandtl's picture of the boundary layer involved some ruthless idealizations. This fact was emphasized by Hermann Schlichting, another of Prandtl's pupils, who would later write an authoritative monograph on the boundary layer.[14] Commenting explicitly on the second of the above equations, Schlichting said:

> Die hieraus folgende Vernachlässigung der Bewegungsgleichung senkrecht zur Wand kann physikalisch auch so ausgesprochen werden, daß ein Teilchen der Grenzschicht für seine Bewegung in der Querrichtung weder mit Masse behaftet ist noch eine Verzögerung durch Reibung erfährt. Es is klar, daß man bei so tief greifenden Veränderungen der Bewegungsgleichungen erwarten muß, daß ihre Lösungen einige mathematische Besonderheiten aufweisen, und daß man auch nicht in allen Fällen Übereinstimmung der beobachteten und berechneten Strömungsvorgänge erwarten kann. (121)

> The disregard of the equation of motion at right angles to the wall that results from this can be expressed in physical terms by saying that, in its transverse motion, a fluid particle in the boundary layer has no mass and experiences no frictional retardation. It is clear that with such far-reaching changes in the equations of motion one must expect that their solutions will show some mathematical peculiarities and that one cannot in all cases expect agreement between the observed and calculated flow processes.

The fluid particles in the boundary layer, as described by Prandtl's equations, have zero mass and zero friction in the direction transverse to the layer. Clearly no one believes that a real, physical object could satisfy these specifications, at least not given all the assumptions about the world taken for granted by physicists. Thus Prandtl portrayed the fluid in his boundary layer in terms that are reminiscent of the idealized fluid of classical hydrodynamics. Euler's equations of inviscid flow generated false empirical predictions, and these errors were usually explained by noting that the equations neglected friction, whether between the fluid elements themselves or between the fluid and solid boundaries. One might therefore expect that a determined effort would be made to remove all such idealizations and unrealities concerning friction in the course of producing the improved, more realistic, boundary-layer equations. This appears not to have been the case. As far as friction is concerned, the particles of fluid in the boundary layer are hardly less exotic than the particles of an ideal fluid. More will be said later about the way in which idealization is an enduring feature of scientific progress in fluid dynamics.

Not only did Prandtl's boundary-layer equations involve physical unrealities, but the reasoning that generated them involved mathematical assumptions for which no justifications were given. Certain mathematical questions had been passed over, for example, questions about the existence and uniqueness of solutions to the equations and the convergence of the approximation techniques that were employed. This left the precise relation between Prandtl's equations and Stokes' equations unclear. As one mathematician noted, even fifty years after the introduction of the boundary-layer equations, this deductive obscurity had still not been dispelled. But, he added, there has been a tendency to disregard it because of the great, practical success of Prandtl's contribution.[15]

The boundary-layer equations, as such, played no explicit part in the mathematical apparatus employed in the early Göttingen aerodynamic work. The mathematics that Prandtl actually used for his theory of the finite wing was confined to the Euler equations of inviscid flow, but the idea of the boundary layer was always in the background and undoubtedly played a heuristic role.[16] The interpretation of the theoretical results depended on qualitative reasoning that appealed to boundary-layer theory. For example, postulating the existence of the boundary layer effectively divided the fluid into two parts. One part demanded recognition of its viscosity, while the other could be treated as if it were an inviscid fluid. If the flow sticks closely to the surface of a solid body, and there is no separation, then the bulk of the flow can be treated as an exercise in ideal-fluid theory. This was the basis of Prandtl's claims, discussed earlier, that for streamlined bodies the theory of perfect fluids had been dramatically confirmed. The viscosity assumed to be present in the boundary layer also provided a resource for explaining the origin of the circulation around a wing. The viscous fluid in the boundary layer possesses vorticity, so that if fluid from the layer were to diffuse into the free stream, this occurrence might modify the overall structure of the flow and introduce a component of circulation, even if the circulating flow were then attributed to a perfect fluid.

The model of the boundary layer was itself subject to development both theoretically and experimentally. At first it had been assumed that the flow within the layer had a laminar character. Later, Prandtl relaxed this assumption and explored the idea of a turbulent boundary layer. Because turbulence implied an increased exchange of energy between the slower-moving boundary layer and the faster-moving free stream, a turbulent boundary layer would possess more energy than a laminar boundary layer because it would have absorbed energy from the free stream. The increased energy delays the separation that occurs when the boundary layer runs out of energy and brakes away from, say, the surface of the wing. The delay means the flow conforms more closely to the surface of the wing. This lowers the pressure drag and thus brings the behavior of the air closer to that of a perfect fluid. The idea of boundary-layer turbulence also explained some intriguing disparities between the wind-channel measurements of the resistance of spheres made in Göttingen and those from Eiffel's laboratory in Paris. Strangely, in Paris resistance coefficients for spheres were about half the value of those in Göttingen: 0.088 compared with 0.22. In the course of a review of Eiffel's wind-channel results, which were otherwise comparable with those in Göttingen, Otto Föppl concluded that, in the case of the resistance of spheres, there was obviously some mistake in the French work: "Bei der Bestimmung des Widerstands einer Kugel ist offenbar ein Fehler unterlaufen."[17]

Prandtl, however, was able to explain the result without attributing a mistake to Eiffel. Rather than a trivial error, the anomaly indicated the presence of something deep. Prandtl argued that in Göttingen the flow in the wind channel was less turbulent than in Eiffel's channel. He deliberately increased the turbulence in the Göttingen channel by means of a wire mesh and reproduced Eiffel's results. What is more, Prandtl argued that the boundary layer itself may have been laminar in Göttingen, whereas in Paris it had been turbulent. This analysis was then subject to an ingenious experimental test in the Göttingen wind channel. Just as Prandtl had introduced the original idea of the boundary layer alongside a demonstration of how to remove the layer by suction, so he now showed how to manipulate the turbulence of the layer. He (counterintuitively) *reduced* the resistance of a sphere by wrapping a trip wire around it to render the boundary layer turbulent. Photographs taken by Wieselsberger provided further corroboration. Not only was the measured resistance reduced, but the introduction of smoke into the flow showed the separation points pushed toward the back of the sphere. The turbulent boundary layer must be clinging to the sphere longer than the laminar layer. In both of these cases, that of the laminar and the turbulent boundary layer, Prandtl's engineering mind linked a novel theoretical idea to a novel technology of intervention.[18]

Klein and the General-Staff Officers

Prandtl wrote his paper on the boundary layer while holding the chair of mechanics at the *technische Hochschule* in Hanover. Felix Klein (fig. 7.3) soon arranged for him to be called to Göttingen. In subsequent years Klein continued to use his contacts with powerful government ministers, such as Friedrich Althoff, to support Prandtl and his work. Although his greatest mathematical achievements were now behind him, it was during the period 1890–1914 that Klein was at the height of his influence as an academic politician.[19] The institutional structures created in Göttingen by Klein provided the context for Prandtl's aeronautical work. It is hardly surprising that Klein always remained a figure whom Prandtl revered.[20]

In the years from the turn of the century and through the period of the

FIGURE 7.3. Felix Klein (1849–1925). Klein, one of the greatest mathematicians of his time, has been called a "countermodernist," but he was a tireless academic reformer and the mentor of Ludwig Prandtl. Klein brought Prandtl to Göttingen and encouraged his aerodynamic research. (By permission of the Niedersächsische Staats- und Universitätsbibliothek Göttingen)

Weimar Republic, Göttingen was a place of extraordinary intellectual brilliance in the fields of mathematics and physics. The university was the home of not only Felix Klein but David Hilbert, Hermann Minkowsky, and Hermann Weyl. They all played leading roles in the sometimes tense discussions that took place over the mathematical foundations of general relativity and the geometrical nature of space, time, and matter. It is right that these aspects of the Göttingen scene have attracted the attention of historians of science and have been subject to detailed analysis.[21] Nevertheless, for my purposes it is important to ensure that this aspect of Göttingen does not dazzle us. It must not obscure the very different character of the work done by Prandtl and his school.

The two greatest Göttingen mathematicians of that time, Klein and Hilbert, represented divergent intellectual tendencies. Hilbert's work can be seen as formal and abstract, whereas Klein's was more concrete and intuitive. Hilbert has been described as a "modernist," while Klein has been presented as a "countermodernist."[22] The differences between the work of the two men are a matter of continuing discussion among historians of mathematics. The sharpness of the contrast between them has, perhaps, been blunted by recent scholarship, and the degree of overlap in their interests is now better appreciated.[23] Despite this, something of the older polarity still remains. Both men wanted to increase unity in the field of mathematics, but they sought unity in different ways and aspired to a qualitatively different kind of unity. Hilbert looked for unity through powerful methods and results that would radically reconfigure what had gone before. Klein sought unity through an encyclopedic and cumulative arrangement of results. He was opposed to the bitter sectarianism of the competing "schools" of mathematicians in the German universities and sought to cure it not by the victory of one tendency, but by an ordering of mathematical knowledge that was at once inclusive and hierarchical. The historian of mathematics David Rowe was surely right when he treated Klein's great, collective project—the *Encyklopädie der mathematischen Wissenschaften*—as a revealing expression of Klein's conception of mathematical knowledge and its proper organization.[24]

Klein's tireless organizational activities have been subject to detailed analysis by, for example, Manegold, Pyenson, Rowe, Schubring, and Tobies.[25] The main outlines of the story, as it bears on Göttingen aerodynamics, can be quickly sketched. After his visit, as a representative of the Prussian government, to the World's Fair in Chicago in 1893, Klein became convinced of the need for German universities to attach more importance to applied science and technology. By 1897 he had been able to persuade the government and the University of Göttingen to create an institute for applied physics. In 1898 Klein and Henry von Böttinger, the vigorous and persuasive chairman of the Bayer chemical company, founded the Göttinger Vereinigung zur Förderung der angewandten Physik und Mathematik. This was a novel institutional vehicle by which commercial firms could finance applied research in the university.[26] On this basis, in 1905, the original Institute for Applied Physics grew into the Institute for Applied Mathematics and Mechanics under Prandtl and Runge.[27]

When Klein was pressing the authorities to appoint Prandtl, he was explicit about the engineering connection.[28] He introduced Prandtl as someone "who combines the expert knowledge of the engineer with a mastery of the apparatus of mathematics and has a strong power of intuition combined with a great originality of thought" ("der mit der Sachkenntnis des Ingenieurs und der Beherrschung des mathematischen Apparatus eine starke Kraft der Intuition und eine große Originalität des Denkens verbindet"; 232). It was Klein who prompted Prandtl to bring problems connected with airships and aerodynamics within the scope of his new institute.[29] The Motorluftschiff-Studiengesellschaft, the Society for the Study of Motorized Airships, which had been founded in July 1906, provided the finance for setting up a model-testing facility at Göttingen, the Modellversuchsanstalt, with its wind channel designed by Prandtl and constructed by Fuhrmann. The channel was given a test run in December 1908 and went into operation in January 1910. In October 1913 the model-testing facility was taken over by the university.[30]

There is no doubt that Prandtl was in close contact with all the leading mathematicians and physicists at Göttingen. He went with them on their afternoon walks; he plotted with them over local academic politics; he sat at the "high table" along with Klein and Hilbert during the intimidating meetings of the Göttingen Mathematical Society.[31] He was a friend and neighbor of Karl Schwarzschild, who was at work solving the field equations of Einstein's general theory of relativity.[32] Unlike that of his colleagues, Prandtl's intellectual life was not dominated by the struggles over the nature of gravity or the structure of matter. His mathematical heritage lay elsewhere, in the *technische Mechanik* of the engineering tradition and August Föppl's textbooks.[33] The significance of this tradition is conveyed with some force by the lengthy account that Runge and Prandtl wrote in 1906 describing the history and work of their Institute for Applied Mathematics and Mechanics.[34] It is full of enthusiastic and detailed specifications of pumps, boilers, condensers, dynamos, diesels, turbines, and electrically driven ventilators (which were used in the aerodynamic experiments). No wonder their colleagues made sly jokes about the Department of Lubricating Oil.[35] This was not the world of the philosopher-mathematician or the physicist-as-cosmologist, any more than it was the world of Horace Lamb, Augustus Love, or Lord Rayleigh. Certainly the Cambridge mathematicians would not have chosen the words with which Runge and Prandtl ended their article:

> Die technischen Wissenschaften sind reich an Kapiteln, deren volles Verständnis eine tiefe mathematische Bildung erfordert. Der Unterricht setzt sich zum Ziel, die Entwicklung der mathematischen Methoden zu vereinigen mit dem vollen Verständnis der praktischen Probleme in dem Umfang und in der Fassung, wie sie sich dem ausübenden Ingenieur darbieten. (280)

> Technical science is rich in material whose full understanding demands a deep mathematical education. The goal of the teaching [at the institute] is to unite the development of mathematical methods and a complete understanding of practical problems in as far as they concern, and in the manner they present themselves to, practicing engineers.

This was neither the voice of Tripos-oriented Cambridge, nor the voice of the mathematical Göttingen that has attracted the lion's share of the historian's attention, but it was the voice of Prandtl's Göttingen. There is no problem in acknowledging this difference as long as it is recognized that the University of Göttingen was not a unified intellectual environment. Of course, there was overlap between its different parts both personally and institutionally: Runge had worked on atomic spectra and von Kármán was a friend of Max Born's and collaborated with him on the quantum theory of specific heats.[36] But it was engineering that defined the orientation of the aerodynamic work, and this orientation had an institutional niche in Göttingen thanks to Klein's efforts.[37]

Institutional plurality was wholly consistent with Klein's vision and practice. Mathematics, for Klein, always had an integrating function in science, but that function was to be discharged in diverse ways. It required coordination and cooperation, but it did not require that everyone have the same preoccupations. That would have run counter to the encyclopedic outlook that informed Klein's organizational plans. Klein had himself lectured on mechanics at Göttingen and had tried to offset the tendency toward excessive mathematical abstraction. In doing this he had, as von Mises put it, restored "the essential but almost lost connection with 'technical mechanics.'"[38] In 1900 Klein gave a general lecture titled "Ueber technische Mechanik" in which he sought to capture the special qualities of the discipline.[39] Like August Föppl, Klein asserted that practitioners of technical mechanics had their own *Fragestellung*, that is, their own way of posing and answering questions. It involved subtle judgments and made unique demands. In particular Klein noted the problematic relation between technologically oriented mathematics and the established knowledge of basic physical principles: "Es ist vielfach nicht möglich, die Erscheinungen mit den Principien oder Grundgleichungen der classischen Mechanik in lückenlosen Zusammenhang zu bringen" (28) (Very often it is impossible to bring phenomena into a rigorous relationship to the principles and fundamental equations of classical mechanics). In the eyes of some of Klein's technological critics, these sympathetic methodological insights were not enough. Nor were Klein's Göttingen seminars on elasticity, graphical statics, and hydrodynamics always deemed a success. The implied commitment to technology, said the critics, was inadequate. This was the line taken by the engineer H. Lorenz, who had been brought to Göttingen by Klein but who had left in disappointment and, in a flurry of anti-Klein criticism, had taken himself off to the TH at Danzig.[40] Klein was always fighting on two fronts. On the one side were those within the universities who argued that technology was not a fit subject for a truly academic institution, and on the other side were those within the *technische Hochschulen* who saw technology as the special preserve of these institutions. Both sides resented Klein's suggestion that the universities involve themselves with applied science and technology.

Klein's attempts to meet these conflicting accusations were not always successful. He eventually got his way, but not without frustrations and compromise. He supported the *technische Hochschulen* in their fight for the right to grant doctoral degrees, but there were delicate issues of status involved. Klein tried to find a division of labor in which different roles were to be played by the different institutions. On one occasion, in 1895, he sought to convey his vision by means of a military metaphor. He spoke of the universities as providing the *Generalstabsoffiziere der Technik* or "general-staff officers" of technology, while the products of the technical institutions would constitute the "frontline officers."[41] Perhaps Klein thought that the heroic image of frontline combat would make the suggestion acceptable to colleagues in the *technische Hochschulen*. If so, he was wrong. The implied disparities of rank and status were too blatant to ignore.

One of those who took offense was August Föppl. Without naming Klein as the source of the military metaphor, Föppl pointedly remarked that if one were going to speak in these terms, then each different arm of the military service (*Waffengattung*) should be accorded equal value. Were they not all equal as comrades in arms? The route to the top, and promotion to high command, should depend on the qualities of the individual, not the particular branch of the service in which they happened to be trained.[42] But while Föppl bridled at the condescension, his son-in-law was a beneficiary, and exemplification, of Klein's rank ordering. Prandtl's institute was devoted to fundamental questions in the field of technology, and he soon occupied a position in which he could influence the strategy of aerodynamic research. Prandtl was indeed a *Generalstabsoffizier*, and the banner under which his army marched was inscribed with the word "Engineer."

The point is confirmed by the qualifications of those whom Prandtl recruited to serve with him. They prefaced their names with their engineering diplomas and styled themselves Dipl.-Ing. Fuhrmann, Dipl.-Ing Föppl, Dipl.-Ing. Betz, Dipl.-Ing. Wieselsberger, and Dipl.-Ing. Munk.[43] Theodore von Kármán (fig. 7.4) also had an engineering training and was a graduate of the Royal Technical University of Budapest. He had originally come to Göttingen to do research with Prandtl on elasticity and the strength of materials but, as he was fond of recounting, turned to hydrodynamics through reflecting theoretically on the experimental difficulties of his colleague Karl Hiemenz.[44] This was the origin of von Kármán's work on the stability of vortex motion and his mathematical analysis of the shedding of vortices that formed a flow pattern called a Kármán vortex street.[45] A number of Prandtl's inner circle

FIGURE 7.4. Theodore von Kármán (1881–1963). Von Kármán trained as an engineer in Hungary and then worked with Prandtl in Göttingen. Later he moved to the *technische Hochschule* in Aachen and built up a rival institute of aeronautics and fluid dynamics. (By permission of the Royal Aeronautical Society Library)

eventually left Göttingen to take up positions in the *technische Hochschulen*. Föppl, Wieselsberger, and von Kármán all went to Aachen, making it a powerful rival to Göttingen. Aachen was not Göttingen, but von Kármán was not a humble man.[46] Without doubt, he and his colleagues still saw themselves as ranking members of the general staff of aeronautical technology.

The Horseshoe Vortex and the Biot-Savart Law

Prandtl's earliest publications on aeronautics did not deal with the circulation theory of lift in either its two- or three-dimensional form. He wrote mainly on airships, the general mechanical problems associated with building an airplane, or the engineering that went into the construction of wind channels for testing models. Thus in June 1909 he lectured to the annual general meeting of the association of German engineers at Mainz on the significance of the model experiments to be done at Göttingen and the equipment that had been developed.[47] September of the same year saw him speaking at the International Aeronautical Exhibition (ILA) in Frankfurt. Here Prandtl concentrated on the principles and preconditions of flight and the practical problems of achieving adequate lift and stability.[48] In a series of articles in the new *Zeitschrift für Flugtechnik* he laid out for the reader's benefit those parts of mechanics of special relevance to aeronautics.[49] His topics included gyroscopes, stability, and air resistance, with a mention of his own boundary-layer theory in connection with the phenomenon of separation and vortex formation. One of Prandtl's concerns was to remove false ideas that continued to cause trouble in the field. Many "inventors," he explained, tried to achieve automatic stability in aircraft by some sort of pendulum device whose rationale was based on a misunderstanding. Prandtl also stressed that resistance depends as much on the suction effects behind a body as it does on what happens at its front face. From the outset he had no time for Rayleigh-Kirchhoff flow as an account of lift. He viewed it as wrong at the level of general principle.

In these articles for the *ZFM*, Prandtl recommended some of the relevant literature (64). Significantly he included both Lanchester's book on aerodynamics and Lamb's book on hydrodynamics. He thus symbolically conjoined what, in the homeland of those two authors, was proving so resistant to unification. On the first page of the initial article in the *ZFM* series, Prandtl also nailed his colors to the mast regarding the principle of the unity of theory and practice. Not for him the slogans of the antimathematical movement, which treated theory, and especially mathematically based theory, as something in fundamental opposition to practice. Using words that closely echoed the formulations adopted a decade earlier by August Föppl, in the preface to his *Vorlesungen über technische Mechanik,* [50] Prandtl asserted:

> Um nun meine Absichten näher zu kennzeichnen, möchte ich vorweg betonen, daß ich das viel ausgesprochene Schlagwort vom Gegensatz zwischen Theorie und Praxis nicht gelten lassen will; der Gegensatz liegt für mein Empfinden zwischen guter und schlechter, richtig und unrichtig angewandter Theorie; eine gute Theorie is in Übereinstimmung mit den Ergebnissen der praktischen Erfahrung, oder sie gibt zum mindesten wesentliche Züge der Erfahrungstatsachen wieder. (3)

> In order to characterize my intentions more precisely, I should emphasize at the outset that I do not want to endorse the much used slogans about the opposition of theory and practice. In my experience the contrast lies between theories that are good or bad for application and between correctly or incorrectly applied theories. A good theory is in agreement with the results of practical experience or, at least, captures the essential thrust of the facts of experience.

Prandtl was not just striking attitudes. His work on the three-dimensional wing would meet the demanding requirement of applicability to engineering practice.

Prandtl tells us that it was in the winter semester of 1910–11 that he began to develop his own mathematical theory of the finite wing.[51] It was treated in a course of lectures that he gave in Göttingen on the theme of aeronautics. The lectures were attended by Otto Föppl and, though the text of the lectures appears to have been lost, hints and mentions of their content are to be found in the early papers coming from the Göttingen group. The basic picture with which Prandtl worked was similar to the one already proposed by Lanchester, namely, a finite wing with circulation around it and with two vortices emerging from the wingtips. Prandtl's aim was to make this picture amenable to a mathematical analysis, but to do so he had to resort to extreme simplification. He treated this complex three-dimensional, physical system as equivalent to three connected-line vortices. One vortex, called the bound vortex, represented the wing and was assumed to run along the span of the wing. The other two vortices, called free vortices or trailing vortices, extended back from the wingtips and were at right angles to the wing. The arrangement, as shown in figure 7.5, first appeared in a brief, qualitative account of the theory that Prandtl published in 1912.[52] The two diagrams, which show the progressive abstraction of the picture, reappeared in Prandtl's published lectures in the form shown here.

Physically the trailing vortices are assumed to extend right back from the wing in flight to the point where the aircraft left the ground. Mathematically

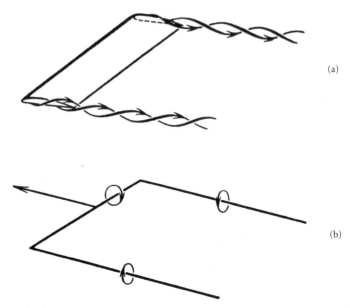

(a)

(b)

FIGURE 7.5. Trailing vortices from the wingtips represented diagrammatically (*a*) and by the even more simplified "horseshoe" vortex system (*b*). Tietjens 1931, 197. (By permission of Springer Science and Business Media)

they are said to extend back "to infinity." The three connected vortices thus form a single bent line of vorticity. A vortex line in an ideal fluid (whether straight or bent) has the same strength everywhere along its length, so the circulation Γ around the wing, that is, around the bound vortex, also gives the vortex strength along the trailing vortices. For reasons examined later, this model was called, rather implausibly, the *Hufeisen Schema*, or "horseshoe schema."

Prandtl's task was to see how the presence of the two trailing vortices modified the flow in the vicinity of the wing. If the trailing vortices are ignored (as they are in the two-dimensional case), the law of lift is (using the usual notation) Lift = $\rho V \Gamma$. Is this law preserved in the three-dimensional case? Surely some significant modification of the flow must occur, and this must have some consequences for the behavior of the wing—but what modifications and what consequences? Before I go into mathematical details, it may be useful to sketch the outcome of Prandtl's analysis in qualitative terms. Using his simplified model, Prandtl was able to predict two unexpected and important effects.

First, Prandtl found that the effect of the trailing vortices ought to be the

creation of a downwash at, and behind, the wing. The swirling air of the trailing vortices would (according to his mathematical analysis) influence the air
in the vicinity of the wing in such a way as to give it a downward velocity
component. The downward component would have the effect of tilting the
airflow. The way that the tilt arises from the introduction of the downwash
component is shown in figure 7.6a. The presence of the tilt means that the
wing operates at a slightly lower angle of incidence relative to this new local
flow. The effective angle of incidence is therefore reduced. Because (over the
working range of the aerofoil) lift is proportional to the angle of incidence,
the lift should be reduced. The same point may be expressed the other way
round. In order to get the same lift per unit length in a finite aerofoil as in an
infinite aerofoil of the same cross section, the angle of incidence relative to
the main flow has to be increased.

The second prediction followed from the first. If the resultant lift force
on the wing is at right angles to the *local* flow, and the local flow is tilted in
the manner shown in figure 7.6b, then the resultant aerodynamic force will
be tilted back relative to the *main* flow and hence to the direction of motion
of the wing. The effect will be that the resultant force will now have a component that opposes the motion. There will be a drag. It is important to recall
that this analysis was all done using ideal-fluid theory. The two-dimensional
analysis showed how a flow involving a vortex and circulation could yield a
lift *but no drag*; the three-dimensional analysis now predicts that it can produce *both a lift and a drag*. For reasons that will become clear in a moment,
Prandtl and his co-workers came to call this novel drag an "induced drag,"
and the tilt of the local flow due to the downwash was called the "induced
angle of incidence." Induced drag was a form of drag that did not result from
viscosity or skin friction or turbulence. It was produced by the very same
inviscid mechanisms that generated the lift.

Having sketched these two initial predictions, I now look at their mathematical derivation. How did the idealization of the "horseshoe" vortex help
Prandtl to develop a mathematical description of the flow which led to these
results? The answer lies in an analogy that exists between the hydrodynamics of a perfect fluid and electromagnetic phenomena. A line vortex is like a
wire carrying an electric current which sets up, or "induces," a magnetic field
around it. The vortex in an ideal fluid does not, of course, induce a magnetic field; it induces a velocity field. The velocity was exactly what Prandtl
wanted to understand because once he had a mathematical expression for the
velocities, he could deduce the pressures, and thus the forces on the wing.
Prandtl's use of the analogy was explicit. "Für die Verteilung der Geschwindig-
keit in der Umgebung irgendeines Wirbelgebildes besteht eine vollkommne

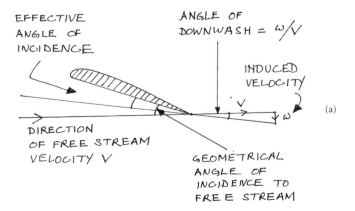

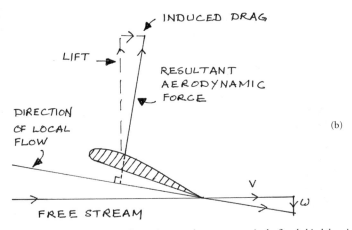

FIGURE 7.6. In (a), the trailing vortices induce a downward component w in the flow behind the wing. This tilts the airflow and effectively lowers the angle of incidence of the wing. The new "effective" angle of incidence is the original geometrical angle of the wing to the free stream minus the angle of the downwash produced by the tailing vortices. In (b), the tilt in the airflow means that the resultant aerodynamic force, which is at right angles to the local flow, is no longer at right angles to the free stream, that is, the direction of motion. This produces a drag component called the "induced drag."

Analogie mit dem Magnetfeld eines stromdurchflossenen Leiters"[53] (For the velocity distribution in the neighborhood of any such vortex formation there exists a complete analogy with the magnetic field around a current carrying conductor).

The analogy to which Prandtl referred was well known before the work on aerodynamics and was discussed in standard textbooks in both electromagnetic theory and hydrodynamics. It was to be found in the books on Maxwell's theory and on technical mechanics written by Prandtl's own teacher and father-in-law, August Föppl.[54] The analogy is not a superficial one but exists at the level of the fundamental equations that can be used to describe the two different areas. Experts in electromagnetism, such as August Föppl, could easily describe the magnetic effects of a current-carrying wire shaped in the form of the idealized "horseshoe," and Prandtl carried these results over to the corresponding system of line vortices.

The law of induction common to the hydrodynamic and electromagnetic cases is called the Biot-Savart law.[55] The law may be explained by considering figure 7.7, which shows an infinite, straight-line vortex of strength Γ. A small segment of the vortex is identified as ds. The point P lies at a distance r from the vortex element, and the line joining ds and P makes an angle θ (theta) with the vortex. Following the electromagnetic analogy, there is a small component of velocity dw "induced" at the point P by the small element ds. According to the Biot-Savart law, the component is

$$dw = \frac{\Gamma ds.\sin\theta}{4\pi r^2}.$$

The velocity component is perpendicular to the plane determined by ds and the line r. This formula links infinitesimal quantities, and the causal relation between such infinitesimals that seems to be implied by the law occasioned some puzzlement. August Föppl had no time for such subtleties. Taken by itself, said Föppl, the formula has absolutely no meaning: "Es hat überhaupt keinen Sinn."[56] Föppl's point was that the real significance of the law only

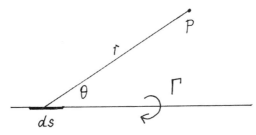

FIGURE 7.7. According to the Biot-Savart law, the infinitesimal amount of induced velocity dw at a point P due to the infinitesimal vortex element ds of a vortex of strength Γ is given by the formula $dw = (\Gamma/4\pi r^2)\, ds\, \sin\theta$.

emerges when it is integrated in order to give the *finite* effect of a *finite* length of vortex or, by extrapolation, the finite effect of a very long (and effectively infinite) vortex. This is what Prandtl needed in order to compute the effect of the, effectively infinite, trailing vortices.

As Prandtl first employed it, the Biot-Savart law was used to give the finite velocity component w at an arbitrary point P in the vicinity of a finite, straight-line segment AB of a vortex (see fig. 7.8). If the strength of the vortex is Γ, and the point P is at a perpendicular distance h from the line AB, then, after integration, the law now reveals that the contribution to the velocity of the finite segment is

$$w = \frac{\Gamma}{4\pi h}(\cos\alpha + \cos\beta).$$

The angles α and β are the angles made by the lines joining the point P with the ends of the finite vortex segment under consideration. The direction of the velocity w at P is at right angles to the plane that passes through the points A, P, and B. Whether the velocity vector faces downward, into the page, or upward, depends on the sense of the circulation around AB. (If an observer who looks from B to A is confronted with a clockwise circulation, then the induced velocity vector points into the page and vice versa.) Notice that if the finite line segment AB is extended to infinity in both directions (so the angles α and β get smaller and smaller as the line gets bigger and bigger), then the velocity at point P should correspond to the velocity at a point situated a distance h from the center of a *two-dimensional* vortex, that is, a point vortex, of circulation Γ. This is exactly what the formula provides. The expression $(\cos\alpha + \cos\beta)$ assumes the value 2 for $\alpha = \beta = 0$, so that $w = \Gamma/2\pi h$.

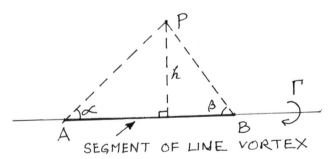

FIGURE 7.8. The induced velocity w at a point P due to a finite vortex segment AB of a vortex of strength Γ is, according to the Biot-Savart law, $w = (\Gamma/4\pi h)$ $(\cos\alpha + \cos\beta)$.

Prandtl's aim was to apply the Biot-Savart law to the "horseshoe" vortex because he was interested in the effect of the trailing vortices. The trailing vortices count as "semi-infinite" lines because they start from the wingtip and go to infinity in one direction. To understand Prandtl's reasoning when he applied the law to his horseshoe system, think of the arrangement in figure 7.8 modified in two ways until it turns into that in figure 7.9. First, A is moved to infinity so that $\alpha = 0$ and $\cos \alpha = 1$. Second, the point B is moved inward until it coincides with the base of the perpendicular from P. This makes $\beta = 90°$ so that $\cos\beta = 0$. The formula then gives the value for the induced velocity w:

$$w = \frac{\Gamma}{4\pi h}.$$

Interpreted in terms of the horseshoe vortex model of the wing, this formula gives the contribution of *one* of the trailing vortices to the flow at a point on the wing that is distance h from the wingtip generating the vortex. The full downwash at any given point on the wing needs the contribution of both trailing vortices to be added together, but the formula reveals the mechanics of the process that generates the downwash. Prandtl spoke of a *zusätzliche Abwärtsgeschwindigkeit*, an additional downward speed. Max Munk called the quantity w the "induced velocity," and this name was taken over by the Prandtl school.

Sufficient has now been said to show how Prandtl was able to reach his predictions about the general effect of the trailing vortices, that is, (1) the

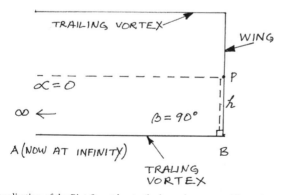

FIGURE 7.9. Application of the Biot-Savart law to the horseshoe vortex. The point P is now a point on the wing. B is now the wingtip. A is at infinity, and AB is the semi-infinite trailing vortex from one wingtip. The Biot-Savart formula gives the induced velocity created by the trailing vortex at point P of the wing.

creation of a downwash, (2) the tilt that the downwash creates in the local flow, and (3) the resulting (induced) drag. One final point to notice about the formula for the induced velocity w of the downwash is that it takes the form of a fraction with h in the denominator. It contains a singularity at the point $h = 0$. For this value of h, the formula requires that the velocity w be infinite, which is physically impossible. Recall that h refers to the distance from the wingtip. This means the application of the Biot-Savart law to the horseshoe model of the three-dimensional wing breaks down for points close to the wingtip. Prandtl had made novel and important predictions, but, because of the singularity, the predictions carried with them a problem. Even if in many cases they were proven correct, they were based on a physically impossible model.

The First Test: Downwash on the Elevator

The first published application and test of Prandtl's approach was provided by Otto Föppl in the *Zeitschrift für Flugtechnik* of July 19, 1911.[57] Föppl had already produced a series of experimental reports using the Göttingen wind channel to test the resistance and lift of flat and curved plates.[58] These studies indicated that the laws of resistance depended in a complicated way on the effects at the edges of the plates. It was clear that the move from an infinite to a finite wing would introduce significant new factors into the account of lift and drag. Prandtl had begun to identify these factors in his lectures. Föppl's aim now was to test a quantitative prediction made by Prandtl on the basis of the new theory he was developing. The prediction kept away from the problematic singularity at the wingtips and concerned the angle at which the air would be moving downward at a specified distance behind the wing. It concerned the induced angle of incidence.

Föppl took for granted the qualitative picture of the two (straight-line) trailing vortices or "vortex plaits" (*Wirbelzöpfe*). The reader of the *Zeitschrift* was assured that these had been rendered visible in the Göttingen wind channel by introducing ammonia vapor into the flow (184). This "really existing flow" ("tatsächlich vorhandene Strömung"), said Föppl, was the empirical basis on which Prandtl had built his theory (184). The question was how the downwash and the tilt in the local flow could be generated and measured in the wind channel. First it would be necessary to introduce a model wing to generate the vortex system that was under study. It should then be possible to detect the downwash by introducing a flat plate at a distance behind the wing. The angle of the plate could be adjusted until it was aligned with the tilt of the airflow. When the plate was correctly aligned with the flow, there should be no lift force on it. This is the empirical clue giving the angle of the flow. It is important that the plate should be flat, because if it were curved or had the cross section of a normal aerofoil it would still generate a lift even when pointing directly into the local flow. The zero-lift position would not reveal the angle of the flow. With a flat plate, however, the observed angle of zero lift gives the actual tilt of the flow, and this can be compared with the predicted angle.

Having reviewed the logic of Föppl's experimental design, we can now look at the details of the experiment and the connections that Föppl made with the realities of aircraft construction. Consider the choice of the distance between the wing and the flat plate. The distance used in the prediction and test was selected on the basis of practical considerations about current aircraft design. Increasingly, and unlike the early Wright machines, aircraft were being built with a control surface, called the elevator, located at some distance behind the main wing. The elevator controls the pitch of the aircraft. In the Wright Flyer, the elevator was at the front and the propellers at the rear. By 1911 designers typically put the propeller at the front and the elevator at the tail end of the fuselage. As Prandtl explained, according to the "horseshoe" model, if the elevator is in a horizontal position behind the wing, it will experience a definite downthrust or negative lift (Abtrieb). This will only disappear if the elevator is rotated by a specific amount which depends on the circulation around the wing. If the elevator is a flat surface (in effect a moving tailplane), then for the reasons just given it will experience a zero-lift force when it is aligned with the downward inclination of the flow behind the wing.[59] Föppl therefore built a model airplane with exactly this kind of adjustable elevator. The model is shown in figure 7.10.

The main wing was 60 × 12 cm and had a camber of 1/18, while the elevator was a flat plate of 20 × 8 cm. Both were made of 2.3-mm-thick zinc. The elevator, which was rigidly attached to the wing by two struts, could be pivoted about its leading edge and fixed at different angles relative to the airflow. There was a distance of 34 cm between the elevator and the main wing, that is, the line running along the span of the wing on which the bound vortex was supposed to be located.

Föppl's experimental procedure involved four steps, each of them using the Göttingen wind channel. First, Föppl removed the elevator from the model, leaving the main wing still connected to the two struts. He placed the wing and struts in the wind channel at a realistic angle of incidence of 4.6°. The channel was run at a single, fixed speed V, and the lift on the wing was measured. The next step was to reposition the wing (still without an elevator) at a different angle of incidence. This time he chose 7.6°. Again the lift

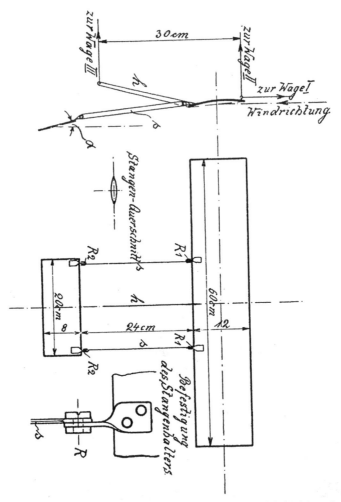

FIGURE 7.10. Föppl's model for testing Prandtl's prediction of downwash behind the main wing. The dimensions of the model are entered into the formula for the Biot-Savart law to yield the predicted angle of flow near the tailplane. From Föppl 1911a, 183. (By permission of Oldenbourg Wissenschaftsverlag GmbH München)

was measured at the speed V. In both cases Föppl expressed the lift as a coefficient ζ_A. (This involved dividing the lift force by the density, the area of the wing, and the square of the speed.) He now had two lift coefficients, one for each of the two angles of incidence. In preparation for the next part of the experiment, Föppl reattached the elevator in order to carry out two sequences of measurements on the whole model. In one sequence the elevator-wing system was suspended so that the angle of incidence of the main wing was

4.6°, while for the other sequence the main wing was at 7.6°. For each of these angles Föppl measured the overall lift of the combined system for a range of different elevator settings. He gave the elevator seven different settings, that is, seven different angles relative to the direction of the free flow. The angles of the elevator to the free airstream ranged from +30° to −10°.

To find the forces on the elevator alone, Föppl subtracted the lift measurement for the wing in isolation from that of the wing plus elevator. The remaining lift force (that is, the lift force on the elevator) was then cast into the form of a lift coefficient. This gave Föppl data that could be expressed in terms of two graphs in which the lift on the elevator was plotted against the angle of incidence of the elevator—one curve for each of the angles at which the main wing had been set to the free stream. The most important feature of these graphs was the point at which the curves passed through the x-axis, that is, the angle of the elevator when its lift coefficient was zero. This was the angle at which the elevator should be parallel with the downwash, that is, the local, downward flow of the air induced by the vortex system. The graphs indicated that when the main wing was at an angle of 4.6°, the zero-lift position of the elevator was 2.8°. When the main wing was at 7.6°, the zero-lift position of the elevator, and hence the angle of the downwash, was 4.3°. The question now was whether Prandtl's theory could predict these angles of downwash from the main wing at the two angles of incidence that Föppl had selected for his test.

Föppl duly announced the predicted value of the angles that had been deduced from the theory—but he did not say on what basis the prediction had been made. He simply informed his readers that in his lectures Prof. Prandtl had derived a formula that gave the tangent of the predicted angle of downwash. The formula was stated, but the deduction that led to it was withheld. The tangent, Föppl said, was given by the ratio w/V, where, according to Prandtl,

$$\frac{w}{V} = \frac{b\zeta_A}{\pi l}\left(1 + \frac{\sqrt{x^2 + (l/2)^2}}{x}\right).$$

As Föppl explained, the coefficients of lift ζ_A to be entered into the formula were the ones that had been determined experimentally for the isolated wing. All the other dimensions could be taken from the model itself. Thus, b was the chord of the main wing (12 cm), l was the span of the wing (60 cm), and x was the distance along the longitudinal axis of the model from the middle of the main wing to the middle of the elevator (34 cm). With these values for the tangent, the predicted angles themselves came out at 3.3° and 4.2°. Given that the two measured angles of the downwash (derived from the zero-lift position of the elevator) were 2.8° and 4.3°, Föppl concluded that the result amounted to "a very good confirmation of the theory"—"eine sehr gute Bestätigung der Theorie" (184). The prediction derived from the horseshoe model was correct.

The force of this claim must have been somewhat diminished because the theory used to make the prediction was not revealed. Readers of the *Zeitschrift* would have known that something was afoot in Göttingen, but Föppl was not going to anticipate Prandtl and expound the theory. He merely said that Prof. Prandtl would soon publish his derivation of the formula in the *ZFM*. No such derivation was forthcoming, but, with the benefit of hindsight, an examination of the formula makes its origin easy to guess. The formula was simply the result of applying the Biot-Savart law to each of the three straight-line parts of the horseshoe vortex and then doing the trigonometry necessary to relate the formulas to Föppl's model.[60]

From Ground Effect to Biplanes

The law of Biot and Savart received a number of further aerodynamic applications before the outbreak of World War I. All of these were published in the *Zeitschrift für Flugtechnik* and came from the Göttingen group. Four of them were by Albert Betz and one by Carl Wieselsberger. I describe them briefly, keeping to the chronological order of their appearance.

In September 1912 Betz published some wind-channel results that showed that a wing operating in the vicinity of the ground would experience an increase in lift.[61] Betz showed this by testing a model wing in a channel fitted with a false floor that could be raised or lowered. The phenomenon was an important one. Aircraft necessarily fly near the ground on landing and takeoff. Pilots were aware that there was a change in flying characteristics produced by these circumstances, but the nature of the change was little understood. This "ground effect" explains why an overloaded aircraft can sometimes take off with apparent success and then fail to gain height, with disastrous consequences. It also explains why some early aircraft could "fly" but never got more than a few feet above the ground.[62] Betz also wanted to get a quantitative estimate of the effect of the walls of a wind channel on the measurements that were carried out in the course of experimentation. He showed that Prandtl's new theory could lead to rough but quantitative predictions that were confirmed by experiment. (The results were approximate, Betz suggested [220], because the "horseshoe" model ignored the downward motion of the trailing vortices.) Both of the subjects that Betz broached in his brief paper were to become a matter of enduring concern and research in subsequent years.

In January 1913, Betz published a second study, this time of the lift and resistance of a biplane.[63] Whereas Föppl had used the Biot-Savart law to study the effect of the induced velocity on the tail wing, Betz now used the same approach to study the mutual interaction of wings that were positioned one above the other. The central point about the application of Prandtl's approach to a biplane is that the trailing vortices from the upper wing will generate an induced resistance not only in the upper wing itself but also in the lower wing, while the trailing vortices from the lower wing will likewise affect both wings. Furthermore, if the wings are not located directly one above the other, the bound vortex corresponding to the wing itself (and not just the trailing vortices) will have to be taken into account when computing the induced velocity and induced drag on the other wing.

With the exception of Kutta's second, 1911 paper, this work represented the first serious engagement with the theoretical aerodynamics of the biplane and the difficult problem of the mutual interaction of the different parts of an aircraft. It will be recalled that the "practical men" in Britain stressed holistic effects to justify their conviction that only the intuition of the engineer could cope with the problems of airplane design. Scientists and mathematicians, they said, simplified problems by studying one part at a time, which doomed them to failure. Such a procedure ignored the all-important effects of inter-action. Perhaps (had they known about it) the "practical men" would have been impressed to be told of the progress that was being made in Göttingen. Here engineers, such as Betz, were using the Biot-Savart law to put the study of interaction on a mathematical as well as an experimental basis.

Betz carried out wind-channel measurements of the lift and resistance of a set of two wings rigidly fastened into a biplane configuration. He studied (*a*) the effect of varying the distance apart of the wings, (*b*) the effect of giv-ing the wings different angles of incidence from one another (*décalage*), and (*c*) the effect of placing one wing ahead of the other (stagger). He found that the effects were small within the range he studied, though the most signifi-cant variable was the stagger of the wings. One of his practical concerns was to form some idea of the relative merits of monoplanes and biplanes. He summed up his results in four propositions: (1) A biplane arrangement with wings of equal span always has a less favorable ratio of lift to resistance than one of the wings taken separately. (2) A biplane can have advantages over a monoplane when the rest of the resistance of the aircraft, for example, a bulky fuselage, is taken into account. (3) A biplane is at an advantage if a high lift at low speeds is required. (4) The greatest maximum lift is obtained when the upper wing of a biplane is placed ahead of the lower wing and is given a slightly smaller angle of incidence than the lower wing. All of these results, said Betz, were rendered intelligible by Prandtl's theory, and the empirical graphs of lift and resistance were duly accompanied by theoretical curves calculated from the theory.[64]

In neither of his papers did Betz specifically mention, or illustrate the use of, the Biot-Savart law. He alluded to the horseshoe model but revealed none of the mathematics involved in his calculations. Like Föppl he promised the reader that a fuller account was to follow from the pen of Prandtl himself. The Great War began in July 1914, but there seemed no immediate concern with secrecy. In a paper that appeared in August 1914, Wieselsberger preempted Prandtl and stated the Biot-Savart law explicitly and illustrated its application.[65] He asked why birds often fly in a V formation. He did not manage to answer the question, but he did succeed in laying out the basic ideas, and the basic mathematics, of Prandtl's theory. In approaching the problem of formation flying, Wieselsberger ignored the beating wing motion involved in bird flight and treated birds as small airplanes. He then followed Prandtl and treated the airplane as a horseshoe vortex. By the use of the Biot-Savart law he showed that on either side of the horseshoe vortex there would be an updraft. This, he argued, allowed another wing, positioned to one side of the first wing, to operate at a more favorable angle of attack. This lowered the component of induced resistance in the direction of flight. On the basis of some plausible numerical assumptions, he made a quantitative estimate of the advantages to be derived from flying in the updraft of neighboring birds. His overall model, however, led to the conclusion that side-by-side flight would be just as efficient as the V formation.

In September 1914 Betz produced a study of wings with a sweepback and a twist at their ends,[66] a configuration frequently used by designers of German aircraft at that time. The name *Taube*, or "dove," was given to such machines. In Betz's paper there was a passing reference to yet another formula attributed to Prandtl and his new theory, though again no derivation was given. The formula concerned the minimum glide-angle that could be expected for a wing of given span and lift. The main result of Betz's experiments on a range of Taube-style wings was to confirm the near optimum character of very simple, rectangular wings. Having neither twist nor sweepback, such wings also had an economic and practical advantage: they were easy to construct. The glide coefficient (given by the ratio of resistance over lift) was not significantly improved by sweepback or twist, though Betz did find they improved longitudinal stability.

Perhaps because the promised theoretical paper from Prandtl was not forthcoming, Betz finally published his own account of the mathematics underlying his papers. Titled "Die gegenseitige Beeinflussung zweier Trag-flächen" (The mutual influence of two wings),[67] the work appeared in the *Zeitschrift für Flugtechnik* for October 1914. Betz concentrated on the case of the staggered biplane with wings of equal span where the upper wing was positioned ahead of the lower wing. Because the analysis proceeded on the assumption that each wing and vortex system could be represented by the simple "horseshoe" schema, the only real novelty in the paper lay in the more complex geometry of the computations, but the explicit development of the mathematics of the theory demonstrated its applicability to what was then a vitally important form of aircraft. It was clear that Prandtl and his colleagues now had a theory that could be used to predict the induced resistance of bi-planes, or triplanes, using only the wind-channel data for a single wing.

`In the same year, 1914, Wieselsberger also published a survey article that described the state of knowledge in German aerodynamics with respect to lift and drag. It did not appear in the *ZFM* but in an Austrian journal, the *Österreichische Flug-zeitschrift*.[68] The article covered both two-dimensional and three-dimensional theory and took the reader through the work of Kutta, Joukowsky, Deimler, and Blumenthal and up to Prandtl's horseshoe vortex. Wieselsberger's survey effectively brought up to date an earlier survey by Reissner, of the TH in Aachen, which had laid stress on questions of stability and propeller theory.[69]

The international situation had been deteriorating throughout 1914, and British statesmen, such as Lord Haldane, became increasingly worried about the "war party" surrounding the German kaiser.[70] With the threat of war, it was ever more important for European countries to monitor the technology of their potential enemies. If anyone had wanted to keep an eye on German aviation, the papers of Föppl, Betz, and Wieselsberger would have given them all they needed to know about the general state of scientific knowledge in the field of aerodynamics. These publications would have made clear that the circulation theory of lift was wholly taken for granted in Göttingen and the German-speaking world. Collectively, the publications showed that the theory had been developed to the point where it was being applied to problems of practical importance. Betz's theoretical analysis of the biplane, however, was the last of the Göttingen research papers to appear in an open and accessible format. Thereafter they would be hidden away from public view in the *Technische Berichte*, published in individually numbered copies by the military authorities and marked *Geheim*—"secret." In the meantime, the Göttingen results were in the public realm and were available to anyone in Cambridge or London who cared to study them.

Making the Horseshoe Model More Realistic

Prandtl never produced the promised article in the *Zeitschrift für Flugtechnik*. This was not because he harbored reservations about the approach. On the contrary, he was happy to produce accounts for general surveys, for example, in volume 4 of the *Handwörterbuch der Naturwissenschaften* published in 1913. The handbook was an encyclopedic survey of the state of the natural sciences and contained articles by both Fuhrmann and Prandtl. Fuhrmann wrote on hydrostatics, and Prandtl wrote on fluid dynamics.[71] In his contribution Prandtl gave an explicit account of the circulation theory and presented a graph contrasting Kutta flow with Kirchhoff-Rayleigh flow (136). He also cited Lanchester's work and gave a diagram (112) that laid out the qualitative basis of the horseshoe model, though the Biot-Savart law was not mentioned by name. Why the hesitation? The simple horseshoe model was clearly in a provisional state and was still undergoing revision. It contained formal features that compromised both its empirical adequacy and its practical utility. Despite the successes of the theory, it would have been understandable if Prandtl had wanted to remove these limitations before presenting the approach to a specialist readership. The time was hardly ripe for an authoritative presentation, which may explain the non-appearance of the article. Then the war intervened, and the form and level of presentation at which he seems to have been aiming were not achieved until 1918.

The problems with the "horseshoe" vortex were both mathematical and physical and were closely interconnected. Mathematically there was the difficulty arising from the singularity in the Biot-Savart formula which has already been remarked on, that is, the problem that arises when $h = 0$. The formula implied that the velocity of the downwash at the wingtips became infinite. The formula yields this result because of the uniformity of the vortex distribution implied by the model, that is, the constant value of the circulation along the bound vortex and hence along the span of the wing. This was a physically false picture. The existence of lift implies that there must be a greater pressure beneath the wing than above it, but the finite length of a real wing allows the air at high pressure beneath the wing to move round the tip to occupy the lower-pressure region above the wing. Such freedom of movement ensures that the pressure difference between the upper and lower surface will be zero at the tips. There will therefore be no lift at the tips and hence no circulation. Circulation cannot be constant along the span in the way that was assumed in the simple horseshoe model; it must fade away to zero at the tips.

Prandtl's problem was to find a model with a more realistic lift distribution along the span of the wing. His response was ingenious. He complicated the simple horseshoe model by introducing a number of horseshoe vortices laid out in the fashion indicated in figure 7.11. (A similar figure was used in an early article by Betz.)[72] Starting from a single "horseshoe" whose span coincided with the full span of the wing, he added others of smaller span. The parts of the vortex that lie along the span are to be thought of as piled on top of one another. In this way the constant distribution of circulation along the span is replaced by a variable, stepwise distribution with a maximum at the midpoint. The arrangement had the consequence that vortices now trailed from a number of points along the rear edge of the wing, rather than merely at the wingtips. This stepwise model, however, was only the starting point of Prandtl's line of reasoning.

Prandtl did not simply introduce a number of horseshoe vortices such as the five in the diagram, or even 50 or 500. He introduced an infinite number. He postulated an infinite number of vortices of infinitesimal strength. The vortices were infinitesimal for two reasons. First, an infinite number of vortices of *finite* strength would result in the absurdity of a wing with infinite circulation and infinite lift. Second, he needed the circulation and the lift at the tips to approach zero. A stepwise model with finite vortices would merely reproduce the problem that dogged the original. The vortices had to become infinitely small at the wingtips. Along the span of the wing the infinitesimal vortices were assumed to be compressed into a single line of bound vorticity (of varying strength) called the lifting line. These refinements made it possible to imagine a smooth, rather than stepwise, lift distribution that was amenable to mathematical treatment. To accord with the known facts, the

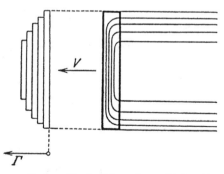

FIGURE 7.11. Stepwise complication of the simple horseshoe model. Prandtl made the horseshoe model more realistic by multiplying the number of horseshoe vortices and imagining them stacked on top of one another. From Tietjens 1931, 209. (By permission of Springer Science and Business Media)

smooth lift distribution had to have a maximum lift at the midpoint of the span and approach zero lift at the wingtips.

Having described Prandtl's refined model in qualitative terms, I now show how he expressed these ideas in mathematical terms. This account will prepare the ground for the next two sections, which describe the technical and mathematical heart of the Göttingen achievement.

Suppose the wing has a span b and lies along the x-axis of a coordinate system so that it runs from $x = -b/2$ to $x = +b/2$. The distribution of the circulation can then be represented by $\Gamma(x)$. The symbol indicates that, for every value of x along the axis between the wingtips, there corresponds a specific value of Γ, the circulation. Thus $\Gamma(0)$ is the value at $x = 0$, the origin, which, following convention, is taken as the center of the wingspan. It is known from experiment that the lift is at its maximum value at this central position. Because it plays an important role, it is customary to give the circulation at this point a special designation and write $\Gamma(0) = \Gamma_0$. The lift and hence the circulation is zero at the tips, so that $\Gamma(-b/2) = 0$ and $\Gamma(+b/2) = 0$. For the moment, and for the purpose of conveying the main outlines of Prandtl's theory, the actual shape of the lift distribution need not be given in more detail than this. The mathematical shape described by the function $\Gamma(x)$ will, for the moment, remain unspecified, but it will be some smoothed-out version of the shape made by the stepwise lift distribution. The details are reserved for the next section. For the remainder of this section, the distribution is simply referred to as $\Gamma(x)$ so that the general structure of the mathematical reasoning can be rehearsed. My aim is to show, in general terms, how Prandtl used the Biot-Savart law to calculate the lift, the induced velocity, and the induced drag.

The first step was to relate each of the infinitesimal horseshoe vortices to the Biot-Savart law. The relevant version of the formula for a vortex of finite strength has already been stated, namely, $w = -\Gamma/(4\pi h)$. Because the analysis was now to be applied to infinitesimal vortices, the formula became $dw = -d\Gamma/(4\pi h)$. The goal was to calculate the downwash at some specified point on the wing with the coordinate, say, $x = x'$. All of the infinite number of trailing vortices (each coming away from the wing at some point with its own specific x-coordinate) will contribute to the downwash at the point x'. The Biot-Savart law gave the (infinitesimal) contribution dw made by each of these infinitesimal vortices. The perpendicular distance h in the formula needed to be re-expressed as $(x' - x)$. This was the distance between the point on the wing from which the infinitesimal vortex emerges and the point x' at which the downwash was to be found. A process of integration that adds the contribution of all the infinitesimal trailing vortices would then give the total downwash at x'. A further calculation, and a further integration, was needed to get the downwash for the entire wing, that is, for all the points like x' which lie along the span between $x = -b/2$ and $x = +b/2$.

The procedure that has just been sketched was based on the assumption that the quantity $d\Gamma$ used in the Biot-Savart formula corresponded to the strength of the infinitesimal vortex at the arbitrary point x. How was this infinitesimal strength to be expressed? The answer was that the strength of the element of trailing vorticity issuing from a point x was equal to the *change of vorticity* on the wing at that point. This can be explained by going back to the stepwise model of a finite number of finite vortices that was shown in figure 7.11. First the outer horseshoe is put in place. Suppose this has strength Γ_1. Then the second horseshoe is added, which has strength Γ_2 and a slightly shorter span, then Γ_3 is added, which again has a slightly shorter span, and so on. Consider the two points on either side of the origin of the x-axis from which the trailing vortices of strength Γ_2 emerge. These are the points at which the distribution of circulation changes by an increase of the amount Γ_2. Thus the *strength* of vorticity trailing from the wing at that point equals the *change* in vorticity around the wing at that point.

This "strength equals change" rule holds even when there are an infinite number of infinitesimal horseshoe vortices. The distribution of circulation along the span is given by the curve $\Gamma(x)$, so the change in circulation is the slope of the graph of $\Gamma(x)$ multiplied by the distance over which the slope reaches. The slope is $\partial\Gamma/\partial x$, and the distance is dx, so the change whose value is sought is $d\Gamma = (\partial\Gamma/\partial x)\, dx$. This expression gave the strength of the circulation or vorticity to be entered into the formula for the Biot-Savart law. The infinitesimal contribution of the vorticity at x to the downwash at x' was therefore

$$dw = -\frac{(\partial\Gamma/\partial x)\, dx}{4\pi(x'-x)}.$$

The total downwash at the point x', designated by $w(x')$, is the integral of all of these infinitesimal contributions, summed over all the vortices issuing from the whole span of the wing. Thus,

$$w(x') = -\frac{1}{4\pi}\int_{-b/2}^{+b/2}\frac{(\partial\Gamma/\partial x)\, dx}{x'-x}.$$

The above integral has a singularity at $x = x'$, when the denominator becomes zero, but the integration could be carried out in such a way as to avoid this problematic point.

Given the downwash it was then possible to calculate the induced angle
of incidence at x'. This angle, φ, follows from the value for $w(x')$ because it
was simply the angle made by combining the downward induced velocity
with the free-stream velocity. The ratio of the two speeds gave the tangent of
the angle φ, but because the angle was small, the angle and tangent could be
equated. The induced angle of incidence was

$$\varphi(x') = \frac{w(x')}{V}.$$

The lift distribution could now be related to the overall lift and induced drag.
Recall that for an infinite wing the flow at every cross section resembles that
at every other cross section. The lift per unit length is constant and is given by
the Kutta-Joukowsky formula as $L = \rho \Gamma V$. Prandtl took this formula to apply
to each separate, *infinitesimal* element of a three-dimensional wing, with the
proviso that the circulation would vary from element to element according
to the distribution $\Gamma(x)$. The overall lift could then be represented by the
integral of all the elementary lifts: $dL(x) = \rho\, V\Gamma(x)dx$. Thus,

$$Lift = \rho V \int_{-b/2}^{+b/2} \Gamma(x)dx.$$

Each point on the wing would generate an element of downward velocity and
would thus be subject to a slight downward slope in the local flow. The element of lift $dL(x)$ at that point would be tilted backward (relative to the main
flow) so that the resultant force possesses a component opposing the motion.
This was the induced drag. The induced drag at a given point x depended on
the induced angle of incidence φ at that point. The component of induced
drag resulting from the backward tilt equals $dL(x) \sin\varphi(x)$. For small angles
the sine of φ is equal to φ itself, so the element of induced drag was $dL(x)$
$\varphi(x)$. Thus the total induced drag was given by the integral

$$Drag = \rho V \int_{-b/2}^{+b/2} \Gamma(x)\varphi(x)dx.$$

This relation could be expressed in terms of a coefficient of induced drag by
dividing the value of the drag force itself by $\frac{1}{2}\rho V^2 F$, where F is the area of the
wing. This gave the coefficient of induced drag as

$$C_{D_i} = \frac{2}{VF} \int_{-b/2}^{+b/2} \Gamma(x)\varphi(x)dx.$$

It will be evident from these formulas that a closely knit structure of theoretical relations was emerging in Göttingen which connected lift, drag, span, and the distribution of circulation along the span of a wing. For the purposes of exposition I have only presented this structure in a schematic form. The mathematical formulas just given all depend on the distribution of the circulation, $\Gamma(x)$, but the actual character of the function governing the distribution has remained unspecified. All that the above formulas entail is that *if* the distribution $\Gamma(x)$ is given, *then* the lift, the induced angle of incidence, and the induced drag can be calculated. Only when the distribution is specified will the theory will have real content. The next question is: How was the distribution of lift and circulation found? How is $\Gamma(x)$ to be defined?

The Elliptical Distribution of Lift

Ideally the shape of the graph of $\Gamma(x)$ showing the distribution of circulation, and hence lift, along the span of the wing, would be deduced from first principles. The deduction would start with the governing equations of fluid motion and, by inserting data about the shape of the wing and the angle of attack, the mathematics should yield the function $\Gamma(x)$ relating the circulation to the x-coordinate along the span. This, said Prandtl later, was the first question that he and his group posed for themselves but the last one to be answered. (During the war the problem was solved for a rectangular wing by Betz. His analysis formed the substance of a 1919 inaugural dissertation submitted to Göttingen.)[73] Initially, however, it was necessary to proceed by trial and error and under the guidance of experiment. The character of the lift distribution along the wingspan could be established empirically by pressure measurements made on a model of the wing in a wind channel. If a mathematical representation could be found for the distribution, and if that function could be integrated, then the equations of the theory (given in the last section) could be employed to deduce further characteristics of the wing. The function $\Gamma(x)$ governing the distribution of circulation had to be (1) empirically plausible and (2) mathematically tractable. Experimentally it transpired that most of the wings used in practice had a similar distribution of lift and hence circulation along their span. There was a strong "family resemblance" between their distributions, and the family in question was well known.[74] The *distribution typically resembled the upper half of an ellipse*. The expression $\Gamma(x)$ is essentially nothing more than the equation for an ellipse.

The equation for an ellipse is simple. Using standard x- and y-coordinates, the ellipse that has a major axis of length b in the x-direction and a minor axis of length a in the y-direction is represented by the equation

$$\left(\frac{x}{b/2}\right)^2 + \left(\frac{y}{a/2}\right)^2 = 1.$$

The ellipse is shown in figure 7.12a. If the y-axis is used to represent the circulation, then the formula describing an elliptical distribution of circulation of the kind shown in figure 7.12b is, by analogy,

$$\left(\frac{x}{b/2}\right)^2 + \left(\frac{\Gamma}{\Gamma_0}\right)^2 = 1.$$

The ellipse has one semi-axis of length Γ_0 (the maximum circulation) and the other semi-axis of length $b/2$ (the half span). This formula can be manipulated

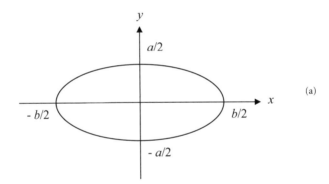

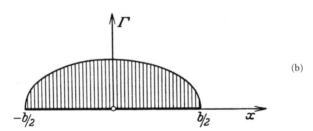

FIGURE 7.12. The geometry of an ellipse provides a model for the lift distribution (a). In (b) there is continuous distribution of lift and circulation $\Gamma(x)$ along a wingspan of length b. The distribution shown here is elliptical. Note zero lift at tips located at $-b/2$ and $+b/2$ and a maximum at the origin $x = 0$. From Tietjens 1931, 213. (By permission of Springer Science and Business Media)

to give an expression for Γ (as a function of x). An elliptical lift distribution is thus given by

$$\Gamma(x) = \Gamma_0 \sqrt{1 - \left(\frac{x}{b/2}\right)^2}.$$

With this formula at hand, the reasoning set out in general terms in the previous section can be reworked to produce quantitative predictions about lift and induced drag. Following this line of reasoning, Prandtl was able to generate three important results. First, he showed that the induced drag along the span of a wing should be constant if the distribution of circulation is elliptical. Second, he showed that under these conditions the induced drag should increase according to the square of the lift coefficient. Third, he predicted a relation between the induced drag and the planform of the wing. He showed that induced drag should be inversely proportional to the aspect ratio. The narrower the wing, the lower the induced drag. This result had immediate implications for the aircraft designer. I shall now show how he reached these conclusions mathematically and then say more about their importance.

Differentiation of the formula for $\Gamma(x)$ gives the expression $d\Gamma/dx$, and the result can be substituted into the formula for the induced velocity, or downwash, that was arrived at in the previous section. The differentiation of the elliptical distribution gives

$$\frac{d\Gamma}{dx} = -\frac{4\Gamma_0}{b^2} \frac{x}{\left(1 - 4x^2/b^2\right)^{1/2}}.$$

The induced velocity is then

$$w(x') = \frac{\Gamma_0}{\pi b^2} \int_{-b/2}^{+b/2} \frac{x}{\left(1 - 4x^2/b^2\right)^{1/2}(x'-x)} dx.$$

The next problem was to evaluate the complicated-looking integral in order to give an actual value to the downwash. It turned out that the integral reduced to a very simple expression. It was equal to $-\pi b/2$. The induced velocity at the point x' of the wing was then

$$w(x') = -\frac{\Gamma_0}{2b}.$$

The induced angle of incidence for this point on the wing follows immediately:

$$\varphi(x') = -\frac{w}{V} = \frac{\Gamma_0}{2bV}.$$

Two features of these formulas for w and φ deserve notice. First, inspection shows that all the quantities that enter into them are constants. The span of the wing, the speed of the free stream, and the value of the circulation at the center of the wing do not change as different positions along the span come under consideration. Both expressions are therefore independent of x'. It follows that for any given wing, provided it has an elliptical lift distribution, both the induced velocity and the induced angle of incidence are constant along the span. The unreal, infinite, induced velocities at the wingtip have been avoided. This was progress.

The second point of note is that both of the formulas have b, the span of the wing, in the denominator. Thus, as b approaches infinity, both w and φ approach zero. The theory therefore implies that for an infinite wing there will be no downwash, that is, no induced velocity, and thus no induced drag. This fits perfectly with the previous work of Kutta and Joukowsky. For an infinite wing moving through an ideal fluid, the absence of induced drag means the absence of all drag, and this was one of the more disconcerting consequences of their analysis. However, the results of the two-dimensional theory turn out to be a limiting case of the more realistic, three-dimensional theory. On Prandtl's approach, Kutta and Joukowsky were not studying the unreal aerodynamics of an imaginary world; they were studying the aerodynamics of the real world but dealing with limiting cases.

The value of the circulation $\Gamma(x)$ given earlier by the formula for the elliptical distribution can be inserted into the Kutta-Joukowsky law. This gives an expression for the total lift of a wing with an elliptical distribution:

$$Lift = \rho V \Gamma_0 \int_{-b/2}^{+b/2} \left(1 - \frac{4x^2}{b^2}\right)^{1/2} dx.$$

A change of variable simplified the integration and gave

$$Lift = \rho V \Gamma_0 \frac{b}{4} \pi.$$

If the lift is now expressed as a coefficient and the equation is rearranged to give an expression for Γ_0, the maximum circulation, it becomes

$$\Gamma_0 = \frac{2VFC_L}{b\pi},$$

where F is the area of the wing. This value of Γ_0 can be inserted into the previously derived expression for the induced angle of incidence, which gives

$$\varphi = \frac{FC_L}{\pi b^2}.$$

This expression yields a relation between the induced angle of incidence and the planform of the wing.

Because of its importance it is worth making this relation explicit and restating the formula. For a rectangular wing of, say, span b and chord a, the area $F = ab$ and the aspect ratio is b/a. This definition of the aspect ratio can be generalized for more complicated shapes. For wings that do not have a constant chord, the chord length can be replaced by F/b, that is, the area divided by the span, thus giving the aspect ratio as b^2/F. If the aspect ratio is represented by the symbol A_R, then the above formula for the induced angle of incidence becomes

$$\varphi = \frac{C_L}{\pi A_R}.$$

This expression reaffirms the point made previously—that for a wing with an elliptical distribution, the induced angle of incidence will be constant along the span. The utility of the new formulation, however, is that it leads to a revealing expression for the induced drag. To arrive at this result it is only necessary to insert the above expression for φ into the formula for the coefficient of induced drag given previously. In the last section it was shown how Prandtl's theory had given the following general result for the coefficient of induced drag:

$$C_{D_i} = \frac{2}{VF} \int_{-b/2}^{+b/2} \Gamma(x)\varphi(x)dx.$$

Because an elliptical distribution for the circulation has been assumed, all of the component parts of this integral are now known. The expression for φ is a constant whose value has just been expressed in terms of the aspect ratio. It can thus be taken out from beneath the integration sign. The remaining integral of the elliptical shape $\Gamma(x)$ has already been evaluated. These results can be combined so that (again using A_R to signify the aspect ratio) the coefficient of induced drag can now be written in the form

$$C_{D_i} = \frac{C_L^{\,2}}{\pi A_R}.$$

This formula expressed a highly significant result. It indicated two things. First, it showed that the induced drag increases rapidly with increased lift. The drag grows with the square of the lift coefficient. Second, it implied that to reduce the drag it was necessary to increase the aspect ratio of the wings. It therefore carried an important lesson for the aircraft designer because it linked a specific design feature of a wing to definite aerodynamic effects. The significance of the aspect ratio of a wing had long been recognized at an empirical level, but now a fundamental, theoretical understanding was emerging.

This deeper understanding had a typically engineering character to it. It identified the need to trade one advantage against another advantage. It pointed to the costs that had to be paid and the compromises that had to be made to get the benefits of increased lift and decreased drag. Increased lift brought increased induced drag. Induced drag could be reduced by increasing the aspect ratio, but an engineer would immediately see a problem. High aspect ratio may be desirable, but a long, narrow wing is not easy to build. Such a wing confronts the designer with the problem of how to make it strong enough without making it too heavy.

Two Practical Conversion Formulas

The formula given earlier linking the induced drag C_{D_i} to the lift C_L led Prandtl to two further results that were significant for engineering and indicative of the power of the analysis he was developing. He discovered two practical formulas that allowed him to reduce drastically the amount of time spent conducting wind-channel tests. It allowed him to generalize results about induced drag and induced angle of incidence from a wing that had been tested to other wings with a different aspect ratio that had not been tested. All that was necessary was that the wings have the same cross-sectional profile. In this section I explain how Prandtl was able to do this.

The formula (deduced above) for the coefficient of induced drag, namely,

$$C_{D_i} = \frac{C_L^{\,2}}{\pi A_R},$$

is analogous to the equation $y = ax^2$, where $y = C_{Di}$ and $x = C_L$, while a is the constant $1/\pi A_R$. Equations having the form $y = ax^2$ generate curves in the shape of a parabola. The constant a determines the curvature of the parabola. According to Prandtl's theory, then, the curve linking the coefficients of induced drag and lift will be parabolic, and the curvature will depend on the aspect ratio. Is this theoretical deduction confirmed experimentally? The "polar curves," as Lilienthal called them, which link the observed coefficients of lift and drag are, indeed, roughly parabolic, but some distinctions need to be drawn before such results can be related to the parabolic formula for the induced drag.

Induced drag arises in an ideal fluid but also plays a role in real fluids. There are, however, other sources of drag that are present in real fluids. A real wing in real air will experience some degree of skin friction due to the viscosity of the air. As Betz had previously argued, the flow of air over a real wing will also generate eddies, and the flow over the upper and lower surfaces will not join together smoothly at the trailing edge. This, too, is a source of drag. Prandtl grouped both of these latter sources under the name "profile drag." Thus the empirical phenomenon of drag is the combined effect of induced drag and profile drag. The empirical coefficient of drag can be understood as the sum of the coefficient of induced drag and the coefficient of profile drag. Thus, using obvious notation,

$$C_D = C_{Di} + C_{Dp}.$$

If the theoretical polar curve relating induced drag and lift is plotted on the same diagram as an empirical curve relating total drag to lift, it turns out that the curves are similar but not identical. They lie close to one another but do not overlap. Their relative position shows that most of the drag on the wing is induced drag. This is particularly true at high angles of incidence (see fig. 7.13).

Prandtl realized that the contingent relation between the two drag curves could be exploited. He spoke of a "fortunate circumstance that had not been suspected at all at the outset"—"einen glücklichen Umstand, der von vornherein keineswegs vermutet wurde."[75] It was a stroke of good fortune that allowed the results of his wing theory to be cast into a form that was highly useful for practical purposes ("gelang es nun, die Ergebnisse der Tragflügeltheorie auch für die Praxis noch wesentlich fruchtbarer zu gestalten"). As Prandtl explained:

> Man trug nämlich die Polarkurven bei sonst gleichen Tragflügeln, aber von
> verschiedenen Seitenverhältnissen auf und erkannte, daß der Unterschied der
> gemessenen Widerstandszahl von der theoretischen in allen Fällen nahezu

der gleiche war. Daraus war zu schließen, daß der Profilwiderstand unabhängig vom Seitenverhältnis ist, woraus sich weiterhin die Möglichkeit ergab, gemessene Polarkurven von einem Seitenverhältnis auf ein anderes Seitenverhältnis umzurechnen. (219–20)

The polar curves of wings that are the same, apart from having different aspect ratios, were laid out, and one could see that the difference between the measured and the theoretical resistance coefficient was approximately the same in all cases. It had to be concluded from this that the profile resistance was independent of the aspect ratio. This raised the possibility that the measured polar curve for one aspect ratio could be converted into that for another aspect ratio.

Prandtl's compressed reasoning can be broken down into two steps. First, the total drag is the sum of induced drag and profile drag (and the larger and more important of these two quantities is the induced drag). Second, if the profile drag C_{Dp} is roughly the same for wings of all aspect ratios, then the *difference* between total and induced drag would be the same if the wings were operating at the same coefficient of lift. In other words, the quantity $(C_D - C_{Di})$ would be a constant. This quantity could therefore be equated for wings of different aspect ratio, provided they both had the same profile and both had an elliptical lift distribution. The implication was that given two wings, wing$_{(1)}$ and wing$_{(2)}$, the total drag of wing$_{(2)}$ could be predicted once

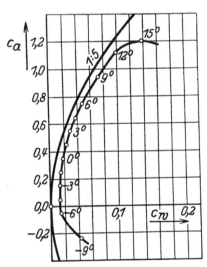

FIGURE 7.13. Two polar curves relating induced drag and lift. The curve at the front on the left is derived from theory, the other is plotted from measurements. From Tietjens 1931, 219. (By permission of Springer Science and Business Media)

it was known for wing$_{(1)}$. If the aspect ratios and the coefficients of total and induced drag for the respective wings are also distinguished by the labels (1) and (2), Prandtl was able to write

$$C_D(1) - C_{Di}(1) = C_D(2) - C_{Di}(2).$$

It had already been established that

$$C_{Di} = \frac{C_L^{\,2}}{\pi A_R},$$

so this result could be substituted into the above equation to give

$$C_D(1) - \frac{C_L^{\,2}}{\pi A_R(1)} = C_D(2) - \frac{C_L^{\,2}}{\pi A_R(2)}.$$

Rearranging the equation gives the drag of one wing in terms of the drag of the other, at the same value of the lift coefficient. The formula thus allows knowledge about one wing to be converted into knowledge about the other wing. The conversion formula was thus

$$C_D(2) = C_D(1) + \frac{C_L^{\,2}}{\pi} \left(\frac{1}{A_R(2)} - \frac{1}{A_R(1)} \right).$$

A second conversion formula was then deduced. This formula dealt not with drag but with the angle of incidence. Once again it converted knowledge gained from one case into knowledge applicable to other cases. The second formula implied that if the angle of incidence associated with a given lift is known for a wing of one aspect ratio, then the angle at which that lift was produced could be predicted for another wing of the same profile but a different aspect ratio. In this case the reasoning depended on the relation between finite wings and an infinite wing with the same profile.

Prandtl argued as follows. Suppose an infinite wing, of a given profile, meets a horizontal airstream at an angle α_0. Let the lift coefficient be C_L. What would happen if this profile were to meet the air at the same speed but now as part of a *finite* wing, not an infinite wing? Prandtl had shown that the effect of the vortices, which now trail from the tips, is to induce a downward flow of the air that presents itself to the wing. This induced angle φ reduces the effective angle of incidence of the wing. If the finite wing is to generate the same amount of lift per unit length as the infinite wing, then it must be restored to the same angle relative to the local flow that it originally had to the free stream. This can only be done if the angle to the horizontal is increased.

The angle of attack α will have to be made equal to the original angle α_0 plus the induced angle of incidence φ. Thus, $\alpha = \alpha_0 + \varphi$. Rearranging the equation leads to $\alpha_0 = \alpha - \varphi$. This expression implies that, for all wings of the same profile, the difference between the angle of attack and the angle of the induced flow will be the same when they are delivering the same amount of lift. Thus for two such wings, using obvious notation,

$$\alpha_0 = \alpha_1 - \varphi_1 = \alpha_2 - \varphi_2.$$

Suppose that wing$_{(1)}$ has the lift coefficient C_L at α_1 and wing$_{(2)}$ has the same lift coefficient at α_2. Prandtl had already arrived at an expression for the induced angle of incidence φ, so he could write

$$\alpha_0 = \alpha_1 - \frac{C_L}{\pi} \frac{1}{A_R(1)} = \alpha_2 - \frac{C_L}{\pi} \frac{1}{A_R(2)}.$$

This gives the second conversion formula

$$\alpha_2 = \alpha_1 + \frac{C_L}{\pi} \left(\frac{1}{A_R(2)} - \frac{1}{A_R(1)} \right).$$

If these two formulas stood up to test, they would fulfill the desiderata for work in technical mechanics identified by August Föppl. I described in chapter 5 how Föppl had insisted that the role of time in the economy of knowledge was different for the engineer compared to the physicist. The value of the conversion formulas was that they would enormously lighten the work load of the engineer engaged in wind-channel research.

The formulas were first published by Betz in 1917 in the confidential *Technische Berichte*.[76] They were then tested in Göttingen by taking wings of different aspect ratios in order to see if the measurements for drag and angle of incidence could be converted into the values for one, arbitrarily chosen, aspect ratio. The first such test was performed by Munk and was also reported in the *Technische Berichte*.[77] Munk used just three different aspect ratios and, in order to keep the section of the wings as constant as possible, simply started with a long span of wing and sawed off the ends to produce the shorter wings. In this way he produced wings of aspect ratio 6, 5, and 4. Munk verified the formulas by calculating the results for the wing of aspect ratio 6 from the other two aerofoils and plotting the three sets of points in the same graph. Later the experiment was repeated with seven different aspect ratios and produced the same positive result. Except for the measurements taken on one wing of very low aspect ratio, the predictions worked well. The conversion formulas did what they were meant to do, that is, collapse all the experimental results into one and the same curve.[78]

Any direct test of the conversion formulas was also an indirect test of the theoretical assumptions on which they were based. As well as sanctioning a practical shortcut that avoided much time-consuming work with the wind channel, the positive results of the test were a corroboration of Prandtl's overall analysis. But all of the reasoning rested on the assumption that the lift distribution was elliptical. This facilitated the calculations but made the result a special case. Could the result be generalized? In November 1913 Prandtl and E. Pohlhausen had established that the induced drag for an elliptical lift distribution was not only constant along the span but represented a minimum value.[79] Any deviation from an elliptical distribution would give a higher value for this form of drag. It was also soon established that the actual planform that produced an elliptical lift distribution was itself of an elliptical shape. This was not because there is any simple rule to the effect that wings generate distributions that mirror the shape of their planform. In general, the shape of a wing does not immediately correspond to that of the resulting lift distribution. A rectangular wing does not yield a rectangular lift distribution. But, despite having the character of special case, it turned out that all the results derived for the elliptical wing could be generalized. The mathematical apparatus that has just been sketched could be applied, without significant loss of accuracy, to non-elliptical wings, for example, to the simple-to-construct rectangular wing that was used as the baseline or "norm" (the *Normalflügmodell*) in the Göttingen profile tests.[80]

The empirical basis for the generalization has already been mentioned. It rests on the "family resemblance" between the lift distributions of all typical wings. Though their lift distributions are not strictly elliptical, they are, mostly, roughly elliptical. As Prandtl pointed out, while the true ellipse gives the minimum possible induced drag, many mathematical functions change their values slowly in the vicinity of a minimum. Results that hold for the minimum are often found to hold, at least approximately, in the neighborhood of the minimum.[81] Thus the Göttingen results had a practical applicability, and a predictive power, that went beyond what might have been expected, given the specialized, and often unreal, assumptions on which they were based. Looking back, some quarter of a century after the creation of Prandtl's theory, Richard von Mises summed up the situation as follows: "It seems appropriate to stress the fact that . . . the parabolic form of the polar diagram and the dependence of this form on the aspect ratio, and the relation between lift coefficient, angle of attack, and aspect ratio, were not known as empirical facts before the wing theory was developed. These facts . . . have been predicted by the theory. Experiments carried out a posteriori have confirmed these theoretical predictions to a degree that is remarkable in view of the numerous idealisations of the theory."[82] Von Mises was not a wholly unqualified admirer of the Göttingen group—he thought they cited one another too much—but the word "remarkable," applied to the success of the Göttingen theory, was reiterated in his book *Theory of Flight*.[83] The repetition attests to the striking and, it would seem, almost baffling power of Prandtl's work. Let us look a little more closely at some of the methodological features that were associated with this success.

Idealization as the Route to Realism

Prandtl took a significant step toward greater realism when he went beyond the idealization of the infinite wing. But, as von Mises emphasized, Prandtl's own work rested on numerous idealizations. Prandtl was fully aware of this. He explained, for example, that the lift force was assumed to be small so that changes in the direction of the airflow would also be small. Mathematically this justified the neglect of all but the lowest order of the quantities under consideration and made the theory linear. As we have seen, the wing was replaced by a bound vortex, a lifting line, and was treated as if it had no chord. Central to the process of idealization was the now familiar horseshoe vortex. The metaphor of the horseshoe is strained because the vortices in Prandtl's model were in the form of straight lines with right-angled bends, whereas horseshoes are curved. How did the schematization acquire this inappropriate name? The answer links together some of the sparse facts about the relation between Prandtl's work and Lanchester's book. It also provides material for reflecting more generally on the role played by idealization.

In Lanchester's *Aerodynamics* there is a drawing of the vortex system around, and behind, a wing (175). Lanchester's sketch is reproduced here as my figure 7.14. The likeness is not exact, but Lanchester drew the vortices in a way that looked roughly like a horseshoe. They are certainly much more horseshoe-like than the vortex system made up of the three straight lines that Prandtl used. Although Lanchester himself called the shape a "hoop or half-ring" (174), this description must have been the origin of the "horseshoe" metaphor. The *Hufeisen* label presumably arose in Göttingen as a natural response to Lanchester's truly horseshoe-like figure.

Prandtl knew Lanchester's book, and he knew Lanchester's drawing. He mentioned it explicitly in one of the few reflective pieces he wrote

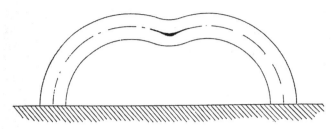

FIGURE 7.14. Lanchester's trailing vortices spread out from the wingtips and reach back to the ground. In doing so they make the space around the wing doubly connected. This picture was probably the origin of the Göttingen label of "horseshoe vortex." From Lanchester 1907, 175.

about his methods of work. In a talk he gave in 1948, called "Mein Weg zu hydrodynamischen Theorien" (My route to hydrodynamic theory), he remarked that it was frequently his doubts about existing treatments of a problem that spurred him to new ideas—and he instanced this particular diagram in Lanchester's book as an example.[84] Unfortunately, he did not specify exactly what it was about the figure that struck a discordant note. A probable answer is that the figure looked wrong because Prandtl took it to be a consequence of Helmholtz's theorems that the trailing vortices would be carried along by the streamlines and, to a first approximation, these would be the straight streamlines of the free flow. The free-vortex lines would not coincide exactly with a prolongation of the original, straight streamlines (because they would have a slight downward movement), but they would not have the marked, outward curving, horseshoe-like shape attributed to them by Lanchester.[85] The defect in Lanchester's figure was removed in the better, though still approximate, straight-line diagram that Prandtl subsequently used. This showed the trailing vortices going straight back from the wingtips.

Was the error in Lanchester's figure obvious to Prandtl the moment he set eyes on the original diagram, or did it take some time before the problem emerged into view? There are grounds for thinking that the error may not have been immediately obvious. The horseshoe-like diagram was not modified in the German translation of Lanchester's book made in 1909 by Prandtl's friends Carl Runge and his wife. Had the diagram seemed obviously wrong from the outset, Prandtl would have mentioned it to his friends, and the matter would have then been raised with Lanchester in the discussions that took place in Göttingen over the translation. The opportunity would have been taken to modify the text in the same way that an opportunity was taken to add a mention of Prandtl's 1904 boundary-layer paper.

Although Prandtl introduced his quantitative theory in 1910, in the summer semester of 1909 he had already given a series of lectures on the scientific basis of airship flight in which, in addition, he had touched, qualitatively, on the circulation round the finite wing of an aircraft. Some of Otto Föppl's notes of those earlier lectures have survived and are reproduced in Julius Rotta's beautifully illustrated book *Die Aerodynamische Versuchsanstalt in Göttingen*.[86] Föppl's lecture notes include a diagram that he presumably copied from one of Prandtl's own blackboard drawings. The diagram shows the vortices curving out from the wingtips in the way Lanchester had originally presented them. It seems that in 1909 Prandtl had drawn the vortex system so that it did indeed still look like a real horseshoe. Föppl's diagram also contains a cross section of the trailing vortices that clearly shows the core of the vortices separated by a distance considerably greater than the wingspan, thus confirming the idea that the vortex lines were not meant to go straight back from the tips. The mere presence of the diagram does not prove that Prandtl had drawn it on the board as an example of truth rather than error, but a probable sequence of events would be this: Prandtl started by accepting the (curved) horseshoe picture, as did Finsterwalder, but within a year realized that it was wrong. Henceforth his model had straight lines. Despite this change of mind, the name *Hufeisen* appears to have stuck and was used, somewhat incongruously, for the simple, straight-line vortex schema that replaced the original curved horseshoe.[87]

I shall now comment on the important transition from the simple, straight-line horseshoe schema to the refined version involving an infinite number of infinitesimal horseshoe vortices. The infinity of vortices coming away from the trailing edge creates a "vortex sheet" spread across, and trailing behind, the span of the wing. In the simple schema there was a vortex line coming from each tip; now there is something like a continuous train of vorticity attached to the rear of the wing. This changes the picture considerably. It also poses a problem. If this picture is right, the earlier picture was wrong, but the supporters of the circulatory theory claimed to have actually *seen* the simple horseshoe structure. In the first publication to use Prandtl's theory, Föppl said that the two vortices trailing from the wingtips had been made visible in the wind channel by introducing ammonia vapor. Nor was it just the members of the Göttingen group who claimed to have seen the horseshoe-like vortices. A similar, though more guarded, claim had been made by Lanchester, who had moved a model aerofoil under water and claimed to have "traced experimentally" the vortices that were postulated in his theory.[88] But if these two trailing vortices are now discarded as theoretical fictions, what was it that had been made visible? One possible answer, according to later versions of the theory, was that the phenomenon reported was really

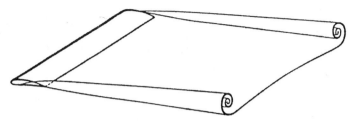

FIGURE 7.15. Prandtl's picture of the rolling up of the vortex sheet behind a finite wing. From Tietjens 1931, 204. (By permission of Springer Science and Business Media)

the rolling up of the vortex sheet. Prandtl argued that the sheet was unstable and rolled up at the edges in the way shown in figure 7.15. The rolled-up sheet then decayed into something resembling the two trailing line vortices. Perhaps this is what had been seen.[89]

There remains a further and deeper question about the move from the single, horseshoe vortex to the infinite number of trailing vortices that now constitute the vortex sheet. I have explained that the single, horseshoe schema could not do justice to what was known experimentally about the distribution of lift along a wing. Greater realism required a non-uniform lift distribution across the span, with zero lift at the tips. Accordingly, Prandtl replaced his single, highly abstract, horseshoe model with an infinite number of similar models. No fundamental principle of the original model was changed in the course of producing the more refined version. In fact, those principles were reproduced an infinite number of times. Can this be right? Can an unrealistic construct be made *more* realistic by repetition? The refined horseshoe model shows that the answer to this question must be yes.

The earlier discussion of Prandtl's boundary-layer theory showed that the realism of a theory may be increased even though physically impossible idealizations were still present. Now the point can be taken further. Realism may be increased by increasing the number of idealizations. It may sound wrong to say that "realism" is increased, while attributing that increase to the increased use of highly "unreal" instruments of thought, such as ideal fluids and infinitesimal vortices, but the discomfiture must be overcome. The essential point is that there is no valid inference from the desirability of greater realism, as that word is normally understood, to the undesirability of idealization. If a theory has been made more realistic, it does not follow that abstractions and idealizations must have been removed or their number diminished. This might, on occasion, be part of the story, but the move to greater realism bears no *necessary* relation to a reduced number of abstractions and idealizations.

Prandtl and his colleagues were not inclined to be apologetic about the abstractions they deliberately introduced into their theory, nor were they in any doubt that they were grasping reality. As Prandtl insisted to his Berlin friend von Parseval, the Göttingen work on vortex theory was successful *because* of its abstractions, not in spite of them. In discussing a paper that von Parseval had given on the formation of vortices on a wing, Prandtl complimented von Parseval on his treatment but contrasted their approaches.[90] He put it like this: "Herr Professor v. Parseval hat der Wirbeltheorie, die bei unseren eigenen Arbeiten immer etwas Abstraktes behalten hat (die allerdings gerade durch die bewußt eingeführten Abstraktionen zu ihren Erfolgen führen konnte), eine anschauliche Deutung gegeben" (63) (Prof. v. Parseval has given an intuitive significance to the theory of vorticity. In our own work it has always been treated rather abstractly [though it is, nevertheless, precisely because of these consciously introduced abstractions that it has led to success]). Prandtl's assistant Max Munk surely spoke for the Göttingen group as a whole when he insisted that the formulas of Prandtl's wing theory represented "die wirkliche auftretenden Vorgänge"—"the actual processes that occur."[91] The consciously introduced abstractions were the means by which the real and actuality occurring processes were described.

The stance of Prandtl and Munk, and the striking achievements of the Göttingen approach suggest a bold generalization. Perhaps successful work of this kind will always be based on idealizations and abstractions. If this is correct, then what is really at issue is not *whether* abstractions are to be used but *which* abstractions are to be used. Which are to be counted as having a role in the laws of nature and which not? Scientists and engineers themselves, collectively, have the responsibility of according or withholding that status and of saying which abstractions and idealizations best describe the actual processes that occur in nature. Different groups may discharge this responsibility in different ways. This fact has already been encountered in the different positions adopted by British and German experts with regard to the Stokes equations. Now we have another example. For the German aerodynamic community, unlike the British, the pragmatic success of the circulatory theory of lift, even within a limited technological domain, was evidence enough that the gulf between thought and reality was being overcome.

Einstein's Folly

The wartime activities of the Göttingen group, and their colleagues in the *technische Hochschulen*, proceeded on a much broader front than I have so far described.[92] On the theoretical side, building on Betz's early papers, there were studies by Betz, Munk, and Prandtl on the aerodynamics of biplanes and triplanes. Using the apparatus of the Biot-Savart law they produced some general theorems that helped to guide the aircraft designer through the maze of possible multiplane configurations. Betz had proven that, for an unstaggered biplane, the induced drag effects of the wings on one another would be equal. For a staggered arrangement it was now shown that the sum of the mutually induced drags was constant and independent of stagger, provided that the lifts, and their distribution, were not changed. (This condition could be satisfied by changing the angle of attack.) In general, it was shown that the best biplane configuration was one with wings of equal length, with the upper wing ahead of the lower. The biplane work also confirmed the important finding that elliptical lift distributions provided a good approximation for wings with non-elliptical planforms.[93]

On the empirical side, the war effort called for wind-channel studies of the drag generated by different aircraft components such as undercarriages and machine-gun mountings, engine-cooling systems, and the ubiquitous struts and bracing wires of the period.[94] Work was also done on the lift and drag of the fuselage, the interaction between the fuselage and the wing, the effect of dividing the wing, the forces on fins and rudders, and the empirical properties of the triplane configuration.[95] Experiments were done to test the resistance of the nose shapes required by different engine types, for example, rotary as compared with in-line engines, and attempts were made to add rotating propellers to the wind-channel models to achieve realism.[96] Some wind-channel tests were also done on models of complete aircraft.[97] Munk and Cario continued the studies initiated by Föppl of the downwash behind a wing. Whereas Föppl had worked with the overall force exerted by the downwash on the elevator, Munk and Cario studied the downwash in much more empirical detail, using fine silk threads to trace the local variations. They uncovered significant complexities in the flow and made clear the need for a more extended program of work.[98]

Numerous studies were carried out to measure the lift and resistance of individual wing profiles. These were mainly overseen by Munk and his collaborator Erich Hückel.[99] Significant efforts were made to ensure that the results were intelligible to those who might use them in practice.[100] An attempt was also made to introduce order into the vast amount of data that had accumulated for different aerofoils, though the classification remained largely at the empirical level.[101] One trend was toward an interest in thicker rather than thinner aerofoils, something that surprised the British when they examined captured German aircraft.[102] An example of these thick aerofoils was the Göttingen 298 used on the famous Fokker triplane. The use of the 298 profile by

the designer Anthony Fokker does not, however, appear to have been a consequence of Prandtl's recommendation or scientific knowledge of its good lift
and drag characteristics (characteristics that, a priori and wrongly, the British
designers doubted). In fact, the aerofoil was introduced into the Fokker production line by their chief engineer Reinhold Platz on the basis of trial-and-
error knowledge. Later, and unknown to the people at Fokker, it was tested
in Göttingen, where it was given its designation.[103]

The Fokker episode indicates that there was a continuing gap between the
"practical men" of Germany and those self-consciously developing science-
based procedures and working in academic and government institutions. The
alienation of the practical men was not purely a British phenomenon, though
it seems to have been less acute as a problem for German aviation than for
British. Evidence in the technical reports indicates that the members of the
Göttingen school were themselves aware of this gap and found it frustrating.
Max Munk addressed the issue directly in a brief report of October 15, 1917,
titled "Spannweite und Luftwiderstand" (Span and air resistance).[104] Referring to the practical conversion formulas linking wings of different aspect
ratio, Munk complained:

> Die kürzlich von Betz veröffentlichen Prandtlschen Flügelformeln werden
> wohl, da sie auf theoretischen Grundlagen beruhen, in der Praxis nicht so
> freundlich aufgenommen werden, wie sie verdienen. Das ist sehr schade, denn
> die Formeln enthalten mehr und leisten Besseres als der Praktiker geneigt ist,
> ihnen zuzutrauen. (199)

> The formulas of Prandtl's wing theory that Betz has recently published will
> probably not be welcomed in the realm of practice as much as they deserve
> because they rest on theoretical grounds. This is a great shame because the
> formulas offer more and give better service than the practical man is inclined
> to believe.

Munk went on to give an explanation of the significance of the formulas for
the aircraft designer and some simple, general rules for the rapid calculation
of the induced resistance and angle of incidence. Despite this evidence of
skepticism in certain quarters, there was no shortage of contract work to be
done for individual aircraft firms during the war years. This is attested by the
frequency with which such names as AEG, Aviatik, Rumpler, Siemens and
Schuckert, and Zeppelin were mentioned in the technical reports. Despite
the problems of communication between the representatives of theory and
practice, Prandtl's institute had achieved a central position in what would
now be called the military-industrial complex of Wilhelmine Germany. If this
development brought frustrations as well as the advantages of government

support, it is clear that striking progress had been made in aerodynamics, both empirically and theoretically.

One of the more unusual aerofoils whose properties were reported on by Munk and his colleagues was designated as profile 95. Visually it stood out from the usual run of aerofoil shapes (see fig. 7.16). The aerofoil looked like the back of a cat when the animal stretched, and it was duly given the nickname *Katzenbuckelfläche*. The Göttingen tests showed that the performance characteristics of the "cat's-back" profile 95 were notably poor. It was tested by Max Munk and Carl Pohlhausen in the course of a run of work on nearly one hundred aerofoils. The results were listed together in the *Technische Berichte* of August 1917 and showed that the maximum-lift coefficient for each wing in this sequence was typically in the region of 130 or 140. The maximum lift coefficient for profile 95, by contrast, was given as 95.2. Again, the maximum lift-to-drag ratio was typically 14 or 15, while the ratio for profile 95 was 10.8.[105] The designer of the cat's-back wing was the celebrated physicist Albert Einstein.[106] In retrospect Einstein felt that his excursion into aerodynamics had been irresponsible—he used the word *Leichtsinn*. From 1915 to

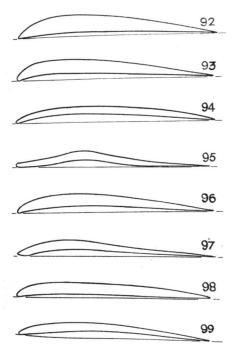

FIGURE 7.16. Profile 95 stands out because of its unusual shape. It is the "cat's-back" wing designed by Albert Einstein. From Air Ministry 1925, publication no. 1120.

1917, Einstein had been a consultant to two aircraft firms, LVG and Merkur, and an aircraft had been equipped with the Einstein wing. The test pilot for LVG, Paul Ehrhardt, barely managed to get the machine off the ground and gave his professional opinion on the wing by saying that the airplane flew like a pregnant duck.[107] The Göttingen tests made the same point in more scientific terminology.

No account remains of how Einstein actually designed the wing, but some insight into his thought processes may be gained from an article he published in 1916 in *Die Naturwissenschaften*. Here he set out to explain, in elementary terms, the basic principles of lift.[108] How does a wing support an aircraft and why can birds glide through the air? Einstein declared, "Über diese Frage herrscht vielfach Unklarheit; ja ich muß sogar gestehen, daß ich ihrer einfachsten Beantwortung auch in der Fachliteratur nirgends begegnet bin" (400) (There is a lot of obscurity surrounding these questions. Indeed, I must confess that I have never encountered a simple answer to them even in the specialist literature). This is a striking claim, given that Einstein was writing a number of years after the publications of Kutta, Joukowsky, and Prandtl.

Einstein drew an analogy between the flow of fluid through a pipe of variable cross section and its flow around a wing. As fluid passes along a pipe that gets narrower, the fluid speeds up. By Bernoulli's law the pressure will be lower in the fast, narrow section than in the broader section. Einstein then invited the reader to consider a body of incompressible fluid with no significant viscosity (that is, a perfect fluid) which flowed horizontally but where the flow was divided by a thin, rigid, dividing wall. The wall was aligned with the flow except that it also had a curved section where the wall bulged upward. (See fig. 7.17, which is taken from Einstein's paper.) The curved section looks a bit like a cat's back and would appear to have the same shape as the underside of the Einstein wing tested at Göttingen. Einstein argued that the fluid above the dividing wall will behave like the fluid in a pipe when it encounters a narrowing of the pipe and will speed up and exert a diminished pressure on the wall. The fluid below the wall will behave like the fluid in a pipe when it encounters a widening out of the pipe so it will slow down and increase the pressure on the wall. The fluid pressure pushing upward on the curved section of the dividing wall will thus be in excess of the pressure pushing downward, so there will be a resultant force upward All that is necessary now is to imagine that most of the dividing wall has been removed, leaving behind just the curved section. This procedure, argued Einstein, will retain the features of the flow that generate the pressure difference and hence will represent a wing with lift.

Einstein's argument rested on the assumption that the removal of all but

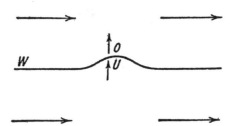

FIGURE 7.17. Flow through a variable cross section. Below the wall W, the fluid slows down so the pressure increases; above the wall the flow speeds up so the pressure decreases. This theoretical argument appears to be the basis of the cat's-back wing. From Einstein 1916. (By permission of the Albert Einstein Archives, Jerusalem)

the curved portion of the dividing wall would leave the flow unchanged at the leading and trailing edge of the remaining arc. He appeared to take this as obvious: "Um diese Kraft zu erzeugen, braucht offenbar nur ein so großes Stück der Wand realisiert zu werden, als zur Erzeugung der wirksamen Ausbiegung der Flüssigkeitsströmung erforderlich ist" (510) (To generate this force it is obviously only necessary for part of the wall to be real. It need only be sufficiently large to produce the effective curvature of the flow). It is puzzling that Einstein made no mention of circulation. Was he aware of the circulation theory of lift? This remains unclear, but given that he assumed the air to be a perfect fluid, it makes it all the more important to ask how he proposed to circumvent d'Alembert's paradox. How did Einstein expect to get a lift force rather than a zero resultant?

Although Einstein wrote in a dismissive way about the aerodynamic literature, he had, in effect, taken the discussion back to where Kutta started it in 1902. When, following Einstein's instructions, all of the wall dividing the flow is removed except for the curved piece, what is left is essentially Kutta's arc at a zero angle of incidence. Einstein did not specify that the curve he discussed in his article was precisely the arc of a circle, but his argument was offered as a general one. If it were right it would apply to Kutta's arc. But it does not. The arc is a counterexample to what seems to be Einstein's argument, that is, to any argument that depends on ideal fluids but does not make provision for circulation. In order to generate lift, and to generate the requisite speed differential between the upper and lower surfaces of the arc, a circulation must be postulated. In a continuous perfect fluid flow, without an independently specified circulation, such an arc would *not* produce lift. D'Alembert's paradox would come into play. Such an arc would not have its stagnation points on the leading and trailing edge. The formula for the complex velocity has singular points indicating infinite velocities for the ends of the arc.[109]

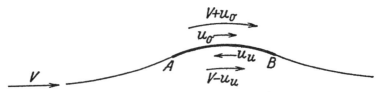

FIGURE 7.18. Flow of an ideal fluid with circulation around an arc. As Kutta showed in 1902, a circulation is necessary to place the stagnation points on the leading and trailing edges, that is, at *A* and *B*. As shown here, Prandtl's diagram, unlike Einstein's, makes the role of circulation explicit. From Tietjens 1931, 174. (By permission Springer Science and Business Media)

In his diagram Einstein put the stagnation points on the leading and trailing edges of the arc even though he did not explicitly invoke a circulation. He put them where they *would* have been, given the appropriate amount of circulation needed to avoid infinite velocities at the edges. He did this by supposing that the guiding effect of the dividing wall, smoothly leading the fluid toward, around, and then away from the arc, would still be present even when the wall was removed and only the arc was left. There are no grounds for this assumption. Perhaps the best that can be said is that Einstein had, in fact, made provision for circulation in his analysis, but had done so tacitly and by questionable means. In his published lectures, a few years later, Prandtl gave a diagram that was almost identical to Einstein's but, in Prandtl's case, the component of circulation in the flow was properly identified and made explicit.[110] Prandtl's diagram is reproduced as figure 7.18 for purposes of comparison with Einstein's figure.

It would seem that Einstein had little knowledge of current developments in the field of aerodynamics. This episode is a salutary reminder of the difference between fundamental physics and technical mechanics. Eminence in the former does not guarantee competence in the latter.

At the Eleventh Hour

At eleven o'clock, on November 11, 1918, a cease-fire was declared on the western front. The "war to end all wars" was over. European civilization would never be the same again, nor would aerodynamics. The military situation did not take Prandtl wholly by surprise, and he had already begun to explore the possibilities for a program of peacetime research.[111] The comprehensive collapse of the German military effort meant that financial support for the Göttingen institute would all but disappear. At the height of the war Prandtl's institute had employed around fifty people, but now some 60 percent of its staff had to be dismissed.[112] At the very moment when the institutional and financial arrangements that had sustained German aerodynamic work were crumbling, the full scope of the Göttingen achievement was coming into view.

Toward the end of the war Prandtl finally brought together the overall theoretical picture that had been hinted at, and promised, but never produced, for the readers of the *Zeitschrift für Flugtechnik*. Even then it was not the readers of the *Zeitschrift* who would be the immediate beneficiaries. On April 18, 1918, Prandtl gave a comprehensive, and confidential, lecture on the theory of the lift and drag of an aircraft wing to the annual meeting of the Wisssenschaftlichen Gesellschaft für Luftfahrt in Hamburg.[113] A few months later, on July 26, 1918, Prandtl presented the first part of his classic paper "Tragflügeltheorie" to the Göttingen Academy of Science.[114] It is the theory of the wing as laid out in these papers that I have described in this chapter.

In presenting his theory to the Academy, Prandtl prefaced the aerodynamic work with a highly abstract sketch of fluid-dynamic principles. Perhaps in deference to Hilbert and the Göttingen fashion for formal axiom systems he even offered two new "axioms" to be added to classical hydrodynamics. (Axiom I stated that vortex layers can arise at lines of confluence. Axiom II was that infinite speeds cannot arise at protruding sharp edges of the body, or, if they do, only in the most limited way possible.) It is difficult to avoid the suspicion that these embellishments were added because Prandtl was conscious of addressing a high-status, scientific audience rather than an audience of engineers. Moritz Epple points out that the "axioms" Prandtl introduced do not justify the approximation processes that he used, nor do they operate as axioms in the way that Hilbert would understand them.[115] The reference to "axioms" appears to have more to do with style than substance. Furthermore, the general principles of fluid mechanics presented at the outset needed the help of drastic approximations before the theory of the wing could be presented in a recognizable and useful manner. As Prandtl introduced these approximations into his exposition, so the tone of the talk to the Academy changed. There was a shift from abstract principle to concrete practice; from science to engineering; and from classical mechanics to *technische Mechanik*.[116]

"We Have Nothing to Learn from the Hun": Realization Dawns

When I returned to Cambridge in 1919 I aimed to bridge the gap between Lamb and Prandtl.

G. I. TAYLOR, *"When Aeronautical Science Was Young"* (1966)[1]

Oscar Wilde declared that if you tell the truth you are bound to be found out sooner or later.[2] There is a corresponding view that applies to scientific theories. Given good faith and genuine curiosity, a true theory will eventually prevail over false ones. These sentiments make for good aphorisms but the epistemology is questionable. Even if it were right, there would still be the need to understand the contingencies and complications of the historical path leading to the acceptance of a theory. My aim in the next two chapters is to describe some of the contingencies that bore upon the fortunes of the circulatory theory of lift in Britain after the Great War. I shall come back to the philosophical analysis of theory acceptance in the final chapter of the book, when all the relevant facts have been marshaled. I begin the present discussion with some observations about the flow of information between German and British experts before, during, and after the Great War.

Scientific Intelligence: Fact and Fiction

Looking back to the period of the Great War, after some sixty years, Max Munk expressed the belief that the aeronautical work he had carried out in Göttingen had rapidly fallen into the hands of the Allies. According to Munk, the secret *Technische Berichte* "were translated in England a week after appearance and distributed there and in the U.S."[3] Exactly how this feat of espionage was performed Munk did not say. Similar stories have been related about the flow of sensitive information in the other direction, from the Allies to the Germans. I have already mentioned the secret testing of the Dunne biplane in the Scottish Highlands before the war. This was said to have attracted the attention of numerous German "spies," though these stories surely owed more to John Buchan than to reality.[4] A more sober counterpart to Munk's beliefs is provided by J. L. Nayler, one of the secretaries to the Advisory Committee. Also speaking retrospectively, he said that the wartime Reports and Memoranda produced in Farnborough and Teddington eventually found their way into German hands. Nayler, though, suggested that this took months rather than weeks.[5] Perhaps British spies were just superior to German spies.[6]

The truth was almost certainly more pedestrian than these claims suggest. There is no evidence that agents acting on behalf of the British government got their hands on any information about the wartime Göttingen work and passed it on to their masters in Whitehall or their allies in Paris and Washington. There appears to have been no successful espionage activity. It is not the speed with which information traveled that is striking but its slowness. When information did travel, the channels were overt and obvious rather than mysterious.[7] The war had the predictable effect of attenuating the flow of technical information between different national groups, but even during the prewar years, with no military or diplomatic impediments, the flow was surprisingly limited. It is important to identify where the restriction lay. It did not arise because of what might be called material or external factors, such as censorship, but because of more subtle, cultural constraints. It was not the physical inaccessibility of reports, journals, or books that caused the problem. What counted was the response, on the intellectual level, even when they were accessible. For example, both Sir George Greenhill and G. H. Bryan were present at the congress in Heidelberg, in 1904, when Prandtl presented his revolutionary, boundary-layer paper.[8] Bryan explicitly mentioned Prandtl's contribution in his postconference report for *Nature*, but he ignored its mathematical content entirely and confined his comments to the experiments and photographs.[9] It is difficult to resist the conclusion that if such important matters can be passed over in these circumstances, then even if there had been "spies" reporting back to the British Advisory Committee, their efforts would have been wasted.

To reinforce this claim I start with some other prewar events and look at the information that members of the Advisory Committee had available to them about their German counterparts. From the outset the committee, and the Whitehall apparatus that supported it, accepted the principle that it was important to monitor the work of foreign experts. Haldane stressed the point in Parliament, and the theme was picked up by the aeronautical press.[10] The commitment to gathering intelligence was made apparent in three ways. First, the preliminary documentation of the committee, when it was established in 1909, included what was, in effect, a reading list for the committee members. The list cited some twenty-two works by French, German, Ital-

ian, and American writers. The German authors included Ahlborn, Finsterwalder, and Lilienthal.[11] Second, the sequence of Reports and Memoranda
issued by the committee began with a description of the program of German
airship research. It was presented by Rear Admiral Bacon at the very first
meeting of the Advisory Committee on May 12, 1909.[12] R&M 1 consisted of
translated extracts from the publications of the German Society for the Study
of Airships and included a lengthy quotation from Prandtl.[13] There was mention of Prandtl's wind channel, his experiments on model airships, and, in-
triguingly, a passing reference was made to his "hydraulic machine" (shown
earlier in fig. 7.1). This was the apparatus used to take his boundary-layer
photographs. There was, however, no mention of the mathematical theory.
Third, and most important of all, the committee was provided with a series of
summaries of foreign papers from leading journals such as the *Zeitschrift für
Flugtechnik*. A steady stream of these summaries was published in the period
between the founding of the committee and the outbreak of the Great War,
when such material was immediately withdrawn from public circulation.[14]

A measure of the size of the intelligence initiative can be gathered by counting the number of such abstracts published yearly in the annual report of the
Advisory Committee. Such a procedure can only provide an approximate
measure of the potential flow of information because it does not take account
of the different scope of the individual publications, but it gives some guide.
Figure 8.1 charts the year-by-year production of summaries and abstracts of
foreign-language publications that were made available to the committee.[15]
Two things stand out. First, the size of the effort put into tracking foreign
work was clearly considerable. Second, there was a consistently high level of
attention given to German work, amounting on average to identifying and
abstracting some eighteen items per year for a period of six years.

Moving from the quantitative to the qualitative character of the information, it is important to know which authors the committee deemed interesting. The answer is that Prandtl and his collaborators were prominent among
them. In December 1910, Glazebrook, as chairman, explicitly drew the Göttingen work to the attention of the members of the Advisory Committee.[16]
In August 1913, in preparation for a forthcoming visit to the laboratory in
Teddington, Prandtl sent a number of his papers to the National Physical
Laboratory (NPL) and received acknowledgment from Selby, the secretary.[17]
Thus, by one route or another, all of the major prewar work of the Göttingen school had been made available, including accounts of the wind channel
and the airship work but also material directly concerned with the circulation
theory of lift. In addition there were abstracts of papers of indirect interest

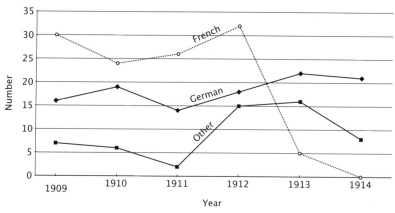

FIGURE 8.1. The number of abstracts of foreign works made available to members of the Advisory Committee for Aeronautics in the years before the Great War. Data from the committee's annual reports.

because of their significance for fluid dynamics in general. More specifically, among the papers summarized, sometimes at length, were those of Föppl on the resistance of flat and curved plates (abstracts 93, 94, 97, 98, 118, and 131), Fuhrmann on the resistance of different airship models (abstracts 95, 96, and 127), and Prandtl's classic study of the flow of air over a sphere in which he introduced turbulence into the boundary layer by means of a trip wire (abstract 234). Of those explicitly related to the idea of circulation and Prandtl's wing theory, accounts were given of Föppl's 1911 study of the downwash behind a wing (abstract 128, but incorrectly attributed to Fuhrmann); Wieselsberger's 1914 study of formation flying in birds (abstract 276); the 1914 paper by Betz on the interaction of biplane wings (abstract 279); Joukowsky's pioneering 1910 article (abstract 299); Blumenthal's 1913 paper on the pressure distribution along a Joukowsky aerofoil (abstract 301); and Trefftz's 1913 graphical construction of a Joukowsky aerofoil (abstract 302).

The principal mathematical formulas associated with the circulation theory in both its two- and three-dimensional forms were also to be found in the abstracts. Thus the basic law of lift, linking density, velocity, and circulation, $L = \rho V\Gamma$, was stated, as was the law of Biot-Savart, which was the basis of the three-dimensional development of the theory. The abstracts provided everything that was needed to show that the circulation theory was capable of mathematical development and was more than a mere collection of impressionistic ideas. The abstracts gave clear, documentary evidence of the progress that the German engineers were making. It would appear that the circulation theory was there for the taking. Nevertheless, the availability of the abstracts generated no more enthusiasm for the theory of circulation in its mathematical form than did Lanchester's original publication with its more intuitive treatment of the subject.

Why might this be? Obviously, the abstracts had no power to force themselves on anyone's attention. They were things to be used selectively and were subject to the filtering effects of interpretation, both in their composition and their evaluation. Thus Glazebrook's act in drawing attention to the Göttingen work was probably indicative of his enduring concern with discrepancies between the results of different wind channels and the fundamental problems shared by the NPL and Göttingen in the interpretation of their findings. Glazebrook was acutely aware that such problems would be grist to the mill of the "practical men" and was anxious lest they be used to persuade the government to cut the budget of the NPL.[18] Furthermore, the precise content of the abstracts reveals the way that reported work may be glossed so that certain aspects of it are given salience at the expense of other readings. Take, for example, the account given in abstract 131, which was devoted to Föppl's 1910 paper in the *Jahrbuch der Motorluftschiff-Studiengesellschaft*.[19] This paper contained a comparison of Rayleigh flow with Kutta's theory of circulatory flow. After summarizing the contents of the paper, the abstract writer drew the conclusion that neither approach to the flow over an inclined plate was satisfactory. What was needed was an understanding of certain subtle, viscous effects. "It is suggested that Kutta's theory throws some light on the experimental results, and in some respects, qualitatively, is in fair agreement with the experiments. At present, however, no entirely satisfactory theory seems to be possible until more is known of the nature of the air flow, the main differences being due to the difficulty of including the frictional effects" (257). The need to include frictional effects was, of course, an abiding theme in the British work. The abstract writer then went on to single out, as the "most striking result" of Föppl's investigation, "the discontinuity in lift and drift coefficients within the region from 38° to 42°" (258). All the attention was thus directed toward extremely difficult, fundamental, and unstable features of the flow that lay far outside the typical working range of an aerofoil. Once again, the British were drawn to the phenomenon of stalling. The focus was on all the things that could not be understood on the basis of inviscid flow at small angles of incidence rather than on what could be achieved using perfect fluid theory over a limited range. Thus the abstract and summary itself prefigured the selective tendencies and implicit evaluations that worked against the circulatory theory.

The prewar information about German thinking on aerodynamics was rich but unexploited, whereas during the war, the pressure of short-term work added to the tendency to pass over the significance of the German theoretical approach.[20] What of the pattern of information flow, and the reaction to it, immediately after the cessation of hostilities? In some quarters in Britain, the outcome of the war produced a jingoistic complacency. Such sentiments were exemplified by C. G. Grey, editor of the *Aeroplane*, when he said in 1918: "We have nothing to learn from the Hun in aerodynamics."[21] This boast was a continuation of a commonplace theme in the aeronautical press, which, throughout the war, dismissed German inventiveness, originality, and skill.[22] Such vulgarity was largely absent from the writing of the more technically sophisticated members of the British aerodynamic community, who had, if not an admiration, at least a healthy respect for German achievements.[23] Among the members and associates of the Advisory Committee there was an understandable degree of self-congratulation as the war drew to a close, but it was modest in tone.[24]

The first reaction to the outbreak of peace by the scientists at the Royal Aircraft Establishment (formerly the Royal Aircraft Factory) was to poke fun at themselves and their critics. The period immediately after the cease-fire, between November and December 1918, saw the production of a light-hearted work titled "The Book of Aeron: Revelations of Abah the Experimenter."[25] This undergraduate-style spoof was composed by the remarkable group of young men who had been recruited by Mervin O'Gorman. Many of them were billeted in a large house in Farnborough called Chudleigh (see fig. 8.2).

FIGURE 8.2. The Chudleigh set. Hermann Glauert is seated on the plinth on the extreme left. George Paget Thomson, in uniform, is standing behind Glauert. David Pinsent (*left*) and Robert McKinnon Wood (*right*) are seated on the lower steps at the front of the group. W. S. Farren, in uniform, is directly behind McKinnon Wood. F. A. Lindemann is standing behind Thomson, and Frank Aston is seated on the right-hand plinth. David Pinsent was killed in a flying accident on May 8, 1918. From Birkenhead 1961.

The house acted as a mess for the RAE and had originally been organized by Major F. M. Green, who, until 1916, had been the engineer in charge of designs.[26]

The "Book of Aeron," circulated as an internal report, had been written by a committee that included R. V. Southwell and Hermann Glauert. It was couched in mock, Old Testament terms, with ancient Egyptian overtones, and told the story of the Land of Rae (the RAE), its ruler Bah Sto (Bairstow), and the wicked scribe Grae (C. G. Grey). The Abah of the subtitle appears to be a reference to the designation "Department BA" in which the experimental work was conducted at Farnborough. Naturally, the clash between the aircraft manufacturers ("the merchants") and the Farnborough scientists ("the men of Rae") played a significant role in the story. The following passage conveys the spirit of the enterprise:

> 2. And the men of Rae built air chariots for their king, and brought forth new chariots of diverse sorts; and to each chariot did they give a letter and a number, that the wise men might learn their habits:
> 3. but the multitude comprehended it not.
> 4. Then murmured the merchants one to another saying, Why strive the men of Rae so furiously against us? For the king goeth to war with their chariots, and behold our chariots are cast into the pit. (3)

After the relief of the armistice, and the lessening of tension, came the serious business of taking stock. Just what had been achieved during the war? What had been sacrificed, scientifically, because of the demands of the war effort? What, if anything, was to be learned from newly accessible German literature? Bairstow and other committee members rapidly let it be known that they deplored the cutback in basic research that had been caused by the war. In terms of fundamentals, they argued, the period of rapid achievement in aerodynamics had been before the war. It was now time to get back to deeper questions and that meant solving Stokes' equations of viscous flow. Pure mathematicians may have given up on this task, but new techniques, perhaps using graphical methods, might be developed for this purpose. In a confidential report of November 1918,[27] mapping out a program of work for 1919–20, the argument was put like this: "General research in fluid motion has been discontinued during the war and it is very desirable that it should be resumed at the earliest possible moment. It is proposed as soon as the opportunity occurs to continue the study of the motion of viscous fluids to which considerable attention has already been given" (9–10). These sentiments represented the beginnings of a campaign by Bairstow and Glazebrook to channel more resources from short-term to basic research. They argued, in letters to the *Times*, that government figures for the aeronautics budget conflated the spending on development with that on research proper. This made the expenditure on fundamental work, which was vital for future technology, appear larger than it really was.[28]

What had the Germans being doing during the war? At first, information filtered through to the British via the French and Americans. In 1919 the American National Advisory Committee for Aeronautics, the NACA, established an office at 10, rue Victorien Sardou in Paris. Their representative, William Knight, actively pursued a policy of information gathering. To the irritation of military and diplomatic circles in Paris and Washington, he made contact with Prandtl and suggested that information on recent developments be shared. Knight contacted Prandtl on November 15, 1919, by letter and managed to get official agreement to visit him in April 1920. He made a second visit in the autumn of 1920.[29] It appears to be due to the efforts made in Paris that the British Advisory Committee was furnished with translations of reports by A. Toussant and by Colonel René Dorand. Toussant was an engineer at the Aerotechnical Institute of the University of Paris at St. Cyr. In November 1919 he had produced a résumé of the theoretical work done at Göttingen based on the *Technische Berichte*. Extracts of Toussant's work were translated by the NACA, and these surfaced in February 1920 as a report designated Ae. Tech. 48 for consideration by Glazebrook and his colleagues.[30] There was, however, little that Toussant could add to the information that was already at hand from the prewar abstracts. He gave mathematical formulas but none of the background reasoning. Dorand's report was titled "The New Aerodynamical Laboratory at Göttingen." It was translated as report T. 1516 in October 1920 and reached the Advisory Committee via the Inter-Allied Aeronautical Commission of Control.[31] In Dorand's report Prandtl was referred to throughout as "Proudet," but it included three pages of blueprints of the Göttingen wind channels and provided technical details of the automatic speed control and the measuring apparatus. Bairstow felt that there was nothing new in Dorand's report.[32] Others, however, noticed that Prandtl seemed to have solved certain problems that plagued the NPL channels and had "managed to achieve good velocity distribution in the working sections."[33]

Both the British and the Americans were keen to locate and translate copies of the *Technische Berichte*. At a meeting that took place on July 13, 1920, at the Royal Society, minute 28 records Treasury authority for the "employment of abstractors to make abstracts of German technical reports."[34] Across the Atlantic, Joseph Ames, who was the chairman of the Committee on Publications and Intelligence, and one of the founding members of the American National Advisory Committee for Aeronautics, wrote from the NACA headquarters to J. C. Hunsaker at the Navy Department in Washington, D.C., on October 15, 1920:

> It is with great pleasure that I am informing you that the National Advisory Committee for Aeronautics has been successful in obtaining a number of sets of "Technische Berichte" and we are mailing you under separate cover volumes No.1, 2 and 3. The Committee is also forwarding a carefully prepared translation of the index of the first three volumes with a list of the symbols used. . . . The importance of the information contained in the "Technische Berichte" cannot be over-estimated and it is the desire of the Committee that all research laboratories and individuals interested in aeronautical research should become familiar with the results of the aeronautical research carried on in Germany during the War.[35]

The success of this search appears to have been due to Knight's persistence. He used Prandtl's good offices to approach the German publishers but had to overcome the numerous obstacles arising from the immediate postwar currency and customs restrictions.[36] Meanwhile the U.S. National Advisory Committee continued negotiations with the British about the translation and abstraction of the reports.[37] The enthusiastic terms of Ames' letter, and the continuing efforts by both the British and the Americans, provide a sufficient basis for rejecting Max Munk's claim that the reports were in the possession of the allies soon after their completion. Had this intelligence already been gathered, all the postwar concern would have been unnecessary.[38]

Postwar Contact with Göttingen

A number of significant changes, both organizational and personal, took place in the higher reaches of British aeronautics at the end of the war. The size of the aeronautical section at the National Physical Laboratory had grown considerably during the conflict. Starting from three or four active workers in 1909, the section had expanded to around forty by the time of the armistice.[39] Predictably, the return of peace meant that the budget was now to be cut back. Lord Rayleigh had died in 1919, and the Advisory Committee he had guided for a decade was formally dissolved and reconstituted as the Aeronautical Research Committee (ARC). The new committee held its first meeting on May 11, 1920.[40] Glazebrook was given the job of restructuring it and preparing it for its new peacetime role. The National Physical Laboratory and the Aeronautical Research Committee now came under the aegis of the newly formed Department of Scientific and Industrial Research (DSIR).[41] Horace Lamb had been appointed to the Aerodynamics Sub-Committee in July 1918 and later joined the full committee.[42] In an attempt to avoid the old hostility between the scientists and the manufacturers, there were now to be representatives of industry on the committee. J. D. North, of Boulton Paul, was appointed to the Aerodynamics Sub-Committee to represent the Society of British Aircraft Constructors. Bairstow left the National Physical Laboratory in 1917 and took up a post with the Air Board, the precursor to the Air Ministry, though he continued to serve on the new committee.[43] Bairstow then moved again and become the Zaharoff Professor of Aviation at Imperial College, London. Sir Basil Zaharoff, who financed the chair, was an international arms dealer.[44] Shortly afterward Emile Mond provided the money to set up a chair in aeronautics at Cambridge in memory of his son killed flying on the western front. This chair was taken by Melvill Jones.[45] Bairstow's post as superintendent of the Aerodynamics Department at the NPL was taken over by Southwell, who moved from Farnborough to Teddington. Lanchester, who sometimes felt that Rayleigh was the only sympathetic member of the committee, left a year after Rayleigh's death. Lanchester had been assiduous in his duties but had already resigned from the Aerodynamics Sub-Committee in December 1918.[46] Now that the emergency of the war was over he felt able to cite pressure of work as a basis for leaving. In his letter to the chairman of the Aerodynamics Sub-Committee he expressed "great pleasure in having been able to serve the committee," but the retrospective account he gave of his departure from the full committee was very different in tone. He complained that he had been sidelined, snubbed, and deliberately edged out by Glazebrook.[47]

At the moment that Lanchester left the Whitehall scene feeling, justifiably, that his ideas had been ignored, moves were under way that would eventually lead to the triumph of the circulation theory he had pioneered. Two things happened. First, on November 13, 1920, Southwell received a letter from Prandtl, who sent him some up-to-date papers on wing theory and material from the *Technische Berichte*. Prandtl explained that his action had been prompted by his meeting with William Knight. Knight had apparently told Prandtl that Southwell wanted to get hold of information about developments at Göttingen. Southwell replied on November 29 with thanks and tentatively asked Prandtl for details about his wind tunnel and the techniques for keeping the flow steady. He also stressed that the exchange with Prandtl had to be considered personal rather than official because of the British government's policy of restricting formal contact with German institutions. Prandtl sent Southwell the required data about the air flow and said that more information would soon be published in a volume to be titled *Ergebnisse der Aerodynamischen Versuchsanstalt zu Göttingen*.[48]

The second development was that two Farnborough scientists, Robert McKinnon Wood and Hermann Glauert, both members of the Chudleigh

FIGURE 8.3. Hermann Glauert (1892–1934). Glauert, an Englishman of German extraction, was a Cambridge mathematician and fellow of Trinity. He worked at the Royal Aircraft Factory in Farnborough during the Great War and visited Prandtl in Göttingen soon after the war's end. He became an advocate of the circulation theory of lift and Prandtl's theory of the finite wing. (By permission of the Royal Society of London)

set, were sent to Germany to report on the situation. McKinnon Wood, a product of the Cambridge Mechanical Sciences Tripos, was deputy director of the Aerodynamics Department at Farnborough and worked on propellers and the experimental side of aerodynamics.[49] Hermann Glauert (fig. 8.3) had studied mathematics at Trinity College, Cambridge, where his circle of friends included David Pinsent, G. P. Thomson, and Ludwig Wittgenstein.[50] He graduated with distinction in the first class of part II of the Tripos in 1913 and won an Isaac Newton studentship in 1914 and the Rayleigh Prize for mathematics in 1915.[51] Glauert was born in Sheffield. His mother was an Englishwoman who had been born in Germany, while his father, a cutlery manufacturer, was German but had taken British citizenship. Originally specializing in astronomy, Glauert published a number of papers on astronomical topics at the beginning of the war and then, through a chance meeting with W. S. Farren, he was appointed to the staff of the Royal Aircraft Factory in 1916.[52]

Based in the Hotel Hessler in Charlottenburg, Glauert wrote on January 21, 1921, to make contact with Prandtl and ask if was possible to arrange a visit to Göttingen.[53] The approach could not have been more different from Knight's. Knight had sent a typed letter, in English, on elaborately headed NACA notepaper. The letter was replete with office reference numbers and subject headings and was introduced with a flourish of (questionable) diplomatic credentials.[54] Glauert penned his note in German on a modest sheet of unheaded paper. He introduced himself as a fellow of Trinity who worked at Farnborough and explained that he was very interested in the reports shown to him by his friend Herr Southwell. Could he and his friend Herr Wood, also of the Royal Aircraft Factory, come along next Monday? The visit duly took place but cannot have been an extended one because on February 2 Glauert was writing from Farnborough (again in German) to say that he and Wood were safely home after a thirty-six hour journey. All of the technical material that Prandtl had given them, he reported, had been carried over the border without difficulty. In return Glauert sent Prandtl a copy of Bairstow's new *Applied Aerodynamics*.[55] It was, he said, currently the best English book on aerodynamics.

Bairstow's *Applied Aerodynamics*, published in 1920, offered a massive compilation of design data from the aerodynamic laboratories which, for Bairstow, primarily meant from the National Physical Laboratory. It was comprehensive and detailed but, as far as lift and drag were concerned, heavily empirical. The circulation theory of lift was placed on a par with the discontinuity theory and rapidly dismissed. Both theories were said to be based on "special assumptions," that is, ad hoc devices designed to get around the fatal zero-drag result. Kutta's work, according to Bairstow, offered no more than a "somewhat complex and not very accurate empirical formula" (364). No account, he complained, was given by Kutta or Joukowsky of the critical angle of stall. Bairstow admitted that Joukowsky had found a way to avoid the infinite speeds at the leading edge of a wing profile, but he spoke of the Joukowsky transformation as if it were little more than a mathematical trick. Bairstow called it a "particular piece of analysis" and did not deem it sufficiently important to explain it to the reader. Prandtl must have perused Glauert's gift with mixed feelings. He would have agreed with all of Bairstow's facts but none of his evaluations. In particular, he would surely have dissented from Bairstow's conclusion that it "appears to be fundamentally impossible to represent the motion of a real fluid accurately by any theory relating to an inviscid fluid" (361). Wasn't this exactly what Prandtl and his colleagues had just done for the flow of air over a wing?

It is clear from the subsequent letters exchanged between Glauert and Prandtl that, during the visit, their conversations had been confined to technical matters. Prandtl was now keen to discuss politics as well as aerodynamics. He was distressed by the economic and political situation, particularly the stance of the French and the severe reparations that were being demanded. Glauert expressed agreement with much that Prandtl said on these topics, for example, with the "absurd restrictions that have been placed on the development of German aviation," but in his replies he encouraged Prandtl to see things in a less pessimistic light. Glauert explained that not everyone in Britain agreed with these policies, and he thought there were strong economic reasons why, sooner or later, they would be modified. Writing now in English, he drew Prandtl's attention to the influential arguments of the Cambridge economist John Maynard Keynes and mentioned that opposition to punitive sanctions was also part of the official policy of the Labour party.[56] He even included a reassuring cutting from the correspondence columns of the *Times*.[57] The concern was not just personal; it was also professional. Glauert was worried that Prandtl would become so antagonized by allied policy that he would cease to take part in scientific exchanges. The anxiety was more than justified. The academic atmosphere in the immediate postwar years was a poisonous mixture of bitterness, intransigence, boycott, and counterboycott.[58]

The immediate result of the Glauert-Wood visit to Germany was the production of two confidential reports. In February 1921 McKinnon Wood produced Technical Report T. 1556, "The Aerodynamics Laboratory at Göttingen."[59] He argued (contrary to government policy) that Göttingen should be included in any future international trials that were envisaged to clarify the discrepancies that existed between the results of different laboratories.[60] When the results of the British and Germans were compared, McKinnon Wood noted that the Göttingen channel gave the same lift and drag, but always for an angle of incidence smaller by about one degree. He also studied and reported on the complicated "three-moment" balance mechanism used to measure the aerodynamic forces. A blueprint of the balance was included in the report. He judged that the Göttingen balances were "very inferior" in sensitivity to the British but then conceded that "our balances and manometers are unnecessarily sensitive for most of the work for which they are required" (14). McKinnon Wood concluded by saying that "Mr. Glauert discussed Prandtl's aerofoil theories with him and obtained some further papers. A discussion of these will be embodied in a separate paper (T. 1563) which Mr. Glauert is writing" (21). The "discussion" to which McKinnon Wood alluded turned out to be a piece of brilliant advocacy that was undoubtedly a major factor in undermining resistance to the circulation theory in Britain. Glauert had a gift for clear exposition and for seizing the essentials of the subject. Once he had accepted that Prandtl's theory represented the path to follow, Glauert produced a notable series of papers and reports explaining, testing, and developing the achievements of the Göttingen school. He corrected inadequate formulations and produced important extensions of the theory, as well as confronting the skeptics.

What made Glauert special? Why did he, with his impeccable Tripos background, strike out in a direction that had hitherto been unattractive to experts who shared with Glauert the intellectual culture of the "Cambridge school"? The "practical men" had always been divided over Lanchester's theory of lift, but the "mathematicians" had been unanimous in their skepticism. Why did Glauert break ranks? No definitive answer can be given to this question, though one fact stands out and invites speculation. The German name, the German father, the command of the German language may have generated some affinity with the body of German work that was under consideration. These facts distinguished Glauert from British experts such as Bairstow who did not have the command of German that would have enabled them to read the literature or converse with Prandtl.[61] (At that time Prandtl had little knowledge of English.) Of course, given the bitterness of the war, personal links to Germany might have had quite a different effect. Such links could be sources of difficulty, and there is evidence that they caused problems, and some inner turmoil, for Glauert. Referring to the outbreak of war in 1914, Farren and Tizard said of Glauert: "His German descent was an embarrassment to him, and he wisely decided to stay where such trivial matters did not assume the importance that they did elsewhere, and where he could work in peaceful surroundings, though not with a peaceful mind. His friends were far afield, and as time went on he became more and more restless and concerned with the difficulty of his position."[62]

Some people in Glauert's position would have kept their distance from all things German, and even shed their German name.[63] One might speculate that, in Glauert's case, the balance in favor of the circulation theory was tipped by the opportunity to meet members of the Göttingen group. His visit to Germany enabled him to explore the mathematics of the circulation theory with Prandtl face-to-face. Others had visited Göttingen before the war, but British experts did not have a great deal of direct contact with their German counterparts.[64] Even here it is necessary to be cautious about the impact of personal contact. Glauert was impressed by the circulation theory *before* he went to Germany. Doubts and qualifications dropped away after the visit to Göttingen and his advocacy became more confident, but he had begun to explore the theory through what he read in Southwell's copies of the *Technische Berichte*. Farren and Tizard suggest that exposure to engineer's shoptalk in the Chudleigh mess at Farnborough during the war made Glauert sympathetic to the needs of engineers and hence (one may suppose) to the theoretical approaches adopted by the engineers. Beyond this, little can be ventured in terms of explanation. It must simply be accepted that, equipped with the recent papers, Glauert made it his business to explore the Göttingen theory in great detail. He became committed to it, even when this led him to diverge from such authorities as Leonard Bairstow, Horace Lamb, R. V. Southwell, and G. I. Taylor.[65]

Glauert Makes the Case

On March 30, 1920, before he went to Germany, Glauert had presented his "Notes on the German Aerofoil Theory" to a meeting of the Aerodynamics Sub-Committee which included Glazebrook, Greenhill, and Lamb. The notes amounted to a brief overview and assessment of two conversion formulas. Part 1 of the paper dealt with the formula linking the aerodynamic characteristics of monoplane wings of different aspect ratios, while part 2 concerned the link between monoplane wings and biplane configurations. Glauert stated the relevant formulas without proof and simply said that they were taken from "German Technical Reports." In his comments to the subcommittee, when introducing the notes, he added that he had not yet been able to locate the papers giving the theory on which the formulas were based. His aim was to marshal some empirical data to find out if the formulas gave the right answers. He concluded that in some cases they did but in some cases they did not. In general, the transformation formulas discussed in part 2 of the notes seemed problematic, while those in part 1 worked well for predicting the induced drag but badly for predicting the induced angle of incidence. Because the (good) result for drag is theoretically dependent on the (bad) result for angle of incidence, Glauert declared himself puzzled.

These notes provided the basis for an article that Glauert published soon afterward in the short-lived journal *Aircraft Engineering*.[66] The article gives further insight into the status he accorded to Prandtl's theory before the Göttingen visit. Glauert put it like this: "Good agreement is not obtained for the angle of incidence, and as the theory estimates the change in drag from the effective change in incidence, it is evident that the basis of the theory cannot be regarded as quite satisfactory. The form of the expression found for the induced drag has a certain theoretical justification, but it is probably safer to regard the results as empirical formulae which are confirmed by experimental results" (161). The tone of this conclusion, in which Prandtl's results were accorded the status of mere "empirical formulae," contrasted with that adopted after his visit with McKinnon Wood to Göttingen and his talk to Prandtl.

In February 1921, after his return from Göttingen, Glauert produced his report T. 1563 on the outcome of his talk with Prandtl. The report was titled simply "Aerofoil Theory" and was based on six sources, all by Betz, Munk, and Prandtl.[67] These sources included Prandtl's *Trägflügeltheorie* and the Göttingen dissertations of his two assistants. Glauert's aim was to give "an account of the development of the theory and of the main results contained in the original papers" (2). He divided his report into five sections: (1) aerofoils of infinite span, (2) the finite monoplane wing, (3) special cases of the monoplane wing, (4) biplane wing structures, and (5) the influence of walls and the free boundaries of a stream on the flow in a wind channel. What followed was one of the most lucid accounts that has ever been given of the basics of the subject. Farren and Tizard refer to the "faultless style" of Glauert's exposition.[68] Although some of the same reservations were carried over from the earlier "Notes on the German Aerofoil Theory," for example, the empirical weakness of the prediction of the induced angles of incidence, these were not deemed to be of great practical importance compared to the accurate predictions of induced drag. Furthermore, the fuller treatment of the relation between monoplanes and biplanes had removed some of the earlier doubts. In the light of further analysis, Glauert now concluded that "the theoretical formulae may be accepted as giving a reasonably accurate method of predicting the biplane characteristics from those of the monoplane" (26).

How was Glauert's report received? What, for example, did the Aerodynamics Sub-Committee make of it? At meeting 38 of the subcommittee on April 5, 1921, minute 375(b) records that "Prof. Lamb remarked that he had read the report with great interest and considered it a very valuable addition to aerodynamic theory." Lamb did, however, say that he found the vortex lines difficult to visualize, and J. D. North suggested that the relevant diagrams were to be found in Lanchester's book.[69] (Whether Lamb found those diagrams acceptable or whether, like Prandtl, he thought they were wrong, is not recorded.) Although one may wonder about the identity of the (implied) prior theory, to which Prandtl's theory was a "valuable addition," Lamb's response may seem positive enough. There is, however, a second version of Lamb's reaction which must put a question mark over this positive interpretation. The second version is given in the minutes of the full Aeronautical Research Committee that met for its tenth meeting a few days later, on Tuesday, April 12, 1921, at the Royal Society. (Lamb now served on both the Aerodynamics Sub-Committee and the full research committee.) Minute 111 of the full committee meeting deals with the business of the subcommittee and refers to Glauert's "Aerofoil Theory" as "report (ii)." It reads as follows: "The report (ii) was stated by Professor Lamb to form a good basis for the commencement of work on the development of an aerofoil theory. Professor Bairstow expressed his dissent."

Had Lamb moderated an earlier, more positive response or did the later minutes simply capture nuances that were lost in the earlier summary? When one recalls the highly qualified wording that Lamb had used in his *Hydrodynamics*, when describing Kutta's work, the later minute seems closer to the authentic voice of this cautious spokesman of the Cambridge school. Either way, the full Aeronautical Research Committee did not receive Glauert's account of the Göttingen work with open arms. Bairstow was clearly not impressed by what he was hearing of Prandtl's achievements, and Lamb's apparent support now had so many qualifications that it is difficult to decide whether he was really being supportive or not. To say that something is a "basis" for a "commencement" of a "development" is not to say a great deal.

Undeterred by this response Glauert presented a second report in May 1921 called "Some Applications of the Vortex Theory of Aerofoils," which dealt with both wing theory and propeller theory. (I confine myself to the former.) Glauert was clearly in no mood to compromise and began by asserting that his previous paper had led to "a satisfactory theory for correlating the lift and drag of different wing structures and for determining the effect of changes of aspect ratio." His aim now was to see whether it gave an accurate picture of the flow of air in the vicinity of the wing. Glauert's talk of the theory "correlating" data suggests he may have still been concerned lest Prandtl's approach merely provided empirical formulas rather than a physically true account of the actual air flow. His intention was to address this anxiety by comparing the calculated and observed "downwash" of air at three locations in the vicinity of a wing: (1) above or below the center of the wing, (2) behind the (main) wing in the region of the tailplane, and (3) at the wingtips.[70]

Before making the comparison Glauert entered a caveat. Prandtl's theory rested on drastic simplifications, and these would necessarily preclude it giving an accurate picture of certain features of the flow. First, the wing was replaced by the abstraction of a "lifting line." For both the simple horseshoe model and the refined model, with a varied distribution of circulation along the span, the chord of the wing was neglected. So the flow close to the wing could not possibly be described accurately. Second, where a vortex sheet was assumed to be issuing from the trailing edge, the sheet would roll up, so the flow behind the wing would have a different character at different distances. As a partial response to this second problem Glauert performed his calculations of the downwash in two different ways: (1) on the assumption of a constant distribution of lift (the simple horseshoe model) and (2) on the assumption of an elliptical distribution of lift (the refined horseshoe model). He argued that the rectangular wing used in the experimental tests would have a lift distribution somewhere between these two extremes. Furthermore, the trailing-vortex system near the wing would be more like the refined model, whereas the system at a distance would be more like the simpler model. Glauert argued that provided the tests were not carried out too near the wing, or too far behind it, the theory ought to give a reliable picture of the surrounding airflow.

The first of the three tests used downwash data taken from a BE2E biplane with its wings at an angle of incidence of 6°. Measurements were made along an axis that was normal to the wing at its midpoint. For distances away from the wing of greater than one and a half times the chord, it was found that the predictions based on a uniform loading agreed fairly well.[71] Other results, however, using wind-channel data from a monoplane wing with an RAF 6 section at an incidence of 3°, showed a downwash that was much greater than predicted. The second test measured downwash along the longitudinal axis, that is, at a number of points toward, and beyond, where the tailplane is typically located. Observed values of the downwash were progressively smaller than those predicted by the elliptical distribution of lift but larger than those to be expected on uniform distribution. The trend of the results was roughly right but not the numerical values. The third test concerned the flow at the wingtips, and theoretical calculations were compared with wind-channel measurements made on a model Bristol fighter. This time, only calculations based on the elliptical distribution were used. (A uniform distribution was ruled out because it failed to represent the fact that lift falls to zero at the tips.) The predictions agreed with the observations in showing that, as one moves along the span of the wing, downwash decreases toward the tips and turns into an upwash beyond the tips.[72]

Glauert admitted that he was perplexed by the mixed results of the first test but deemed the results for the flow round the wingtips "quite good." The theory represented the flow "with reasonable accuracy," especially given all the approximations involved.[73] The results for downwash on the longitudinal axis obviously took Glauert into the area studied by Föppl in the very first published test of Prandtl's theory. Glauert concluded that the theory "cannot be used in any simple manner to predict the angle of downwash behind the wings"—which is exactly what Föppl had tried to do. The operative words, though, are "in any simple manner." Glauert pointed out that the inaccuracy probably arose because the vortex sheet behind the wing was unstable and so the theory must be made more complicated to allow for this effect. He then noted that Prandtl had offered some suggestions about how to describe the rolling up of the vortex sheet. These promised to bring calculation and observation back into alignment.

The study of the downwash behind a wing structure had held the promise of giving "a direct method of testing the underlying assumptions of the theory," but it proved to be a complicated phenomenon and generated a lengthy and ramified program of experimental and theoretical investigation.[74] Glauert was clearly sensitive to the problematic character of the empirical data and the complex relation between theoretical calculation and experimental measurement. It is also clear that he did not treat the empirical difficulties confronting Prandtl's theory as refutations of the theory. He saw them as challenges that called for its further development. In a quiet but determined way Glauert shouldered the burden of developing the theory mathematically, and he did so, for a while, almost single-handedly. Farren and Tizard said Glauert was a "bonny fighter" in argument and worthy of any opponent, but they remembered him as a man of "essential modesty and gentleness."[75] This characterization accords with the calm and nonpolemical character of everything he wrote. Not all of those who came to support Prandtl shared these character traits. One who did not was the redoubtable Major Low.

Confrontation at the Royal Aeronautical Society

From 1922 the Royal Aeronautical Society (RAeS) became the main public forum in London for informed, and sometimes sharp, debate over the merits of the Prandtl theory. This was a matter of deliberate policy. Bairstow had become president of the RAeS and, at an ordinary general meeting on November 2, 1922, reported that the council of the society felt that there should be more opportunity for the expert discussion of technical subjects. Talks to the society need not be kept accessible to a general audience and could be prolonged over more than one session. An obvious topic for such treatment, said Bairstow, was Prandtl's theory. To start the ball rolling Bairstow invited Major Low to give his "Review of Airscrew Theories."[76]

Major Low was Archibald Low, the designer from Vickers, who has already been mentioned in relation to the conflicts between the manufacturers and the Royal Aircraft Factory. Low's role in that dispute showed that he was not a divisive man, but he always had definite opinions and was prepared to speak his mind. He belonged to the section of practical men inclined to be sympathetic rather than hostile to Lanchester. Having constantly defended the National Physical Laboratory and the Factory from its detractors, he said he now felt justified in offering some outside criticism. In the journal *Aero-*

nautics Low had earlier expressed the view that "there was a tendency on the part of the official circle of aero-dynamic science in this country to think they were absolutely 'it' and that there was very little outside. . . . That was a dangerous attitude of mind to get into. He believed we no longer had anything like the supreme position of advantage."[77] Low had acquired his rank of major during the war and was now employed by the Air Ministry. He was based in the library of the ministry and was engaged on translation work. Low was later to become a member of the Fluid Motion Panel of the Aeronautical Research Committee. Although his mathematical expertise was not comparable to that of, say, Glauert, his contributions were deemed interesting by authorities such as Taylor and Southwell.[78] As Bairstow said, when introducing Low, their speaker had earned the reputation of being "very interesting and very contentious." Laughter greeted this remark, but it may have been nervous laughter.

Low used his talk as an excuse to lay out the basis of the circulation theory and Prandtl's work. He had a command of the German literature and could not resist taking Bairstow to task for the inadequacy of the foreign references in the latter's recently published *Applied Aerodynamics*. Low described for his audience some of the German papers that had been available for a number of years but had lain neglected. He described the basic geometry of conformal transformation and sketched the main results of the work on the infinite wing. He then gave a qualitative account of Prandtl's theory of the finite wing and reported that the transformation formulas, linking wings of different aspect ratio, had been confirmed experimentally. Here Low quoted the first volume of the *Ergebnisse* that Prandtl had mentioned in his exchange of letters with Southwell.

In the course of the talk it became clear that Low wanted to force Bairstow and others to acknowledge their culpability for neglecting the circulation theory. They had disregarded Lanchester and left it to the Germans to develop insights that Lanchester had published in 1907. Lanchester had shown "remarkable insight into the physics of a problem that had baffled scientists of the last century. Had our physicists followed up his ideas, this country might have shared in the work" (43). Low went on to make a comparative observation. He noted that Lanchester's work on the theory of lift had been ignored in this country while being known in Germany. By contrast, G. H. Bryan's work on the theory of stability had been fully appreciated in Britain but had made much less impact in Germany. To illustrate his claim Low cited Joukowsky's acknowledgment of Lanchester in the *Zeitschrift für Flugtechnik* and contrasted it with a reproach by Reissner, directed at his fellow countrymen, for their neglect of Bryan. As Low put it:

Although not till recently honoured in his own country, Lanchester has had very full recognition in Germany, unlike Bryan, who is generally ignored. In Joukowsky's words, "Lanchester's distinguished service is the elucidation of the transition from plates of infinite span . . . to finite span in simply connected space" (*Z.f.F.u.M.*, 1910, p. 282). Compare this with Reissner's reproof to German writers, "Bryan's highly distinguished service in first (1904) putting the problem of aeroplane stability in complete mathematical form should not be ignored in citing names" (*Jahrbuch d. Wiss. Gesell. f. Luftfahrt*, 1915–16, p. 141). (43–44)

Knowing how these two quoted sources should be interpreted is obviously no easy matter, but Low's point is an interesting one. Perhaps the strengths and weaknesses of the two nations complemented each other. Any overall assessment of British and German aerodynamics should take this possibility into account.[79]

In the lengthy discussion after Low's talk, Bairstow declared that he would speak "mainly as a critic of the Prandtl theory" (62). Bairstow admitted that he was impressed by the way Prandtl had brought experiments on aerofoils of different aspect ratio into agreement and by Betz's success in bringing calculated and measured pressure distributions into alignment. Overall his position was that Prandtl's theory connected together a great number of facts. It was "a very good empirical theory," but, he told his audience, they should not think of "scrapping all their previous work." Prandtl's theory "was not sufficiently well established" (62). Bairstow declared himself surprised that Low had got through the whole of his lecture "without mentioning a fundamental property of air on which its motion depends, viz., its viscosity" (63). This brought Bairstow to what he called his fundamental objection to Prandtl's theory: "They could have various theories which were good or defective in various proportions, but ultimately if they were going to deal with a real physical problem they must come back as the basis to physical ideas. They had in the equations given by Stokes, and the experiments of Poiseuille and Stanton, very strong experimental indications that these equations were sufficient to account for the phenomena, whether it was a steady flow or an eddying flow. These equations did not appear in the Prandtl theory" (63).

In what was presumably a reference to the boundary-layer equations, Bairstow said that Prandtl gave "other equations" but that nobody knew what relation they had to the Stokes equations. The Stokes equations were currently the subject of research by a group at Imperial College. The members of this group "naturally looked for the source of the circulation of which the Prandtl theory makes use, [but] without finding it. In the solution of Stokes' equations it appeared there was no circulation, i.e., the motion of a viscous fluid around a body moving in it was free of circulation. He knew of no natural mechanism which could produce circulation in a viscous fluid and that seemed to him to make a great difference to one's appreciation of the Prandtl theory" (63). Prandtl's theory, said Bairstow, apparently speaking of both the theory of the boundary layer and the aircraft wing, was not a "fundamental theory" in the way that Stokes' equations were fundamental. He concluded by suggesting that both Lanchester and Prandtl were aware of these limitations and knew that they had not provided the last word in aerofoil or propeller theory. The "ultimate solution," insisted Bairstow, must be along other lines.

There was no way in which Low could match the technical authority of this attack, but he was not lost for a tart rejoinder. Casting himself in the role of the "engineer," responding to Bairstow the "pure scientist," he said he had no objection to providing scientists with endowments and facilities to allow them to pursue their "strictly abstract studies." But who knows when, if ever, these studies will bear fruit? As an engineer "he did not intend to wait for them on this occasion" (65). With the benefit of hindsight one cannot deny that Low had a point.[80]

Saying and Showing

In accordance with the Royal Aeronautical Society's policy of sustained, technical discussion, Low's paper was followed, later in the year, by other material relating to the Prandtl theory. On November 16, 1922, R. McKinnon Wood gave a paper titled "The Co-Relation of Model and Full Scale Work."[81] Like Low, McKinnon Wood also took the opportunity to describe the basis of the circulation theory and, like Low, found himself confronting Bairstow. It transpired in the discussion that Bairstow was engaged in experiments at the NPL to work out where viscous and nonviscous flows differed in the case of aerofoil shapes. Bairstow said he had no doubt that the theory of nonviscous flow would yield results that could be tested, but "he did not expect to find the circulation at all in the experiments" (499).

There were also two talks given to the society which were devoted to wind-tunnel studies of the vortex system behind a wing both by N. A. V. Piercy of the East London College.[82] Piercy had been a colleague and collaborator of Thurstone but worked at a much more sophisticated level both empirically and theoretically. Using the college's wind tunnel, Piercy produced detailed measurements of the airflow both behind the wing and in the region of the wingtips. It was clear that there were vortex structures to be mapped, and these corresponded, at least qualitatively, to the expectations created by

Lanchester's and Prandtl's work. Although they were broadly supportive of the circulatory picture, the results were actually understood by Piercy to support Bairstow's suspicion that too little weight had been given to the role of viscosity.

Piercy was very conscious of the empirical variability of the phenomena under study in his wind tunnel. He argued that the vortex effect behind a wing sometimes achieved its maximum value after the wing had stalled and thus after the lift (and, presumably, the circulation) had dropped away. How could this be explained on the Lanchester-Prandtl theory? He made three suggestions, none of which could be easily accommodated within the circulatory theory as it stood. First, he wondered whether, during a stall, the wingtip vortices continue to exist but are not joined together by a vortex that lies along the span of the wing. This would produce the effect to which he was referring, namely, wingtip vortices without lift or with diminished lift. But if the vortices can exist without circulation around the wing at, and beyond, the angle of stall, surely "it is not necessary for them to be so joined at a smaller angle" (502). Second, even if it were the case that the two wingtip vortices are still joined (in some fashion) after the stall, "they may be joined in such a manner as not to give cyclic lift" (502). "As a third alternative," said Piercy, we may "suppose that cyclic lift may be destroyed to a considerable extent by viscous effects," but then, "it seems reasonable to conclude that cyclic lift is not immune from viscosity at smaller angles" (502).

Piercy was well aware that supporters of the circulation theory had always sought to draw a line between normal flight at low angles of incidence and the phenomenon of stall at high angles of incidence. Their position was that a good theory of the former did not have to explain the latter. An explanation was desirable but not necessary. This had been a central part of the argument between Lanchester and Bairstow in 1915. Piercy sided with Bairstow on this matter. He dismissed the defense as an evasion. It was, he said, "beside the point." "The question is whether we can afford to neglect at 8 deg. incidence, say, a factor so powerful as to be able to overthrow the vortex system at, say, 16 deg. Should we not rather conclude that at any angle viscosity is playing an essential and important role in the whole system of flow?" (502). The long-standing British concern with stalling, and the desire for a unified and realistic theory of broad scope, was still in play.

The next paper in the 1923 volume of the Royal Aeronautical Society journal which discussed the circulation theory appeared under the name Glauert, but it came from Muriel Glauert, Hermann Glauert's new wife. The paper was called "Two-Dimensional Aerofoil Theory" and was based on a technical report written some two years earlier for the Aeronautical Research Committee, but under the name Muriel Barker, not Muriel Glauert.[83] Muriel Barker worked for the Royal Aircraft Establishment and was the holder of the post-graduate Bathurst Studentship in Aeronautics at Cambridge. Her notes on Kutta, which were mentioned and used in a previous chapter, were probably made when gathering material for writing the original technical report. The discussion of two-dimensional aerofoil theory for the RAeS journal was based on the assumption, rejected by Bairstow and many others, that inviscid methods are legitimate. To develop this starting point, Muriel Glauert introduced a general theorem due to Ludwig Bieberbach.[84] The theorem showed that there was one and only one conformal transformation of the form:

$$\zeta = z + \frac{b_1}{z} + \frac{b_2}{z^2} + \dots$$

(where b_1, b_2, ... were complex) which would map the space around a shape, such as an aerofoil, in the z-plane into the space round a circle in the ζ-plane, leaving the region at infinity unchanged. She then worked through, in mathematical detail, the special case of this theorem provided by the Joukowsky transformation and dealt with circular arcs, Joukowsky aerofoils, double circular arcs, struts, Kármán-Trefftz profiles, von Mises profiles, and Trefftz's graphical methods. Muriel Glauert's paper makes it clear that the British had been doing their homework. They had now brought themselves up to date and absorbed all the mathematical techniques and results of the German work on the two-dimensional wing that I described in chapter 6.

Next to be published in the sequence of Prandtl-oriented discussion papers was one by Hermann Glauert himself, titled "Theoretical Relationships for the Lift and Drag of an Aerofoil Structure."[85] At first glance Glauert's paper has the appearance of being no more than an elementary treatment of the circulatory theory—far less mathematical, for example, than Muriel Glauert's paper. Unlike his wife's paper, or his own technical reports for the Aeronautical Research Committee, the present paper was not replete with mathematical formulas. The appearance of simplicity, however, is misleading. The paper may have been essentially qualitative in its argument but it was in no way elementary. It was sharply focused on difficult problems, but the problems in question were methodological ones. It dealt with the orientation that was needed to appreciate Prandtl's approach—the very thing that divided Glauert from his mathematically sophisticated British contemporaries. They did not need to be convinced of the mathematics; they needed to understand the mathematics in a different way.

The solution of a physical problem in aerodynamics, said Glauert, can be analyzed into three steps. First, certain assumptions must be made about what quantities can be neglected, for example, gravity, compressibility, and viscosity. Only rarely is it necessary to take into account the full complexity of a phenomenon. Second, the physical system, in its simplified form, must be expressed in mathematical terms, for example, a differential equation and its boundary conditions. Third, the mathematical symbols must be manipulated until they yield numerical results that can be tested experimentally or used for some practical purpose. This third step, said Glauert, must not be misunderstood. It is where some of the greatest difficulties arise because the mathematical problems may be insurmountable. At this stage it may be necessary to simplify further the initial, physical assumptions or to confine attention to a limited range of cases, such as small deviations from known motions. It is important to remember, said Glauert, that "in no case are these assumptions absolutely rigid" (512). Glauert's three steps are not simply sequential: what happens during the third step can feed back into what was called the first step.

Glauert then rehearsed the assumptions that were made by proponents of the circulation theory, that is, the assumption that the air could be represented as a perfect fluid with neither compressibility nor viscosity; the assumption that the fluid flow is irrotational; and the need to postulate a circulation to avoid a zero resultant force or to resort to the theory of discontinuous flow. "In view of this discussion," said Glauert, "it appears that no satisfactory solution of an aerodynamic problem is to be expected when the effects of compressibility and viscosity are neglected, and it becomes necessary to consider the effect of these two factors" (513). This sentence is a striking one. It appears to concede all the points made by the critics of the circulation theory. Is this not exactly what Leonard Bairstow would say? Is not Glauert here following the line that led the young Taylor to dismiss Lanchester? Given Glauert's accomplishments as a stylist, however, both the import and the impact of these words would have been carefully weighed. He would not have inadvertently conceded too much or expressed himself inaccurately on such an important question.

How could Glauert grant that no satisfactory solution can be expected if viscosity is neglected without also granting the dismissive conclusions drawn by the critics of the circulation theory? The answer hinges on what it is to "neglect" compressibility and viscosity and what it is to "consider" their effects. Is this something done at the outset, in step 1 of the methodology? Or is it done at step 3, not as a sweeping assumption but as a technique for making the mathematics tractable? The vital but subtle methodological point that Glauert was making can be expressed like this: viscosity cannot, indeed, be wholly neglected but, contrary to first appearances, that does not preclude the use of perfect fluid theory. *There were ways of operating with the mathematics of a perfect fluid that involved consideration of viscosity.* The acts of consideration that were in question could not be *stated* in the inviscid equations themselves but would be *shown* in how they were deployed and interpreted.

Glauert explained that two important facts about viscosity must be accommodated. First, there is the no-slip condition, which stands in contrast to the perfect fluid property of finite slip. Second, viscous forces are proportional to the rate of change of velocity and hence are important close to a body such as a wing but become negligible at large distances. These are the physical facts for which approximations must be found. They cannot be dismissed in step 1 of the sequence of steps Glauert had described. That would indeed amount to a decision to "neglect" them, and it is known that this produces the empirically false result of a zero resultant force. Rather than neglecting these two facts, their reality must be taken into account by a justifiable approximation, an approximation of the kind introduced in step 3. Glauert's development of this point deserves to be quoted in full. Notice the specific meaning he attached to the word "ignore" in the quoted passage and the implied contrast between ignoring something (in step 1) and approximating its properties (in step 3):

> It is known that the solution obtained by ignoring the viscosity is unsatisfactory, but it is by no means obvious that the limiting solution obtained as the viscosity tends to zero is the same as the solution for zero viscosity. In particular, in the case of a body with a sharp edge, there is a region where the velocity gradient tends to infinity, and where the viscous forces will be of the same order of magnitude as the dynamic forces, however small the viscosity. On the other hand, the layer round the body in which viscosity is of importance can be conceived as of zero thickness in the limit, and this conception is equivalent to allowing slip on the surface of the body. It appears, therefore, that the non-viscous equations will be the same as the limit of the viscous equations, except in the region of sharp edges. (514)

The argument was that under the right conditions the equations of inviscid flow are legitimate approximations to the viscous equations and their use does not amount to "ignoring" or "neglecting" the viscous properties of the flow. The crucial requirement is that the inviscid flow must be one that can be understood as a limiting case of a viscous flow. Glauert appears to have carried this crucial lesson away with him from his conversations with Prandtl. The Royal Aeronautical Society paper was therefore not merely an elementary exposition of the theory of lift; rather, it was an attempt to confront the habits of thinking that had justified the systematic neglect of the inviscid approach by British experts. Up to this point the conviction of British mathematical experts that ideal-fluid theory was false, and ultimately useless for aerodynamics, had carried almost everything before it. Only the theory of viscous flow dealt with reality. Ideal-fluid theory may provide some residual mathematical challenges, and some suggestive analogies, but it could not be taken seriously as a means for directly engaging with reality. Glauert was challenging this assumption. He sent a copy of his paper to Prandtl along with copies of Piercy's two papers. Of Piercy's work he remarked that the experimental results were interesting but expressed doubt about the theoretical interpretation: "His experimental results are of considerable value, but his interpretation of them leaves a good deal to be desired." Glauert described his own piece as a "short note I wrote in justification of the principles underlying the vortex theory of aerofoils." [86] The question was: could Glauert shift the way his contemporaries understand those underlying principles?

The International Air Congress of 1923

The International Air Congress for the year 1923 was held in London. It provided a further occasion for assessing the advances that had been made in aeronautics during the war years and for addressing unresolved problems. It was a highly visible platform on which the supporters and opponents of the circulatory theory could express their opinions and, in some cases, air their grievances. In the morning session of Wednesday, June 27, there were three speakers: Leonard Bairstow, Hermann Glauert, and Archibald Low.

The first to speak was Bairstow, whose talk was titled "The Fundamentals of Fluid Motion in Relation to Aeronautics." [87] Bairstow was explicit: his aim was nothing less than the mathematical deduction of all the main facts about a wing from Stokes' equations and the known boundary conditions. The work of Stanton and Pannell had shown that eddying motion did not compromise the no-slip condition and had established kinematic viscosity as the only important variable. [88] "These experiments appear to me," said Bairstow, "to remove all doubt as to the correctness of the equations of motion of a viscous fluid as propounded by Stokes and the essential boundary conditions which give a definite solution to the differential equations. The range of these equations covers all those problems in which viscosity and compressibility are taken into account, and from them should follow all the consequences which we know as lift, drag etc. by mathematical argument and without recourse to experiment. Such a theory is fundamental" (240–41). The boundary conditions were empirical matters, but thereafter everything should follow deductively: lift, drag, changes in center of pressure, the onset of turbulent flow and stalling characteristics, along with a host of other results important to the designer of an aircraft. Confronted by an aerodynamic problem the response would not be "let us experiment" but "let us calculate." This was what it meant to possess a "fundamental" theory.

Bairstow was not being naïve. He, as well as anyone, knew the problems standing in the way of any such employment of the Stokes equations. But he insisted that these difficulties had to be confronted because only in this way could aerodynamic theory be given a proper foundation in physical reality. Until aerodynamic results could be derived from the Stokes equations, they lacked a true and reliable foundation. They could be no more than makeshift approximations combined with ad hoc appeals to experimental findings. For Bairstow this was not an intellectually acceptable state of affairs. Bairstow did not deny that the success of Prandtl's theory was "striking." What worried him was that Prandtl made this "start without reference to fundamental theory" (241). If Prandtl's approach was legitimate, then it *must* be the case that it can be related to the Stokes equations.

If the successes of the circulation theory could no longer be denied, Bairstow now said *it was those very successes that constituted the problem.* The theory worked, but why did it work? What might, at first, have appeared to be the strength of the circulation theory—that it worked—was now identified as a source of worry. "The questions which naturally arise," said Bairstow, are "(i) Why does the circulation theory apply with a sufficient degree of approximation in some cases and what is the fundamental criterion of its applicability? (ii) Is further progress possible along the same lines?" (242). There might seem to be an obvious response to the second question. Why not just try and see what happens? In Bairstow's opinion, however, the rational thing to do was to seek guidance from the fundamental equations in advance, rather than resort to trial and error. But it was the first question that provided the most characteristic expression of Bairstow's position. The inviscid model of air was physically false. The appearance of truth must be explained away by showing why a viscous fluid sometimes behaves like an inviscid fluid. This capacity to appear inviscid should be deducible from the Stokes equations.[89] Bairstow therefore proceeded to lay out for his audience some of the mathematics of viscous flow.

In the course of his discussion Bairstow remarked that the postulation of a boundary layer was an attempt to respond to the "essential failure" of the (inviscid) theory to meet the boundary conditions, that is, the condition of no slip. He conceded that this move, that is, postulating a boundary layer, "does not present an impassable barrier to acceptance," but he insisted that difficulties begin "when the region is defined as of infinitesimal width" (244). If Bairstow was going to countenance a boundary layer at all, it had to be an empirically real layer with a finite depth, not the mathematical fiction of an infinitesimally thin layer. Experimentally, he said, the infinitesimally thin boundary layer was "unacceptable at the trailing edge," where there was a clear wake; and, in any case, the inviscid approach to lift that it appeared to sanction (that is, the Kutta-Joukowsky formula making lift proportional to circulation) "leads to an estimate of lift which is 25 per cent too great" (244).

At the time he gave his talk to the Air Congress, a program of work and publication was under way designed to carry Bairstow toward his fundamental goal. Money had been acquired from the Department of Scientific and Industrial Research to pay for two assistants, Miss Cave and Miss Lang, to work under Bairstow's guidance at Imperial College. One paper from the team had already appeared: Bairstow, Cave, and Lang's "The Two-Dimensional Slow Motion of Viscous Fluids."[90] In the same year as the Congress these three authors also published "The Resistance of a Cylinder Moving in a Viscous Fluid."[91] In the latter paper Bairstow explained that the purpose "was to prepare the ground for a solution of the complete equations of motion for very general boundary forms, and steps are now being taken toward that end" (384). In the event, he did not actually address the complete equations but followed Stokes and Lamb in using an approximation. Stokes had simplified his own equations by doing the opposite of Euler and had neglected everything but viscosity. All the inertial terms had been dropped and only the viscous terms retained. This had enabled him to arrive at equations that described, for example, the very slow motion of a very small sphere in a very viscous fluid. The formula was accurate near the sphere but failed at a large distance from the sphere. Stokes also drew attention to the fact that his approximation could not be applied to two-dimensional flow. It could not be made to work for the two-dimensional case such as a circular cylinder.[92]

The Swedish mathematician Carl Wilhelm Oseen had proposed another approximation for the full viscous equations in 1910.[93] These produced the same results as Stokes' analysis in the neighborhood of a sphere but differed at large distances. In 1911 Lamb had published a paper in which he drew attention to Oseen's approach and had simplified the working.[94] Lamb also showed how to apply Oseen's approximate form of the Stokes equations to the case of a circular cylinder. He showed how it could be extended to the two-dimensional case in a way that had proven impossible using Stokes' own simplification. It was Lamb's work that provided Bairstow and his team with their method. "The line of attack adopted by us," said Bairstow, "was suggested by Lamb's treatment of the circular cylinder" (385–86).

Bairstow was able to generalize Lamb's result for the circular cylinder to an ellipse. Most of the resulting formulas could be evaluated without resort to graphical or mechanical methods, but these could not be avoided when analyzing shapes such as cross sections of wings and struts. In the case of a wing Bairstow was not able to reach a determinate result, but, as he put it, at least the problem "has been attacked and a method of solution indicated" (384). In their second paper Bairstow, Cave, and Lang offered a complicated, general formula for the lift of a wing shape, that is, an expression for the vertical component R_y of the resistance. The implications of the formula, however, were not clear. Bairstow could only say, "Except in the case of symmetry it is not obvious that R_y will vanish, but rather that a lift may be expected" (419). The computations needed to get this result were considerable, but despite all the expenditure of effort he had done no more than demonstrate the *possibility* of a lift.

Bairstow had sent a copy of the collaborative 1923 paper, on the resistance of a cylinder in a viscous fluid, to Prandtl. Prandtl replied on October 17, in German. He expressed polite interest but said that he had certain doubts about Bairstow's calculations. There followed two closely typed pages of technical objections. Prandtl demonstrated that Oseen's approximation, and Bairstow's use of it, was only acceptable in the vicinity of the cylinder when the Reynolds number was small, that is, when the flow was very viscous. He identified the precise equations in Bairstow's paper that were inadequate and explained why they failed when the Reynolds number was large and the viscosity therefore relatively small, that is, in the cases that were relevant to aerodynamics. He signed off, rather abruptly, after recommending that Bairstow acquaint himself with the theoretical work of Blasius and the experimental measurements of Wieselsberger.[95]

Evanescent Viscosity

Glauert's contribution to the London Congress was a paper titled "Some Aspects of Modern Aerofoil Theory."[96] It covered much of the same ground as the methodological paper given to the RAeS, but with the addition of technical results about propeller theory and wind-tunnel corrections. Though deeply opposed to Bairstow's view, Glauert did not directly attack what had just been said. Instead he quietly sought to outflank it by demonstrating that the concern with "fundamentals," as Bairstow conceived it, was out of touch with events at the front line of active research.

Glauert began by pointing out that the study of the forces and moments on a body in motion through a viscous fluid was beset by complexity and progress had been slow. But a "modified form of the classical hydrodynamics" was proving successful. "The present paper," Glauert went on, "is concerned only with the problem of aerofoil structures, whose essential characteristic is that they give a relatively large lift force at right angles to the direction of motion at the expense of a relatively small drag force retarding the motion" (245). A few minutes earlier Bairstow had drawn attention to the "limited" scope of application of the circulatory theory as a point of criticism and as an unacceptable feature of the work of Kutta, Joukowsky, and Prandtl. Right at the outset of his talk Glauert was doing the opposite. He was drawing attention to the limited focus of the work as a wholly-taken-for-granted feature that in no way told against it. Glauert was implicitly making an engineering-style response to Bairstow of exactly the kind that Lanchester had made, explicitly, in the 1915 confrontation.

Having led his audience, nonmathematically, through the main developments in aerofoil theory, Glauert concluded by saying that the most important feature of the "modern" approach was that it "presents us with a point of view" with which to examine new problems. It provides us with a small number of theoretical conceptions "which serve to bind into a single unity a multitude of experimental results" (255). Clearly this point of view was different from that adopted by Bairstow—and Glauert's idea of unity was not Bairstow's. For Bairstow unity meant deducibility from the Stokes equations; for Glauert it meant linking experimental results by adopting the modern, methodological standpoint. Glauert expressed himself in an interesting way. Referring to the moment when the boundary layer became infinitely thin, he said, "The effect of the evanescent viscosity is represented in the non-viscous solution by the possibility of a circulation round the aerofoil" (246).

The word "evanescent" means "passing quickly from sight or memory." It was also the old Newtonian word for describing the infinitesimal quantities that entered into the differential calculus. Infinitesimals were "evanescent quantities" that were neither zero nor nonzero but poised on the very brink of vanishing. Newton and his followers spoke in this way because they did not possess the modern concept of a limiting process.[97] Glauert, of course, did possess it. His idea was to contrast the limiting value of, say, $f(\mu)$ as $\mu \rightarrow$ o with the value of $f(o)$, that is, the value of the function at $\mu = o$. These can be different. Glauert's "evanescent viscosity" was thus viscosity on the point of reaching the limit zero, but his language was meant to register a methodological as well as a mathematical point. It signalizes the difference between deciding that viscosity is zero in step 1 of his three-step methodology, and accepting that it may be treated as zero in step 3.

Bairstow wanted to know why ideal-fluid theory sometimes worked. Glauert's answer was that it works when the ideal flow is a limiting case of a flow that would take place in a fluid of small viscosity. This answer was not one that Bairstow was prepared to accept. Recall that Bairstow saw no "impassable barrier" to the idea of a boundary layer, but, he said, difficulties began when the region was said to have an infinitesimal width (244). Bairstow did not want the relationship between inviscid and viscous flows to hinge on a limiting process, and certainly not on a limiting process that was carried out informally. He did not want Glauert's "evanescent" viscosity; he wanted what, in his own mind, would count as *physically real* viscosity.[98]

The Albatross Wing

Major Low's paper, the third Congress paper of the morning, was titled "The Circulation Theory of Lift, with an Example Worked out for an Albatross Wing-Profile."[99] The "albatross" of his title was not the bird but the name of a German aircraft company that had played a prominent role in the war.[100] Low's aim was to apply the circulation theory to an actual aircraft wing and to show the relation of the theory to drawing-office practice. He also wanted to straighten out one or two points of recent history. He began by reminding his listeners that the origin of the circulatory theory was grounded in the work of British physicists. Fifty years ago, said Low, Rayleigh had published his paper on the spin of the tennis ball and explained the force that made it veer by reference to the circulation around the ball. A similar idea, he said, was to be found in the work of P. G. Tait of Edinburgh (where Low had himself been a student, graduating with an honors degree in mathematics and natural philosophy in 1903).[101] Low regaled his audience with a story about experiments, done in the dark cellars of the old Edinburgh University buildings, on the spin of golf balls. Tait was helped by his son, "the lamented Freddie Tait." Freddie had been a professional golfer and, in the name of science, was required by his father to shoot golf balls through screens in order to trace their trajectory. On one occasion, in the gloom, he missed the screens. This resulted in the experimenters dodging around as the ball "ricocheted interminably off the walls of the cellar" (255).

This story was merely the disarming prelude to a point that was not intended to be amusing. Lanchester, Low went on, had boldly applied the idea of circulation to the wings of an aircraft and had given a thorough, descriptive account of the mechanism of flight. That was nearly twenty years ago. Why was it only now that the circulation theory was being taken seriously in the land of its origin? Low had an answer, and it was not a flattering one: "Had Rayleigh put forward the theory, how we should have vied with each other in the will to believe it, if not in power to understand it! But when it was offered by a man outside the circle of recognised physicists it was ignored" (255).

Leaving his audience to ponder this sociological point, Low went on to expound some of the basic techniques associated with the theory, confining himself to "strictly graphical and descriptive" methods. First, he gave a graphical method for transforming a circle into Joukowsky profiles and then tackled the more difficult, inverse task of going from a given aerofoil back to a circle or a close approximation to a circle. As before, Low was conveying to his audience the content of recent German material, this time using a postwar publication by Geckeler.[102] Low showed how to start from the Albatross wing and, using drawing-office methods, map it back to an approximate circle by a series of trial-and-error steps. "There now remains only a routine of laying off and measuring straight lines on the drawing board to determine the velocity and hence the pressure at every point of the field" (273). Low assumed a velocity $U = 10$ m/sec and an air density of $\rho = 1.2$ kg/m³. Using the formula $L = \rho U \Gamma$, he derived the data to construct a theoretical curve for the Albatross wing relating lift to angle of incidence. Because the formula was based on the assumption of an infinite span, the curve could not be compared directly with wind-tunnel data derived from a finite wing. Low then appealed to the Göttingen transformation formulas relating wings of the same section but different aspect ratio. This allowed him to recast known experimental data on the Albatross wing into its equivalent for an infinite wing. Low now had two curves that linked lift and incidence for the Albatross wing, one curve coming from wind-tunnel tests, the other derived from the circulation theory. For the range of $-5°$ to $+10°$ the two graphs were close together. The theory was supported by experiment. Given an arbitrary wing, a designer could now predict from the circulation theory the curve relating lift to angle of incidence at least up to the point of stall.

Having achieved his main goal, Low then returned to the theme with which he had begun. "In conclusion," he said, "it is desired to call attention to the fact that this fundamental physical theory was first stated by an English writer, and then allowed to fall into complete neglect in the country of its origin, largely owing to the attitude taken up by some of Lanchester's fellow members of the Advisory Committee for Aeronautics" (275). On this note the talk ended.

Calling the circulation account of lift a "fundamental physical theory" can only have been meant as a thrust at Bairstow, who had just explicitly denied it the status of being a fundamental theory. But the remarks blaming the Advisory Committee for Aeronautics for the neglect of Lanchester were even more pointed. The austere figure of Professor Sir Richard Glazebrook, who had been the chairman of the Advisory Committee, and who was thus the main focus of Low's complaint, was present at the talk. In fact, he was more than present. He was presiding over the session at which Low had just delivered his paper.[103]

Professor Glazebrook's Excuse

R. V. Southwell, from the NPL, opened the discussion after the talks and sought to defuse the situation with good-natured praise for all of the speakers. Southwell wondered if the Stokes equations were quite as secure as Bairstow assumed. He raised the possibility that the underlying physics might involve even more complications than those already expressed in the equations. He also reported that wind-tunnel experiments under way at the NPL seemed to be finding a value for the circulation around a wing that was similar to that predicted by Prandtl, though he, Southwell, doubted if the flow near the wing would correspond to that assumed by the circulatory theory. On the other hand, he was enthusiastic about Bairstow's fundamental research program and fully supported the need to explain the success of Prandtl's approach. Bairstow's own contribution to the discussion was a bland response to a Dutch speaker from the audience who had sketched some of the recent work at Aachen and Göttingen. Bairstow said he was glad to hear that Continental workers were taking viscosity seriously. The discussion ended on a bizarre note when Sir George Greenhill proceeded to inform the audience that the modern approach to aerodynamics was based on a paper that Rayleigh had written fifty years ago on the irregular flight of the tennis ball. This intervention was remarkable for two reasons. First, Greenhill was rewriting history and was expecting his audience to have forgotten all about the discontinuity theory of lift and his own, and Rayleigh's, contribution to it. Second, it is unclear whether Greenhill had come into the lecture late or whether he had failed to register what had been said in his presence. Glazebrook had to draw Greenhill's attention to the fact that he was repeating a version of what Major Low had just said. This done, Glazebrook thanked all of the speakers and promptly declared the meeting closed.

Despite the pointed criticism of the Advisory Committee, Glazebrook had chosen not to respond to Low. He might have been distracted by Greenhill's odd behavior but, leaving psychology aside, there is another possible explanation for Glazebrook's nonresponse. The Wednesday session was not the first time the issue of Lanchester had been raised at the Congress. Low and Glazebrook had crossed swords on the previous Monday, June 25. It is possible that Glazebrook had decided he had said all he was prepared to say and was not going to be drawn out on the subject again. On that Monday, Glazebrook had given a paper titled "Standardisation of Methods of Research."[104] In the discussion that followed he had encountered some criticism by Major Low about the reliability of wind-tunnel results. Low cited some negative remarks from G. P. Thomson's book on aerodynamics and argued that wind-tunnel data needed to be corrected. The "Lanchester-Prandtl theory," said Low, had shown how to make the corrections, and this theory would soon be the subject of his own talk. Glazebrook, who did not like to air the problems of wind-tunnel research in public, suggested that Prof. Thomson had surely changed his mind. Then, perhaps alerted by Low's mention of his forthcoming talk, Glazebrook added a comment that was not a direct response to anything that had actually been said. As if to head off trouble, Glazebrook launched into an apologia for the way Lanchester had been treated: "With regard to the reference to the Prandtl theory, I trust there is no one here who will in any way depreciate the enormous value of the work done by Mr. Lanchester and of the suggestions he has made. But it was not until Prandtl put some such suggestions into mathematical form that it was possible to attach to them the kind of value they have now gained, or to give Mr. Lanchester all the credit and praise that we should desire to give for his work" (65). This preemptive statement may explain why, on the following Wednesday, Glazebrook remained silent. He had no wish to go round the issue again.

Glazebrook's desire to give due, if belated, credit to Lanchester may be accepted at its face value, but as an excuse for the neglect of the circulation theory, his claim has three, obvious weaknesses. (1) To say that we can now see that Lanchester was doing something valuable because Prandtl has made it clear to us does not explain why the British could not have worked it out for themselves. (2) In reality, as I have argued, British mathematicians had no difficulty in seeing the underlying mathematical form of Lanchester's ideas. It was not the obscurity of the relation to mathematics that was the cause of the trouble, but the opposite. British experts such a G. I. Taylor were very familiar with the mathematical form of the circulation theory. It was actually the underlying mathematical form (the potential flow of an inviscid fluid) that they rejected on the grounds that it could not refer to processes that were physically real. (3) When British mathematicians were presented with a developed mathematical expression of Lanchester's theory, they still experienced difficulty in coming to terms with it—witness, for example, G. H. Bryan's negative review of Joukowsky, the responses of Lamb to Kutta, and Bairstow's response to Prandtl and Betz. In all cases the work struck them as a problem rather than a solution. None of these three points is accounted for by Glazebrook's version of the events. It is not difficult to see why a well-informed practical designer such as Low might have felt less than convinced by Glazebrook's answer. No wonder he could not resist raising the matter again and putting Glazebrook on the spot.[105]

I have now looked at some of the discussions about aerodynamics that took place in Britain in the immediate postwar years. It is clear that the mathematically sophisticated British experts did not take the view that "there was nothing to learn from the Hun." They were learning and learning quickly, but there was disagreement about what, and how much, was to be learned. How, and on what terms, was the Göttingen work to be assimilated? While the arguments at the Royal Aeronautical Society and the International Congress were conducted in the public realm, there had been other arguments that were still running their parallel course behind the closed doors of committee rooms. It is to these that I now return. In the next chapter I pick up the story of the discussions initiated in the Aeronautical Research Committee by Glauert's resolve to champion the merits of the circulation theory and Prandtl's theory of the finite wing.

The Laws of Prandtl and the Laws of Nature

Prandtl was not vastly outstanding in any one field, but he was eminent in so many
fields. He understood mathematics better than many mathematicians do.

MAX MUNK, *"My Early Aerodynamic Research"* (1981)[1]

After Glauert and McKinnon Wood had presented the reports on their Göttingen visit, discussions continued in the Aeronautical Research Committee
as the British experts sought to mobilize a collective response to the German wartime achievements. These (sometimes sharp) exchanges took place
in the monthly meetings of the committee and its subcommittees that were
held in London. The Cambridge contingent made the journey to London
together by train and engaged in lively aeronautical debate en route. "I fear
we must have been a pest to our fellow travellers," recalled one.[2] The upshot
of the committee meetings are recorded not only in the minutes of their discussions but also in the confidential technical reports circulated among the
participants. The content of the technical reports sometimes surfaced in the
published Reports and Memoranda issued by the committee and sometimes,
in the case of especially important results, in leading scientific journals. A
number of the main experiments done in this period appeared in the *Philosophical Transactions of the Royal Society* and in the *Proceedings of the Royal
Society*. There were some significant and perplexing changes in the analysis
of the experimental material as the data made the journey from the private to
the public realm.

I have described how Taylor, in his 1914 Adams Prize essay, had dismissed
Lanchester's idea that the flow of air over a wing was describable in terms of a
perfect fluid in irrotational motion with circulation. If Prandtl was right, then
Taylor had been wrong. Led by Glauert, the postwar argument in the Aeronautical Research Committee seemed to be going in Prandtl's direction. The
circulation theory was gaining ground. By 1923 Glauert felt able to write to
Prandtl to tell him that his "aerofoil theory has certainly aroused much interest here and it would not be an exaggeration to say that it has revolutionised

many of our ideas."[3] But Taylor (see fig. 9.1) was not to be easily convinced that his earlier reservations had been misplaced. In the postwar discussions, he made it his job to scrutinize Glauert's reasoning and to oppose it whenever he detected a logical gap or a questionable premise.

Glauert versus Taylor

There was clearly a desire by the members of the Aeronautical Research Committee to put the theory of circulation and Prandtl's analysis of the finite wing to the test, but disagreement emerged about how to proceed. This gave rise to a sequence of technical reports in which Taylor and Glauert crossed swords. Part of the problem concerned experimental technique. A further difficulty was that Glauert was sensitive to the fundamental distinction between the ideas underlying the two-dimensional picture of Kutta flow (that is, flow that is smooth at the trailing edge) and Prandtl's three-dimensional picture of the wing as a lifting line with trailing vortices. Glauert wanted these ideas kept distinct, while other participants in the discussion ran these two ideas together and counted them as forming one single theory whose basic assumption was the irrotational character of the flow.

To explain what was at issue it is necessary to go back to December 1921 and the mathematical report submitted by Muriel Barker.[4] She had suggested that the theoretical streamlines she had plotted for the flow over a Joukowsky aerofoil with circulation could be the basis for an experimental test: "it would be most instructive," she had written, "if these same quantities could be obtained practically" (3). Miss Barker's report and the question of what to do next were discussed by the Aerodynamics Sub-Committee and by the full Research Committee during February and March 1922.[5] Should they follow her suggestion and place a model of a Joukowsky aerofoil in a wind channel or should they use a more practical aerofoil, for example, the RAF 15? If they used a real section then should they ask Miss Barker to generate the theoretical streamlines by tedious computation or could a quicker method be found? Were mechanical or electrical methods of generating the theoretical streamlines of comparable accuracy to those produced by the laborious calculations that would be needed? Lamb was in favor of using the Joukowsky profile and direct calculation. Southwell wanted to use a more realistic profile and a mechanical method. He mentioned that Taylor had developed a piece of apparatus that enabled him to use a soap film to model the potential surfaces of ideal fluid flow. Bairstow added that he and Sutton Pippard had devised graphical methods for solving Laplace's equation.[6] Then there was the possibility of using the techniques developed by Hele-Shaw derived from photographs of creeping flow. It was decided that Southwell and Taylor would report back on different analogue methods of producing theoretical streamlines.

Southwell started with his report T. 1696.[7] He supported Muriel Barker's suggestion that comparisons be made of theoretical and empirical streamlines for an infinite wing, that is, where the model wing would reach right across the tunnel to exclude the effect of flow around the tips. In this way, said Southwell, "a direct check can be imposed upon one of the fundamental assumptions of the Prandtl theory" (2). Southwell then described the method developed by Taylor for simulating the streamlines and the bench-top apparatus that had been built.[8] A soap film was stretched between the walls of a box while precise measurements were made of the position of the film. The film connected the outline of a small wing profile to other boundaries within the confines of the box. (These boundaries represented the walls of the wind tunnel.) Southwell explained how this technique could take into account the circulation as well as automatically correcting for the effect in the flow of the tunnel walls. "Using orthodox mathematical methods," said Southwell, "it would appear that the problem thus presented is one of extreme difficulty" (2). Taylor, however, followed this up with a brief note, designated T. 1696a, in which he said that he had actually applied the soap-film method to a model aerofoil but had not taken the matter further.[9] The small size of the apparatus prevented the measurements being made with the required accuracy. Taylor therefore backed the use of an electrical method, and eventually such a method was developed by E. F. Relf and formed the basis of the experimental comparisons that were later published.[10]

At this point Glauert intervened. In May 1922 he submitted his "Notes on the Flow Pattern round an Aerofoil" (T. 1696b).[11] First, he took issue with Southwell's claim that it would be difficult to allow for the influence of the channel walls by use of analytical methods. Glauert said that the effects could be represented in a simple way using standard mathematical techniques, the so-called method of images. He then went on to make some comments about the proposed experimental comparison involving an infinite wing and two-dimensional flow. It was important "to have a clear understanding of its bearing on the general question of aerofoil theory" (2). The implication was that some of the thinking behind the proposal lacked the requisite clarity. Not every test of the two-dimensional work was automatically a test of the three-dimensional claims, for example, the hypothesis that the flow over a wing is smooth at the trailing edge is not a necessary presupposition of Prandtl's work. Prandtl used the idea that lift is proportional to circulation and that the circulation around a wing can be replaced by the circulation around a line vortex, that is, that the chord is negligible. But, said Glauert, no assumption is made "as to the relationship between the form and attitude of the aerofoil and the circulation round it, the analysis always being used only to estimate the behaviour of one aerofoil system from the known behaviour of another system of the same aerofoil section" (2). Taken in its own terms, he went on, the Prandtl theory has been applied "with considerable success" to three cases: (1) the effect of changes of aspect ratio, (2) the estimation of the behavior of multiplane structures on the basis of monoplane data, and (3) the description of flow patterns such as downwash. The comparison of predicted and observed data shows that the "agreement is reasonable." This, Glauert insisted, constitutes "a satisfactory check of the fundamental equation" (3).

Glauert acknowledged that the hypothesis that the rear stagnation point is on the trailing edge overestimates the circulation and therefore the lift. It does so because of departures from the idealized condition of irrotational flow. The real flow detaches itself from the top surface of a wing before reaching the trailing edge and forms a "narrow, eddying wake behind the aerofoil." Glauert had discussed this in his earlier report, "Aerofoil Theory," but the committee seemed to be using the well-known facts about the existence of a turbulent wake as an objection to Prandtl's work. If the wake really was to be a focus of interest, it would be necessary to make assumptions about the distribution of vorticity associated with "the contour of the aerofoil and inside the wake region." Prandtl's aim was to give a first-order approximation for the flow at a distance from the aerofoil, and at points outside the wake. The vorticity of the aerofoil can then be concentrated at a point or, in the three-dimensional case, in a line, just as Prandtl assumed. It is legitimate under these circumstances to "ignore completely the series of alternative small vortices in the wake" (4). Glauert concluded by saying that the proposed experiment on an infinite wing would, indeed, illuminate the relation between aerofoil sections and the circulation round them, "but will not have any bearing on Prandtl's aerofoil theory" (4).

Taylor did not agree. He produced a written reply, designated T. 1696c, in which he challenged both Glauert's response to Southwell about mathematical techniques and Glauert's claim that the experiment would be irrelevant to Prandtl's theory.[12] On the latter point, Taylor declared that all the reasons Glauert "brings up to support his view were well known to most of the Committee which discussed the proposed experiments and some of them were actually brought up in the discussion. It is curious, therefore, that Mr. Glauert should come to a view which is different from that of the members who proposed the experiments" (1).

Taylor said that the experiment on the infinite wing *would* constitute a test of Prandtl's theory because the theory was based on the assumption

that the flow at a distance from the wing was irrotational. Glauert's position, it seemed to Taylor, was that this assumption can be made a priori, but it cannot. It is an empirical matter, and the proposed experiment was designed to test it. Second, Glauert had said that the experimental evidence gathered so far had provided a satisfactory check on the fundamental equations of the theory. Taylor replied that if "satisfactory" meant "sufficient" he could not agree. The fundamental equation $L = \rho V\Gamma$, relating lift and circulation, might hold true for some body of data, and some experimental arrangement, *but not for the reason that Prandtl had given*, that is, not because the flow was irrotational. In fact, said Taylor, "there are an infinite number of kinematically possible distributions of velocity for which this is the case, but only certain of them will correspond with irrotational motions" (3). Finally, Taylor turned to Prandtl's assumption that the chord of the wing could be neglected. Again, insisted Taylor, this could not be assumed a priori. "The assumption can only be justified by experiment or by calculation of the type indicated by Miss Barker or by the purely empirical method of comparing the results of Prandtl's calculations with observed lifts and drags"

(4). For these reasons, said Taylor, "I do not agree with the conclusions reached by Mr. Glauert."

The Experiment of Bryant and Williams

It was not until July 3, 1923 (after the confrontations at the Royal Aeronautical Society and the International Air Congress), that the Aerodynamics Sub-Committee took a formal decision to test the Prandtl theory. The minutes read as follows: "That the investigation of the air flow behind aerofoils of finite and infinite span be carried out to ascertain how far the Prandtl theory of circulation can be substantiated by experiment." The decision was passed to the full committee and ratified on July 10.[13]

After some delays the much-discussed experiment went ahead at the National Physical Laboratory. L. W. Bryant and D. H Williams performed a large number of measurements to build up a picture of the speed and direction of the air at points around a vertically positioned aerofoil that stretched across the full seven-foot depth of the NPL's largest air channel. They did not use a Joukowsky aerofoil but a thick, high-lift section, chosen by Bairstow, that they positioned at approximately 10° to the air flow, which moved at about 49 feet per second. From a study of the velocity data it was possible to evaluate the circulation around various, selected contours. Some of these were chosen to loop round the wing itself, while some were closed contours that did not include the wing but were merely located in the space around it. If the flow was irrotational, the latter contours should yield a zero circulation. By contrast, according to the circulation theory, all contours enclosing the wing should show the same, nonzero circulation whose magnitude was related to the lift by the fundamental equation $L = \rho V \Gamma$. The lift was established by taking pressure measurements. Inserting this empirical value for L in the equation gave a predicted value for the circulation Γ, which could be compared with the circulation computed from the velocity measurements.

In November 1923 the committee received Glauert's technical report T. 1850 titled "Experimental Tests of the Vortex Theory of Aerofoils."[14] This was a general summary of evidence in support of Prandtl's approach but began with a preliminary analysis that Glauert had made of data provided by Bryant and Williams. The measurements, carried out at what Glauert called a "relatively high angle of incidence," enabled him to present a graph that showed the circulation plotted against the area of the contour around the wing along which the circulation had been measured. The graph indicated that the circulation was roughly independent of the area. Different-sized contours were indicating the same value for the circulation—just as the theory implied.

What is more, the numerical value of this circulation was almost exactly the value predicted from the formula. Things looked promising for the supporters of the theory. There had yet to be any direct test of the zero circulation in the contours that did not enclose the wing, to check that the flow was indeed irrotational, but the indirect evidence gathered so far seemed to confirm the assumption of irrotational flow.

Taylor set about to show that this evidence was not as good as it looked. In January 1924, he submitted his "Note on the Prandtl Theory."[15] It was meant to block the inference that Glauert was making on the basis of the preliminary data. Taylor argued as follows. If the flow is irrotational, then *all* the contours around the wing will have the same circulation. This condition was consistent with the experimental results, but no experiment can test all contours and, so far, only a few had been checked. It does not follow that, because *some* contours show the same circulation, the flow must be irrotational. In the present case Taylor was convinced that the flow could not be irrotational. His suspicions were aroused by the use of a wing at high incidence, that is, approaching the stalling angle. Previous experiments had shown that the wing would be experiencing high resistance, although it should have zero resistance in an irrotational flow. The high resistance means the flow cannot be everywhere irrotational. There would be a significant wake, and the fluid elements in the wake would be rotating—some this way, some that. If the experimental readings showed that various contours all had the same circulation, then there must be something peculiar about the contours. "Mr. Glauert's result would have to be attributed to a fortunate choice of his contour rather than to an irrotational type of flow" (3).

How could contours of different sizes exhibit the same circulation without this indicating the irrotational nature of the flow through which the contour had been drawn? Taylor gave an example to show how this result might arise. Imagine a wing profile surrounded by two circular contours, A and B, which have a common center on the wing. Both are large compared to the wing, but B has a greater radius than A. (See fig. 9.2, which is taken from Taylor's report.) Suppose that the wing generates a wake with, say, positive vorticity issuing from the upper surface and negative vorticity from the lower surface. The wake is shown in the diagram contained between the dotted lines. The circulation around the circle A is, by Stokes' theorem, equal to the total vorticity within the area enclosed by the contour. Here there will be two sources of vorticity: the vorticity along the surfaces of the wing and the vorticity provided by the wake. Now consider the circulation around the larger circle B. Any *difference* between the circulation around A and B must come from the vorticity in the wake that lies in the area between the two circles.

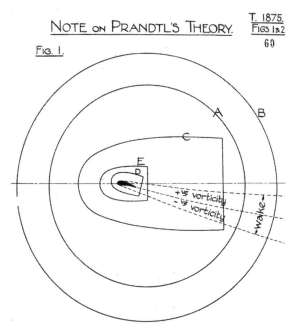

FIGURE 9.2. G. I. Taylor's counterexample to the conclusion drawn from Bryant and Williams' experiment. Circulation around contour B is the same as around contour A, if equal amounts of positive and negative vorticity are generated by the wing. The irrotational character of a flow cannot be deduced from the sameness of circulation around a number of different contours. From ARC Technical Report 1875. (By permission of the National Archives of the UK, ref. DSIR 23/1889)

Taylor, however, was interested in the conditions that would make them have the *same* circulation, that is, the conditions that might lead the unwary into supposing that the contours passed through a flow that was wholly irrotational. Such a condition would arise if the amounts of positive and negative vorticity issued into the wake were equal. The flow would not be irrotational, but it would yield the same circulation for the different contours. The conditions of this example may be special, but they are sufficient to expose the fallacy of inferring the irrotational nature of the flow simply from the equality of circulation around a number of different contours.

This was the essence of Taylor's argument but not the end of it. Glauert's preliminary analysis of the experiment not only suggested that the circulation was the same on contours of different sizes, but it also revealed that the circulation's magnitude was almost precisely that predicted from the basic law relating lift to density, speed, and circulation. Could that be explained by the unwitting selection of a "favorable" contour? Taylor showed that the answer to this question is yes. His argument was based on the equations that, respectively, express the conservation of momentum perpendicular to, and parallel to, the direction of motion of an aerofoil. Consider a body in a stream of fluid of velocity U parallel to the x-axis. Taylor wrote down, in what is now a standard way, the momentum flux equations across an arbitrary contour C which encloses the body. After some manipulation, and disregarding small quantities, he arrived at the following two expressions. One concerned the drag forces lying in the direction of motion; the other concerned the forces perpendicular to the motion and defined the relation between lift and circulation:

$$D = \rho \int_C \left(\frac{p}{\rho} + \frac{1}{2}q^2 \right) l \, ds \text{ and}$$

$$L = -\rho \int_C \left(\frac{p}{\rho} + \frac{1}{2}q^2 \right) m \, ds + \rho U I,$$

where D is drag, L is lift, U is the velocity of the free stream and I is the circulation, ρ is density, p is pressure, q is velocity, and m and l are direction cosines. These indicate the slope of the contour. Thus m expresses the orientation of the contour at a given point by giving the cosine of the angle between the normal at that point and the y-axis, while l functions in the same way but gives the cosine of the angle between the normal and the x-axis. The expression in brackets, under the scope of the integral sign, is proportional to what is called the "total head" and (as explained in chap. 2) is a quantity measured by a Pitot tube facing the oncoming stream of air.[16]

For Taylor's immediate purposes it was the second of these equations, dealing with the lift L, that was most important. The equation is obviously similar to the Kutta- Joukowsky equation. The familiar product of density, speed, and circulation (here expressed as $\rho U I$) is clearly visible, but with the addition of the integral on the right-hand side. Taylor drew attention to the following features of the equation. If the flow is irrotational, the expression in the brackets, the total head, is a constant, and as a result the value of the integral around the contour is zero. This removes the term from the equation and leaves $L = \rho U I$, the basic law of lift. Thus the simple law follows from his analysis, given the assumption that the flow is irrotational. But suppose there is a wake. Within the wake the flow is not irrotational. Along the part of the contour that passes through the wake, the total head will not be a constant but will vary in magnitude. In general, under these conditions, the value of the integral will not be zero. The simple proportionality between lift and circulation will then no longer hold. But, Taylor noted, the bracket under the integral is multiplied by m, the direction cosine. If the contour is so chosen that the part of it that cuts the wake is perpendicular to the main flow, that is, perpendicular to the x-axis and parallel to the y-axis, then m will be zero. (The angle between the normal to the contour and y-axis will be 90°, and the cosine of 90° is zero.) This feature of the mathematics makes the integral zero once again. For such a contour, the integral term in the equation will disappear and leave an expression that coincides with the simple law of lift.

It follows that one cannot infer that the flow is irrotational just because the data are related by the law $L = \rho UI$. Even though some of the flow is not irrotational, this relation between lift and circulation can hold because of the choice of the contour. Taylor summed up his argument in the following words: "We have now seen that the relation $L = \rho UI$ may be expected to hold even when the motion is not irrotational provided that the circuit used in calculating I is chosen in a particular manner. If Mr. Glauert's circuits are in fact chosen in this manner his result though interesting in itself cannot be regarded as being in any sense confirmatory of Prandtl's fundamental hypothesis that the motion at a great distance from the aerofoil is irrotational" (7).

Taylor's deduction depended on neglecting small quantities. To justify this step he had assumed that the disturbances caused by the aerofoil declined as $1/R$, where R is the distance from the aerofoil. Taylor was conscious that this step might hide a problem for his argument. Could the $1/R$ assumption be tantamount to admitting that the flow was irrotational? Taylor seemed unsure but circumvented the problem by showing that there were other counterexamples that would block Glauert's inference, and these did not depend on the $1/R$ assumption.

Here Taylor produced his pièce de résistance. He reanalyzed Rayleigh's 1876 paper on discontinuous flow around an inclined plate and showed that the circulation in the flow conformed to the law $L = \rho UI$, provided the contour cut the dead air of the wake perpendicular to the direction of the main motion. (See the contour marked C' in fig. 9.3, again taken from Taylor's report.)

Rayleigh flow clearly bears no resemblance to the flow assumed by Prandtl. Although the relevant integral can be specified, the idea that Rayleigh flow has a "circulation" is not physically well defined because there is no unique value attributable to it. The value will depend on the contour. This was Taylor's point. With the "right" sort of contour, he was able to show that the value of the integral will tend to the same limit, $L/\rho U$, as that derived from the assumption of an irrotational flow with circulation. Armed with these remarkable results Taylor drew his final conclusion: "It appears, therefore, that if Mr. Glauert's contours were taken in the special way described his results would be expected whatever the type of flow round the aerofoil may

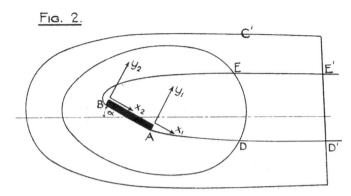

FIG. 2.

FIGURE 9.3. Taylor's reanalysis of Rayleigh flow. The circulation around contour C′ conforms to the Kutta-Joukowsky law of lift, $L = \rho U\Gamma$. Note that the contour cuts the wake of "dead" air at right angles. From ARC Technical Report 1875. (By permission of the National Archives of the UK, ref. DSIR 23/1889)

be. It cannot be taken as confirmation of Prandtl's hypothesis" (14). Were "Mr. Glauert's contours" taken in this way? Were they perpendicular to the main flow where they cut the wake? The answer is yes. Bryant and Williams had used just such contours to establish the circulation around the wing, and those were the ones Glauert employed in his preliminary analysis. Taylor had pulled the rug from beneath Glauert's feet.

Despite Taylor's criticisms, Bryant and Williams' experiment was written up as a technical report (T. 1885) in February 1924.[17] It was eventually published, in an expanded form, in the *Philosophical Transactions of the Royal Society* in 1926 along with an appendix from Taylor.[18] Taylor's appendix gave essentially the same mathematical and methodological argument as his "Note on the Prandtl Theory," described earlier. The combination of paper and appendix makes uncomfortable reading because no attempt had been made to modify the contours in the light of Taylor's comments. The only significant differences between the original technical reports presented to the Research Committee and the published papers lay in the tone of Taylor's contribution. He had made two alterations. First, the published version contained none of the sharp words directed at Glauert. Second, Taylor chose to reformulate his methodological point and express it in different words. Instead of saying that the purpose of his argument was to show what type of contour should *not* be used, Taylor now said his purpose was to show what sort *must* be used if, that is, one desires to get the result that $L = \rho UI$ *for a flow that is not irrotational*. Thus, "the object of the present note is to find out which type of contour must be chosen in order that the lift-circulation relation may be satisfied when the motion is not irrotational" (238).

Logically, this declaration is equivalent to what Taylor said to Glauert, but it is no longer framed as an objection. In the unpublished note Taylor had uncompromisingly stated that the result of the experiment "cannot be taken as confirmation of Prandtl's hypothesis." In the published appendix his conclusion was reexpressed as follows: "The relationship between lift and circulation may hold when the motion is not even approximately irrotational, provided that large contours are chosen so that they cut the wake in a straight line perpendicular to the direction of motion of the aerofoil" (245).

An anonymous report in *Nature*, commenting on the published account of the experiment, correctly summarized the position and noted that Bryant and Williams had used contours of a special type, "so that the accuracy with which the observed lift force agrees with that predicted from measurements of circulation is no indication that the flow is in fact an irrotational motion with circulation."[19] Although *Nature* got it right, it is not surprising, given the rewording, that the significance of Taylor's note sometimes proved elusive. For example, von Kármán and Burgers, writing in 1935 in their chapter in Durand's *Aerodynamic Theory*, appeared to miss this point.[20] They correctly noted that, when the contour cuts the wake at right angles, "then the value of the circulation is the same for all curves under consideration" (8). But they went on: "If the condition that the line must cut the wake at right angles were not stipulated, the magnitude of the circulation would become indefinite, as an arbitrary number of vortices, 'washed out' from the boundary layers along the surfaces of the aerofoil, and rotating either in one sense or in the other, could be included. The necessity of this condition was pointed out by G. I. Taylor"(8). This is true, but it turns the original argument on its head. What Taylor had sought to prohibit has now become a stipulation.[21]

Assessing Taylor's Argument

What is the correct understanding of Taylor's "Note on the Prandtl Theory" and the mathematical arguments he put forward? What might explain the inversion of its perceived significance? Why did Taylor himself begin that process when he reworked the conclusion to the original note in his published appendix? Clearly, Taylor was not trying to rehabilitate the old discontinuity theory; he was merely using it to embarrass the supporters of Lanchester and Prandtl. He was showing that there was no unique explanation of the phenomena summarized by the Kutta-Joukowsky law.[22] Taylor can therefore be seen, at least initially, as putting forward counterexamples in order to make a logical point. He was concerned with what follows, or does not follow, from the standard formulation of the circulation theory of lift. He had caught Glauert out in a hasty inference, but to show that proposition A does not

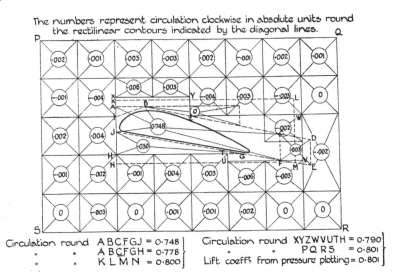

The numbers represent circulation clockwise in absolute units round the rectilinear contours indicated by the diagonal lines.

Circulation round   A B C F G J = 0·748
       "   "   A B C F G H = 0·778
       "   "   K L M N = 0·800

Circulation round   X Y Z W V U T H = 0·790
       "   "   P Q R S = 0·801
Lift coefft from pressure plotting = 0·801

FIGURE 9.4. The contours and circulation studied in Bryant and Williams' experiment. Notice the parts of the contour line around the wing, such as CF and LM, which cut through the flow behind the wing and do so at right angles to the main flow. From Bryant and Williams 1926, 209. (By permission of the Royal Society of London)

entail proposition B does not prove that B is false. Taylor's argument did not establish the falsity of the conclusions that Glauert drew about the results that were emerging from Bryant and Williams' experiment.

As well as measuring the circulation along contours that enclosed the wing, Bryant and Williams also measured the circulation around contours that did not enclose the wing. This part of their argument remained intact. They divided the space around the wing into zones, like tiles on a bathroom wall, and then measured the circulation around each zone (see fig. 9.4). The question they posed was the following: Were these local circulations all zero? Zero circulation was required for these contours (that is, contours not surrounding the wing) because that would indicate that the main flow was irrotational, namely, an irrotational motion but with circulation around the wing. Not all of these local circulations turned out to be zero. As can be seen in figure 9.4, there were a number of anomalous values particularly among those recorded near the leading edge. Despite this, enough of the readings were sufficiently close to zero for the measurements to be seen as a vindication of the theory. This result had emerged after Glauert's preliminary analysis and after Taylor's original note. It was therefore too late to have played any role in the exchange in the Aerodynamics Sub-Committee, but it was acknowledged in the published version of Taylor's note attached to Bryant and Williams' paper in the *Philosophical Transactions.*

The new data may have contributed, in some measure, to the change in Taylor's tone between the unpublished and the published versions. Despite limitations in the experimental design, the overall picture that emerged from the experiment favored the circulation theory. Taylor therefore had little choice but to begin the published version of his comments by accepting that, after all, the flow *was* mainly irrotational. As he put it: "In their paper 'An Investigation of the Flow of Air Round an Aerofoil of Infinite Span,' Messrs. Bryant and Williams show that the flow round a certain model aerofoil placed in a wind channel is not very different from an irrotational flow with circulation. There are, however, differences which are considerable in the wake, a narrow region stretching out behind the aerofoil" (238).

The acknowledgment that the flow outside the wake was "not very different" from irrotational gave the supporters of the circulation theory all they really needed. Taylor had himself identified this as "Prandtl's fundamental hypothesis," but though he was now conceding the point he did not linger on the concession. Taylor immediately drew attention to the flow inside the wake, which was nonirrotational. This point was the real focus of his interest. The Aerodynamic Sub-Committee discussions of the preliminary data coming from Bryant and Williams, in December 1923, had charged Taylor with the task of differentiating "between the effect due to circulation and that due to eddying on the forces as measured on a complete aerofoil."[23] The "Note on the Prandtl Theory" was his response to this request. In the same spirit, when Bryant and Williams gave their technical report to the committee, Taylor drew attention to the apparent presence of local areas of significant circulation where there should have been none. Glauert, by contrast, wondered if this result could be an artifact produced by the compounding of small errors elsewhere in the data.[24]

In taking the line he did in his note, Taylor showed that his thinking fell into the familiar pattern that was characteristic of British work—with the exception of Glauert's. Taylor's counterexamples were an expression of the old argument that perfect fluid theory must be false because it predicted zero drag for an infinite wing. The wake was the physical source of the drag, and drawing attention to it and exploring the consequences of its presence, merely underlined the standard objection. If, along with a wake and a viscous drag, the Kutta-Joukowsky law of lift turned out to be approximately true of the flow, then some other reason had to be found to explain the law than the one originally advanced. In a real, viscous fluid, there was no well-defined quantity that could be called "the" circulation of the flow. The value of the relevant integral would not be contour-independent but would, in general, vary from contour to contour. This was why Taylor stressed the contour dependence of the experimental results. Taylor, however, was not asserting anything that the defenders of Prandtl's theory had not granted long ago. In 1915 Betz had made a correction to allow for the role of the wake. There was no inconsistency between what Taylor said in the passage quoted earlier, in which he conceded the generally irrotational nature of the flow, and what Glauert had said originally in his technical report "Aerofoil Theory," in which he had conceded a viscous wake. Both men acknowledged the presence of rotational *and* irrotational flow in the phenomenon before them. The difference between them lay in their reaction to this agreed fact. It was a difference of emphasis and preferred method. Taylor wanted to know where the inviscid approach failed, whereas Glauert wanted to know where it worked. Taylor's eyes were directed to the viscous wake, whereas Glauert's gaze was on the nonwake.

This concern with the wake may help explain the inversion that took place in the perception of Taylor's argument, and even why Taylor himself reexpressed his original, negative point in an oddly positive way. The ideas Taylor used, negatively, to construct the counterexamples to the circulation theory became resources that could be used, positively, to study the wake. This study rapidly became a subject of research in its own right. The ideas Taylor originally advanced as counterexamples found a new use. Perhaps it was the transition to this new role that gave rise to the later misunderstanding. The new studies of the wake did not displace the circulation theory of lift but came to complement it. The viscous flow inside the wake found a place alongside the inviscid flow outside the wake. Nor did the two merely coexist. Rather, the latter could be seen as the limiting case of the former. As the wing increases in efficiency, so the wake gets smaller. In the limit the wake is simply the vortex sheet behind the wing, which was central to Prandtl's analysis. Taylor's viscous, rotational wake becomes, to use Glauert's word, "evanescent." Understood in terms of Glauert's methodology, the reality of the wake was not being ignored but was allowed for in the limiting process by making the right choice of the inviscid flow. The counterexample then becomes identical with the phenomenon it was meant to contradict. Perhaps Glauert had begun to convince Taylor that the seeming contradiction between his counterexamples and the theory were not as logically sharp as it first appeared. Glauert would surely have discussed the problem with Taylor in the time between Taylor's (unpublished) note and his (published) appendix. If minutes had been taken of these discussions, it might have been possible to trace the process by which Taylor came to reformulate his original doubts.

What is a matter of public record is how Taylor's argument was deployed for the purpose of studying the wake. Recall that Taylor's analysis of momentum relations concerned not only lift but also drag. The physical basis of Taylor's calculation of drag was the idea that drag arises from a loss of momentum in the fluid flow behind the obstacle. His analysis showed that this loss implied a pressure reduction in the wake, namely, a diminution of the quantity called the "total head" or the "total pressure." (These terms were explained in chap. 2 in connection with Bernoulli's theorem.) This account of drag was taken up by Fage and Jones at the National Physical Laboratory in a paper published in 1926 in the *Proceedings of the Royal Society*.[25] They cited Taylor's comments on Bryant and Williams' work, but they did not read them negatively. They understood them as positive suggestions about the nature and measurement of drag and proceeded to do the experiments needed to test them. Bryant and Williams, they said, had explored the velocity of the flow in the wake of a model aerofoil spanning the wind tunnel and had shown that, for all practical purposes, it was two-dimensional. They went on: "In an Appendix to the above paper [by Bryant and Williams], Prof. G. I. Taylor shows that there is good reason to believe, on theoretical grounds, that the drag of an aerofoil can be determined with good accuracy from observation of total-head losses in the wake, provided that these observations are taken in a region where the velocity disturbances are relatively small" (592). Fage explained that the drag under discussion was not Prandtl's "induced drag," which was a by-product of the lift, but "profile drag" associated with the shape and attitude of the wing section (592).

Fage and Jones' experiment was to be carried out on an infinite wing with two-dimensional flow for which the induced drag should be zero. Using the symbol H to represent the total head $\left(p + \frac{1}{2}\rho q^2 \right)$ and, like Taylor, neglecting small quantities, Fage and Jones rewrote Taylor's drag equation in a simplified form as

$$D = \rho \int_C H.ds \, ,$$

where "the integration is taken along a line passing through the wake at right angles to the undisturbed wind direction" (594). Outside the wake, H will be constant from streamline to streamline but will vary as the line of integration passes through the wake. For a contour that cuts the wake parallel to the y-axis, the value of the direction cosine l is unity, which explains its apparent absence from Fage's simplified expression. If Taylor's analysis was right, the experimenter could measure the drag on a two-dimensional or infinite wing by summing up the losses in the total head across the span. This was what Fage and Jones did using a model wing of 0.5-foot chord mounted, with only a small clearance, right across the 4-foot wind tunnel. The wind speed was 60 feet per second. They calculated the drag from pressure measurements taken in the wake and performed the required integration using graphical methods. They found that most of the loss of total head pressure (H) came from a loss in velocity (q) rather than a reduction in the static pressure (p) of the wake. Finally they compared the predicted drag with the result of their direct drag measurements when the wing was suspended on wires and attached to scales. The two methods they concluded were "in close agreement" (593).

A further feature of Taylor's argument that was given a new employment was his picture of the equal discharge of positive and negative vorticity into the wake (respectively from the upper and lower surfaces of the wing). This also helped to integrate the circulation theory into the study of viscous flow.[26] The theorem that the circulation in all circuits enclosing the aerofoil had the same value, if the contour cut the wake at right angles, was, as one later researcher put it, "of fundamental importance in the calculations of the lift of aerofoils allowing for the boundary layer."[27] An important sequence of papers starting in the mid-1930s was devoted to this theme, and they all traced their approach back to Taylor's appendix. New ways were sought to generalize the old Kutta condition in order to quantify the circulation under more complex and realistic conditions.[28]

As independent evidence in favor of some version of the circulation theory increased, the original significance of Taylor's analysis, as a source of counterexamples, decreased. The ideas became consolidated in a new context. This may explain why von Kármán and Burgers expressed themselves as they did. Perhaps they were not misreading Taylor so much as rereading him. That is, they were reinterpreting his original contribution in the light of later concerns and assimilating his ideas to the new preoccupations of a research agenda in which the circulatory theory of lift was taken for granted. Taylor's line of thought was now being used to supplement rather than undermine the theory of circulation.

The Experiment of Fage and Simmons

A significant part of the experimental evidence for the circulation theory came from Arthur Fage (1890–1977). Fage was a retiring man who had trained as an engineer at the Royal Dockyard School in Portsmouth. His father had been a coppersmith in the dockyards. Fage won a scholarship to the Royal College of Science in London and then moved to the National Physical Laboratory in October 1912 as a junior assistant in the aeronautics section. In 1915 he published *The Aeroplane: A Concise Scientific Study*.[29] The book was "written to meet the requirements of engineers" (v). It embodied a wholly empirical approach and made no mention of either the discontinuity theory or the circulation theory. With his "infinite capacity for taking pains," Fage "became progressively a better and better research scientist" and acquired the reputation of being one of the NPL's most meticulous experimenters.[30]

After Bryant and Williams' work on the infinite wing, the next step in the Research Committee's plan was to test Prandtl's account of the flow around a finite wing. This work was undertaken by Fage and L. F. G. Simmons.[31] They tested a rectangular model wing with a 3-foot span and a chord of 0.5 foot that was set at an angle of incidence of 6° to a wind of 50 feet per second. The wing had a cross section known as the RAF 6a. This particular wing was chosen because it had been studied in earlier experiments at the NPL, and the lift coefficient and the distribution of lift along the span were already known. Fage and Simmons used a speed-and-direction meter to probe the space round the wing to build up a detailed, quantitative picture of the flow. They measured the flow as it cut a number of transverse planes across the wind channel at various distances behind the wing (see fig. 9.5). If the x-axis is taken as the longitudinal axis of the channel, then positions on these transverse planes would be defined in terms of y- and z-coordinates where the y-axis lay parallel to the span of the wing and the z-axis indicated positions above and below the level of the wing. These coordinate axes are shown in the figure. Notice that the origin is located at one of the wingtips. The transverse planes chosen for study were called A, B, C, and D. Distances were expressed in terms of the chord of the wing. Plane A, which featured very little in the subsequent discussion, lay at a distance of about half a chord in front of the wing. Planes B, C, and D, the main interest of the experimenters, lay respectively at distances of $x = 0.57$, $x = 2.0$, and $x = 13.0$ chords behind the wing. The aim was to measure the properties of the trailing vortices as they cut through planes B, C, and D. On the basis of previous work (such as Piercy's), Fage and Simmons were by now confident of the existence of the vortices but their question was: Did these vortices behave quantitatively in the way that Prandtl had assumed?

For convenience the test wing was actually mounted vertically and the speed-and-direction meter was inserted through a hole in the floor of the wind channel. The meter could be moved up and down, parallel to the leading and trailing edges. The tips of the wing were fastened to transverse runners on the roof and floor of the wind channel so that the wing could be made to slide from one side of the channel to the other without altering its angle of incidence. These two degrees of freedom (the wing going from side to side and the meter going up and down) allowed measurement to be made within

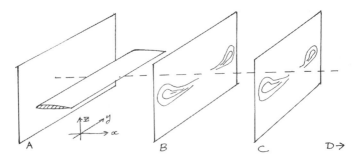

FIGURE 9.5. A schematic representation of Fage and Simmons' experiment to test Prandtl's theory of the finite wing. The flow was determined by detailed measurement in three planes, B, C, and D behind a wing (plane D, which is at a large distance behind the wing, is not shown in the diagram).

a given, transverse plane. After taking measurements in plane A, they had to remount the wing at a different distance from the meter to measure the flow in plane B, and similarly for planes C and D. The measurements were made with an instrument of a standard type developed at the NPL (see fig. 9.6). It consisted of four open-ended manometer tubes forming a head that could be tilted on its mounting as well as turned from side to side. The orientation of the head of the meter could be manipulated from outside the wind channel. When the head was directly aligned with the local flow, the pressure in the four tubes was equal, which allowed the direction of the air at that point to be read off. The speed of the air was given by the difference in pressure between one of the four tubes and a fifth open-ended tube pointing in the direction of the free flow.[32]

Fage and Simmons took hundreds of measurements in order to build up a quantitative picture of the flow, and they were able to draw a number of important conclusions. First, they found that the velocity components (u) in the direction of the main flow (that is, parallel to the x-axis) were very little changed by the air having passed over the wing. The circulation theory implies that the speed immediately above and below the wing was modified, but measurement showed that soon after its passage over the wing the u-component settled down to a steady value that only varied by about 1 percent. This (approximately) constant velocity could therefore be ignored when considering the flow in the transverse planes. Exploiting this fact, Fage and Simmons argued that the flow within each plane could be considered as a two-dimensional flow. This two-dimensional flow was taking place in the (y, z) plane, so the two components of speed in the flow were v (along the y-axis) and w (along the z-axis). Fage and Simmons proceeded to compute the stream functions, streamlines, and vorticity of this flow at a large

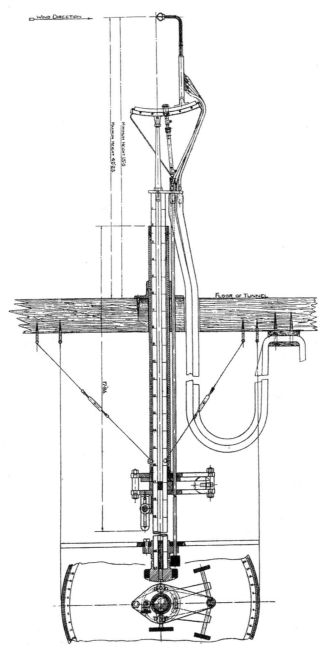

FIGURE 9.6. Apparatus used by Fage and Simmons for detecting the speed and direction of the flow behind a wing. As shown, the neck of the apparatus is inserted through the floor of the wind tunnel. The head of the apparatus can be raised or lowered, turned from side to side, and also rotated. From Fage and Simmons 1926, 306. (By permission of the Royal Society of London)

number of points in each transverse plane by using the values taken from their measurements.

By the term "vorticity" Fage and Simmons meant $(\partial w / \partial y - \partial v / \partial z)$. This is the quantity previously introduced and defined (in chap. 2) as the measure of the rotation of a fluid element. The two terms "rotation" and "vorticity" were used interchangeably. Each of the two parts of the expression represents a rate of change and hence the slope of a graph. Thus the value of $\partial w / \partial y$ at a given point refers to the rate of change of the speed w with distance along the y-axis at that point. A corresponding definition applies to $\partial v / \partial z$. Fage and Simmons had a sufficient number of velocity readings of v and w at a sufficient number of points to find the relevant slopes and rates of change. They were thus in a position to compute the rotation or vorticity at each of these points. Once numerical values of this expression were established for a large number of points on the planes B, C, and D, it was possible for the experimenters to draw curves linking up the points of equal vorticity. The result was a striking picture of the flow.

Fage and Simmons' figures show the contours of equal vorticity in each of the transverse planes behind one-half of the wing (see fig. 9.7). The wing is not shown on these diagrams but lies horizontally, obscured, as it were, by the vorticity. The wingtip would be on the right, and the center of the wing, which is off the diagram, would be on the left. The top figure refers to plane B immediately behind the wing; the other two refer to planes C and D, which are set farther back. Examination of the figures shows that initially (at about half a chord behind the wing) vorticity is spread in a narrow sheet along the trailing edge but becomes more concentrated near the tip. At a distance of two chords behind the wing, the vorticity at the tip has becomes less intense and more spread out. These changes, said Fage and Simmons, demonstrate the unstable character of the vortex sheet and show that it rolls up in the way predicted by Prandtl. The third picture shows the rolling up to be almost complete. The vorticity is now confined to the area near the wingtip. Citing Lamb's *Hydrodynamics*, Fage and Simmons state that the diagrams show that, in accordance with the classical mechanics of vortex motion, the vortices behave as if they are attracted to one another. Their centers are progressively displaced toward the center and away from the tips.

Fage and Simmons then used their data to test the quantitative aspects of the theory and to construct detailed connections between their observations and the mathematics of Prandtl's picture. The central questions were whether the flow (outside the trailing vortices) was irrotational and whether the lift and circulation around the wing were connected to the trailing vorticity

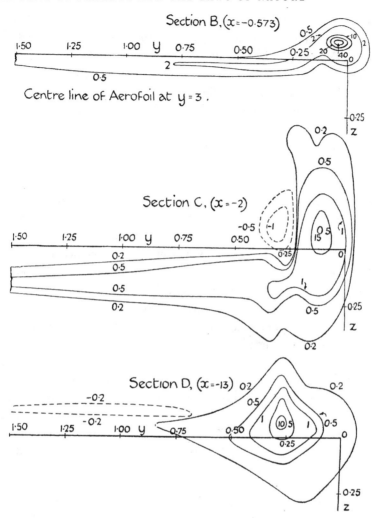

FIGURE 9.7. Lines of equal vorticity at three distances and across three planes, section B, section C, and section D, behind a finite wing. The wing is not shown but its centerline is off the diagram on the left and the wingtip is close to the origin O, which in this figure is positioned on the right. From Fage and Simmons 1926, 132. (By permission of the Royal Society of London)

according to the laws identified by Prandtl. To address the question of irrotational flow, Fage and Simmons computed the circulation around a sequence of rectangular contours of increasing size drawn on their transverse planes. They chose rectangular contours, with sides parallel to the y- and z-axes, because they already had the relevant velocity components needed for the calculation. The circulation they sought was given by the quantity $(vdy + wdz)$ calculated around the contour. The chosen rectangles all enclosed the same areas of vorticity in the wake of the wing. According to Stokes' theorem the circulation around these contours gave a measure of the vortex strength they enclosed. The calculated values for the circulation were close to one another, and this, argued Fage and Simmons, showed that outside the vortex wake the flow was irrotational—that is, as the contours got bigger, no further rotating fluid elements, and hence no further vorticity, can have been enveloped by the contour.

This argument for the irrotational character of the flow outside the vortex wake had exactly the same form as Glauert's inference when he gave his preliminary analysis of Bryant and Williams' results. A number of contours of increasing size display the same amount of circulation, ergo the contours pass through an irrotational flow. This was the inference to which Taylor had taken exception. Had Fage and Simmons fallen into the same trap as Bryant and Williams? The answer is no. As a deductive inference the argument put forward by Fage and Simmons is not compelling (for the reasons that Taylor gave), but as an inductive inference the conclusion is plausible. Given their other computations dealing with vorticity in the flow, and the contours showing its distribution, these results reinforced the conclusion that the flow outside the wake was, indeed, irrotational.

The most important quantitative question was whether the strength of the vortices trailing from the three-dimensional wing had been correctly predicted by the proponents of the circulation theory. Fage and Simmons deduced that, according to Lanchester and Prandtl, the strength of vorticity leaving the semi-span of the wing should equal the circulation around the median section of the wing. If these two quantities could be found and compared, then the prediction could be tested. Fage and Simmons developed this argument mathematically, but the intuitive basis of the deduction can be seen immediately by inspecting figure 9.8.

The figure shows a version of the refined horseshoe vortex. The vortices that run along the span of the wing all pass through a contour around the median section of the wing. The value of the circulation around this contour is determined by the vortices that run through it (from Stokes' theorem). All of these vortices then peel off the trailing edge of the wing. Their total strength can be assessed by the circulation around another contour, namely, any contour that captures and includes them. For example it can be measured by the circulation around the rectangular contours drawn in the planes B, C, and D that have been previously discussed. Fage and Simmons compared these empirically based values with the predicted circulation around the median section.

How did Fage and Simmons evaluate the predicted circulation? It is the

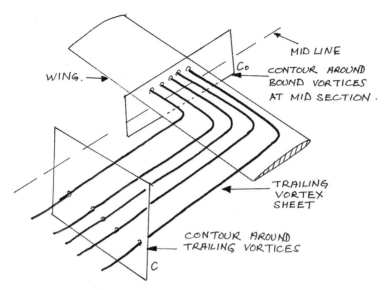

FIGURE 9.8. Circulation around contour C_0 at the midsection of the wing is created by the bound vortices running along the span of the wing. According to Prandtl the combined effect of this circulation should equal the circulation around contour C created by the vortices which trail behind the half span of the wing.

quantity that has previously been called Γ_0. Recall that in chapter 7, it was shown that by making the assumption that the lift distribution was elliptical, Prandtl had given an expression that predicted the value of Γ_0 from known properties of the wing and the flow. Prandtl had shown that

$$\Gamma_0 = \frac{2VSC_L}{b\pi},$$

where C_L is the coefficient of lift, S is the area of the wing, b is the span, and V is the free-stream velocity. All of these were known quantities for the wing used in Fage and Simmons' experiment. In this way they were able to compare the predicted value of Γ_0 and the observed results for the circulation around the vorticity coming away from the trailing edge. The two values should be the same. The theoretical prediction was corroborated for the flow up to two chord lengths behind the wing, that is, for planes B and C. On plane D, at a distance of thirteen chord lengths, the agreement was not so good. The measured vorticity was around 18 percent too small. Some of the vorticity appeared to have dissipated, but, within the limits of experimental error, Fage and Simmons declared that the match between theory and observation was a good one.

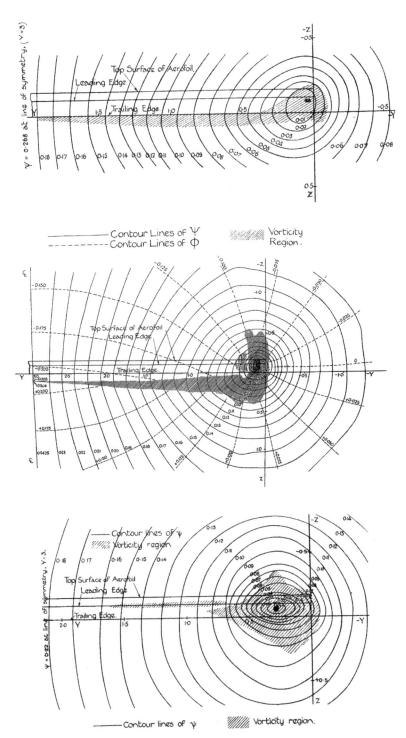

FIGURE 9.9. The streamlines of a trailing vortex at three distances behind the tip of a finite wing. The position of the wing is shown on the diagrams with the wingtip on the right, behind the center of the vortex. The centerline of the wing is on the left. From Fage and Simmons 1926, 320–22. (By permission of the Royal Society of London)

Fage and Simmons not only drew the lines of equal vorticity (shown in fig. 9.7) but also drew the actual streamlines of the vortices that issued from the wings and passed through the planes B, C, and D. This, too, was an exercise that involved considerable computation. First they had to check that the preconditions for the existence of a stream function, ψ, were fulfilled. Was the continuity equation satisfied? Laboriously, they concluded that it was. They then identified the values of the stream function at a large number of points in the (y, z) plane in order to draw the streamlines ψ = constant. On grounds of symmetry they took the line $y = 3$, the center line of the wing, as the streamline ψ_0. The pictures they produced from their measurements and computations were convincing. "The diagrams," they said, "illustrate clearly the changes in the character of the airflow behind the aerofoil: close to the aerofoil, the contour lines are ovals with the ends pointed inwards; as the distance behind the aerofoil increases, the contour lines become more and more circular—indicating that the vorticity is becoming more concentrated" (320). The results are shown in figure 9.9.

Fage and Simmons had significantly strengthened the case for the circulation theory. They had effectively picked up the argument at the point where Pannel and Campbell had left it in 1916 when curiosity seems to have been swamped by collective incredulity. Unlike the conclusions of Bryant and Williams, the work of Fage and Simmons was not challenged by Taylor.[33] Others, such as Lanchester and Föppl, had made qualitative observations of trailing vortices and, more recently, photographs had been taken, but now the vortices had been directly measured and linked to the mathematically defined relations of the circulation theory. Could it now be said that the entities that were subject to empirical measurement in the laboratory had been clearly identified as one and the same with the entities referred to in the circulation theory? If so, the work of Fage and Simmons would represent a qualitative change in the status of the circulation account of lift. What had previously been a *theory* could now, arguably, be counted as a directly verifiable *fact*.

Footprints in the Snow?

I want to dwell for a moment on the significance of this transition from theory to fact. It is a phase-change that has often taken place in the history of science. Assertions to the effect that, say, the blood circulates in the human body, or that water is made of hydrogen and oxygen, may once have been speculations but today can be taken as matters of fact. It would be a questionable use of language to keep calling them theories. How is this change of status, from theory to fact, best described? One attractive answer was provided in the form of a striking metaphor used by the Cambridge mathematician William Kingdon Clifford.[34] Clifford, who had been a second wrangler in 1867, a fellow of Trinity, and a friend of Rayleigh, had wide-ranging interests in mathematical physics. In a lecture he gave in Manchester (some fifty years before the experiments that concern us), Clifford took as his example not hydrodynamics but the wave theory of light. This conception of light, he said, must now be accepted as fact. The difference between a theory and a demonstrated fact, he went on, "is something like this":

> If you suppose a man to have walked from Chorlton Town Hall down here say in ten minutes, the natural conclusion would be that he had walked along the Stretford Road. Now that theory would entirely account for all the facts, but at the same time the facts would not be proved by it. But suppose it happened to be winter time, with snow on the road, and that you could trace the man's footsteps all along the road, then you would know that he had walked along that way. The sort of evidence we have to show that light does consist of waves transmitted through a medium is the sort of evidence that footsteps upon the snow make; it is not a theory which merely accounts for the facts, but it is a theory which can be reasoned back to from the facts without any other theory being possible. (117)

The thought is that if you can track a process in great detail, and see it develop step by step, then you can reach a stable understanding that is unique, unchallengeable, and enduring. Such an understanding, said Clifford, deals with facts, not theories.

Clifford's metaphor may be a tempting one, but it cannot be wholly right, and it led Clifford himself astray. The development of physics showed that what he thought of as a demonstrated fact was actually a theory. The alleged impossibility of any alternative to the wave theory was refuted by the emergence of an alternative. In Clifford's time the wave theory had superseded an older particle theory, but in 1905 Einstein once again postulated light particles. These new-style particles or "photons" were invoked to explain the photoelectric effect, something that was proving difficult to understand in terms of waves. The photoelectric effect takes place when light is incident on a metal surface and releases electrons from the surface. How could the energy spread across a wave front be concentrated in the way necessary to release a charged particle? This was the problem Einstein's theory was designed to answer. The light energy, he argued, was concentrated because light consists not of waves, but of particles, albeit particles with unusual properties.[35] These developments took place after Clifford's death. He cannot be blamed for not anticipating them, but they amount to a counterexample to Clifford's argument and show the need to introduce qualifications into his overconfident picture.

What was Clifford's error? It was that of assimilating fundamental scientific inquiry to commonsense knowledge. While there are many similarities and connections, Clifford ignored a crucial difference. Everyone has seen, or could see, a man creating footprints in snow. The cause and the effect can be conjoined in experience, and both are open to inspection. This is the basis of subsequent inferences from footprints to their human causes and the basis of the conclusions that can be drawn about, say, the route someone had taken from Chorlton Town Hall to the location of Clifford's lecture. The physicist, on the other hand, did not come to the wave theory of light by seeing light waves creating diffraction patterns or rainbows. The two things were not conjoined in experience in the way people and footprints have been conjoined. The inference to light waves did not have the same inductive basis as the commonsense inference with which Clifford was comparing it.[36]

Clifford's metaphor may have broken down for light waves, but it might still be applicable to Fage and Simmons' achievement. It could be argued that Fage and Simmons *were* confronting the vortices *and* observing them bringing about their effects. Was not this conjunction precisely what the experiment was designed to expose? Even if the experimenters could not actually see the flow of air, they could have made it visible, and others had done exactly that. In any case, the diagrams showing the streamlines of the vortices and the contour lines of equal vorticity allowed them to follow the path and development of the postulated vortices. The experimenters could set these diagrams side by side with the measured lift forces on the wing. Causal connections and correlations of phenomena that were originally speculative had, in a sense, been exposed to view, and the step-by-step progress of the circulation had been traced. Perhaps tracking the vortices through the pattern of measurements registered in Fage and Simmons' diagrams was, after all, similar to tracking footprints in the snow.

If Clifford's metaphor is applicable to the aerodynamic work, does this mean that the question "How does a wing produce lift?" can now be answered with the same level of certitude as the question about the man walking down the Stretford Road? In a sense, yes, it does. The phenomena of circulation, vortices, and lift had been made, or were on the way to being made, part of the routine and reality of daily life. At least, this was true for the laboratory life of some of the experimentalists working in this area. They were becoming increasingly familiar with the patterns in the data and the range of effects to be accounted for. Expectations were crystallizing, and experimenters were learning what they could take for granted. Techniques of calculation and prediction were becoming more confident and refined. What was once strange was becoming familiar and part of predictable, daily experience—like getting to know a new town. Learning to live with the theory of circulation was like learning to live in a new environment with new architectural styles and a new street plan. You want to get to Prof. Clifford's lecture starting out from Chorlton Town Hall? Then go down the Stretford Road! You want to calculate the induced drag? Then use Prandtl's formula![37]

Significantly, this was not yet how some of the most influential British experts saw the issue. They acknowledged that Fage and Simmons' results represented a triumph of sorts for Lanchester, Prandtl, and Glauert, but they did not accept that the answer was now known to the question How is lift produced? On the contrary, they maintained that, despite the experimental advances and the increase in empirical knowledge, the answer to this question remained wholly unknown. Many questions, they acknowledged, had now been answered, but not *this* one. These experts were not simply being stubborn or blind in the face of mounting evidence, and their reaction underlines just how careful one must be in applying Clifford's metaphor. It must be accepted that what looks like demonstrated fact from one point of view may appear less compelling or revealing from another point of view. This skeptical response to the mounting experimental evidence was articulated with great clarity by Richard Vynne Southwell (fig. 9.10). Southwell has already been mentioned in connection with the postwar contact with Prandtl and Göttingen, but it is appropriate to look more closely both at the man and at his response to the growing experimental literature.

Our Ignorance Is Almost Absolute

Southwell entered Trinity in 1907 to read mechanical sciences. He was an engineer, but an engineer with impressive mathematical skills.[38] In 1909 he was placed in the first class of part I of the Mathematical Tripos and in 1910 graduated with first-class honors in the Mechanical Sciences Tripos. He was coached by Pye and Webb, two of the best mathematical coaches of the time. On graduation he began research on elasticity theory and the strength of materials and in 1912 became a fellow of Trinity. In 1914 Trinity offered Southwell the post of college lecturer in mathematics but he did not take up the offer because of the outbreak of war. He volunteered for the army and was sent to France. In 1915, however, he was brought back to work on airships for the

FIGURE 9.10. Richard Vynne Southwell (1888–1970). Southwell was a product of the Mechanical Sciences Tripos but held a lectureship in mathematics. He was superintendent of the Aerodynamics Department at the National Physical Laboratory after the Great War before returning to Trinity. Despite the experimental support for the circulation theory, Southwell argued that ignorance regarding the cause of lift was almost absolute. (By permission of the Royal Society of London)

navy. In 1918 he was transferred to the newly created Royal Air Force, with the rank of major, and was sent to Farnborough in charge of the aerodynamic and structural department. After demobilization, and a brief return to Trinity, in 1920 he went to the National Physical Laboratory as superintendent of the Aerodynamics Department. He stayed at the NPL for five years and then returned again to Cambridge, where (unusual for an engineer) he was a faculty lecturer in mathematics.

It was in the field of applied mathematics, rather than practical engineering, that Southwell made his outstanding contribution. He developed novel mathematical techniques for the analysis of complex structures of the kind used in the building of airships. The technique was called "the relaxation of constraints" and depended on replacing the derivatives in the equations and boundary conditions by finite differences.[39] Though the technique was initially developed to deal with engineering problems, Southwell later demonstrated its power as a general method of solving differential equations. Referring to the unavoidable complexities of practice, and the uncertainties in data of whatever kind, he called his own Relaxation Method "an attempt to construct a 'mathematics with a fringe.'"[40] He was not only interested in elasticity and the strength of materials but also worked on viscous flow. Like Bairstow, Southwell started from Oseen's approximation to the full equations of viscous flow and the developments provided by Lamb.[41] In 1929 Southwell was offered the chair in engineering at Oxford, which he accepted after some hesitation but where he stayed until his retirement. Southwell had a lively sense of the different demands confronted by engineers and mathematical physicists, but it may be revealing that Glazebrook said of him that, although he was an Oxford professor, he was still a Cambridge man.[42]

As superintendent of the Aerodynamics Department at the NPL, South-well played a prominent role in the discussions that took place in the Aero-nautical Research Committee after the war when plans for future work were thrashed out. Southwell always placed great emphasis on fundamental scien-tific research. It was the long term, not the short term, that counted. Though an engineer by training, he defended the value of academic research of the kind so often attacked by the practical men. This came out clearly in the pol-icy discussions that took place in February 1921, devoted to the topic of "The Aeroplane of 1930." The participants were invited to anticipate the character and needs of aviation in ten years' time. Southwell wittily subverted the discussion by posing the question If we could know where we would be in ten years' time, why wait? His point was that fundamental advances could not be predicted. He suspected that, whatever we said, we would be wrong.[43] The most we can do is to be conscious of the gaps in existing knowledge and try to fill them. Consider, he said to the committee, the fundamental cause of the lift and drag on an aircraft wing: "We have much empirical data in regard to aerofoils, but our ignorance of the mechanisms by which their lift and drag are obtained has hitherto been almost absolute." Here was a worthy focus for research: the true mechanisms of lift and drag must be identified.[44]

One might assume that Bryant and Williams' experiments, as well as those of Fage and Simmons, were performed to identify the mechanisms that Southwell had in mind. But if this were so, we would expect that the results of the work (give or take Taylor's reservations) would have been seen by South-well as furnishing the desired account of lift and drag. This was not how he saw them. The same sense of ignorance about fundamental causes still per-vaded Southwell's thinking *after* this experimental work had been completed and *after* Glauert had begun to provide his superbly clear exposition and de-velopment of the circulation theory. The same pessimism that was expressed privately in committee in 1921 was expressed again, and publicly, some four years later in two lectures that Southwell gave in 1925. One of these lectures, on January 22, was to the Royal Aeronautical Society; the other, on August 28, was to the British Association meeting in Southampton.

The lecture to the RAeS was titled "Some Recent Work of the Aerodynamics Department" and was meant as a summary of the achievements of the department during the years of Southwell's superintendence.[45] His return to Trinity was an opportunity to take stock. Southwell began by welcoming the change from ad hoc wartime experimentation to programs of research guided by theory. Two main lines of theoretical concern were identified. First, there was the classical theory of stability, and Southwell described in detail the recent work of Relf and others. This had taken the experimental determination of the damping coefficients for roll, yaw, and pitch to new levels of sophistication. The second set of theoretical concerns dealt with the fundamentals of fluid flow. For aerodynamics, said Southwell,

> I suppose no problem is so fundamental as the question—why does an aerofoil lift? We can hardly rest satisfied with the present position—which is, that we have next to no idea. To answer the question completely would involve no less than the solution of the general equations of motion for a viscous fluid, and attacks on these equations have been made from all angles. Considering the energy expended, the results have been very small; but then, these are about the most intractable equations in the whole of mathematical physics. (154)

Southwell mentioned the role played in this (so far fruitless) endeavor by Bairstow, Cowley, and Levy and then moved on to the approach adopted by Prandtl, namely, using the inviscid theory of the "hydrodynamic textbooks" informally conjoined with the idea of a viscous boundary layer. In this way the "once discounted" classical theory of the perfect fluid had been "reinstated" and could provide a close approximation to the truth when used "*under proper control*, and aided by assumptions based on physical intuition" (156). At the NPL, said Southwell, every opportunity had been taken to check the validity of Prandtl's theory, and "in the main one must say, I think, that it has passed the ordeal with flying colours" (156). The most important tests "are those which Messrs. Fage, Bryant, Simmons and Williams have made" (156). Southwell explained that at the time of his lecture this work had not yet been published but it had confirmed the most important result, namely, "the theoretical relation between lift-coefficient and the circulation" (156).

At this point Southwell's audience might have been puzzled. They were being told that Prandtl's theory had passed the tests to which it had been subject with "flying colours," and yet a moment before, Southwell had declared that experts had "next to no idea" how a wing produced lift. Didn't these claims contradict one another? The answer is that Southwell's argument was consistent but depended on a suppressed premise. For Southwell, the experiments of Fage and Simmons only justified the use of inviscid theory as a way of *representing* the real flow. They did not show that it *truly described* the flow. As far as Southwell was concerned, Fage and Simmons were not tracing footprints in the snow. In their experiments the imprint of reality had not been made in some familiar and reliable medium. Their analysis had used ideal fluid theory. The nature of the beast that left the footprints was still under discussion. The inviscid approach left it an open question whether the "actual flow" corresponded to the representation, and the most plausible answer was that it did not. The no-slip condition was violated by the inviscid representation, and Prandtl had assumed that the flow was steady. The eddies in the wake were neglected. The place to look to resolve these issues, Southwell concluded, was the boundary layer. It was this aspect of Prandtl's work that really engaged Southwell. As he put it, "the conditions in this layer are the ultimate mystery of aerodynamics: somehow or other, in a film of air whose thickness is measured in thousandths of an inch, that circulation is generated which we have just seen to be the essential ingredient of 'lift'" (158). Research should concentrate on the boundary layer. Theoretically this required a deeper understanding of the equations of viscous flow; experimentally it called for the development of special instruments such as microscopic Pitot tubes to probe the boundary layer. Southwell mentioned that Muriel Glauert was working mathematically and experimentally on the calibration of such an instrument.[46]

Here was the explanation of Southwell's apparently conflicting claims. Prandtl's theory of the finite wing "worked," but it could not be true because the mathematical analysis depended on false boundary conditions. This was the suppressed premise, which rendered the argument consistent. Although Prandtl's wing theory could pass many tests, and even pass them with flying colors, it could not, by its very nature, answer the question that Southwell wanted to answer. In a very British way, he wanted to know how a *viscous* fluid generates lift. In the discussion after the lecture, in response to Major Low, Southwell said: "The really interesting part of Prandtl's work was the work he had been doing subsequently in his study of the 'boundary layer,' because that work might ultimately explain why the assumptions which could not be correct could make such amazingly true predictions" (166).

In a lecture titled "Aeronautical Problems of the Past and of the Future," delivered later in the same year, Southwell insisted that the aim of research was *"not so much to achieve, as to understand."*[47] Scientists should not be content with "achievement," "*unless it be the result of understanding*"—something of which the "practical man" would never be persuaded (410). Understanding meant understanding based on a sound theory. Southwell identified three triumphs of British aeronautics that, in his opinion, met this condition. They were (1) the ability to build stable aircraft, (2) the analysis of the dangerous maneuver of spinning and its avoidance, and (3) the achievement of control in low-speed flight even after the aircraft had stalled. In all three cases, he argued, the end result had enormous practical value but the driving force had been the aim to understand. And it was mathematical analysis that had furnished the understanding.

The theory of lift was conspicuous by its absence from this list of triumphs. For Southwell, Prandtl's wing theory was an achievement that was not yet informed by an adequate theoretical understanding. Bryant, Williams, Fage, and Simmons were mentioned by name, and Southwell used diagrams taken from their papers. The role that he accorded the work, however, was that of showing that the effects of viscosity can be ignored as far as the sliding of air on air is concerned but cannot be ignored very close to the surface of a wing or in the wake behind the wing. It is what happens in these regions that constitutes "the ultimate problem of hydrodynamics" (417). It was this "ultimate" problem that Southwell had in mind when he asked: Why does a wing generate lift? He was not denying the role of circulation, nor was he belittling the insights of Lanchester, Prandtl, or Glauert as they continued to develop the inviscid theory of lift. His point was that no one, following this route, could hope to explain the origin of circulation.[48] Within inviscid theory, circulation had to be a postulate not a deduction.

Southwell's skeptical position was endorsed by H. E. Wimperis, the quiet but influential director of scientific research at the Air Ministry.[49] Wimperis had trained as an engineer in London and Cambridge and had sat the Mechanical Sciences Tripos in 1890. During the Great War he had served as a scientist with the Royal Naval Air Service and had designed a bomb sight that carried his name. After the war he worked at Imperial College in a laboratory financed by the Air Ministry. Along with Tizard, he was later to play an important role in the development of Britain's radar defense system. In 1926 Wimperis, in his role as director of research, published a survey article in the *Journal of the Royal Aeronautical Society* called "The Relationship of Physics to Aeronautical Research."[50] One of Wimperis' aims was to send the message that the Air Ministry and government were aware of the need for fundamental research. What, he asked, was engineering but applied physics? Government scientists at the National Physical Laboratory and Farnborough must have the freedom to pursue basic, physical problems. A second aim was to argue that this policy had already produced significant results. Here Wimperis cited, among other examples, the mathematical work that had been done on fluid flow and, in particular, the flow around a wing. It rapidly became clear, however, that in Wimperis' view, the approach based on inviscid theory was not an exercise in real physics but a mere preliminary to a genuine understanding of lift. On a classical hydrodynamic approach, he noted, the circulation must be added in an arbitrary way to the flow, and this only provides an "analogy with the lift force experienced by an aerofoil" (670). Admittedly there have been some successful predictions made "by the employment of this convention" (670), but the theory becomes "somewhat far-fetched" in its account of what is happening on the surface of the wing. "Circulation," said Wimperis, "must have a physical existence since velocity is greater above the wing than below; though this real circulation is a circulation with no slip, whereas the mathematical circulation has slip. Hence the rather amusing situation arises of adding to the mathematical study of streamlines a conventional motion which could not really arise in an inviscid fluid!" (670). Southwell was right, said Wimperis, in insisting that the real problem lay in discovering what was actually happening in the very thin, viscous layer close to the wing. This was a problem in physics rather than something that could be evaded by the use of mathematical conventions and unreal boundary conditions.

Glauert's Textbook

In the German-speaking world the circulation theory of lift and Prandtl's wing theory found their way into the textbooks during the Great War. Richard Grammel, who taught mechanics at the *technische Hochschule* in Danzig, led the way in 1917 with his *Die hydrodynamischen Grundlagen des Fluges*.[51] In 1919, immediately after the war, Arthur Pröll of the TH in Hannover, published *Flugtechnik: Grundlagen des Kunstfluges*.[52] H. G. Bader published his *Grundlagen der Flugtechnik* in 1920, and in 1922 Richard Fuchs of Berlin and Ludwig Hopf of Aachen produced their comprehensive *Aerodynamik*.[53] The content and level of these books contrasted markedly with what was available on the British textbook scene. As described earlier, both Cowley and Levy's *Aeronautics in Theory and Experiment* of 1918 and Bairstow's *Applied Aerodynamics* of 1920 dismissed the circulatory theory, whereas G. P. Thomson's *Applied Aerodynamics* of 1919 was almost purely empirical. Thomson spoke of the "complete failure" (26) of mathematical hydrodynamics to account for lift. He concluded that an account of lift required an understanding of eddies and turbulent motion: "This is the solution we want for aerodynamics, and not that found by the ordinary mathematical method" (32). In 1926 the situation changed radically when Cambridge University Press published Hermann

Glauert's *Elements of Aerofoil and Airscrew Theory*.[54] This work showed the power of the "ordinary mathematical method" of which Thomson, like most of his Cambridge companions, had despaired. Glauert's *Elements* proved to be an outstanding work of exposition which, even today, some eighty years later, is still confidently recommended to students.[55]

The book consisted of seventeen brief chapters and surveyed all the main themes of modern aerofoil theory for a reader with no previous knowledge of fluid mechanics (though a significant degree of mathematical competence was presupposed). The first chapter described the main facts to be explained, while chapters 2–5 outlined the theory of perfect fluids. Chapters 6 and 7 introduced the theory of conformal transformation and the specific properties of the Joukowsky transformation. Chapter 8 dealt with viscosity and drag. Here Glauert introduced the Stokes equation for viscous flow and informally derived Prandtl's boundary-layer approximation. Chapter 9 was called "The Basis of Aerofoil Theory," and in it Glauert sought to bring together the apparently antithetical ideas of viscous and inviscid flow into a practical synthesis. His aim was "to obtain the true conception of a perfect fluid" (127). The form of the desired synthesis was described by Glauert as follows: "The viscosity must be retained in the equations of motion and the flow of a perfect fluid must be obtained by making the viscosity indefinitely small" (117).

When analyzing the motion of an object in a fluid, the concept of a perfect fluid must be deployed in a way that retains the effects of viscosity. If a perfect fluid is defined as a fluid devoid of viscosity, such a requirement is contradictory: how can the viscosity, which is excluded by definition, also be "retained"? Here, once again, Glauert sought to convey his novel, Göttingen-style methodology. He argued that the requirement he formulated can be satisfied even though the boundary conditions, the way the fluid behaves at a solid boundary, are wholly different for a perfect fluid and for a viscous fluid. (For a perfect fluid the boundary conditions are that it cannot penetrate the solid boundary but it can slide smoothly along it; for a viscous fluid the conditions are zero penetration of the boundary and zero slip along it.) The trick needed to retain viscosity in the equations of motion is to start with a viscous boundary layer of finite thickness around the object and imagine it to become an infinitely thin sheet of vorticity. The boundary layer is a real phenomenon belonging to viscous fluids, and a vortex sheet is an idealization appropriate to perfect fluids. The connecting link that allows viscosity to be "retained" is that

> in the limit the boundary layer becomes a vortex sheet surrounding the surface of the body and the vortices of this sheet act as roller bearings between

the surface of the body and the general mass of the fluid. The conception of a perfect fluid with a vortex sheet surrounding the surface of the body therefore represents the limiting conditions of a viscous fluid when the viscosity tends to zero, and the existence of the vortex sheet implies that the perfect fluid solution need not satisfy the condition of zero slip at the boundary. (117–18)

If the cross section of a wing is drawn in two dimensions, the line that traces its profile is to be thought of as made up of an infinite number of points acting like infinitesimal roller bearings. These rollers are said to be rotating fluid elements, that is, "fluid elements in vortical motion" (119), and they stand in for the boundary layer. The rotating fluid elements pass along the surface of the body and finally leave it to pass downstream in the wake. Glauert linked this picture to von Kármán's work on the so-called vortex street that exists behind a bluff body placed in a fluid flow. Clearly Glauert was introducing a significant degree of idealization, but even this degree would soon be surpassed. Having replaced the boundary layer on the wing surface by a sheet of vorticity, the chord and profile of the wing was then ignored altogether. In two dimensions the wing was reduced to a single point, that is, to the "cross section" of a line of vorticity imagined to be perpendicular to the page on which the figure is drawn.

How did Glauert explain the origin of the circulation around a wing? The discussion of this sensitive and difficult question was located in a short section of chapter 9. Combining candor with British understatement, Glauert introduced the issue as follows: "The process by which the circulation round an aerofoil develops as the aerofoil starts from rest presents certain theoretical difficulties, since the process would be impossible in a perfect fluid, and it is again necessary to consider the limiting condition as the viscosity tends to zero" (121). Circulation is impossible if the analysis starts with $\mu = 0$ and confines itself to this condition. To overcome the difficulty it is necessary to start by considering $\mu \neq 0$ and then make the transition from viscous to nonviscous flow by imagining that $\mu \to 0$. Here within the one sentence we see the British and German conceptions of an ideal fluid directly juxtaposed. Glauert then proceeded to offer a qualitative account of the required transition. The analysis effectively hinged on two diagrams of the flow at the trailing edge of a wing. I reproduce the diagrams in figure 9.11.

At low speeds, as the aerofoil starts from rest, the air behaves like an ideal fluid and curves round the trailing edge as shown in (a) in figure 9.11. There is a stagnation point S on the upper surface not far from the trailing edge. Glauert argued that, as the velocity increases, the streamlines coming from the undersurface are unable to turn round the trailing edge "owing to the large viscous forces brought into action by the high velocity gradient" (121).

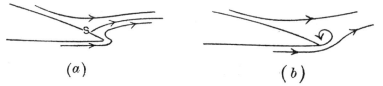

(a) (b)

FIGURE 9.11. The initial moment in the creation of a vortex (a). The vortex detaches itself from the trailing edge and floats downstream (b), leaving behind an opposing circulatory tendency. From Glauert 1926, 121. (By permission of Cambridge University Press)

The flow thus breaks away from the trailing edge in the manner shown in (b). The result is that "a vortex is formed between the trailing edge and the old stagnation point S" (121). When the vortex has reached a certain stage of development, it breaks away and floats downstream in the wake of the wing.

Although Glauert does not make the point explicitly, it is obviously important for the argument that the vortex that detaches itself and moves downstream is rotating in the correct direction. Its direction of rotation determines the direction of the circulation in any contour that surrounds it. The information about the direction in which the vortex rotates is contained in Glauert's diagram rather than his text. The diagram shows that the vortex rotates in a counterclockwise direction. Why counterclockwise? The presumption must be that the flow that initially went round the trailing edge and then progressively failed to navigate the sharp corner is moving more rapidly than the flow that comes away from the stagnation point. The speed difference of the adjacent bodies of fluid would constitute a surface of discontinuity and hence a surface of vorticity—and the differences, in this case, would produce a counterclockwise vortex.

The next step in the argument was equally crucial. In the diagram a vortex has detached itself from the wing, and this, the argument goes, generates an equal and opposite circulation around the wing. The detached vortex had a counterclockwise circulation, so the circulation around the wing will be clockwise, and this produces the speed differential postulated by the circulation theory. But why does this process create an equal and opposite circulation? The answer given by Prandtl and Glauert was that such an outcome is required in order that Kelvin's theorem is satisfied. Kelvin's theorem initially looked as if it would rule out the onset of circulation entirely. If there is zero circulation at the onset of movement, there will be zero circulation at all later times. But the enemy is now converted into an ally. Glauert explained that the circulation around a large contour, enclosing both the wing and the detached vortex, will indeed stay zero. This is necessary to satisfy Kelvin's theorem, but the theorem can be satisfied by virtue of two opposing vortices whose

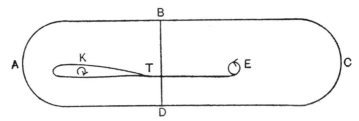

FIGURE 9.12. The creation of circulation around a wing (around contour ABD) counterbalances the circulation (around the contour BCD) created by the vortex which detaches itself from the trailing edge. The two opposing circulations sum to zero, and thus the flow is said to conform to Kelvin's theorem. Kelvin's theorem entails that if there is no circulation in the flow of an ideal fluid when motion begins, then there can be no circulation at a later time. From Glauert 1926, 121. (By permission of Cambridge University Press)

respective circulations cancel each other out. The argument is represented diagrammatically in figure 9.12, again taken from Glauert. The large contour is called ABCD. The vortex is called E, and the circulation around the wing is K, while the circulation around the vortex is, accordingly, −K. As Glauert put it: "the circulation round any large contour ABCD which surrounded the aerofoil initially was and must remain zero, and as this contour includes the vortex E there must be a circulation K round the aerofoil which is exactly equal and opposite to the circulation round the vortex E" (121). The detached vortex floats away on the free stream with velocity V, leaving the wing with circulation K and lift given by the Kutta-Joukowsky law $L = \rho KV$. Glauert's argument was moving rapidly at this point, but though potentially puzzling, this sequence of steps is now to be found in all standard textbooks.

Having prepared the ground in the first nine chapters, Glauert moved on to discuss the aerofoil in three dimensions. In chapter 10 he introduced the mathematics of the simple and refined horseshoe vortex system as well as the important concepts of induced velocity and induced drag. In chapters 11–13 he dealt with the effects of varying the aspect ratio of a wing and generalized the results to biplanes. In chapter 14 he discussed wind-tunnel corrections, while in the final chapters, 15 and 16, he applied the circulation theory to the airscrew.

Two Reviews and Two Perspectives

Glauert would have known that, however cogent he made his book, he could not meet all the demands of his intended audience. The diversity of interests that would inform the response would pull in opposing directions. The structural tensions, present in British aerodynamics from the outset, were still at work. By its very nature, Glauert's *Elements of Aerofoil and Airscrew Theory* could not satisfy the prejudices of both the practical engineer and the mathematical physicist. Indeed, the book was designed to bring about a change of approach in both parties. Until it had worked its effect it was bound to be viewed with a certain reservation from both sides, even if, at the same time, its virtues were acknowledged.

The review that appeared in the *Journal of the Royal Aeronautical Society* was signed "A.R.L." With its mixture of bluff praise and barbed comment, the review was such that no regular reader would have failed to recognize it as the work of Major A. R. Low.[56] Glauert was identified as one (but only one) of the leading exponents of the Lanchester-Prandtl approach, and the *Elements* was welcomed as (perhaps) the first full-length book on the subject in the English language. There was the hint that engineers might find Glauert's discussion too advanced, and Low strongly recommended that students read a work by H. M. Martin "as an introduction to the volume under review, and, as well, for Mr. Martin's mastery of elementary exposition" (167).[57] Low simultaneously praised Glauert for the adequacy of his general references, which would "introduce the reader to the most important German work," and criticized him for not citing Fuchs and Hopf's *Aerodynamik* of 1922, which "quite evidently influenced the author, both as to selection and arrangement of materials" (168). The book by Fuchs and Hopf was broader in scope than the *Elements*, covering both lift and stability, and Low had reviewed it in the *Aeroplane*. While he had praised their treatment of lift, he had been scathing about their treatment of stability.[58] Low acknowledged Glauert's originality in using the method of images to arrive at a formula for tunnel interference effects in the case of rectangular (as distinct from circular) tunnels but quibbled at an "unnecessary reference to Hobson's Trigonometry" (168) to deal with a mathematical point that Low considered elementary.[59] Low concluded his review by describing the *Elements* as "an important contribution to English aerodynamic literature" and (high praise indeed) as a book that "should be of the greatest value to all designers of aircraft" (168).

Approaching the *Elements* from the direction of the mathematician rather than the engineer, R. V. Southwell reviewed it for the *Mathematical Gazette*.[60] He praised Glauert's "power of concise exposition" (394) and said the book gave "an admirable account of a fascinating theory." It could be recommended as "indispensable to every student of modern aerodynamics" (395). But, as well as offering praise, it is clear that Southwell was taking care to locate Glauert's achievement in a particular way. He began by noting that when Mr. Asquith first appointed the Advisory Committee for Aeronautics in 1909, "nothing seemed more certain than that aerodynamics must develop as a purely empirical science" (394).[61] Theoretical hydrodynamics was not sufficiently developed to take account of both inertial and viscous forces. Today, said Southwell, that "difficulty is not yet overcome, but it has been turned" (394). Prandtl's wing theory is not "an exact theory," but Prandtl "has supplied what for practice is almost as useful—a theory which can predict" (394).

Southwell then clarified his distinction between exact and predictive theories. Consider Glauert's chapter 8, which contained an account of skin friction and the origins of viscous drag. Southwell granted that Glauert's brief treatment was appropriate given the limited purposes of the *Elements* but expressed the hope that Glauert would produce a follow-up volume to expand the "somewhat slender outline" of the present chapter. The follow-up volume would call for a "slightly altered arrangement" of the material. It appears to me, said Southwell, that

> we ought to recognise not one "Prandtl theory," but two. The first forms the main subject of the present work; its methods are those of the classical theory, and its assumptions, based on Lanchester's picture of the flow pattern, are justified, ultimately, by its success in prediction. The second, which develops the notion of the "boundary layer" is more fundamental, more difficult, and (probably) less productive of concrete results than the first; its aim is to explain the circulation round a lifting wing in terms of the known equations of viscous flow. (395)

The implication was that Glauert had adopted an "arrangement" of his material that did not adequately recognize the difference between the two theories. Glauert ran these two distinct theories together and aspired to a "combined" presentation. This may help the student, said Southwell, but it cannot do justice to either theory because "their methods are too distinct to permit a really satisfactory blend" (395). He did not believe that they could be combined in the way that Glauert wanted. "For the combined theory seeks to bring phenomena, in their very essence dependent on the viscosity of the fluid and its interaction with the solid boundary, within the scope of analysis which he knows is strictly applicable only to vortex motions existent throughout all time in a fluid devoid of all viscosity" (395). Southwell thus insisted on keeping apart what Glauert had sought to bring together. What the author had aspired to unify, the reviewer saw as incompatible. The themes invoked in Southwell's review were familiar and characteristic of the British experts: there was the desire for a "fundamental" theory based on Stokes' equations, a commitment to the "essential" difference between real and perfect fluids, and the appeal to the eternal character of vortices in a perfect fluid, that is, to Kelvin's theorem. That Glauert, like Prandtl, was deliberately trying to overcome the idea that there is an "essential" difference between real and perfect fluids finds no recognition. Southwell acknowledged in Glauert's unified presentation not a principled methodological stance but a mere pedagogical expedient, an "arrangement" of material to help students—and an arrangement that could not be sustained in the face of reality or in the pages of a more advanced treatise.

Southwell said that the views he expressed "imply no criticism" of Glauert's book. The claim may look disingenuous but I think it should be accepted as authentic. The words would make sense if Southwell were reading Glauert's book as an exercise in technology rather than physics. Once Prandtl's wing theory was understood as no more than an instrument of prediction, as something that could be assessed using purely pragmatic criteria, then the real business of science could be thought of as proceeding in parallel to the technology. There would be no need for any quarrel between those engaged in the two distinct sorts of activity, provided they were kept apart and not confused with one another. Thus Southwell could honestly declare that he was *not* criticizing Glauert's book but simply making it clear what manner of book it was, and what criteria were appropriate for its assessment.

This left just "one small detail" (395) that Southwell certainly wanted to criticize in an explicit way. He was worried about the imaginary roller bearings that Glauert interposed between a fluid and a material body or between two layers of fluid moving in different directions or with different speeds. References to roller bearings cropped up at a number of points in Glauert's book, for example, on pages 95, 100, and 117, and were represented diagrammatically on page 131. Southwell thought such talk was misguided, and he implied that Glauert should know better. Vortices don't behave like roller bearings, and it won't help the beginner to understand "the purely mathematical concept of vorticity" (395), that is, the technical definition of the rotation of a fluid element. Southwell's point was that "vorticity," as the term is used in fluid dynamics, can be present when nothing in the flow behaves like a "vortex," as that term is used in common language, that is, nothing is swirling, rolling, or rotating. For example, mathematically, "vorticity" is present when two immediately adjacent layers of ideal fluid move horizontally with uniform but different speeds. All the fluid in the respective layers moves in straight lines, but for the mathematician, this phenomenon is equivalent to an infinitely thin sheet of vorticity between the layers. Talk of "roller bearings," however, will produce an incorrect picture in the mind. The beginner "will misunderstand either the vortex sheet, or the action of roller bearings" (395). The harshest criticism was thus directed at Glauert's engineering imagery. Southwell was not, in general, against visualization.[62] The complaint was against the way that viscous processes and the viscous boundary layer were represented in nonviscous terms.[63]

Despite these reservations, the publication of Glauert's *Elements* in 1926 represented the de facto victory of the circulation theory of lift among British experts. The theory and references to Glauert's exemplary account of it found their way into all subsequent treatises and textbooks, such as Lamb's *Hydrodynamics* and Ramsey's *Treatise on Hydromechanics*. The victory was, of course, underpinned by the steady accumulation of evidence from experimentalists such as Fage.[64] The increasingly secure position of the circulation theory was, however, of a qualified kind. The victory was no simple rout of the opposition. The situation might be described with the use of political metaphors by saying that territory was conceded and new spheres of influence agreed on. The power of the circulation theory had been demonstrated, and a certain zone of occupation was now recognized—though not the full legitimacy of what had taken place. The task now was to get on with life under the new dispensation. In the Great War, Germany may not have prevailed, but in the field of practical aerodynamics a new respect was accorded to the circulation theory and Prandtl's wing theory. In 1927 Prandtl was invited to London to deliver the Wright Memorial Lecture to the Royal Aeronautical Society and to receive the Gold Medal of the society.[65]

There had been a previous suggestion that Prandtl might give a talk, which had been conveyed via Glauert in 1922. Prandtl had felt compelled to turn down the invitation, however, because of his lack of English.[66] The Wright Lecture was a much grander affair, and Prandtl, who clearly appreciated the invitation, now felt better equipped to cope, though he still had some anxieties. In the preparatory exchange of letters with the chairman and the secretary of the society he fussed over what he should wear. Should he be in *Frack*, that is, tailed coat? In hesitant English he announced that "I have at this time English lessons and believe to be able up to the date of the lecture, to read the paper myself."[67] In the event, despite displaying the recommended tails, white tie, and white waistcoat, he only delivered the opening passages of the lecture and then called on the help of Major Low. Low, who had worked with Prandtl to translate the text, read the remainder.[68]

Those opening passages, however, touched on a matter of some delicacy. They concerned the origin of the theory of the aerofoil and the relative contributions of Prandtl and Lanchester. Who invented the theory and who should get the credit? Prandtl was diplomatic but forthright. He said that Lanchester had worked on the subject before he, Prandtl, had turned his attention to it and that Lanchester had independently obtained an important part of the theory. Prandtl insisted, however, that the ideas he used to build up his theory had occurred to him *before* he read Lanchester's 1907 book. This prior understanding, he argued, may explain why "we in Germany were better able to understand Lanchester's book when it appeared than you in England" (721). The truth, Prandtl went on, is that "Lanchester's treatment is difficult to follow." It makes "a very great demand on the reader's intuitive perceptions," and "only because we had been working on similar lines were we able to grasp Lanchester's meaning at once" (721).

Is Prandtl here corroborating Glazebrook's excuse for the British neglect of Lanchester? Surely not, though he certainly shared some of Glazebrook's ideas about Lanchester's work. Like Glazebrook, Prandtl did not countenance the possibility that it was the understanding of Lanchester, rather than the failure to understand him, that lay behind the British response. But, while going along with part of Glazebrook's story, Prandtl's comments actually serve to accentuate the tensions between the different parts of Glazebrook's excuse. They made it even more necessary to explain why the Germans were in a position to grasp Lanchester's meaning when, allegedly, the British had not been able to rise to the occasion. Glazebrook had excused one failure by citing another failure, and what Prandtl had to say aggravated rather than alleviated this logical weakness.[69]

Negotiating Kelvin's Theorem

Prandtl's lecture had the title "The Generation of Vortices in Fluids of Small Viscosity." The choice of subject matter is revealing. Prandtl used the opportunity to address the two problems that most worried the British. First, how did circulation arise? Second, why did perfect fluid theory, though false, work in practice? Prandtl argued that these problems can be resolved by a careful analysis of what is, and what is not, implied by the theorems of classical hydrodynamics.

Consider Kelvin's theorem, which, in Prandtl's words, asserted that "in a homogeneous, frictionless fluid the circulation around every closed fluid line is invariable with time." What did Prandtl mean by a "fluid line"? A closed fluid line is not just a closed geometrical line imaginatively and arbitrarily projected into the fluid. It is meant to be a line that is always made up of the same fluid elements. "Let us suppose a 'fluid line' to be a line composed permanently of the same fluid particles" (722). If the whole of a body of fluid is at rest, then the circulation around any such circuit is zero, and it follows from Kelvin's result that it will stay zero for all time. Prandtl then invited his listeners to imagine a body, such as a wing or strut, surrounded by perfect

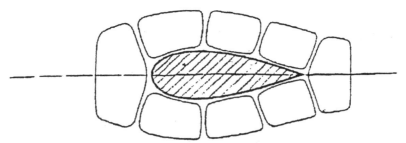

FIGURE 9.13. Stationary fluid around a strut subdivided into a mesh of fluid lines. From Prandtl 1927c, 723. (By permission of the Royal Aeronautical Society Library)

fluid where the fluid is at rest. The fluid around the object is supposed to be subdivided into a mesh of small circuits in the manner shown in figure 9.13, where each circuit is a fluid line.

Kelvin's theorem implies that the circulation around each circuit or fluid line remains zero, and this apparently leads to the conclusion that rotation cannot appear anywhere. From this it would seem to follow that lift is precluded, as the British critics of the circulation theory always argued. For Prandtl, however, "this conclusion is premature" (722). He went on: "We must first ascertain whether every point of the fluid set in motion is actually enclosed by the lines which in the state of rest were closed, if our conclusion is to be permissible. But closer investigation shows that it is possible to give instances in which this is not the case" (722).

How is this possible? Where do the rotating fluid elements come from? Prandtl's answer was based on what happens when the body moves through the fluid "so that the upper and lower streams flow together, or as we shall say become confluent, at the sharp rear edge of the body" (723). Such a motion is shown in figure 9.14.

The network of circuits is divided by a surface "along which our conclusion as to the absence of vorticity is no longer applicable" (723). This surface is indicated by the dotted line in the figure. The vorticity does not arise because some material element of fluid is set in rotation; it comes from the relative motion of two adjacent bodies of fluid. Prandtl did not try to challenge Kelvin's theorem by finding a drop of perfect fluid that somehow escaped the division into closed circuits; rather, he was exploiting the fact that rotation (defined technically) can exist without anything (or any finite thing) rotating. Kelvin's theorem does *not* imply that "rotation cannot appear anywhere"; it has a more specific, and limited, meaning. A vortex sheet, Prandtl insisted, can arise from confluence, and do so "*without contradicting Kelvin's theorem*" (723).

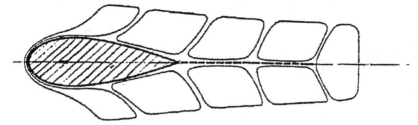

FIGURE 9.14. Strut in motion with confluence at rear edge. Prandtl argued that circulation can be created in an ideal fluid by confluence without violating Kelvin's theorem. From Prandtl 1927c, 723. (By permission of the Royal Aeronautical Society Library)

Confluence, according to Prandtl, can also generate the circulation around a wing. In an attempt to show how it can do this, he set out the sequence of events in the first few moments after a wing has begun to move through still air. Prandtl referred to the lower surface of the wing as the "pressure" side and the upper surface as the "suction" side. "During accelerations," he said,

> the velocity at the rear edge is greater on the pressure side than on the suction side, since the path along the pressure side is shorter. Consequently, after confluence a discontinuous distribution of velocity is set up which effectively constitutes a sheet of intense vorticity; the surface of discontinuity then begins to roll up into a spiral. The circulation for each circuit enclosing the wing and the surface of discontinuity still remain zero, from which it may be inferred that the circulation around the wing is equal and opposite to the circulation of the vortex produced by rolling up of the sheet. This is the method of generation of the circulation around the wing. (723)

It cannot be said that the argument is entirely clear.[70] Prandtl acknowledged that "the mathematician" would object and insist that an inviscid fluid would flow around the trailing edge. There would be a stagnation point on the upper (suction) surface of the wing rather than a vortex sheet coming away from the trailing edge. This would be so even if the trailing edge were sharp and the perfect fluid had to move at an infinite speed to get round the corner. The result would be no circulation and no lift. This must be the correct flow, the mathematicians would say, because it would be "everywhere irrotational in accord with the theoretical laws for a flow produced from a state of rest!" (723). For the mathematicians, this flow alone would be consistent with the theorems of Lagrange and Kelvin. On Prandtl's reading the theorems do not carry this implication. He argued that both his proposed flow (with circulation) and the mathematician's flow (without circulation) were consistent with the classical theorems of hydrodynamics. But, he insisted, only the flow with circulation and the smooth confluence at the trailing edge is physically realizable. This was really all Prandtl needed. Even if the process by which circulation was generated remained obscure, it was the logical possibility of circulation, and the logical right to postulate it, that really mattered for the perfect fluid approach.

Prandtl then turned to his second topic: Why did perfect fluid theory work? Going back to his discovery of the boundary layer in 1904, he explained that the thickness of the layer was inversely proportional to the square root of the viscosity and that it generated a tangential friction proportional to the 3/2 power of the speed and the square root of the viscosity. "But," he said, "there is something of more importance to us here" (725). The boundary layer is the cause of the formation of vortices. Kelvin's theorem only applies to flows devoid of viscous forces, and so, in motion starting from rest, the circulation will remain practically zero in fluid circuits that do not pass, or have not passed, through a boundary layer. This is why real fluids, such as air, behave like perfect fluids in irrotational motion at a distance from a solid body. By means of photographs and a film Prandtl then demonstrated that the boundary layer could be manipulated, for example, removed by suction, and he showed that this procedure had a dramatic effect on the flow. He concluded his lecture with a discussion of the role of turbulence in the boundary layer and explained that an increase in turbulence could reduce drag. This allowed Prandtl to clinch his justification for using perfect fluid theory: "We thus get the unique characteristic that it is precisely these turbulent flows of low resistance around bodies which can be so closely represented by the theory of the perfect liquid" (739).

No one could say that Prandtl had evaded the arguments of his British critics. He had confronted their doubts, but had he dispelled them? The utility of Prandtl's wing theory had been largely conceded, but did the Wright Lecture remove the residual worries about its theoretical basis? The immediate answer was that it did not. Partly this may have been because of the difficulty in following certain steps in Prandtl's line of thought, but a deeper reason lay in the divergent readings of Kelvin's theorem.

Coffee Spoons and Theology

Kelvin's theorem did not categorically preclude circulation in a perfect fluid but asserted, conditionally, that it could only exist under certain circumstances. In Britain effort was put into making sure that the proof of Kelvin's theorem was as rigorous as possible.[71] In Germany the focus was subtly different: it was the scope of the theorem that attracted attention. In 1910 Felix Klein published a paper in the *Zeitschrift für Mathematik und Physik* in which he argued

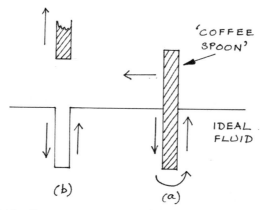

FIGURE 9.15. Klein's coffee-spoon experiment. A surface, the "spoon," is immersed in an ideal fluid and moved forward (*a*). It is then quickly removed (*b*), leaving behind a surface of discontinuity (shown in exaggerated form). The result is a vortex sheet and hence the creation of circulation.

that it was easy to create circulation in an ideal fluid—as easy as stirring a cup of coffee.[72] If a thin, flat surface (the "spoon") is partially inserted in a body of ideal fluid, moved forward, and then briskly removed, the result would be a vortex with a circulation around it—but, said Klein, Kelvin's theorem would not be violated. The mechanics of the process are shown in figure 9.15. The motion of the surface has the effect of forcing the fluid to move down the front face and up the back face, as indicated by the arrows. Removing the surface then leaves two adjacent bodies of fluid moving in opposite directions. The result is a surface of discontinuity, that is, a sheet of vorticity, which then rolls up into a vortex. This does not contradict the theorem, argued Klein, because Kelvin's proof assumed continuity of the fluid, and this precondition is violated by the insertion and removal of the mathematically simplified "coffee spoon." The coffee-spoon experiment was not an exact prototype for Prandtl's confluence argument in the Wright Lecture (because a wing is surrounded by air, not dipped into it), but it surely provided an analogical resource. Klein's argument encouraged a tradition of critical assessment of Kelvin's theorem. Further papers, in which the argument was extended and assessed, were written by Lagally, Jaffé, and Prandtl. Later contributions on this theme came from Betz and Ackeret.[73] By contrast, Klein's coffee-spoon paper received no mention in Lamb's *Hydrodynamics*.

Prandtl was right to anticipate objections from "the mathematicians" or, at least, from some Cambridge mathematicians. For example, his defense of perfect fluid theory failed to convince the Cambridge mathematician Harold Jeffreys, who later became Plumian Professor of Astronomy and Experimental Philosophy at Cambridge.[74] Jeffreys (fig. 9.16) has not previously featured in the story and was not a specialist in aerodynamics. His primary contributions were to geophysics, but he published creative mathematical work in an impressively wide range of subjects. Jeffreys, a notoriously withdrawn man, distinguished himself in part II of the Mathematical Tripos in 1913 and was elected a fellow of St. John's College in 1914. He stayed at St. John's for the rest of his life. During the Great War Jeffreys worked at the Cavendish laboratory on gunnery and then on meteorology with Napier Shaw (who was on the Advisory Committee for Aeronautics). Like his friend G. I. Taylor, Jeffreys originally became interested in circulation and viscous eddies from a meteorological standpoint. In the 1920s, prompted by his lecturing commitments in applied mathematics at Cambridge, Jeffreys began a series of papers on fluid dynamics which made explicit contact with the work that had been done on circulation in aerodynamics.

The first in the series of papers, in 1925, was called "On the Circulation Theory of Aeroplane Lift."[75] Although an outsider in the field, Jeffreys sent a copy to Prandtl and received a somewhat formal reply. Prandtl clearly thought

FIGURE 9.16. Harold Jeffreys (1891–1989). Jeffreys was a powerful and wide-ranging applied mathematician who originally approached fluid dynamics from the standpoint of meteorology. Like Taylor and Southwell, he argued that Kelvin's theorem precluded the creation of circulation in an ideal fluid. (By permission of the Royal Society of London)

that Jeffreys needed to do his homework. He suggested Jeffreys read the 1904 paper on boundary-layer theory and the 1908 application of the theory by Blasius and duly enclosed the references.[76] After something of a delay, Jeffreys acknowledged the response but said, rather untactfully, that he was too busy at the moment to follow up the references. He would get down to them as soon as he could.[77] He added: "Of course it would not in the least surprise me to find that all the ideas in my paper had been anticipated, but they were not in any work I had seen & I thought it well that they should be published simply because they were not well known in this country." It may have been this exchange that gave Prandtl his sense of what topics needed to be addressed in the Wright Lecture and that helped him imagine the archetypal "mathematician" resisting his account of the origin of circulation in an ideal fluid.

In 1930, three years after Prandtl's Wright Lecture, and after discussions with Glauert and Taylor, Jeffreys published "The Wake in Fluid Flow Past a Solid."[78] Jeffreys started by noting that in many cases it was possible to approximate the motion of a real fluid by a "cyclic irrotational motion, with local filaments of vorticity." He instanced the work of Kutta and Joukowsky on two-dimensional flow and that of Lanchester and Prandtl on three-dimensional flow. But, he insisted: "The existence of cyclic motion is in disagreement with classical hydrodynamics, which predicts that there shall be no circulation about any circuit drawn in a fluid initially at rest or in uniform motion, and that there is no resultant thrust on a solid immersed in a steady uniform current" (376).

As far as Jeffreys was concerned, classical hydrodynamics had long "ceased to be a representation of the physical facts" (376). He agreed with the qualitative explanation that Prandtl had advanced to show why a perfect fluid theory could be used to approximate a real flow at a distance from a solid boundary, but he did not accept Prandtl's account of Kelvin's theorem. For Jeffreys, classical hydrodynamics implied that a wing, starting to move from rest in a perfect fluid that was also at rest, could not generate circulation and lift. Prandtl had argued in his Wright Lecture that the generation of circulation and lift was consistent with Kelvin's theorem; Jeffreys said it was not. Zero lift was the clear and inescapable consequence of the theorem in the case under discussion. Understanding the generation of lift required starting out with the theory of viscous flow. For Jeffreys, as previously for Bairstow, the problem was why ideal fluid theory seemed to work. Inquiry should not concentrate on explaining its numerous failures but on its few remarkable successes. "Considerable attention has been given to the reason why classical hydrodynamics fails to represent the experimental facts; but it appears to me that these efforts arise from an incorrect point of view . . . the remarkable thing is not that classical hydrodynamics is often wrong, but that it is ever nearly right" (376).

Jeffreys' way of addressing this question was to anchor the mathematics in physical processes and to make sure that what were really results in mathematics were not treated as results in physics. Their physical application had to be justified, not taken for granted. Consider, for example, Kelvin's theorem and the way it was used to explain the creation of circulation around a wing. The vortex that forms at the trailing edge, and then detaches itself, is said to cause the circulation around the wing. The circulation around the departing vortex brings about the opposite circulation around the wing. How? The answer given by Prandtl and Glauert was that Kelvin's theorem had to be satisfied. Jeffreys was not convinced by this answer, and surely he was right to be suspicious. If Kelvin's theorem prohibits the creation of new circulation, why are *two* violations of the prohibition acceptable merely because they are violations in opposite directions? Things that cannot exist cannot cancel out. If it is illegal to drive down a certain street, two people may not drive down it and plead that the law was not broken because they were driving in opposite directions.

Jeffreys wanted to know why Kelvin's theorem, which was a theorem about inviscid fluids, could be used in the course of an argument in which the role for viscosity had already been granted in order to explain the origin of circulation. This could only be justified if something equivalent to Kelvin's result could be deduced starting from the premise of viscous flow. To explore this possibility, Jeffreys set himself the goal of deriving the rate of change of circulation with time for a viscous fluid. Kelvin's theorem for an ideal fluid is expressed by writing $d\Gamma/dt = 0$, and Jeffreys wanted to know the value of $d\Gamma/dt$ for a real, viscous fluid. The general circulation theorem for viscous flow that Jeffreys derived involved the integral of some five separate expressions, each of considerable complexity. For a uniform, incompressible fluid, however, only one of the five terms survived. For aerodynamic purposes, Jeffreys was then able to replace Kelvin's circulation theorem by the equation

$$\frac{d\Gamma}{dt} = \int_C v \frac{\partial \xi_{ik}}{\partial x_k} dx_i = \int_C v \nabla^2 u_i dx_i,$$

where Γ is the circulation around the contour C moving with the fluid and v is the kinematic viscosity (that is, viscosity divided by density). In Jeffreys' equation the three coordinate axes are represented not by x, y, z, but by x_i

where $i = 1, 2, 3$, and the corresponding velocity components are given by u_i. The summation convention is used for repeated suffixes, and the term ξ_{ik} is the vorticity, which is defined as

$$\xi_{ik} = \left(\frac{\partial u_i}{\partial x_k} - \frac{\partial u_k}{\partial x_i} \right).$$

What did this new expression for $d\Gamma/dt$ mean? Jeffreys followed an earlier discussion in Lamb's *Hydrodynamics* and offered an explanation of the physical significance of the result as follows.[79] The equation linking circulation and time, he said, can be recognized as one that represents a *diffusion* process. It shows that vorticity and the circulation it induces obey laws that are analogous to the laws governing the diffusion of temperature or density. From this analogy it follows that vorticity must diffuse outward from a solid boundary. Circulation cannot arise spontaneously within the body of viscous fluid itself. Before the diffusion process has carried the vorticity to regions distant from the boundary, the fluid in these distant regions shows no rate of change of vorticity with time. The rate of change of circulation around a contour therefore depends on the vorticity *near the contour*. There will therefore be "no appreciable circulation except on contours part of which have passed near a solid boundary: in other words vorticity is negligible except in the wake" (380).

Jeffreys' paper "The Wake in Fluid Flow Past a Solid" covered much of the same ground as Prandtl's earlier but more qualitative treatment in the Wright Lecture, but it is clear that Jeffreys felt that only now had a proper basis been provided for the conclusions that had been advanced. He carefully investigated the orders of magnitude of the quantities involved in the diffusion of the vorticity. This analysis, he said, "constitutes the theoretical justification of the 'boundary-layer' theory of Prandtl and his followers" (380). Jeffreys' treatment converged with Prandtl's but was offered as one "based on the physical properties of a real fluid and not on mathematical conceptions of vortex lines and tubes" (389). In a further paper, "The Equations of Viscous Motion and the Circulation Theorem," Jeffreys made a similar claim about Prandtl's account of the origin of circulation and the starting vortex that detaches from the trailing edge.[80] Only an understanding of viscous circulation, said Jeffreys, can provide the real "physical basis" needed for applying the theory of vorticity to real fluids.

Where did this leave Kelvin's theorem and the (apparently) inconsistent use of that theorem by supporters of the circulation theory? Jeffrey's position was that the diffusion picture showed that it was not really Kelvin's theorem that was being invoked to explain the relation between the circulation around the detached vortex and the circulation around a wing. Rather, it was the theorem for circulation in a viscous fluid that was really in play. Kelvin's theorem dealt with inviscid fluids, but the counterpart theorem for viscous fluids, the diffusion equation, gave the same numerical result for the initial stages of the flow. "Thus," Jeffreys stated, "the conditions assumed by classical hydrodynamics are reproduced, in the specified conditions, by the real fluid" (381).

Jeffreys was not alone in saying that Kelvin's theorem clearly ruled out the creation of circulation by a wing in an ideal fluid. This had been Taylor's position in 1914, and it was still Southwell's position in 1930 when he gave the prestigious James Forrest Lecture.[81] Southwell asserted that classical hydrodynamics left the existence of circulation around a wing "an altogether amazing coincidence" (360). He added that the assumption of circulation was "rather theological" (361). The allusion was to Kelvin, for whom the eternal character of circulation and vortex rings indicated a divine origin. Southwell, like Jeffreys, was unmoved by the first part of Prandtl's Wright Lecture, dealing with Kelvin's theorem and perfect fluid theory, but he was enthused by the second part on the boundary layer and the creation of vortices. Southwell reproduced Prandtl's photographs showing the control of the boundary layer by suction and showing how to make a divergent nozzle "run full." He selected and emphasized the places where Prandtl's concerns came closest to the long-standing British interest in viscous fluids and eddying flow. Southwell further assimilated this aspect of Prandtl's work to the British tradition by arguing that the analysis of backflow in the boundary layer was similar to Mallock's work on reverse flow and eddies that was done in the early years of the Advisory Committee for Aeronautics.[82]

Lamb had also made gentle fun of the theory of circulation by exploiting the theological overtones of Kelvin's theorem. In his Rouse Ball Lecture of 1924, titled "The Evolution of Mathematical Physics," Lamb had said of perfect fluid theory that "this theory cannot tell us why an aeroplane needs power for its propulsion; nor, indeed, can it tell us how the aeroplane obtains its sustentation, unless by assuming certain circumstances to have been established at the Creation which, in all reverence, we find it hard to believe."[83] The "certain circumstances," of course, were the provision of suitably adjusted values of the circulation. Every takeoff and landing, Lamb hinted, would require divine anticipation and intervention. But if the tone was joking, the point was serious. Perfect fluid theory predicts zero drag and makes a mystery out of the origin of circulation. In the last edition of his *Hydrodynamics*, in

1932, Lamb returned to the problem of the origin of circulation and of understanding how it resulted in a smooth flow being established at the trailing edge of a wing. He clearly felt that no satisfactory account had been given of this. He cannot have been convinced by what Prandtl and Glauert had to say, and his reference to Jeffreys' efforts was noncommittal. Jeffreys may have deepened the discussion and clarified some of the physical principles, but it was still mathematically incomplete. Lamb summed up by saying: "It is still not altogether easy, in spite of the attempts that have been made, to trace out deductively the stages by which the result is established when the relative flow is started. Fortunately, some beautiful experiments with small-scale models in a tank come to our help. A vortex with counter-clockwise sense is first formed, and detached from the edge, and then passes down stream, leaving a complementary circulation around the aerofoil in the opposite sense" (691).

No one would have been deceived by the understatement. Lamb was saying that, by his standards, no one had given a satisfactory mathematical analysis of the processes by which a circulation was created. Prandtl may have been able to offer "beautiful photographs," but that only meant that the analysis was still confined to the empirical level.[84] Lamb was surely right. The circulatory theory was accepted by the British without its supporters being able to offer a rigorous account of the origin of circulation. This had been a source of difficulty for the British from the outset. It was still a worry, but now, with varying degrees of unease, they appeared ready to live with the problem.

Lamb also remained skeptical when Prandtl and Glauert represented the surface of a body, such as a wing, by a sheet or line of "bound" vorticity. Lamb did not claim that the concept of a bound vortex was logically defective, but, in responding to a paper that Glauert submitted to the Aeronautical Research Committee in 1929, he deemed it "artificial." He succeeded in deducing Glauert's results, concerning accelerated motion in two dimensions, by other more usual routes.[85] Once again, there was tension between the advocates of two different approaches to applied mathematics: those who insisted that the mathematics described what they took to be physically real processes (and described them in a rigorous way) and those content with mathematical descriptions that were acknowledged to be expedient rather than physically true. Lamb never shifted from the view that he expressed in his presidential address to the British Association meeting in 1925.[86] He spoke for many in British aerodynamics when he said that Prandtl had provided "the best available scheme for the forces on an aircraft" (14). The choice of the word "scheme" was meant to imply that Prandtl had failed to give a *fundamental* account of the physical reality of the process.[87]

Plus Ça Change

Prandtl's boundary-layer theory provided the material that might give substance and depth to the "scheme" of the wing theory. It suggested that inviscid approximations might be replaced by a more realistic account of the physics of viscous processes. The boundary layer became the focus of a sustained British research effort organized by the Fluid Motion Panel of the Aeronautical Research Committee. The original intention was that Lamb would be the editor of the volumes that would draw the results together, although Fage, who clearly found Lamb's work very demanding, put in a request that the mathematics should be kept as simple as possible. "Lamb's *Hydrodynamics*," said Fage in the course of a discussion of the proposed monograph, "was more suitable for the professional mathematician and was very difficult."[88] In the event, Lamb did not live to complete this task and it fell to Sidney Goldstein. Goldstein had been a pupil of Jeffreys' but he had also gone to Göttingen after the war to study with Prandtl.[89] Despite continuing resistance, it is clear that the overall strategy that Prandtl had adopted in his Wright Lecture had been an appropriate one. He had engaged with the preoccupations of the British experts with viscous and eddying flow while reminding them of the intellectual resources that Göttingen had to offer.[90] Writing to Prandtl, after the Wright Lecture, Major Low said that he had spoken to many mathematicians and physicists and they all said that "your paper will give a new direction to aerodynamic research in this country." Low identified the transition from laminar to turbulent flow as the special point of interest for the British audience.[91]

This concern with the boundary layer and turbulence became the new research front, and it was congenial territory for the British even though their head-on assault on the Navier-Stokes equations had proven frustrating. If the battle for circulation in the theory of lift was over, the war on turbulence in the boundary layer was about to begin.[92] But even here the old worries were not far beneath the surface. On February 6, 1930, members of the Royal Aeronautical Society discussed a report titled "Modern Aerodynamical Research in Germany."[93] The report was presented by J. W. Maccoll, who had visited Göttingen and Aachen.[94] Maccoll, who had a command of German, was a government scientist and was to hold the post of research officer in the Department of External Ballistics at the Woolwich Arsenal. He described in mathematical detail the original work on the laminar boundary layer and then the more recent work on the transition to turbulence. In the discussion that followed Maccoll's paper, Bairstow identified what he saw as two fundamentally different approaches to the current problems in fluid dynamics.

Bairstow declared that he had "been impressed by the extreme complication of the whole subject and the apparently little connection between the German methods of solution and the equations of motion of a viscous fluid. All would have noticed how often new variables were introduced into the equations to deal with failures of the original hypothesis. It seemed that the Germans were making an engineering attempt to get solutions of practical value and had little hope of solving the equations of motion in a sense that would satisfy Professor Lamb" (697).

Bairstow was describing, albeit in a one-sided way, the difference in approach between a mathematically sophisticated engineer, adopting the methods of *technische Mechanik,* and that of a mathematical physicist drawing on the finely honed traditions and research strategy of the Cambridge school. Bairstow might not have sat the Tripos, but he still took Prof. Lamb as his reference point.[95] The difference in approach to which Bairstow was alluding, between the Cambridge and Göttingen traditions, has been present in one form or another throughout the story I have been telling. It was implicated in the original British dismissal of the circulatory theory, and it was central to the manner in which the theory was finally accepted by the British.[96]

In an article titled "Twenty-One Years' Progress in Aerodynamic Science" which Bairstow published in 1930, the same year as the remarks just quoted, he surveyed the work that had been done since the creation of the Advisory Committee for Aeronautics in 1909. Bairstow invoked a revealing comparison to describe the discomfort that still surrounded the relation between the theory of viscous and inviscid fluids in aerodynamics. He likened the problem of reconciling the viscous and inviscid approaches to the problems that British physicists were experiencing in reconciling the wave and particle conceptions of light and of the electron. Two fundamentally different models were in use, but it was impossible to see how they could both be true.[97] Bairstow quoted the exasperated response to this situation of one of the country's leading physicists, a response that mirrored, perhaps, the frustrations of Bairstow's own work on the Navier-Stokes equations. "Aerodynamic theory," said Bairstow,

> is now rather like the physical theory of light; Sir William Bragg recently said that physicists use the electron theory on Mondays, Wednesdays and Fridays, and the wave theory on alternate days. Both have uses but reconciliation of the two ideas has not yet been achieved. So it is in aeronautics. In our experimental work we assume that viscosity is an essential property of air and the building of a compressed-air tunnel is the latest expression of that belief. The practically useful theory of Prandtl comes from considering air as frictionless or inviscid. (29)

At the end of his survey Bairstow returned to this theme and defined his view of the prospects of aerodynamics in terms of this ambiguous and problematic image. We can be assured, he said, that aerodynamics has "a future comparable with that in electron theory" (30).

Despite Glauert's efforts to renegotiate the conceptual distinction between perfect fluid theory and the theory of viscous fluids, it is clear that the leading British mathematical physicists were in no hurry to abandon their view that the distinction was fundamental. The boundary separating the objects of the two theories was treated as ontologically rigid rather than methodologically flexible. Eventually, though, by the mid- and late 1930s, what Glauert called the "true conception of a perfect fluid" appears to have filtered into British mathematical and experimental practice. It was not acknowledged explicitly, but it was implicit in the use of potential irrotational flow as an engineering ideal. By the 1940s its use for this purpose had become routine, for example, in estimating the role played by the viscous boundary layer.[98] By this time the circulation theory of lift, and Prandtl's wing theory, had already become an established part of British aerodynamics. The earlier insistence on a rigid conceptual boundary between ideal and real fluids nevertheless helps to explain why, when Prandtl's wing theory was finally accepted by the British, there was still a note of reservation. Prandtl's theory may have been, as Bairstow conceded, "the best and most useful working hypothesis of modern times"—but it was still a working hypothesis.

For many years, one of the standard British textbooks in the field was Milne-Thomson's *Theoretical Aerodynamics*.[99] The book ran through four editions between 1947 and 1966 and contained the following, revealing observation on the lifting-line theory. Following an explication of Lamb's contrast between a scheme and a fundamental theory, Milne-Thomson said, "The student should be warned, however, that the investigation on which we are about to embark is one of discussing the deductions to be made from schematization of a very complicated state of affairs and that the 'laws of Prandtl' which will be used as a basis are not necessarily laws of nature" (191). Contrasting the Laws of Prandtl with the Laws of Nature was just a picturesque way of saying what most British experts had felt all along. Prandtl's work on the aerofoil was an exercise in engineering pragmatism rather than a contribution to a realistic and rigorous mathematical physics.

A Conclusion and a Warning

My question at the beginning of this volume was: Why did British experts in aerodynamics resist the circulatory theory of lift when their German counterparts embraced it and developed it into a useful and predictive theory? My answer has been: Because the British placed aerodynamics in the hands of mathematical physicists while the Germans placed it in the hands of mathematically sophisticated engineers. More specifically, my answer points to a divergence between the culture of mathematical physics developed out of the Cambridge Tripos tradition and the culture of technical mechanics developed in the German technical colleges.

This abbreviated version of my argument and my conclusion is correct, but a condensed formulation of this kind carries with it certain dangers. It may invite, and may seem to permit, assimilation into a familiar, broader narrative, which destroys its real significance. Thus it may appear that the "moral" of the story is that (at least for a time) certain social prejudices encouraged resistance to a novel scientific theory and led to scientific evidence being ignored or overridden by social interests and cultural inertia. According to this stereotype the story came to an end when "rational factors" or "epistemic factors" eventually overcame "social factors" and science was able to continue on its way—a little sadder and wiser, perhaps, but still securely on the path of progress.

Is there really any danger of the episode that I have described in so much detail being trivialized in this way? I fear there is.[100] In one form or another, the narrative framework I have just sketched is widely accepted. It has numerous defenders in the academic world who confidently recommend it for its alleged realism and rectitude. It is deemed realistic because no one who adopts this view need deny that science is a complicated business. Scientists are, after all, human. Sometimes the personality or the metaphysical beliefs of a scientist may imprint themselves on a historical episode. Sometimes political interests and ideologies will intervene to complicate the development of a subject and perhaps even distort and corrupt a line of scientific inquiry. What worldly person would ever want to deny that this can happen? But who could approve of these things or, after sober reflection, think that they represent the full story of scientific progress? The intrusions of extra-scientific interests must therefore be exposed as deviations from an ideal that is characteristic of science at its rational, impersonal, and objective best. As well as personal and social contingencies (the argument goes on), it is vital to acknowledge that there are rational principles that, ultimately, stand outside the historical process and outside society. These represent the normative standards that science must embody if it is to achieve its goal. Fortunately the norms of rational thinking are realized with sufficient frequency that science manages to do its proper job. The norms ensure that the Voice of Reason and the Voice of Nature are heard. With due effort, and a degree of good fortune, this is how science actually works. The rest (the deviations and failings) merely provide a human-interest story of which, perhaps, too much has been made.[101]

Doesn't the episode I have described fit into this stereotype? The dispute over the circulation theory ended because the evidence had become too strong to resist. Isn't that really all there was to it? The British experts were initially too impressed by the great name of Rayleigh, and their resistance to the circulation theory was not a credit to their rationality. Eventually, though none too soon, they came round. Ultimately, therefore, evidence and reason triumphed over prejudice, tradition, and inertia. Reality stubbornly thwarted vested interests, and rationality subverted conventional habits and complacent expectations. Knowledge triumphed over Society. Isn't this how my story ends?

The answer is no. This is not the story, and it is not how the story ends. Such a framework does not do justice to even half of the story I have told. In reality the end of the story is of a piece with its beginning and its middle. There was continuity both in the particular parameters of the episode I have described as well as in the general epistemological principles that ran through it. The supporters of the circulation theory never provided an adequate account of the origin of circulation, and the critics never deduced the aerodynamics of a wing from Stokes' equations. Nor were there any qualitative differences in the relations linking knowledge to society and to the material world at the end of the story compared with the beginning of the story. There were changes of many kinds throughout the course of the episode, but they were not changes in the fundamentals of cognition or the modes of its expression. Fundamental social processes were operating in the same, principled way before, during, and after the episode described, and they are operating in the same way today. Society was not an intruder that was eventually dispelled or an alien force that had to be subordinated to the norms of rationality or the voice of nature. There was no Manichean struggle between the Social and the Rational.

Trivializing versions of how the story ends may appeal to propagandists who want to spin simple moral tales, but to the historian and sociologist such tales indicate that the complexities of the episode are being edited out and its structure distorted. This danger is amplified if only a summary version of the story is retained in the memory. To offset this tendency I want to make explicit the methodological framework in which the story should be located, and I want to defend this framework against trivializing objections and misguided alternatives. Such is the function of the discussions in the final chapter. The aim is to keep the details of the story alive and its structure intact while, at the same time, reflecting on its broader significance.[102]

Pessimism, Positivism, and Relativism: Aerodynamic Knowledge in Context

Menschen haben geurteilt, ein König könne Regen machen; wir sagen dies widerspräche aller Erfahrung. Heute urteilt man, Aeroplan, Radio, etc. seien Mittel zur Annäherung der Völker und Ausbreitung von Kultur.

Men have judged that a king can make rain; we say this contradicts all experience. Today they judge that aeroplanes and the radio etc. are means for the closer contact of peoples and the spread of culture.

LUDWIG WITTGENSTEIN, *Über Gewißheit. On Certainty* (1969)[1]

The British resistance to the circulatory theory of lift casts light on a number of wider themes, some methodological, some cultural, and some philosophical. Here I take the opportunity afforded by the case study to pose and answer some of the questions of this kind that have been raised. The first thing is to make explicit the methodological principles that have informed my own inquiry. The present study is not only an exercise in the history of science and technology; it is also a contribution to the sociology of knowledge. My approach has been that of the Strong Program in the sociology of knowledge, and so I start this chapter with a brief account and defense of the main features of the program and the perspective it is designed to encourage. It is a perspective very different from the naïve, philosophical narratives I identified at the end of the last chapter and that I mentioned in the introduction.[2] I then address two broader questions. First, was the resistance to the circulatory theory of lift an all-too-typical example of British failings in the field of technological innovation? I argue against this pessimistic reading. Second, what about the controversial topic of "relativism"? Aviation, as a successful and impressive technology, is often cited as a quick and decisive refutation of relativism. I believe this line of antirelativist argument is groundless and obscures the most striking characteristics of aerodynamic knowledge.

In this chapter I make use of the writings of the Viennese physicist Philipp Frank. Frank was a specialist in modern physics, but he wrote some half dozen papers on fluid dynamics and aeronautical topics.[3] He knew many of the leading experts in these fields and was a lifelong friend of Richard von Mises. Frank is relevant to my discussion for two reasons. First, he compared the way a scientific theory is assessed to the way the performance of an aircraft is assessed. I follow Frank in this aeronautical comparison and then close the circle by applying his comparison directly to theories in aerodynamics. Second, Frank's work provides a valuable resource when discussing relativism. I use Frank's simple and forthright definition of the word "relativism" to structure my own discussion.

The Sociological Perspective

It is important to appreciate the difference between the professional perspective of the sociologist and the perspective that prevails, and perhaps comes naturally, to social actors themselves in the course of everyday life. I refer to these as the "analyst's perspective" and the "actor's perspective." The concerns of those engaged in sociological analysis are usually not identical to those of the social actors they study, though, of course, analysts themselves will sometimes occupy the very roles that they investigate professionally. Conversely, sociological perspectives are sometimes invoked in the course of everyday interaction. (Major Low adopted such a stance when he speculated on what would have happened if Rayleigh had backed the circulation theory. He was reflecting on the role played in British aerodynamics by Rayleigh's authority.) Despite this overlap and interweaving, it is the differences in the perspectives of the analyst and the actor that I want to emphasize.

In everyday life much of our curiosity centers on deviations from what normally happens or from (what we feel to be) our justified expectations. We want to know why things go wrong more than we want to know why they went right. Going right tends to be taken for granted. It is the failure of the airline to keep to its schedule that makes irate travelers demand to know the causes of the delay. They do not demand to know how and why a punctual departure was achieved. If they were to pose such a question, it would be heard as a hostile comment rather than a disinterested inquiry. The structure of everyday curiosity can be remarkably one-sided. Using a terminology that has become current in the sociology of science, such everyday curiosity may be described as "asymmetrical." For the sociologist, however, the atypical is not the only thing that needs explaining. The typical is as interesting as the atypical, and the normal or the expected course of events is at least as important as the deviations. The professional curiosity of the sociologist may therefore be called "symmetrical," in contrast to the "asymmetry" of much commonsense curiosity.[4]

If an "asymmetrical" curiosity prompts us to ask for causes for half the story, then a "symmetrical" curiosity must lead us to demand causes for the whole story. If the commuters only want to know the causes of delay, then sociologists must risk the resentment prompted by their wanting to know the causes of nondelay. They must ask the questions others don't ask or don't want to answer. If sociologists were to study the workings of an airline, they would try to grasp its organizational features and see how its various parts related to one another. There are pilots and crew to be trained, maintenance schedules to be established, fuel supplies to be arranged, safety standards to be adhered to, duty rosters and wages to be negotiated, and shareholders to be satisfied. These dimensions of the organization would be common to all or most airlines, so the sociologist could construct a general model of such an organization and note the difference of practice between different instances of the model. This airline might devote twice as much time to safety training as that one; this one might repeatedly demand more flying hours between checks and repairs than that one; this one might meet its schedules by taking more risks. Such an investigative procedure would bring *both* the successful and the unsuccessful, the efficient and the inefficient, the safety conscious and the risk takers under the scope of the same model. By casting both sides of the story in the same terms, it is possible to use the different performances to probe the working of the general model, and hence to explore more deeply what it is to have a social organization capable of producing the range of observed outcomes.[5]

These considerations, drawn from the practice of general sociology, also apply to the sociology of knowledge. The central thrust of the Strong Program is that explanations in the sociology and history of science should be both "causal" and "symmetrical" in the sense that I have just explained. The same type of cause should explain the attractions of both true and false theories, and both successful and unsuccessful lines of work (where the judgments of truth and success derive from hindsight or are the analyst's own). I have said that this approach has informed my case study, but when I asked "Why did the British resist the circulation theory?" I may seem to have adopted the asymmetrical stance of common sense. It is true that the question *could* be posed in a purely commonsense way. This, I suspect, is how Major Low meant it when he asked why Lanchester had been ignored. Despite his sociological insight, he primarily wanted someone to take the blame. Stated in isolation Low's question is worded in a way that is consistent with either a symmetrical or an asymmetrical stance. What differentiates the two stances is the purpose behind the question and the distribution of curiosity informing the answer. The evidence I have presented indicates that it was local cultures, and the institutions sustaining them, that explain the reactions to the theory of circulation exhibited by *both* the British *and* the German experts. These were the causes of the phenomena that I set out to explain, and the causes were of the *same kind* in the two cases. Cambridge was not Göttingen, but both were influential and brilliant research institutions. Mathematical physics was not technical mechanics, but both were based on rich, mathematical traditions. Lamb's *Hydrodynamics* was not Föppl's *Vorlesungen*, but both were much-used textbooks that, respectively, encouraged and transmitted their own characteristic, mental orientation.

The scientific study of a complex phenomenon, or the development of a complex technology, typically calls for the cooperation of specialists from a number of fields. The creation of the atomic bomb in Los Alamos involved physicists, chemists, metallurgists, engineers, and experts in fluid dynamics. An episode in the history of science and technology of the kind I have analyzed likewise counts as a complex, real-world phenomenon, and its study will involve specialists from different fields, for example, historians, sociologists, and psychologists. The psychologist studies the mental capacities necessary for learning about the world and becoming a competent member of society, perhaps even a member of a specialist subgroup—for example, a subgroup whose members are able to read Föppl's textbook or sit the Tripos examination. The sociologist studies the social processes without which, ultimately, there would be no professional identities such as "the engineer" or "the physicist" and no institutions such as "the textbook," "the examination," or "the university."

It is evident that the episode I have described in my case study cannot be called "purely" sociological any more than it is "purely" psychological or "purely" a matter of grappling with the world. Likewise, the desire for a "complete" description or a "complete" explanation of the episode can be dismissed as utopian. But it is not unrealistic to hope for insights into parts of the problem, and some aspects of the episode may call for psychological study, while other aspects may call for sociological study. The one does not exclude the other. My emphasis on the sociological dimension is not a denial of the psychological dimension or any other naturalistic dimension. Rather, the emphasis on society arises because sociological variables are the ones most relevant to the question I am asking. British and German experts did not diverge because their basic cognitive faculties differed or because their personalities were different or because one group engaged with the material world while the other turned its back on it. As far as the present episode is concerned, they differed primarily because their education and professional lives were different. They worked in different disciplines and institutions whose traditions and reward structures diverged from one another.

One final feature of the sociological approach must be emphasized. It is central to my account that the actors involved were not detached intelligences moving in an abstract world of thoughts, theorems, and deductions. Nor did they move exclusively in a world of committee meetings, personal confrontations, status conflicts, and power struggles. These things were part of their world but not the whole of it. The experts in my story experimented in wind tunnels, built models, observed and measured the forces on them, flew airplanes, and sometimes died in them. The sociological variables to which I have drawn attention are not to be conceived in a way that excludes these practical, experimental activities or diminishes their importance. The sociological processes I have identified do not stand between people and the material environment with which they are engaged. Contrary to the claims repeatedly made by their critics, those who follow the Strong Program do not treat the social world as something to which scientists respond *instead* of responding to the natural world. The cultures, institutions, and interests that I have identified did not block the active involvement with material reality but were the vehicle of that involvement and gave a specific meaning to it.[6]

Reasons and Causes

One of the objections that critics have repeatedly directed against the Strong Program is that the commitment to causal, sociological explanation entails neglecting the role of reasons. The critics say that a Strong Program analysis involves disregarding the reasons that social actors themselves offer for their behavior. According to the critics these reasons can, on occasion, provide a sufficient explanation of the behavior and thus render redundant any attempt to construct a causal, sociological explanation. A recent example of this criticism is to be found in a 2006 paper by Sturm and Gigerenzer.[7] The authors say: "Even after a strong sociological explanation has been given for the beliefs of a scientist, it remains sensible to ask: Very fine, but how are these beliefs connected to the scientist's justificatory reasons? Can these reasons perhaps explain better why the scientist acquired the relevant beliefs?" (144). As the wording indicates, these critics assume that the candidate sociological explanation will have been constructed without any significant reference to the agent's reasons. The critics wish to make good this alleged lack, and they put their money on finding cases where nothing more than the scientist's own reasoning is needed to explain some pattern of scientific belief. The agent's reasons, they say, can be the cause of their beliefs, and a proper explanation of these beliefs should be in terms of these reasons (141).

Does the causal explanation that I have put forward to account for my findings in the history of aerodynamics depend on, or result in, a lack of serious attention to the agent's reasons? I hope it will be evident that such a complaint is groundless. The problem I have posed, the central problem of the book, is a problem about the reasons that were used to justify a certain scientific judgment. I have attended closely to the reasons given by the actors in the story and subjected them to a close analysis. I fear, however, that critics will dismiss my discussion of scientific reasoning as mere window dressing. Thus Sturm and Gigerenzer say, "*Mentioning* that scientists claim to employ certain reasons or reasoning standards, and mentioning which ones these are, is not the same as taking these standards *seriously*—seriously in the sense that they are acknowledged as causes of the scientist's acceptance or rejection of claims" (142). For these critics, nothing short of treating reasons as self-sufficient explanations will count as taking them seriously. This I certainly have not done and will not do because it would corrupt the analysis. Despite this, attending to the actor's reasons has played a central role, even though the analysis culminates in a causal explanation. The actor's reasons are not merely mentioned; they are given a substantial role. Consider the stated reasons offered by the British to justify their rejection of the ideal fluid approach, for example, their complaint that the origin of circulation must forever remain a mystery. These reasons certainly illuminate the British rejection. When the reasons are examined, however, it becomes clear they do not adequately explain the behavior of the British experts. This is not because the British had other "real" reasons. The inadequacy of the explanation is because their German counterparts understood the reasons that moved the British as well as the British did, but responded differently. Kelvin's theorem was as familiar to Prandtl as it was to Taylor or Jeffreys. For me, therefore, taking reasons seriously, and assessing their causal role, requires asking why the same facts and the same theorems, that is, the same reasons, caused such divergent responses in the two groups of professionals.

Faced with a situation of this kind, where stated reasons underdetermine the response, should the historian look for a difference in the intellectual presuppositions behind the reasoning of the two groups? Perhaps this approach could uncover hidden premises and lead to an explanation of the different reactions. An appeal to presuppositions might keep the analysis within the realm of self-sufficient reasons in the way that the critics want. A search for presuppositions is certainly important in any historical analysis, and it has been a feature of my own procedure. As I have shown, such a search uncovers subtly different conceptions of an ideal fluid. The British experts treated an ideal fluid as a fluid of zero viscosity ($\mu = 0$), while their German counterparts treated it as the limit of a fluid of small viscosity ($\mu \to 0$). The British drew a strong boundary between inviscid theory and viscous theory.

German-language work involved a weaker and differently positioned bound-ary. Thus von Mises was inclined to treat the objects of both the Euler and the Stokes equations as abstractions, while Prandtl was inclined to treat them both as realities. Despite their differences, both of these German-language thinkers placed viscous and inviscid fluids on a par with one another. This aligned the two against the more literal-minded realism of the British, who treated viscous fluids as real and inviscid fluids as unreal.

There can be no doubt, then, that identifying presuppositions of this kind deepens the analysis, but it still cannot furnish an explanation of the divergent responses of the British and the Germans. Presuppositions are simply *reasons for reasons*, so the real problem is postponed rather than solved. It merely leads to further questions: Where do the presuppositions come from? Why did the British and Germans have different presuppositions?

The point may be made in another way. I identified a sequence of judgments that informed the technical content of the aerodynamic knowledge of lift. At each point in the sequence the British experts jumped in one direction while the German experts jumped in the other. Such was the case regarding (1) the significance attached to the "arbitrary" value of the circulation, (2) the meaning of the zero-drag result, (3) the importance of explaining the critical angle at which a wing stalled, (4) the reaction to the overoptimistic lift predictions derived from the theory of circulation, and (5) the problem of explaining the origin of the circulation around a wing within the confines of the theory of ideal fluids. The divergence of judgment on these questions was systematic, fundamental, and constitutive of the rival understandings of the two groups. It cannot be dismissed as a coincidence, but nor can it be explained by the divergent reasons themselves. The deployment of reason is the problem, not the solution. The phenomenon calls for a causal explanation, and that is what I have given.

So far I have described my procedure in terms of an apparent transition from reasons to causes. I have said, in effect, that my analysis may have started with reasons but it finished by my making an appeal to causes because reasons became equivocal. I justify the claim that the analysis is causal (and conforms to the Strong Program) by saying that an appeal to causes is, in the end, unavoidable. This argument is correct, but as a way of speaking it can generate problems. At best it is a provisional way to state the methodology behind the analysis.

The problem is that two modes of speech and two perspectives are in play: those of the actors and those of the analyst. Keeping both modes of speech in play may suggest that there are two different sorts of cause at work, namely, rational causes and sociological causes. Are we to conclude that reasons cause *some* of the behavior of the scientists under study but not all of it, so that the remainder has to be explained by sociological and nonrational causes? Does rationality provide a *partial* cause alongside other kinds of cause furnished, say, by the social context? Some such view may seem to be underwritten by the historian's own investigative procedure or, at least, by the way the procedure is sometimes presented. First, it seems, reasons are examined and then, and only then, are sociological causes to be invoked (as if they were a residual category). But granted that the work of the historical *analyst* sometimes exhibits such a pattern, it would be wrong for the analyst to project this expository sequence into the picture of the historical *actor* and imagine that actors are, or may be, subject to a corresponding sequence of influences. I do not believe that a dualism of rational and sociological causes, which allegedly compete or alternate with one another, or supplement one another, can be the basis of a satisfactory perspective. It is eclectic and merely encourages baldly posed questions. Something more unified, and hence reductive, is called for. If the Strong Program is correct, then "rational causes," which have so often been treated as sui generis, are really nothing more than a species of social causes.

   Think of the confrontation between Lanchester and Bairstow when they clashed in public in 1915. Bairstow said that Lanchester could not explain why aircraft stalled, so his theory could not be taken seriously; Lanchester said that he did not need to explain this stalling because a theory of narrow scope could still be valuable within its limited domain. My claim is that, in tracing the arguments that Bairstow and Lanchester used against one another, a good historical analyst will, at the same time, be tracing the causal texture of their interaction. There will be no duality of rational and social causes and no transition from one to the other. A properly historical account of the interaction between Lanchester and Bairstow will be in terms of social causation from the outset. A unified, social-causal perspective of this kind can be sustained if the analyst focuses relentlessly on the credibility that the participants and their audience attach to the arguments that are being advanced. Why did the failure to explain the onset of a stall worry Bairstow in a way that it did not worry Lanchester? Why was Bairstow's concern shared by other British experts but not, to the same degree, by Kutta, Prandtl, and other supporters of the circulatory theory in Germany? These are the questions that will expose the sociological basis of the power of reason, and these are the questions to which I have given answers.

   The importance of credibility as a causal category, with its variable and distributed character, is at its most striking when the overall scene is brought into view, for example, the systematic divergence of the German and British responses to circulation. But actual or possible divergences of this kind are not confined to the large scale. They are a feature of *every* act of reason giving and *every* act of responding to reasons, whether interpersonal or intrapersonal, whether public or private. This is because, on its own, invoking and formulating reasons can *never* be sufficient to render a belief causally intelligible or a course of action causally explicable. Things may not look this way from the *actor's* point of view. Sometimes the reasons that are advances in the course of an interaction are accepted by other actors as sufficient justification or explanation. But I am giving the *analyst's* perspective. I am speaking here from the point of view of a historian or sociologist who is committed to giving a causal analysis of a passage of interaction and behavior. Of course, critics say the claim that reasons are never sufficient is mere dogmatism or the result of an irresponsible generalization. Like others before them, Sturm and Gigerenzer see here nothing but a lack of prudence on the part of supporters of the Strong Program.[8] They think it is more judicious to allow that reasons *sometimes* explain rather than *never* explain. In fact my claim is not dogmatic; it is made on the basis of a general and principled argument. The argument comes from Wittgenstein's analysis of rule following and has explicitly informed the Strong Program from the outset.[9] Because the argument is so important, and so often misunderstood or ignored, I rehearse it here and then connect it to my overall analysis.

The General Argument

In paragraphs 185–99 of *Philosophical Investigations*, Wittgenstein imagined a pupil who is being taught to follow a simple rule, namely, the rule of "add 2." The pupil must try to follow the rule and generate the rule-bound sequence of 2, 4, 6, 8, etc., by adding 2 to the previous member of the sequence. The pupil is taught by a familiar mixture of examples and explanations and is then encouraged to go on to produce further numbers in the sequence. Wittgenstein then imagined the pupil deviating from the expectations of the teacher and the other competent rule followers in the surrounding culture. On reaching 2,000, the pupil does not say 2,002, but 2,004, 2,008, etc. Wittgenstein studied the likely reactions of the competent rule followers and the resources at their disposal as they tried to get the deviant pupil to understand what the rule means and what the rule requires. It rapidly emerges from the analysis that just as the original rule might be misunderstood, so too might any further explanation. Such explanations also depend on a small number of illustrative examples that the pupil must use as a pattern to generate the new instances that are required. Thus, the injunction to go on *in the same way* merely pushes the problem back to that of defining "same." Any attempt to furnish an algorithm for producing the next member of the sequence merely provides rules for following rules and the original problem of furnishing an adequate analysis repeats itself.[10]

Ultimately any attempt to furnish reasons, or to ensure that the pupil's behavior is guided by what the rule requires, depends on the pupil reacting normally and automatically to the training provided by those already acknowledged as competent. There is nothing else available, and all appeals to "reason," "logic," "meaning," "implication" (as well as the concepts of "must" and "have to") come down, in the end, to this. The basis of the process is participation in a shared practice. This applies not just to the pupil now learning the rule but to all who have ever learned it and all who now teach others and identify and correct their errors. The meaning of the rule and the compulsion of the rule depend on shared dispositions to react and shared dispositions to interact with one another in ways that ensure that individual responses stay accountable and aligned. In other words, rules are conventions. To obey a rule, said Wittgenstein, is to follow a "custom" and to participate in an "institution."

Wittgenstein did not mean this statement as a criticism of rule following or the practice of citing rules. The claim is not that rules are unreal or that rule following is a sham. He was not saying that rule followers have no reasons for what they do or, in reality, are acting for other reasons. His point was that the institution of rule following is a social reality, and his aim was to expose that reality to view. The institution is vital for, and constitutive of, cognitive and social order. Citing the rule as an explanation of why pupils act as they do, or why they should act thus and so, is the currency with which one rational, social agent interacts with another rational, social agent. The rules and reasons are actor's categories. Invoking rules is not an idiom for *describing* the interaction in causal terms; it is a means of *acting* causally within the situation. What Wittgenstein's argument shows, and was meant to show, was that the actor's account will not suffice if it is taken out of context and treated as if it were self-sufficient, that is, as an analyst's account. The analyst needs to understand an interaction in terms of causes. This is why Wittgenstein adopted the sociological perspective. In his example he focused attention on the process of socialization. More generally he adopted the explanatory stance by invoking the concepts of convention, custom, and institution.[11]

Wittgenstein's genius lay in identifying a simple example that epitomizes all the central points concerning the relation of reasons and causes. Rule following is the perfect example of rational compulsion, and Wittgenstein's analysis can be related directly to the questions posed by critics of the Strong Program. Here, if anywhere, the case could be made for the operation of rational *rather than* social causes. Doesn't the rule provide a *sufficient* reason for the behavior of a competent rule follower? Can't the rule *better* explain the behavior of the rule follower than any social causes can? It seems to give the critics everything they want. But, as Wittgenstein's analysis shows, the critic's case collapses into a regress of rules for following rules. For the purposes of analysis and explanation, the regress must be stopped and so the phenomenon must be seen as a combination of psychological and sociological *causes*. The stated reasons and the appeal to reasons prove insufficient to explain what happens, and, for the scientific analyst, the actor's account needs to be reconfigured in terms that acknowledge an all-pervasive and self-sufficient sociopsychological causation.

Despite the central role played by Wittgenstein's argument in setting up the Strong Program, its critics insist on misreading the sociological approach as the claim that reason giving is spurious. Such a misreading is question begging. It amounts to making the assumption from the outset that "real" reasons and "good" reasons are not to be analyzed in terms of social causes. The categories of the "rational" and the "social" are set in opposition to one another. The critics cannot believe that rationality can be a sociological phenomenon in anything other than a trivial sense. They therefore counterpose the rational and the social and read the symmetry requirement of the Strong Program as an a priori exclusion of the power and presence of *real* reasons. From the critic's standpoint it then seems easy to refute the sociological approach. All that is needed is an example of an action authentically based on good reasons, for example, a case of rule following or a well-founded scientific inference. This, critics assume, refutes the preposterous generalization put forward by sociologists of knowledge. That the sociologist, following Wittgenstein, is actually challenging the dualistic assumptions upon which the critic's argument rests is never considered.[12]

Although Wittgenstein's example deals with a particular case, the lesson that can be drawn is wholly general. The lesson is that the familiar distinction between "cognitive factors" and "social factors" is wrong. The injunction to "disentangle" these two things is incoherent. The social factor cannot be considered "external" to the cognitive process; it is constitutive, and you cannot "disentangle" that which is constitutive. The cognitive factor in Wittgenstein's example is, of course, the rule itself and the orientation to the rule. But what, when properly analyzed, is "the rule"? Is it something that mysteriously exists "in advance" of the acts of following, like a rail stretching to infinity? No, said Wittgenstein, that is just a mythical picture. We must think in a different way. A rule is something that exists solely in virtue of the social practice of following the rule (just as, in economics, a currency exists solely in virtue of the practice of using the currency). The meaning and implications of the rule only exist through being invoked by the actors to correct, challenge, justify, and explain the rule to one another in the course of their interactions. This is what Wittgenstein meant by calling a rule an "institution." The implication (though these are not Wittgenstein's words) is that the rule, that is, the cognitive factor, *is actually itself a social factor.* Those who appeal to a combination of cognitive factors and social factors, as if they are two, qualitatively different kinds of things, are not being prudent; they are being muddled or metaphysical.[13]

The processes that Wittgenstein brilliantly distilled into his example are the same ones that occurred on a larger scale in my case study. That which is recognizably social, for example, the disciplinary identities, the institutional locations, the cultural traditions, the schools of thought, are not "external" to the reasoning processes that I have studied but are integral to them. They are constitutive of the step-by-step judgments by which the different bodies of knowledge were built up. Experts gave reasons to explain and justify their views and found that sometimes they were accepted and sometimes rejected. Facts and reasons that inclined the members of one group to orient in one direction inclined the members of the other group to orient in a different direction. As one would expect from Wittgenstein's example, these acceptances, rejections, indications, and orientations fell into patterns. The patterns form the customs, conventions, institutions and subcultures described in my story.[14]

Subcultures and Status

One of the subcultures I identify in my explanation (German technical mechanics) belongs to the general field of technology, while the other (British mathematical physics) falls more comfortably under the rubric of science. My explanation therefore presupposes a society in which technological and scientific activity are understood to be different from one another. The picture is of culture with a division of labor in which the roles of technologist and scientist are treated as distinct or distinguishable. These labels are the categories employed by the historical actors themselves. Their role in my analysis derives from their prior status as actors' categories.[15] Although the members of the two subcultures interact with one another, my data justify attributing a significant degree of independence to them. To speak of "subcultures" carries the implication that the practitioners within each respective subculture routinely draw upon the resources of their own traditions as they perform their work and confront new problems.[16] A symmetrical stance requires that both science and technology be placed on a par with one another for the purposes of analysis. This injunction is directed at the analyst and is consistent with the historical actors themselves according a very different status to the two activities: for example, some of the actors may see science as having a higher status than technology. The point of the methodological injunction to be "symmetrical" is that it requires the analyst to ask why status is distributed in this particular way by the members of a group and to keep in mind that it could be distributed differently.[17]

Attributions of status can be expressed in subtle ways. They may take the form of assumptions (made by both actors and analysts) about the dependence of one body of knowledge on another. Is technology to be seen as the (mere) working out of the implications of science? Is the driving force of technological innovation typically, or always, some prior scientific innovation?[18] An inferior status may be indicated by an alleged epistemological dependence and a reluctance to impute agency and spontaneity to technology. The symmetry postulate does not assert the truth or falsity of any specific thesis about dependency or independence, but it does require that such a thesis is not introduced into the analysis as an a priori assumption. At most the dependency of technology on science is merely one possible state of affairs among many other possibilities; for example, science may depend on technology rather than technology on science, or the two may be completely fused together or completely separate. The actual relation is to be established empirically for each episode under study. In the case of the theory of lift it is clear that the technologically important ideas worked out by Lanchester, Kutta, and Prandtl were not the result of new scientific developments. On the contrary, they exploited an old science and old results, namely, ideal fluid theory, the Euler equations of inviscid flow, and the Biot-Savart law. The shock engendered after the Great War by the belated British recognition of the success of this approach was not the shock of the new but the shock of the old.[19] The science that was exploited was not only old; it was also discredited science— discredited, that is, in the eyes of Cambridge mathematical physicists pushing at the research front of viscous and turbulent flow.

The advocates of the circulatory theory of lift brought together the apparently useless results of classical hydrodynamics and the concrete problems posed by the new technology of mechanical flight. The theory of lift in conjunction with the theory of stability constituted the new science of aeronautics. Given the way that scientific knowledge was harnessed to technological concerns, the new discipline might be called a technoscience. Some commentators have argued that "technoscience," the fusing of science and technology, is a recent, indeed a "postmodern," phenomenon, exemplified by the allegedly novel patterns of development shown in information tech-nology and computer science. Others have argued that, because the division of labor between science and technology is a relatively recent development, so their fusion into "technoscience" is, in fact, a return to the original con-dition of science.[20] Did not science, in its early modern form, derive from a fusion of the work of the scholar and the craftsman?[21] Whether or not this account of the origins of science is true, identifying early twentieth-century aerodynamics as an instance of technoscience would support the thesis that technoscience is not a novelty.

In the 1930s Hyman Levy, who had earlier coauthored *Aeronautics in The-ory and Experiment*, wrote a number of books of popular science. Along with Bernal, Blackett, J. B. S. Haldane, Hogben, and Zuckerman, Levy belonged to a remarkable group of scientists who played a significant role in British cultural and political life during the interwar years.[22] One of Levy's books was titled *Modern Science: A Study of Physical Science in the World Today.*[23] Aero-dynamics was one of his main examples. He did not call it a technoscience, but he did offer it as an exemplary case of the unity of theory and practice. Writing from a Marxist standpoint, he cited the work of Prandtl and von Kármán and offered the strange transitions from laminar to turbulent flow as evidence that nature embodied the laws of dialectics. While it is plausible to see the later developments of aerodynamics as moving toward a unification of theory and practice, the fact remains that in the early years there was a dis-cernable difference between the stance of the mathematical physicists and the engineers. The history of Levy's contributions, and his own earlier, negative stance toward the circulation theory, underlines this point. When Levy was active in the field and working at the National Physical Laboratory, there was still a significant difference in approach between mathematical physicists and technologists—at least, between British mathematical physicists and German technologists.[24] It is clear that behind the emerging "synthesis" of theory and practice, there still lay the "thesis" of mathematical physics and the "antith-esis" of engineering. Historical contingency rather than historical necessity determined the balance between them. I now look at one such contingency.

A Counterfactual Committee

If Haldane had not followed the advice of a Trinity mathematician when he set up the Advisory Committee for Aeronautics but had, say, recruited Cam-bridge engineers rather than mathematical physicists, he might have got a very different committee. In principle he could have done this because Cambridge had a distinguished school of engineering.[25] Predictably, there had always been a tension between the demands of a practical engineering education and the demands of the traditional, Cambridge mathematical curriculum. There was not time in the day to succeed at both, except for the outstanding few. It was not until 1906 that a satisfactory accommodation was reached when Bertram Hopkinson, the professor of mechanism and applied mechanics, gave a viable and independent structure to the Mechanical Sciences Tripos. This took the engineers out of the competitive hothouse, though inevitably it meant they operated at a somewhat less sophisticated mathematical level. The products of Hopkinson's department played a distinguished role in the development of British aeronautics. Busk (who made the BE2 stable), Farren (who championed full-scale research at Farnborough), Melvill Jones (who worked on low-speed control and gunnery), Southwell (who worked on airship structures), and McKinnon Wood (the experimentalist who went with Glauert to see Prandtl) were all products of the Cambridge school of engineering. Hopkinson himself, though of an older generation, learned to fly during the Great War and did important work on aircraft testing at Martlesham Heath. He met his death at the controls of a Bristol Fighter in a flying accident in August 1918.[26]

Perhaps a counterfactual Advisory Committee, made up of men like Bertram Hopkinson, would have embraced Lanchester and the circulatory theory. This is consistent with my analysis, although the conclusion would only follow if Cambridge engineers were significantly different in their judgments and orientation from their Mathematical Tripos colleagues and significantly similar to the German engineers from Göttingen and Aachen. Such a premise is plausible but it cannot be taken for granted, and there is some evidence that calls it into question. In certain respects Cambridge engineers adopted attitudes that were similar to those of the more traditionally trained mathematicians. This is not surprising in the case of the older products of the Cambridge school of engineering because they too were steeped in the earlier Mathematical Tripos tradition. Bertram Hopkinson's father, John Hopkinson, had been both an engineer and a senior wrangler. After a fellowship at Trinity he become a professional, consulting engineer in London but retained a strongly mathematical bent. He developed a mathematical analysis of the alternating current generator and predicted that it should be possible to run such generators in parallel. Practical engineers knew that, given the available machines, they could not be run in parallel, but none of this blunted the confidence of the mathematically able Hopkinson.[27] Bertram Hopkinson himself had also read for the Mathematical Tripos and had been a highly placed wrangler. Although more practically inclined than his father, he had worked on topics in hydrodynamics and had written a paper on the theory of discontinuous flow.

He extended the work of A. E. H. Love by making allowance for the presence of sources and vortices.[28] None of those who contributed to the mathematical development of the theory of discontinuous flow ever went on to work on the mathematics of the circulation theory of lift.

What about the somewhat younger generation of Cambridge engineers such as Busk, Farren, Melvill Jones, McKinnon Wood, and Southwell? Here, too, the evidence indicates that, although they had some sympathy with Lanchester, that sympathy had its limits. And those limits were characteristic of the milieu of Cambridge mathematical physics. Southwell's role as a lecturer in mathematics at Cambridge (despite his engineering background), and his commitment to a fundamental physics of lift, have already been described. McKinnon Wood was perhaps different. He had come round quickly to the side of the Lanchester-Prandtl theory of circulation, though exactly why is unclear. It is possible that he was taking his lead from Glauert. Writing retrospectively, McKinnon Wood recalled that, as a student, he had borrowed Lanchester's *Aerodynamics* from the Union library but then, after reading it, had thought no more about it.[29] Looking back, all that he could report was regret at this "mysterious blindness." Farren appears to have accepted a version of Glazebrook's excuse and recalled Lanchester as an almost a tragic figure. He was a man who had clear insights but who could not give them expression. "The vision of Lanchester brooding over the irony of a world which at last understood what he had seen so clearly, but had been unable to explain, will not fade from the minds of those who knew him."[30]

Bennett Melvill Jones sought to shed light on these mysteries in a talk he gave at the Royal Aircraft Establishment at Farnborough in 1957.[31] He described how, as a young man, his friend and fellow engineering student at Cambridge, Edward Busk, had introduced him to Lanchester's work. It had deeply impressed them. After studying Lanchester's treatise, "Ted and I solemnly decided to spend the rest of our lives on aeronautics." But, Melvill Jones went on, aeronautics in Britain had then developed in an almost entirely empirical fashion, that is, without the guidance of Lanchester's theory. This may seem surprising, he told his audience, "because the theory of ideal fluid, which now forms the basis of all your work, had already been very fully developed by Lamb and others. But in this theory, no body which starts from rest can, when in steady motion, experience any reaction whatever. And to us, whose business it was to study this reaction, this did not seem very helpful." The source of the trouble, he explained, was the impossibility of accounting for the origin of circulation. This was the very same objection that the young G. I. Taylor had used against Lanchester in 1914. The origin of the circulation and hence the lift could not be deduced within the terms of the theory. If

Melvill Jones' recollections were correct, then the products of the Mechanical Sciences Tripos had been at one on this question with the products of the Mathematical Tripos. Both saw its dependence on the classical hydrodynamics of potential flow as a fatal objection to Lanchester's approach.

Melvill Jones went on to say that now (that is, by 1957) it may seem obvious how this difficulty should have been resolved. He added, ruefully, that perhaps he and his contemporaries had not been very clever in making the response they did, but, he insisted, "our view was, at that time, shared by all engineers." Here Melvill Jones was making a historical mistake. Even if the negative reaction to ideal fluid theory was shared by all Cambridge aeronautical engineers of his generation, or even by all British engineers, the response was not shared by the German engineers who developed the circulatory theory. In an attempt to justify the negative reaction to Lanchester, Melvill Jones said that the theory of ideal fluids, then known as hydrodynamics, "was regarded purely as an exercise for the amusement of students." Again, while this may have been true in Cambridge it was not how things were seen in Göttingen or Aachen. This is not how ideal fluid theory was regarded by Betz, Blumenthal, Föppl, Fuhrmann, von Kármán, Kutta, von Mises, Prandtl, or Trefftz. For them it was more than a source of ingenious examination questions; it was a serious instrument of research and a serious technique with which to gain some purchase on reality.

The Cambridge ethos seems to have enveloped its engineers as well as its mathematicians. It therefore cannot be assumed that a counterfactual committee of Cambridge engineers would have responded as the German engineers did and welcomed the circulatory theory.[32] This answer obviously depends on the range of factors that are allowed into the imaginary picture. If engineers had been given a dominant position on the ACA, they might have gained greater independence of mind. They might then have been less influenced than they actually were by the older Tripos tradition of mathematical physics. No one can know. The fact remains that these were not the men who initiated the research program into the theory of lift, and they did not have a dominant role on the committee that Haldane actually created.[33]

Education for Decline?

The British economy and the character of British culture provide the backdrop to the episode that I have been investigating. I now want to see how the British resistance to the theory of circulation fits into this broad picture. There are two, starkly opposed theories about the economic fortunes of Britain throughout the nineteenth and twentieth centuries and each carries with it a particular image of British society and British science. How does my story bear upon the dispute between the supporters of these two, opposing theories?

The first theory has been called declinism. The declinists hold that, after its initial lead in the industrial revolution, when the country was led by men who were "hard of mind and hard of will," Britain ceased to be the workshop of the world and, ever since, has been in a steady state of decline both culturally and economically. The main causes of the decline, the argument goes, are to be found in the antiscientific, antitechnological, and antimilitary inclinations of subsequent generations of the British elite. The entrepreneurial spirit was drained out of British society, and innovation gave way to inertia. An important role in causing this sorry state of affairs is attributed to the universities. Universities are said to have cultivated the arts and humanities at the expense of science and technology. Literary and genteel inclinations were encouraged rather than more robust industrial and military values.[34]

The historian Correlli Barnett has formulated an influential version of the declinist theory in his book *The Collapse of British Power*.[35] Barnett places great emphasis on national character. He diagnoses a fatal complacency that infected the British character after victory in the battle of Waterloo. Moral principle rather than self-interest became the dominant motive of political activity. Barnett is clear about the causes. The blame lies with the high-flown sentiments of evangelical religion and the ideology of individualism and free trade. Liberal economic doctrines, he concludes, were "catastrophically inappropriate" for a Britain facing the growing economies of (protectionist) America and Prussia (98). By the 1860s, the misguided British faith in the "practical man," along with the weakness of the educational system, had made the country dependent on its commercial and military rivals for much of its more advanced technology (96). By 1914 Britain, with its overextended empire, may have looked like a world power, but it was "a shambling giant too big for its strength" (90). In *The Audit of War*, Barnett acknowledges that the Cavendish Laboratory made Cambridge a "world centre of excellence" in science but insisted that this was "of little direct benefit to British industry."[36] The lack of industrial and commercial relevance was due to the cultural bias of the universities, particularly Oxford and Cambridge, against "technology and the vocational" (xiii): "Here amid the silent eloquence of grey Gothic walls and green sward, the sons of engineers, merchants and manufacturers were emasculated into gentlemen" (221).

Barnett acknowledges the success of the desperate, last-minute efforts that were made during the Great War to overcome the lethargy and inefficiency of British capitalism. In *The Collapse of British Power* he describes how the

Ministry of Munitions brought about a "wartime industrial revolution" (113) by setting up more than two hundred national factories for the manufacture of ball bearings, aircraft and aircraft engines, explosives, chemicals, gauges, tools, and optical instruments. But the effort was not to last, and the full lesson was not learnt. British economic performance and British power, he asserts, continued their trajectory of decline throughout the 1920s and 1930s. The defects of the British national character similarly bedeviled the country's struggle during World War II and, to Barnett's evident disgust, expressed themselves after 1945 in the electorate's desire to build a "New Jerusalem"—a welfare state.

This pessimistic picture has been widely accepted, but is it true? That it captures something is beyond doubt, but the basis of Barnett's account has been challenged by the advocates of a rival picture. They may be called the antideclinists. Antideclinists acknowledge that there was an economic slowdown after 1870 but insist that if this counts as a "decline," it was a relative, not an absolute, decline. It was almost inevitable as new nations industrialized and began to play a role in world trade. Thus in the early 1870s Britain produced 44 percent of the world's steel, whereas by 1914 it produced 11 percent, but, as the economic historian Sidney Pollard put it, "It is evident that a small island with only limited resources of rather inferior ores could not go on forever producing almost half the world's output of iron or steel; that share had to drop."[37]

Pollard is by no means uncritical of British economic policy, but he takes the view that in 1914 the British economy was riding high. Decline, he suspects, is a political myth: "the statistics are against this argument."[38] With regard to the universities, antideclinists draw attention to the significant role of the British educational system in producing large numbers of scientists and technologists, many of whom have gone into government and industrial service.[39] The declinists also tend to overlook the civic universities that emerged in the late nineteenth century and often worked in close conjunction with local industry.[40]

David Edgerton has presented the antideclinist case in a series of publications that includes *Science, Technology and British Industrial Decline, 1870–1970* and *Warfare State Britain, 1920–1970*.[41] He argues that the declinist view rests on the claim that Britain has not spent enough on research and development (R&D). The declinist premise is that economic growth depends on R&D, but, says Edgerton, this premise is false: economic growth is largely independent of investment in R&D. In any case, British spending on R&D has long been comparable to, or greater than, its competitors. Rather than being in the thrall of an antiscientific elite, Britain has been a scientific powerhouse during the twentieth century. Since 1901 Britain has won roughly the same number of Nobel Prizes as Germany. By 1929 well over half the students at British universities were studying science, technology, or medicine. In *Warfare State Britain* Edgerton notes that despite this, "The image of Germans as both militaristic and strong innovators and users of high technology in warfare is still a standard one in popular accounts" (274). It would be closer to the truth, Edgerton argues, to invert the usual stereotypes and apply this description to the British. British military policy has been to invest in the high technology of its navy and air force—and use them ruthlessly against the economy and civilian population of the enemy.

Edgerton has also marshaled the neglected evidence about the massive involvement of British scientists in governmental and military roles. The declinists ignore the historical, statistical, and economic data that support the view of Britain as a technological and militant nation. For Edgerton the declinist theory is not detached, historical scholarship; it is an ideology that expresses the partisan claims of a disgruntled political lobby. The unending complaint that British universities ignore technology and shun the scientific-military-industrial complex is actually the expression of an insatiable demand for ever more engagement. As Edgerton puts it: "I take the very ubiquity in the post-war years of the claim that Britain was an anti-militarist and anti-technological society . . . as evidence not of the theory put forward, but of the success (and power of) the militaristic and technocratic strands in British culture" (109).

How does my study fit into this debate? Does the British resistance to the circulation theory of lift provide evidence in favor of the declinist, or the antideclinist, view? At first glance it would seem to support declinism. I have identified an elite group of academically trained scientists who turned their backs on a workable, and technologically important, theory in aerodynamics. They left it to German engineers to develop the idea of circulation that had originated in Britain. This lapse seems to confirm the familiar, declinist story about the weaknesses of British culture. Rayleigh's strengths lay in research rather than development, and as one commentator has put it, "Rayleigh's weakness in this direction was an example of the British research and development effort in general."[42] The way Lanchester was sidelined surely epitomizes the resistance to modernization. Putting aeronautics in the hands of a committee of Cambridge worthies and country-house grandees reveals the amateurism so characteristic of British culture.[43]

A closer examination of the episode casts doubt on this declinist reading. First, recall how closely the elite British scientists whom I have studied worked with the government and the military. From 1909 they were given a role close to the center of power, and they embraced it readily. The first place that Haldane went for advice, as the government minister responsible for military affairs, was Trinity College. The University of Cambridge provided the mathematical training of a significant number of Britain's leading aerodynamic experts, including Rayleigh, Glazebrook, Greenhill, Bryan, Taylor, Lamb, Southwell, and of course Glauert, who later championed the circulatory theory. The intimate connection between academia and the military evident in the field of aerodynamics does not, in isolation, refute the declinist picture, but it certainly runs counter to it. It exemplifies the policy of high-technology militarism that has long characterized British culture but that is ignored by declinists.[44]

Second, the initial opposition to the circulatory theory must be put in context. The work of the Advisory Committee covered a broad range of other aerodynamic problems as well as lift. Along with the National Physical Laboratory and the Royal Aircraft Factory at Farnborough, the committee was deeply immersed in the theory of stability and control. We should not forget the words of Major Low, who recommended a comparative perspective. He argued that the British and the Germans both had their strengths and weaknesses, and where one was strong, the other was weak. The British strength was stability.[45] In this area the same group of experts who fell behind in the study of lift excelled and led the world. And they did this by using their Tripos, or Tripos-style, mathematics. In this case they did not call on fluid dynamics but on the equally venerable and equally abstract dynamics of rigid bodies. I have described the central role played in this area by the Cambridge-trained G. H. Bryan, who used the work of E. J. Routh, the great Tripos coach.

The conclusion must be that the same cultural and intellectual resources that had failed in the one case, the theory of lift, succeeded in the other, the theory of stability. British experts may have faltered over the theory of lift because they were mathematical physicists rather than engineers, but it was not because they were effete or antimilitary or significantly antitechnological. Stability is as much a technological problem as is lift. Nor did these experts falter over lift because they were torpid and unresponsive in the face of novelty and innovation. On the contrary, they rejected the circulatory theory of lift precisely because it *lacked* novelty. It was novelty that the British experts wanted, not the old story of perfect fluids in potential motion. They were seeking innovations in fluid dynamics and looking for new ways to address the Stokes equations. The theory of decline therefore points the analyst in precisely the wrong direction to appreciate the true nature of the British response.

The lesson to be learned is that broad-brush, pessimistic macro-explanations will not suffice. They do not fit the facts of the case I have studied.[46] In Barnett's words, British cultural life is identified as "stupid," "lethargic," "unambitious," and "unenterprising."[47] Some of it certainly is, but an unrelieved cultural pessimism can shed no light on the specific profile of success and failure that I have been analyzing. Declinist pessimism does not possess the resources to account for the successes of British aerodynamics that went along with the failures. In the event, it cannot even illuminate the failures themselves. Barnett's contemptuous adjectives do no justice to the clever and ambitious British mathematicians and physicists whose work has been examined. His picture of pacifist tendencies hardly accords with the speed with which young Cambridge mathematicians volunteered for active service in 1914. What is needed, and what I have given, is a micro-sociological explanation that addresses both success and failure in a symmetrical manner. My account of the British resistance to the theory of circulation does not appeal to the vagaries of national character or broad, cultural trends. Like other studies related to the Strong Program, it rests on the technical details of the methods used by clearly identified scientific groups and the character of the institutions that sustained them.[48]

Playing Chess with Nature

How can the Tripos tradition be invoked to explain both failure, in the case of lift, and success, in the case of stability? Should not the same cause have the same effect? The answer is that my causal explanation is not meant to explain success and failure. It explains the preconditions of success and failure, that is, what succeeded or failed, and the response to success and failure. It concerns the path that the Cambridge scientists and their associates chose to follow, or declined to follow, rather than the consequences of the choice and what was found along the path that was selected. Neither the British nor the Germans knew what awaited them. Neither group could select their course of action in the knowledge of what would be successful or unsuccessful. Both had to take a gamble. In the case of the theory of lift, the British gamble failed and the German gamble succeeded.

Lanchester did not approve of the theoretical choices made by what he called the Cambridge school, but he had a clear understanding of the methodological gamble involved. He did not use the language of gambling, but as we have seen in his reflections on the role of the engineer, he chose the more cerebral metaphor of playing chess. The scientist and the engineer were like players in a game of chess who were confronting an opponent called Nature. As a player, Nature was subtle and her moves could not be easily predicted. Lanchester expressed the contingency of the outcome by saying that, at the outset, no one could identify which moves in the game were good ones and which were bad. A sound-looking move might turn out to be a mistake, and an apparent mistake might turn out to be the winning move. Only the unknown future would reveal this.

Lanchester's metaphor can, and should, be taken further. It does not just apply to the opening moves in the game of research or to what is sometimes called "the context of discovery" as distinct from "the context of justification." It applies to the entire course of research and development ranging from the origin of ideas to their acceptance and rejection. The game with Nature does not come to an end when results begin to emerge and when the chosen strategy of research starts to generate successes and failures. The uncertainties Lanchester identified at the outset of the game still inform the responses that have to be made to the feedback from experience. If the scientist or engineer scores a success, the question remains whether it will prove to be of enduring significance or short-lived. If there is a failure, does it indicate the need for a revision of the strategy or merely call for more resolve? These radical contingencies and choices can never be removed, and in one form or another, whether remarked or unremarked, they are present throughout science. Indeed, they are present in every single act of concept application. The game never comes to an end.[49]

Policies and Compromises

In 1954 Philipp Frank published an article in the *Scientific Monthly* called "On the Variety of Reasons for the Acceptance of Scientific Theories."[50] He drew the striking conclusion that "the building of a scientific theory is not essentially different from the building of an airplane" (144). I will use Frank's argument to comment on the theories developed in fluid dynamics and aerodynamics, but first I should say a little about Frank himself.[51] From 1912 to 1938 he was the professor of theoretical physics at the German University of Prague. A pupil of Boltzmann, Klein, and Hilbert, Frank had taken over the chair from Einstein when Einstein received the call to Zurich and then to Berlin. He had attended Einstein's seminars in Prague, and Einstein strongly supported his appointment.

In their student days, before World War I, Frank and von Mises talked philosophy in their favorite Viennese coffeehouse and together played a seminal role in the formation of the Vienna Circle.[52] In the interwar years, as established academics, Frank and von Mises jointly edited a book on the differential and integral equations of mechanics and physics, *Die Differential- und Integralgleichungen der Mechanik und Physik*,[53] which brought together a range of distinguished contributors. Von Mises edited the first volume on mathematical methods, while Frank handled the second, more physically oriented, volume which included chapters by Noether, Oseen, Sommerfeld, Trefftz, and von Kármán, who wrote on ideal fluid theory.[54] The Frank-Mises collection, which was an update of a famous textbook by Riemann and Weber, established itself as a standard work in German-speaking Europe.[55] In 1938 Frank was forced to leave Prague because of the threatening political situation in Europe, and he went to the United States. During and after World War II, he taught physics, mathematics, and the philosophy of science at Harvard.

Like that of von Mises, Frank's philosophical position was self-consciously "positivist" in the priority given to empirical data and the secondary, instrumental role given to theoretical constructs. Frank admired Ernst Mach as a representative of Enlightenment thinking, though his admiration was not uncritical, and he did not go along with Mach's rejection of atomism.[56] Much of Frank's philosophical work was devoted to the analysis of relativity theory, quantum theory, and non-Euclidian geometry.[57] He was a firm believer in the unity of science and rejected the idea that there was a fundamental divide between the natural and human sciences.[58] He also insisted on the need to understand science as a sociological phenomenon. The sociology of science was part of "a general science of human behaviour" (140)—a theme central to the *Scientific Monthly* article.[59]

Frank asserted that most scientists, in their public statements, assume that two, and only two, considerations are relevant when assessing a scientific theory. These are (1) that the theory should explain the relevant facts generated by observation and (2) that it should possess the virtue of mathematical simplicity. Frank then noted that, historically, scientists (or those occupying the role we now identify as "scientist") have often used two further criteria. These are (3) that the theory should be useful for technological purposes and (4) that it should have apparent implications for ethical and political questions. Does the theory encourage or undermine desirable patterns of behavior, either in society at large or in the community of scientists themselves? Such questions are often presented in a disguised form, for example, Is the theory consistent with common sense or received opinion or does it flout them? Common sense and received opinion, Frank argued, typically fuse together a picture of nature and a picture of society. The demand for consistency then becomes a form of social control that can be used for good or ill.

In Frank's opinion it is naïve to believe that theory assessment can be confined to the two, internal-seeming criteria. He offered three reasons. First, he noted that no theory has ever explained *all* of the observed facts that fall under its scope. Some selection always has to be made. Second, there is no unproblematic measure of simplicity. No theory has "perfect" simplicity. Simplicity will be judged differently from different, but equally rational, perspectives, depending on background knowledge, goals, and interests. Third, criteria (1) and (2) are frequently in competition with one another. The greater the number of facts that can be explained, or the greater the accuracy of the explanation, the more complicated the theory must be, while the simpler it is, the fewer are the facts that can be explained. Linear functions are simpler than functions of the second or higher degree, which is why physics is full of laws that express simple proportionality, for example, Hooke's law or Ohm's law. "In all these cases," wrote Frank, "there is no doubt that a nonlinear relationship would describe the facts in a more accurate way, but one tries to get along with a linear law as much as possible" (139–40). What is it to be: convenience or truth? Nothing within the boundaries of science itself, narrowly conceived, will yield the answer. This is why scientists have always moved outside criteria (1) and (2), and, consciously or unconsciously, invoked criteria of types (3) and (4).

These unavoidable choices and compromises tell us something about the status of any theory that is accepted by a group of scientists. "If we consider this point," said Frank, "it is obvious that such a theory cannot be 'the truth'" (144). But if the chosen theory is not "the truth," what is it? Frank's answer was that a theory must be understood to be "an instrument that serves toward some definite purpose" (144). It is an instrument that sometimes helps prediction and sometimes understanding. It can help us construct devices that save time and labor, and it sometimes helps to mediate a subtle form of social control. "A scientific theory is, in a sense, a tool that produces other tools according to a practical scheme" (144), he concluded. Like a tool, its connection to reality is not to be understood in terms of some static relation of depiction but in active and pragmatic terms. Its function is to give its users a grip on reality and to allow them to pursue their projects and satisfy their needs—but it does so in diverse ways. It was at this point that Frank produced his comparison between assessing a theory in science and assessing a piece of technology, such as an airplane. Writing, surely, with the performance graphs of von Mises' *Fluglehre* before his mind, he argued:

> In the same way that we enjoy the beauty and elegance of an airplane, we also enjoy the "elegance" of the theory that makes the construction of the plane possible. In speaking about any actual machine, it is meaningless to ask whether the machine is "true" in the sense of its being "perfect." We can ask only whether it is "good" or sufficiently "perfect" for a certain purpose. If we require speed as our purpose, the "perfect" airplane will differ from one that is "perfect" for the purposes of endurance. The result will be different again if

we chose safety. . . . It is impossible to design an airplane that fulfils all these purposes in a maximal way. (144)

It is the trade-off of one human purpose against another that gave Frank his central theme. Only by confronting this fact can the methods of science be understood scientifically. It is necessary to ask in the case of every scientific theory, as one asks in the case of the airplane, what determined the policy according to which these inescapable compromises are made and how well does the end product embody the policy? We must understand what Frank called, in his scientistic terminology, "the social conditions that produce the conditioned reflexes of the policy-makers" (144).[60]

In Frank's terms, Lanchester's metaphor of playing chess with nature as well as my sociological analysis are ways of describing scientific "policies." Just as there were policy choices made over the relative importance of stability and maneuverability, and policy choices about how to distribute research effort between the theory of stability and the theory of lift, so *within* the pursuit of a theory of lift there were policy decisions to be made. My analysis identifies one policy informing the Cambridge school and another policy guiding the Göttingen school. Again using Frank's terms, the members of the respective schools constructed different technologies of understanding, that is, different theoretical "instruments." Their policies, when constructing their theories, maximized different qualities and furthered different ends. The British wanted to construct a fundamental theory of lift, whereas the Germans aimed at engineering utility. Who were the "policy makers"? One might identify, say, Lord Rayleigh as the "policy maker" in Britain and Felix Klein as the "policy maker" in Germany, but there is no need to assume that policy is made by individuals. Such a restriction would not correspond to Frank's intentions; nor is it part of my analysis. Policies can emerge collectively. They can be tacitly present in the cultural traditions and research strategies of a scientific group. One could then say that *everyone* is a policy maker by virtue of their participation in the group, or one could say that the policy maker is *the group itself*. In my example the "social conditions" that determine the "conditioned reflexes of the policy-makers" reside in the division of labor between physicist and engineer.

One implication of Frank's "policy" metaphor is that a *stated* policy need not correspond to an *actual* policy. The devious history of aircraft construction in post–World War I Germany provides some obvious examples. Is this large aircraft really meant as an airliner or is it a bomber? Is this an aerobatic sports plane or a disguised fighter? Is all this enthusiasm for gliding just recreation or a way of training a future air force—and keeping the nation's aerodynamic experts in a job?[61] The difficulty of distinguishing a real from an apparent policy comes from the problematic relation between words and deeds. Sometimes the self-descriptions and methodological reflections of members of the Cambridge school could sound similar to those of German engineers. Both Lamb and Love occasionally invoked the ideas, and sometimes the name, of Ernst Mach, but that did not make Lamb into a positivist, nor turn Love's work on the theory of elasticity into *technische Mechanik*. Their real policy lay elsewhere.

In an address to the British Association in 1904, Lamb acknowledged that the basic concepts of physics, geometry, and mechanics were "contrivances," "abstractions," and "conventions."[62] But Lamb soon left behind this uncharacteristic indulgence in philosophizing and turned the discussion back to the work of his old teacher, G. G. Stokes. He spoke warmly of "the simple and vigorous faith" that informed Stokes' thinking.[63] Lamb then raised the metaphysical question of what lay beyond science and justified faith in its methods. Why, as Lamb put it, does nature honor our checks? He gave no explicit answer, but the theological hint was obvious. Lamb also distanced himself and the Cambridge school from the "more recent tendencies" in applied mathematics. He deplored the fragmentation of the field and regretted the passing of the large-scale monograph, which was a work of art, in favor of detailed, specialized papers. What differentiated the Cambridge school, he went on, related "not so much to subject-matter and method as to the general mental attitude towards the problems of nature" (425). It is this "general mental attitude" that constitutes the real policy.

How is an authentic "mental attitude" to be filtered out from misleading forms of self-description? The answer is: by looking at what is done and at the choices that are made. Words must be supported by actions. Bairstow, Cowley, Jeffreys, Lamb, Levy, Southwell, and Taylor not only gave their reasons for resisting the ideal fluid approach to lift, but they acted accordingly. This is why, in previous chapters, I have identified the mental attitude that informed the work of the Cambridge school and its associates as a confident, physics-based realism rather than a skeptical positivism. Stokes' equations were not only said to be true, but they were treated as true. This was the attitude and policy that Love expressed by invoking the role of the "natural philosopher" rather than the engineer. And this was why Felix Klein, in his 1900 lecture on the special character of technical mechanics, could express admiration for Love's treatise on elasticity and yet pass over it because it could not be taken as an example of *technische Mechanik*.[64]

Simplicity and the Kutta-Joukowsky Law

I now apply Frank's ideas to the Kutta-Joukowsky law: $L = \rho U \Gamma$, where the lift (L) is equated to the product of the density (ρ), the speed (U), and the circulation (Γ). The law is certainly simple, but what is the meaning of this simplicity? Is it a sign of the "deep" truth of the law and hence a quality that should command a special respect? The idea that nature is "governed" by simple mathematical laws is a familiar one—it goes back to the origins of modern science—but positivists have no time for this sort of talk.[65] Frank could have pointed out that the simplicity, and apparent generality, of the Kutta-Joukowsky law derives not from its truth, but from its falsity and from everything that it leaves out of account. The law says nothing about the relation between the shape of the aerofoil and the amount of lift. It contributes nothing to the problem of specifying the amount of circulation and (when used in conjunction with the Kutta condition) gives predictions for the lift that are consistently too high. The law cannot, in any direct or literal way, represent something deep within reality because its individual terms do not refer to reality. They refer to a nonexistent, ideal fluid under simplified flow conditions.

Frank would predict that if an attempt were made to repair the law, and make allowance for some of the factors that have been ignored, then the result would no longer possess the impressive simplicity of the original. This was precisely what happened when, in 1921, Max Lagally of the *technische Hochschule* at Dresden, produced an extension of the Kutta-Joukowsky formula.[66] Lagally exploited a result arrived at previously by Heinrich Blasius, one of Prandtl's pupils, and this result needs to be explained first in order to make sense of Lagally's formula. Blasius had developed a theorem, based on the theory of complex functions, that allowed the force components X and Y on a body to be written down as soon as the mathematical form of the flow of an ideal fluid over the body had been specified.[67] In these terms, for a uniform, irrotational flow U along the x-axis, with a circulation Γ, the Kutta-Joukowsky theorem takes the form

$$X - iY = i\rho U \Gamma.$$

Here X, the force along the x-axis, represents the drag, while Y is the force along the y-axis and represents the lift. The letter i is a mathematical operator. The right-hand side of the equation economically conveys the information that the drag is zero (because $X = 0$) and also that the lift obeys the Kutta-Joukowsky relation (because $Y = \rho U \Gamma$). Blasius' derivation of this result depended on there being no complications in the flow. Lagally added some complications in order to see what effect they would have. In Lagally's analysis the main flow has a horizontal component U and a vertical component V. More important, he assumed that there were an arbitrary number of sources and an arbitrary number of vortices in the fluid around the body. He specified that there were r sources located at the points a_r where each source had a strength m_r, and s vortices located at points c_s where each vortex had vorticity κ_s. When the formula was adjusted to allow for these conditions, it looked like this:

$$X - iY = -i\rho\kappa(U - iV) + 2\pi\rho\sum m_r(u_r - iv_r + U - iV) - i\rho\sum\kappa_s(u_s - iv_s + U - iV),$$

where u_r and v_r are the components of velocity at a_r (omitting the contribution of m_r) and u_s and v_s are the velocity components at c_s (omitting the contribution of κ_s). The original Kutta-Joukowsky formula can be seen embedded in Lagally's formula on the immediate right of the equality sign.[68]

If the original Kutta-Joukowsky relation could be admired for its elegance, like the sleek lines of a modern aircraft, can this be said of Lagally's formula? I doubt if it attracted much praise on this score. But if the long formula really is an improvement on the short one, why shouldn't it be seen as more beautiful? If we do not find it beautiful is it because we can't imagine such complicated mathematical machinery "governing" reality? Frank and his fellow positivists would not want the question to be pursued in these metaphysical terms. They would say: If there is something important about the simplicity of the original formula $L = \rho U\Gamma$, then look for the *utility* that goes with simplicity. What does it contribute to the economy of thought? This question will expose the real attraction of simplicity and explain what might have been lost, along with what has been gained, by Lagally's generalization.

Frank called a theory *a tool that produces other tools* according to a practical scheme. He meant that the simple law provides a pattern, an exemplar, and a resource that is taken for granted in building up the more complex formula.[69] This is how Lagally built his generalization, and if Frank is right, other scientists and engineers, interested in a different range of special conditions, will follow a similar path. This pattern fits what I have found. Recall the way Betz experimentally studied the deviations between the predictions of the circulatory theory and wind-tunnel observations. He sought to close the gap between theory and experiment by retaining the Kutta-Joukowsky law while relaxing the Kutta condition, that is, the understanding that the circulation is precisely the amount needed to position the rear stagnation point on the sharp trailing edge. Again, recall the later episode in which, prompted by the work of G. I. Taylor, the condition of contour independence was relaxed so that a "circulation" could be specified for a viscous flow. In both these examples the development exploited the same resource as Lagally, that is, the simple law was retained as a basic pattern. Simple laws are a shared resource and an accepted reference point. They are used when a group of scientists are striving to coordinate their behavior in order to construct a shared body of knowledge. They are salient solutions to coordination problems, which may explain the obscure "depth" attributed to them. The depth is a social, not a metaphysical, depth.[70]

"The Phantom of Absolute Cognition"

The continuity between Frank's ideas, developed in the 1930s, and the more recent work in the sociology of scientific knowledge was noted by the philosopher Thomas Uebel in his paper "Logical Empiricism and the Sociology of Knowledge: The Case of Neurath and Frank."[71] Uebel concluded (I think rightly) that Frank had anticipated all the methodological tenets of the Strong Program (147), but he insists that there is an important difference: the advocates of the Strong Program are "relativists," whereas Frank "did not accept the relativism for which the Strong Programme is famous" (149). This statement is incorrect. The similarity does not break down at this point. Frank was also a relativist. I first want to establish this fact and then I shall use Frank's relativism to illuminate some examples of aerodynamic knowledge.

Frank's relativism was implicit, but clearly present, in his paper on the acceptance of theories, for example, in his assertion that there was no such thing as "perfect" simplicity. He meant that there is no *absolute* measure of simplicity that could exist in isolation from the circumstances and perspectives of the persons constructing and using the theory. If there is no absolute measure, then all measures must be relative, that is, relative to the contingencies and interests that structure the situation. Recall also the trade-off between simplicity and predictive power. Frank said this meant there was no such thing as "the truth" because there was no absolute, final, or perfect compromise. The relativist stance is epitomized by Frank's comparison between assessing a theory and assessing an airplane. Talk about an "absolute aircraft" would be nonsense. All the virtues of an aircraft are relative to the aims and circumstances of the user. If the process of scientific thinking has an instrumental character, and theories are technologies of thought, then talk about an absolute theory, or the absolute truth of a theory, is no less nonsensical.

Frank made his relativism explicit in a book called *Relativity: A Richer Truth*.[72] Einstein wrote the introduction, and the book contains a number of examples drawn from Einstein's work, but the book is not primarily about relativity theory. It is a discussion of the general status of scientific knowledge and its relation to broader cultural concerns. Frank's purpose is much clearer in the title of the German edition, *Wahrheit—Relativ oder absolut?* (Truth—relative or absolute?),[73] which poses the central question of the book. Does science have any place within it for absolutist claims? Frank said no. No theory, no formula, no observation report is final, perfect, beyond revision or fully understood. The world will always be too complicated to permit any knowledge claim to be treated as absolutely definitive. In developing this argument Frank draws out the similarities between relativism in the theory of knowledge and relativism in the theory of ethics. Are there any moral principles that must be understood as having an absolute character? The claim is often made, but Frank argues that if close attention is paid to the actual employment of a moral principle, it always transpires that qualifications and complications enter into their use. "For this so-called doctrine of the 'relativity of truth' is nothing more and nothing less than the admission that a complex state of affairs cannot be described in an oversimplified language. This plain fact cannot be denied by any creed. It cannot be altered or weakened by any plea or admonition on behalf of 'absolute truth.' The most ardent advocates of 'absolute truth' avail themselves of the doctrine of the 'relativists' whenever they have to face a real human issue" (52).

The book on relativism was written during the 1940s after Frank had left Prague. It was a response to a systematic attack on science by theological writers in the United States. They blamed science for the ills of the time, such as the rise of fascism, the threat of communism, the decline in religious belief, and the loss of traditional values. The critics said that science encouraged relativism and relativism was inimical to responsible thinking. Frank confronted the attack head on. He did not seek to evade the charge by arguing that scientists were *not* relativists (and therefore not guilty); indeed, he said that scientists *were* relativists (and should be proud of the fact). The danger to rational thought and moral conduct came, he said, not from relativism but from absolutism. If we try to defend either science or society by making absolutist claims, we will merely find ourselves confronted by rival creeds making rival, absolute claims. If we take the issue outside the realm of reason, we must not be surprised if it is settled by the forces of unreason (21). Relativism, he argued, is the only effective weapon against totalitarianism and has long been instrumental in the progress of knowledge. It has been made "a scapegoat for the failures in the fight for democratic values" (20).

Frank alluded to the many caustic things that critics said about relativism and then added, "this crusade has remained mostly on the surface of scientific discourse. In the depths, where the real battle for the progress of knowledge has been fought, this battle has proceeded under the very guidance of the doctrine of the 'relativity of truth.' The battle has not been influenced by the claim of an 'absolute truth,' since the legitimate place of this term in scientific discourse has yet to be found" (20–21). Notice that Frank placed the words "absolute truth" in quotation marks because, as a positivist, he would have been inclined to dismiss the words as meaningless. For him they had no real content and no real place in meaningful discourse. The claims of the absolutists were to be seen as similar to the claims of, say, the theologian. But if the best definition of relativism is simply the denial that there are any absolute truths, and if relativism is essentially the negation of absolutism, then relativism is meaningless as well. The negation of a meaningless pseudoproposition is also a meaningless pseudoproposition. Relativism would, likewise, be revealed as an attempt to say what cannot be said. This may explain why Frank also placed the words "relativity of truth" in quotation marks. There is much to recommend this analysis. It might be called the *Tractatus* view of relativism.[74] Where, however, does this analysis leave Frank's book? Does it not render the book meaningless and pointless? The answer is no. The reason is that absolutism, like theology, has practical consequences, and whatever the status of its propositions, the language is woven into the fabric of life. It provides an idiom in which things are done or not done. Even for the strictest positivist this penumbra of practical action has significance.

What *is* done, or not done, in the name of absolutism? The answer that Frank gave is clear. Absolutism inhibits the honest examination of the real practices of life and science. It is inimical to clear thinking about the human condition. The meaningful task of the relativist is grounded in this sphere. It is to be expressed by combating obscurantism and fantasy and by replacing them with opinions informed by empirical investigation. That is the "richer truth" referred to in the title of the English-language edition of Frank's book. This down-to-earth orientation also provides the answer to another problem that may appear to beset Frank's relativist position. What is scientific knowledge supposed to be *relative to*? The answer is that it is relative to whatever causes determine it. There are as many "relativities" as there are causes. That is the point: knowledge is part of the causal nexus, not something that transcends it. Knowledge is not a supernatural phenomenon, as it would have to be if it were to earn the title of "absolute." Knowledge is a natural phenomenon and must be studied as such by historians, sociologists, and psychologists.

Frank's relativism, and the relativist thrust of the positivist tradition, seems to have been forgotten.[75] A number of prominent philosophers paid a moving tribute to Frank after his death in 1966, but they did not mention his relativism.[76] In the course of this forgetting, a strange transformation has taken place. In his *Kleines Lehrbuch des Positivismus*, von Mises spoke of "the phantom of absolute cognition."[77] That phantom still stalks the intellectual landscape, but in Frank's day it was scientists who were accused of relativism, whereas today it is scientists, or a vocal minority of scientists, who accuse others of relativism. From being the natural home of relativism, science has been polemically transformed into the abode of antirelativism and hence of absolutism. A significant role in this transformation has been played by philosophers of science who are today overwhelmingly, and often aggressively, antirelativist in their stance. The involvement of analytic philosophers should have ensured that the arguments for and against relativism were studied with clarity and precision. This has not happened. The philosophical discussion of relativism is markedly less precise today than when Frank addressed it fifty years ago and provided his simple and cogent formulation of what was at stake.

Objectivity and Reality

Time and again critics attempt to refute relativism by drawing attention to the *objectivity* of what is known in both science and daily life.[78] Such attempts are misguided. The only kind of counterexample that could refute relativism would be an example of absolute knowledge. Proof or evidence of objectivity will not suffice unless the objectivity in question can be shown to be an absolute objectivity. The demand for objectivity is legitimate, but it is meant to preclude subjectivity, and subjectivism is not the same as relativism. The subjective–objective distinction is one thing, the relative–absolute distinction is another thing, and the two should not be conflated. Frank was admirably clear on this point and knew that his defense of relativism was not an attack on objectivity. He (rightly) believed in both the relativity and the objectivity of scientific knowledge.[79]

Rather than explore this theme in an abstract way, let me take an example from aerodynamics. The example, which concerns the rolling up of the vortex sheet behind a wing, is designed to show the objective (that is, nonsubjective) character of knowledge at its most dramatic. The question is: Can the example be understood in relativist terms? Here is the example. In the spring of 1944, at a crucial stage of the Second World War, London was attacked by V-1 flying bombs. The V-1 was a large bomb fitted with small wings and a ram-jet engine, and it flew some 300 miles per hour. The bombs were launched from sites on the French and Dutch coasts by means of a shallow ramp that pointed in the direction of the target. After the bomb had traveled a predetermined distance, its engine was switched off by an onboard device that simultaneously altered the trim of the wings, causing the bomb to fall to the ground and explode. In an effort to stem the attacks, the pilots of the Royal Air Force chased after the bombs and tried to bring them down in open country where they would do less harm. It was not possible to close in on the bomb to shoot it down because of the danger that the resulting explosion would destroy the attacking aircraft. Some pilots therefore developed the technique of flying close to the bomb, making use of the airflow behind the wing of their aircraft to flip the missile on its side so that it would drop to the ground.[80] This technique did not involve direct, metal-on-metal contact with the V-1 but, it has been argued, exploited the rolling up of the vortex sheet behind the wing of the aircraft. It was the rotating air of the vortex that turned the missile on its side. According to an article in the *Annual Review of Fluid Dynamics* in 1998, the rolling up of the trailing vortices behind a wing of high-aspect ratio was, for a long time, considered to be a matter of little practical importance by experts in aerodynamics. The experts acknowledged its existence but not its utility. But, says the author, if the theorists ignored the significance of the roll up, "fighter pilots who used their own vortices to topple V-1 flying bombs had another opinion."[81]

The example shows knowledge and skill tested by uncompromising, external criteria. The pilots' subjective feelings had to be mastered and their judgments subordinated to the objective demands of the situation. What, then, is "relative" about this episode and the knowledge and skills involved? The brief quotation from the *Annual Review of Fluid Dynamics* already indicates the lines on which an answer can be given. First, the relevant knowledge and beliefs were distributed unevenly across the groups mentioned in the *Review* article. The experts who worked theoretically, or who experimented with wind tunnels, had one opinion about vortex sheets; the pilots who chased the bombs over the fields of Kent are attributed with another opinion. Second, the character of the knowledge varied. The experts had a mathematically refined understanding; the pilots had a rough-and-ready but practical sense of what they needed to do. What they lacked in rigor they made up in skill. Third, although the experts and the pilots were oriented to the same features of reality, they did not share a common language or common concepts. The article makes no mention of communication between pilots and aerodynamic experts on the matter, but it would almost certainly have been problematic. It had never been easy for pilots and aerodynamicists to talk to one another.[82] Fourth, the range of circumstances that the members of the two groups took into account differed markedly. The experts operated in a world that was artfully controlled, shielded, and simplified; the pilots functioned in an environment saturated with complexity, interaction, noise, vibration, jolting, turbulence, and distraction.

The conclusion must be that although both groups were actively engaged with a reality that was largely independent of their subjective will, the quality of that engagement was different. In both cases their understanding was objective rather than subjective, but it was also to be seen as relative to their standpoints. In neither case did it have an absolute character. In developing the argument of his book, Frank was therefore right to insist that the doctrine of the relativity of truth "does not imperil by any means the 'objectivity' of truth" (21).

It may be objected that the pilots had *causal* knowledge of reality. Theories may come and go, and verbal accounts may vary, but don't actions and interventions put an agent into *direct* contact with reality? This, it may be said, proves that there is a way of grasping reality that is not merely relative. But does it? The critic who takes this line must confront and answer the question What is supposed to be absolute about the knowledge of causes and the exploitation of this knowledge? The correct answer is that there is nothing absolute about causal knowledge. This conclusion ought to be well known because it was established over two hundred years ago by David Hume in his *Treatise of Human Nature.* Hume gave a relativist analysis whose essential points remain unchallenged to this day. His argument was that all knowledge about causes, for example, that A causes B, whether expressed in words or actions, is inductive knowledge based on experience. Inductive inferences, said Hume, can never be given an absolute justification. Inductive knowledge is irremediably relative.[83]

The limited, fallible, and relative character of practical knowledge can be generalized from the example of the flying bombs to my entire case study. The pattern of flow over a wing described by Lanchester, Kutta, Joukowsky, and Prandtl is not the only one that can render mechanical flight possible. Their picture, which is now called "classical aerodynamics," and whose history I have been describing, rests on the principle that the separation of the flow from the surface of the wing must be minimized. Flow separation, it was assumed, always leads to a breakdown of the lift. It has now been discovered that flow separation can be both exploited and controlled in a way that actually generates lift. Leading-edge vortices, and even shockwaves, can be exploited to create lift.[84] This was not realized until many years after the events I have described. As one authority put it, "We must realise, however, that Prandtl's is but one of many possible bases of wing theory and there can be no doubt that more comprehensive assumptions will eventually be developed for this interesting type of physical flow." [85] Until the late 1950s, all of the technical knowledge in aerodynamics concerning lift had been developed on what can now be seen as a narrow basis. What the future holds is always unknowable, but the more recently acquired, broader perspective serves to expose the hitherto unappreciated relativities of past achievements. But we should not allow ourselves to think that, as these historical relativities are exposed, knowledge progressively sheds its relative character and moves closer to something absolute. To cherish such a picture is to indulge in metaphysics.

The Universality of Science

My example of trailing vortices depended for its force on the *difference* between the understanding of two groups of agents, the scientists and the pilots—the one group believing that trailing vortices had no practical significance, the other group knowing in a tacit and practical way that they did. What happens to the arguments for relativism when all parties know the same thing? This question is important because the universality of scientific understanding is often taken to provide an adequate response to relativism. There is only one real world; the laws of nature are the same in London and Berlin; a true theory applies everywhere, and science knows no bounds of nation or race. "It is transnational and, despite what sociologists claim, independent of cultural milieu."[86] If science is independent of cultural milieu, then it cannot be relative to cultural milieu. Granted the premise, the conclusion follows, but my case study shows that the premise is false. The understanding of the phenomenon of lift was not the same in London and Berlin or Cambridge and Göttingen.

Such a response, based upon mere historical fact, is unlikely to satisfy the critics of relativism. It will be said that my study deals with a passing phase. Isn't the important thing what happened *after* the episode that I described— when the truth emerged? The transnational character of science may take time to reveal itself, and progress may be inhibited by unfavorable social conditions, but universality triumphs in the end. It will be insufficient for the relativist to object that the antirelativist has shifted the discussion from what was the case to what *ought* to be the case or to what *will be* the case. The picture of universal knowledge has force because there is, here and now, much science that is indeed transnational. This fact cannot be denied, so what can the relativist say?

Frank argued that relativism is consistent with universality. He said that the conditions leading to the spread of scientific knowledge were the very ones that ought to encourage a healthy relativism. As experience is broadened, the tendency to treat a belief as absolute will be undermined. Dogmatically held theories will encounter challenges, and a growing appreciation of the complexity of the world will undermine their apparently absolute status. Absolutism is parochialism—the cognitive equivalent of parish-pump patriotism. But the central reason why universalism is no threat to relativism is that the extent of a cultural milieu is purely contingent. In principle, a culture could be worldwide. The universal acceptance of a body of knowledge could only serve as a counterexample to relativism if universality indicates, or requires, absolute truth.

Let me explain this by an example. By the early 1930s Hermann Glauert had become a fellow of the Royal Society and the head of the Aerodynamics Department at Farnborough. He was at the height of his powers and had just finished a lengthy contribution on the theory of the propeller for William Durand's multivolume synthesis of modern aerodynamic knowledge. Then tragedy struck. On August 4, 1934, a Saturday, Glauert took his three children for a walk across Laffan's plain near Farnborough. The party stopped to watch some soldiers who were arranging an explosive charge to blow up a tree stump. The party stood, as required, at a safe distance some two hundred feet away, but the instructions they had been given were based on a misjudgment. Glauert was struck by a piece of debris from the explosion. No one else was hit, and his children were unhurt, but Glauert died instantly.[87] In a dignified letter to Theodore von Kármán, who had now moved from Aachen to the California Institute of Technology, Glauert's wife (Muriel Barker) recalled the last time she and her husband had met von Kármán. Along with G. I. Taylor they had all sat together, in the garden of Taylor's Cambridge house, having tea and making plans for the future.[88] Von Kármán replied, in English: "The few people really interested in theoretical aerodynamics always felt as one family, and I am very proud to say that I had the feeling that your late husband and I were really friends, also beyond the common scientific interests."[89] Von Kármán's metaphor of the "family" to describe the relationship between leading members of the profession is striking. It resonates with, and lends support to, the theme of the transnational or universal character of science, though of course allowance must be made for the circumstances in which the expression was used. Perhaps it is the exchange of letters itself, rather than any particular choice of words, that should be considered the salient point. Former enemies in a bitter war are now consoling one another and affirming their solidarity. This epitomizes the increasingly cosmopolitan nature of the scientist's world, at least as it was emerging, in the interwar years, in the field of aerodynamics.[90]

How is this emergence of a transnational science to be interpreted? There is a methodological choice to be made. In framing a response the choice lies between (1) invoking some form of inner necessity governing scientific progress and (2) settling for mere contingency. On the first approach it will be tacitly supposed that a "natural tendency" or telos is at work guiding the development. This idea will not recommend itself to an empirically minded analyst, who would therefore choose the second approach. Internationalism is to be analyzed strategically, not teleologically. The relevant comparisons are with the globalization of markets or the spread of the arms race.

Any move toward transnational knowledge should be interpreted in a wholly matter-of-fact manner. Sometimes scientists will reach out across national boundaries, and sometimes they will not. It will depend on opportunities and on perceived advantages and disadvantages and will vary with time and circumstance. (Recall Prandtl's ambivalent reaction to cooperation in the immediate postwar years.) There will be no inner necessity at work, and references to the "transnational" character of science should not be accompanied by starry-eyed sentiments about Universal Truth. Why was von Kármán in the United States? What was he doing at Caltech, and who was supporting his research?[91] Each case needs to be examined by the historian for its particular features and causal structure. Thus in my study I found there was a phase when the reports of German work prepared for the Advisory Committee for Aeronautics lay gathering dust, and there was another phase when copies of the *Technische Berichte* were sought with urgency. Both should be counted as equally natural. The universality of science and technology, or the absence of universality, depends on familiar, human realities. Some of these will be the brutal realities of war, power politics, and military and diplomatic strategy. Others will be the softer and more agreeable realities of the kind recalled in the exchange between Glauert's young widow and von Kármán— such as taking tea on a Cambridge lawn. These two levels, so different and yet so intimately connected, need to be brought together and linked to the calculations and experiments carried out at the research front. This is what I have sought to do.[92]

Idealization, Abstraction, and Approximation

The "general mental attitude" (as Lamb would have called it) informing the German work on aerodynamics gradually gained ground in Britain in the interwar years and finally became routine during and after World War II. To convey this attitude I have appealed to the work of some of the main practitioners of *technische Mechanik*, and in this chapter I have supplemented these sources by drawing on Philipp Frank's writings. To drive the point home, I introduce one final, representative thinker to characterize the methods of aerodynamics. The "general mental attitude" of modern aerodynamics is well captured in the work and writing of Dietrich Küchemann, yet another of Prandtl's distinguished pupils.[93]

Küchemann came to Britain in 1945, immediately following World War II, after meeting McKinnon Wood in Göttingen. McKinnon Wood was then working for the Combined Intelligence Objectives Sub-Committee (CIOS). He was on a mission similar, but more formal and grimmer in tone, to the one he had undertaken with Glauert in 1921. He questioned Küchemann about the latter's work on swept wings and the flow over engine ducts and fairings.[94] Küchemann then became one of the many German experts who, under varying pressures, began to work for their former enemies in the postwar years.[95] His personal history thus illustrates the transnational character of science and the contingencies that contribute to it. Küchemann subsequently chose to stay in the United Kingdom and eventually took British citizenship. In 1963 he was elected a fellow of the Royal Society and in 1966 became the head of the Aerodynamics Department at Farnborough. His later trajectory was not unlike Glauert's, but whereas Glauert started from Cambridge, Küchemann started from Göttingen.

There is another parallel. At the beginning of my story I described the lectures on aeronautics that Sir George Greenhill had delivered at Imperial College, London, in 1910. In the early 1970s Küchemann also gave a series of lectures at Imperial. Like Greenhill's lectures, these were also turned into a book, but whereas Greenhill was transmitting the old style of Cambridge mathematical physics, Küchemann was carrying with him the style of Göttingen engineering. Küchemann's book was called *The Aerodynamic Design of Aircraft*.[96] In it he offered the following cautionary words about the character of aerodynamic knowledge:

> the most drastic simplifying assumptions must be made before we can even think about the flow of gases and arrive at equations which are amenable to treatment. Our whole science lives on highly idealised concepts and ingenious abstractions and approximations. We should remember this in all modesty at all times, especially when someone claims to have obtained "the right answer" or "the exact solution." At the same time, we must acknowledge and admire the intuitive art of those scientists to whom we owe the many useful concepts and approximations with which we work. (23)

Even the most elementary and pervasive of all the concepts of fluid mechanics, the "fluid element," fits Küchemann's description. It is a tool of analysis that facilitates a mathematical grasp of real fluids, but, taken in isolation, the concept is no more than a convenient fiction shaped by the demands of the differential calculus.[97] Küchemann's message is clear, and it is in no way idiosyncratic.[98] Aerodynamics demands modesty in the status attributed to it. Claims to possess right answers and exact solutions should be viewed with suspicion because they will soon need to be qualified. While acknowledging the highly mathematical character of aerodynamics, we must accept and embrace both the intuitive and utilitarian character of the enterprise.

Everything that Küchemann says about aerodynamics can be found in the examples I have described. It applies to Kutta's mathematical arc, Joukowsky's theoretical aerofoils with infinitely thin trailing edges, Prandtl's bound vortex and his reduction of the wing to a lifting line, Betz's expedient modification of the Kutta condition and his relaxed attitude to infinite velocities, Schlichting's comments about fluid in the boundary layer for which the laws of motion are suspended in directions normal to the layer, and Glauert's evanescent boundary layers. But Küchemann's description applies equally to the work of Stokes, Rayleigh, Greenhill, Bryan, Lamb, Taylor, and Bairstow, even if some of them, on occasion, might not have embraced this characterization of their methods as readily as those trained in the self-consciously engineering tradition of technical mechanics.

In all of these instances, a detailed examination of the practices that were adopted reveals that the whole science of aerodynamics, both British and German, indeed lived on highly idealized concepts and ingenious abstractions and approximations, and the more successful it was, the more radical the idealizations. If Küchemann was right, these characteristics of aerodynamic knowledge were not merely a passing phase. The men and women I have studied did not resort to these expedients because of the immaturity of the field or because they had, for the moment, to be content with second best. Their manner of thinking and knowing was not one that would be left behind as if, today, aerodynamics rests upon a qualitatively different methodological basis than the one it rested on then. It does not.

Let me support this statement by two examples of recent work. First, consider the current status of the Navier-Stokes equations. To this day they have not been solved, and their properties remain shrouded in obscurity. There are no known, general, "closed-form" solutions to the Navier-Stokes equations. There are approximate numerical solutions but none in the form of the analytical equations that were characteristic of classical hydrodynamics and Prandtl's wing theory.[99] The emergence, since the 1990s, of computational fluid dynamics, which is based on powerful digital computers, has not altered this situation. Computational fluid dynamics has made available numerical solutions to the Navier-Stokes equations and graphic representations of flow patterns over the wings and bodies of aircraft. These are enormously useful, but they have all been produced by programming techniques that embody the forms of idealization, abstraction, and approximation to which Küchemann was referring.[100]

My second example shows that Prandtl's wing theory has retained its capacity to inspire novel work. Although a product of the early twentieth century, it lies at the heart of some twenty-first century developments in aviation technology. These developments exploit some new solutions to Prandtl's old equations.[101] The solutions (which are analytic rather than merely numerical) take into account the possibility of creating a variable twist along the span of the wing. The equations show that, with the right distribution of twist, even wings that have (say) a rectangular planform can be made to generate the theoretical minimum amount of induced drag. Previously this minimum had been associated exclusively with an (untwisted) elliptical planform. The necessary twist required for minimum drag depends on gross weight, altitude, speed, and acceleration, but using modern technology, a wing could be built that automatically adjusted itself to these changing conditions. The result, over the course of a long flight, would be significant drag reduction and cost saving. These two pieces of evidence show that Küchemann was not describing a passing phase in the emergence of aerodynamics but the enduring character of knowledge in this field and, perhaps, the unavoidable conditions of all practical cognition.

I have already argued, by reference to Frank, that if the concept of "relativism" is to be used with precision, it can only mean one thing: a denial that there are any absolute truths. This is a necessary and sufficient condition for an account of knowledge to be identified as a form of relativism. A relativist can be comfortable with knowledge that is conjectural, inconsistent, expedient, and partial, that is, with everything that science and technology really is. It is the nonrelativists for whom these facts about science are a cause of trouble. They can acknowledge them, but only as stations on the road to an absolute end point. But the familiar pragmatics of scientific work cannot belong to the realm of the absolute. It would be perverse to use the words "absolute truth" as a label for theories that are approximate or that get this bit right and that bit wrong or which depend on useful fictions and abstractions. It should be clear that Küchemann is denying that practitioners in aerodynamics can ever make any claim to absolute certainty, the absolute status of their concepts, or absolute finality in their knowledge.[102] It follows, given Küchemann's analysis, that aerodynamic knowledge *must* be understood in a relativist manner, namely, relative to all the contingencies of the "intuitive art" that enters into idealization, abstraction, approximation, and inductive inference.[103]

Relativism and Hypocrisy

The conclusion that an understanding of aerodynamics supports, rather than undermines, a relativist analysis of knowledge will sound strange to ears accustomed to the incessant academic rhetoric directed against relativism.[104] Antirelativists confidently use aerodynamics in their attacks on what they see as the debilitating evil of relativism. Airplanes have become a veritable symbol of the absurdity of the relativist position. Consider the following challenge issued by the well-known scientist Richard Dawkins.[105] Dawkins who, after his scientific career, went on to occupy a chair in the Public Understanding of Science at Oxford, has said, "Show me a cultural relativist at thirty thousand feet and I'll show you a hypocrite. Airplanes built according to scientific principles work. They stay aloft, and they get you to a chosen destination. Airplanes built to tribal or mythological specifications, such as the dummy planes of the cargo cults in jungle clearings or the beeswaxed wings of Icarus, don't" (32). Dawkins calls this a "knock-down argument," and I suspect that most philosophers agree with him.[106] In reality it is nothing of the kind. Who are the targets to be knocked down? They are identified as "cultural relativists" and then defined as people who believe that "science has no more claim to truth than tribal myth" (31). There may be persons who cannot distinguish science from myth, but no relativist is committed to such a position simply by virtue of being a relativist. Dawkins has simply chosen an easy target, namely, a foolish version of relativism, and omits to mention that a foolish absolutist might also believe that science has no more truth than a tribal myth. But to reject foolish versions of relativism is not to refute relativism.

Dawkins acknowledges that there are "sensible" people "who, confusingly, also call themselves cultural relativists." The reader is told that these "sensible" relativists believe that "you cannot understand a culture if you try to interpret its beliefs in terms of your own culture" (32). This is indeed a reasonable position, but notice that, on this definition, the sensible relativist stance puts no critical pressure on Dawkins' position. "Sensible" relativism, so defined, can be conceded by an absolutist (as it is conceded by Dawkins) without any inconvenience. An absolutist can allow that *some things* are relative as long as *not everything* is relative. Thus the claim that the *meanings* of concepts are relative to culture can be accepted because (the antirelativist will say) it is the truth-status of the *propositions* conveyed by those meanings that really counts, and this, it will be claimed, is not relative. The real argument is therefore not about meaning, it is about truth—as Frank saw clearly.

Dawkins says that airplanes built according to scientific principles work and stay aloft, while those built according to mythological principles don't. This assertion makes no allowance for the fact that flying machines were mostly developed on a trial-and-error basis by practical men whose stance was often unscientific. Nor does Dawkins allow for cases like the Davis wing. The Davis wing was used on thousands of Consolidated B-24 bombers during World War II. The aerofoil was produced by the inventor David R. Davis according to a procedure he kept secret. The wing section went into production because, to everyone's surprise, it outperformed rivals when tested in the wind tunnel at the California Institute of Technology—von Kármán's home base. When, later, the secret of its design was revealed, it turned out to have no intelligible relation to the laws of fluid dynamics. The procedure was pseudoscientific hocus-pocus.[107]

Nor does Dawkins say *which* scientific principles are supposed to be playing the star role in his version of the history of aviation. The discontinuity theory of lift was based on scientific principles and, at one stage, was supported by no less a figure than Rayleigh, but that wasn't much help. Einstein may have regretted his involvement in aviation, but he was deploying the same formidable scientific intellect that had proven so successful in other areas. On the one side, then, there is the Davis wing, which was unscientific but worked, while on the other side there is the Einstein wing, which was scientific but didn't work. The procedures of science are neither necessary nor sufficient for success.

Dawkins makes mock of what he calls "tribal" science and paints a picture of pathetic, nonflying, cargo-cult replicas of aircraft to drive home the point. He assumes that "tribes" do not have real science, and real science does not have "tribes." The historical episode that I have studied could be read as a counterexample to this questionable assumption. In a nontechnical sense of "tribe," I have examined the different practices and rituals of two, scientific tribes. One of these tribes lived on the banks of the river Cam and was called the Cambridge school; the other lived on the banks of the river Leine and was called the Göttingen school.[108]

At no point does Dawkins grasp the nettle. Relativists deny that humans are in possession of any absolute truths. If Dawkins rejects relativism and uses aerodynamics as his leading example, he must think aerodynamics is a case of absolute truth. Can he really think this? If not, he had better find another example or become a relativist. There is no middle way—other than obscurantism and evasion.[109] Where does this leave Dawkins' challenge? *Show me a cultural relativist at thirty thousand feet and I'll show you a hypocrite.* Dawkins is committed to a proposition of the form "All As are Bs," hence "Show me an A and I'll show you a B." A necessary and sufficient condition for refuting a claim of this form is to produce an A that is not a B. A sufficient reply would therefore be to introduce Dawkins to Dietrich Küchemann. Here is someone who knows all about the reality of flying and yet is a relativist about the very science that it involves. Of course, Küchemann would not normally be called a "cultural relativist," but I have explained why he must be counted as a bold and unequivocal relativist—and that is what the argument is all about.

The authority behind Küchemann's observations about the methods of aerodynamics will be evident. As an aerodynamicist he had the reputation of being one of the best of his generation. His work at Farnborough was devoted to the aerodynamics of transonic and supersonic flight. Would Dawkins really dare to impute intellectual hypocrisy to the man who discovered the novel aerodynamic principles embodied in the remarkable wing of the supersonic Concorde?[110] Of course, while Concorde was an aerodynamic triumph, everyone knows that it was also an economic disaster.[111] This makes it a resonant symbol for many things—but the weakness of relativism certainly isn't one of them. The rise and fall of the Concorde project demands a relativist analysis. It was cases of this kind that Frank had in mind when he said that strengths and weaknesses trade off against one another. This was why he cited the design of airplanes to remind his readers that even the best piece of technology cannot simultaneously meet all human demands at once—and why he then used an airplane as a metaphor for the relativity of scientific knowledge in general.

"The Whimsical Condition of Mankind"

In my study of the difference between the British and German responses to the circulatory theory of lift I have followed out the implications of Frank's comparison. I have tried to dig into what he called "the depths," where, as he rightly said, "the real battle for the progress of knowledge has been fought."[112] I have engaged with the details of the scientific and technical argumentation over the theory of lift because it is here, in these details, that both the social character of knowledge and the consequent relativity of knowledge find their most revealing expression. The story reminds us of the sheer contingency and unpredictability of the outcome of any research enterprise and shows how complex and fine-grained that contingency can be. It shows the vital and ineradicable role played by cultural traditions and the institutions that transmit these traditions. And, as Frank predicted, nowhere in the analysis of scientific discourse was a legitimate place found for the term "absolute truth."

This insight is in constant danger of being forgotten or obscured by the false friends of science. The relativity of all scientific concepts to culture and society is deemed unacceptable by the self-appointed guardians of knowledge who claim to "take reflective responsibility, as it were, for the normativity of our most fundamental cognitive categories."[113] Historians and sociologists, like experimental psychologists and anthropologists, have always known that it is not normative posturing but close and careful *empirical* studies of cognition that are needed. Strange though it may seem, this principle needs special emphasis when the cognition in question is that of scientists and engineers. The practices of scientists and engineers must be studied in a hard, factual light as natural phenomena that belong to the material world of cause and effect. Only studies conducted in this spirit can carry the analysis beyond ideology and propaganda and lay the basis for a proper, public understanding of science and technology.[114]

Detailed empirical studies always need a methodological context, and attention must always be given to the broader framework in which they are understood. Thus one may legitimately ask where all the intellectually brilliant activity of the men and women I have studied is to be positioned in relation to the grand philosophical categories of Progress, Reality, and Truth. Now that the facts of the case study are at hand, the answers are not difficult to supply, though their implications may be disturbing.

Let me take each category in turn. That the work of the German engineers constituted technical progress is beyond doubt, and it is this which eventually had its impact on the British. The German work had utility and practicality relative to goals and interests shared by the experts of the two nations. The British led the way in the study of stability, but when it came to the study of lift and drag, failure and frustration took its toll on the British experts. The practical rewards and opportunities offered by the German approach eventually tempted even the strictest to compromise their principled commitment to theories with a firm basis in physical reality. The theory of circulation and Prandtl's theory of the finite wing allowed the experts to do things that they wanted to do, and that fact alone was, in its own way, rewarding. If the attraction of moving forward could not rationally compel a change of mind on important theoretical questions, it encouraged a pragmatic accommodation.

That everyone in the field of aerodynamics, British and German, was, each in his or her own way, grappling with reality is also evident, and this was wholly taken for granted in all of the reasoning of the actors I have described. Although one may question the extent of Sir George Greenhill's connection

with reality in his notorious Reports and Memoranda No. 19, on discontinuous flow and free streamlines, this work did not set the pattern for the future
reports of the Advisory Committee. As a group the British were no less concerned than their German counterparts with understanding the real performance of real airplanes under real conditions of use. And wasn't Sir George
himself acting as the spokesman of the practical realists when he (correctly)
took G. H. Bryan to task for neglecting the gyroscopic effect of the engine and
propeller in his analysis of stability?

Engagement with reality may be common ground, but my example shows
that there are different ways of grappling with reality. It also shows that these
different modes of engagement are social modes belonging to, and sustained
by, different groups with different local traditions. Such differences can divide groups that otherwise share much by way of a common culture, as did
the British and German experts in aerodynamics. Even more important, the
example shows that there are no independent methodological principles by
which these different forms of engagement could be reliably and usefully assessed. Such principles as emerged in the episode were themselves integral to
the forms of engagement they were used to justify. They were rationalizations
of existing practices and institutions. That there are different ways of engaging with the world may seem obvious; that the only ground available to the
actors for justifying their choices is question begging is perhaps less obvious.
But obvious or not, it follows directly from the fundamentally social character of cognition. This is one reason, though not the only reason, why the so-
ciologically minded David Hume was right when he said that all the sciences
have a relation to human nature and that "however wide any of them may
seem to run from it, they still return back by one passage or another."[115]

What, finally, is to be said about truth? The progress in aerodynamics
made in the *technische Hochschulen* and the University of Göttingen derived
from the use of a theory of perfect fluids in potential motion. The theory
dealt with an idealization and a simplification. This theory was dismissed in
Cambridge and London as physically false and logically self-defeating. It was
false because it denied the viscosity of the air and self-defeating because circulation was unchanging with respect to time, and its origin was beyond the
reach of the theory. The premises of the British objections were true and the
reasoning based on them was sound, but the conclusions led to failure rather
than success.

The German advances in the understanding of lift and the properties of
wings depended on the use of abstract and unreal concepts that were sometimes employed with questionable logic. Progress in aerodynamics thus depended on the triumph of falsity over truth. Everyone knows that false prem-

ises can sometimes lead to true conclusions and that evidence can sometimes support false theories, but the story of the aerofoil involved more than this. The successful strategy involved the deliberate use of known falsehoods poised in artful balance with accepted truths. The supporters of the theory of circulation showed how simple falsehoods could yield dependable conclusions when dealing with a complex and otherwise intractable reality. This is the real enigma of the aerofoil.

The enigma would hold no surprises for Hume. It would simply be an expression of what he called "the whimsical condition of mankind."[116] The lesson Hume learned from the study of history and society was that "the ultimate springs and principles" of the natural world will never be accessible to the human mind. The utmost that reason can achieve is the simplification of complexity. Humans live and operate in a world of limited experience dominated, necessarily but beneficially, by custom, convention, habit, and utility. Hume acknowledged that "the philosophical truth of any proposition by no means depends on its tendency to promote society" but argued that we should be neither surprised nor unduly alarmed if truths (or supposed truths) that lack utility sometimes "yield to errors which are salutary and *advantageous*" (279). The story I have told deals with a technology that may seem remote from the world whose problems exercised Hume, but the central fact to emerge in my story, the fact I have called the real enigma of the aerofoil, can be understood in the humane, skeptical, and sophisticated terms he offered. Among the British it was an accepted truth that the air was a viscous fluid governed by Stokes' equations. In the field of aeronautics that truth, if truth it be, yielded to the erroneous but salutary and advantageous picture of the air as an inviscid fluid governed by Euler's equations.

Are there general lessons to be learned? Not if aerodynamics is a special case, but I do not think it is a special case. The conclusions reached in this case study surely can be generalized.[117] What, then, should be concluded? Individual developments in the sciences will differ in their details, but what Frank had to tell us about the compromises involved in the design of airplanes applies (and was meant to apply) to the technology and instruments of *all* thinking. There are always compromises to be made. The warning given by von Mises against the phantom of absolute cognition will always be relevant. And there will always be a role in science and engineering for the blunt advocacy of a Major Low and the rapier responses of a G. I. Taylor. Above all, what Küchemann had to say about the idealizations of aerodynamics captured the essence of the creative work of Lanchester, Prandtl, and Glauert. But idealizations are salutary and advantageous falsehoods which play a vital role in all science, pure as well as applied. In stressing the role of idealization,

Küchemann may have identified a feature of cognition that is more salient in engineering than in physics—but it is the engine of progress in all fields. Those who point to the airplane as a symbol of the truth of science, the power of technology, and the reality of knowledge are therefore right—but do they know what they are saying? The enigma of the aerofoil is the enigma of all knowledge.

Notes

Acknowledgments

1. Galison and Rowland 2000.

2. Hashimoto 2000; see also Hashimoto 2007.

3. Hashimoto 1990.

4. See, for example, Giacomelli and Pistolesi 1934, 331–33, 348; Kármán 1963, 25–27; Hanle 1982, 55–56; Anderson 1997, 100–108.

5. Most of Hashimoto's discussion concerns a range of other problem areas in aeronautics, namely, (1) the theory of stability, (2) the theory of the propeller, (3) the scale-effect controversy, (4) the use of graphical methods by engineers, (5) the international comparisons that were carried out after World War I on different wind tunnels, and (6) wind-tunnel correction factors. I briefly discuss topic 1 because I use the success of stability research to provide a contrast with the frustrations encountered in the study of lift.

Introduction

1. Hume 1960, xix (first published 1739–40).

2. The question of who has the relevant expertise to analyze and comment on science and technology raises difficult problems. Insiders will know things that outsiders don't, while outsiders may notice things insiders pass over. There are no easy answers. A thought-provoking discussion is found in the methodological reflections at the end of Collins 2004, chaps. 41 and 42.

3. The early pioneers would actually have spoken of "drift" rather than "drag." The latter term became common currency around the time of World War I and is now standard.

4. See, for example, Gibbs-Smith 1960; Dallas Brett 1988; Crouch 1989; Jakab 1990.

5. Hanle 1982; Rotta 1990; Vincenti 1990; Anderson 1997; Galison and Roland 2000; Ackroyd, Axcell, and Ruban 2001; Darrigol 2005; Eckert 2006. Three older works should also be mentioned: Giacomelli and Pistolesi 1934; Kármán 1963; Tokaty 1971.

6. Durkheim 1956.

7. The philosopher Dudley Shapere tells sociologists and historians to "disentangle what is involved in scientific reasoning from the social influences." Shapere, as quoted in Callebaut 1993, 453; see also Shapere 1986. These citations are merely illustrative, and numerous other sources could be named. I come back to these questions in the final chapter.

8. Shapin and Schaffer 1985. This pioneering study exemplifies the thesis that "solutions to the problem of knowledge are embedded within practical solutions to the problem of social order" (15). Some historians still follow the philosophers in their assumption that the social and the cognitive can be "disentangled" (e.g., Brush 1999, 186). I set out my objections in chapter 10.

9. I confine myself to the British and German debates. Not only does this limit keep my story within reasonable bounds, but I have found no evidence that the parallel discussions that must have been taking place in, say, France or Italy impinged on the British–German debate. It is therefore reasonable to isolate the British–German relation for the purposes of analysis. Of course, the assumption that there were parallel debates elsewhere in Europe should be put to the test by further international comparisons. Something should also be said about the notable absence of an American dimension to my story. This absence derives from the historical facts themselves. Today, given the United States' dominance in aviation, it can be difficult to believe that there was ever a time when it was Europe, and not America, that was the driving force in the field of aviation and the theoretical and experimental sciences associated with it. Students of the history of U.S. aviation agree that before 1926 there was little fundamental work done on the theory of lift. There were excellent programs of empirical work, for example, Durand's work on propellers at Stanford. On this see Vincenti 1990, chap. 5. Nevertheless, the United States did not contribute significantly to basic aerodynamic theory. The American side of the story thus begins as mine ends. The theoretical "lag" in the United States is identified in Hanle 1982 and Roland 1985.

10. Kingsford 1960. Although Kingsford accepts Glazebrook's explanation of Lanchester's neglect—that it was a matter of Lanchester's inadequate presentation and the resulting breakdown of communication (1960, 84 and 123)—the book retains its value as a record of the facts of Lanchester's rich life. The book is useful on his years in the automobile industry and conveys the extraordinary scope of Lanchester's interests and activities, ranging from aeronautics to poetry. Kingsford does not, however, go into Lanchester's theory of lift in any technical detail. The theory is given a name but no substance. The result is that when, for example, he briefly touches on Lanchester's confrontation with one of the leading experts at the National Physical Laboratory in 1915, the reader has no idea what is at stake. The one sentence quoted from Lanchester in this context (142) is rendered largely unintelligible. I analyze the content and the background of this important confrontation in chapter 4.

11. Minute Book of the Advisory Committee for Aeronautics, PRO: DSIR (Department of Scientific and Industrial Research) 22/1. The abbreviations used to designate archival material are given at the beginning of the reference list.

12. The word "aerofoil" refers to the characteristic cross section of a wing with its rounded leading edge, cambered upper surface, and thin trailing edge. Modern American usage replaces "aerofoil" with "airfoil," but this change only took place around 1922. Until that date both British and American experts used "aerofoil" in their technical reports, and that remains British usage today. I thank John D. Anderson for information about the timing of this change in the American technical literature (pers. comm.). On historical grounds I retain "aerofoil," but, apart from that special case (and of course quoted material), American spelling is used throughout the book—thus "airplane" not "aeroplane," and "formulas" not "formulae."

13. The classic analysis of the interdependence of theory and experiment is Duhem 1954.

14. I have also set aside the question of the engine that powers an aircraft. The building of a reliable, lightweight source of power was central to the historical development of aviation but falls outside the scope of my discussion. There is an extensive literature devoted to aircraft en-

gines. For an introduction to this theme and for an account of the evolution of modern means of aircraft propulsion, see Constant 1980.

15. The term "relativism" is properly defined in chapter 10.

16. Warwick 2003.

17. Previous studies, such as Goldberg 1984 and Brush 1999, had tended to invoke the hold of the mathematical training in Cambridge in a negative way. It was used to explain the alleged shortcomings of Cambridge science, such as the apparently dogmatic commitment to the theory of the ether and the lack of enthusiasm for Einstein's work. Warwick rejects this one-sided appeal to local cultural traditions as an explanatory resource in the history of science (that is, merely as an explanation of error) and adopts a stance that can be described as "symmetrical" in the sense indicated earlier. See also Warwick 1989.

Chapter One: Mathematicians versus Practical Men

1. Bryan 1916, 510.

2. Wright 1912, 374. Howard Theophilus Wright (1867–1944) was an aircraft builder and designer.

3. An editorial in *Flight*, the journal of the Royal Aero Club, declared that the ACA represented "the nation's brain in aeronautics" (*Flight* 1912b, 958).

4. On the relation of the American NACA to the British ACA, see Roland 1985, 1:3–4, 11, 21–22, 28.

5. For an overview of the role of aviation in British culture, see Edgerton 1991a. Edgerton challenges many of the historical and cultural stereotypes about aviation.

6. For a perceptive and readable description of the social trends and tensions of the Edwardian period, see Hattersley 2006.

7. Gollin 1984, 1989. On the broader literary background and history of British invasion anxieties, see Clarke 1966.

8. Gollin 1989, chap. 3. See also *Aeroplane* 1912b.

9. See Haldane and Kemp's translation of Schopenhauer's *The World as Will and Idea* (Schopenhauer 1883–86.) Haldane gave the Gifford Lectures in 1902–3 and 1903–4, which were published as *The Pathway to Reality* (Haldane 1903–4). Haldane's best-selling philosophical work was *The Reign of Relativity* (1921). He wrote an autobiography, which was published in 1929, after his death. A sympathetic, if somewhat uncritical, appraisal of Haldane is to be found in Sommer 1960.

10. On the army reforms, see Spiers 1980. On Haldane's considerable contribution to the organization of the British university system, see Ashby and Anderson 1974.

11. For a thoroughly hostile portrayal of Haldane, in which he is presented as an insufferable snob devoid of any redeeming qualities, see Driver 1997, chap. 5.

12. F*light* 1901b, 257.

13. See Glazebrook 1902; Mosely 1978. Mosely brings out the important role played by Rayleigh and Glazebrook in setting up the National Physical Laboratory (NPL). On the German model for the NPL, see Pfetsch 1970; Cahan 1982, 1985.

14. "Memorandum of Events between 1906–1915," RBH-NLS: MS 1609 (ii), p. 394. A variant of this passage is in Haldane's autobiography (1929, 234). The reference to the most "stable results" is probably coincidental, but it could be an unwitting echo of the ACA's overriding concern with aerodynamic stability.

15. Hill 1920, 532. Sq. Ldr. Roderic Hill was in command of experimental flying at Farnborough from 1917 to 1923. On Hill's exploits, see Babbington Smith 1961.

16. Jex and Culick 1985.

17. This view was also expressed in the pages of *Nature* and attributed to French aviators, who "while expressing great admiration for the Wright performances are of opinion that the successful balancing of the Wright machine is mainly a feat of skill on the part of the aviator, and that their object has been to construct machines with which anyone can fly" (Bryan 1908, 670).

18. Gollin 1984, 269–77 and 303.

19. Lanchester dismissed Dunne's speculations as "metaphysical nonsense." Letter, December 6, 1935, FWL-RAeS, folder 11a.

20. Dunne persevered, and his machine proved capable of flight. For a diagram of the Dunne machine, see *Aeroplane* 5 (Oct. 30, 1913): 487. For Dunne's own account of his machine and the ideas behind it, see Dunne 1913.

21. *Aeronautics* 1919, 209.

22. "It is rumoured thereabouts to this day that a broken walking-stick is all that remains in existence of a concealed telescopic camera and of the foreign spy who carried it" (*Flight* 1910, 709).

23. For a detailed account of the secret trials, with maps and photographs of the location as it exists today, see Walker 1974, esp. 206–7.

24. "Report and Proceedings of a Sub-Committee of the Committee of Imperial Defence," January 28, 1909, PRO: CAB 16/7.

25. C. S. Rolls was the financial side of the Rolls-Royce partnership. Henry Royce provided the engineering skill. Rolls flew a Wright machine and became the first man to fly the English Channel in both directions without alighting. He was killed in a flying accident in 1910 (Dallas Brett 1988, 47–50). In the 1890s Hiram Maxim experimented with a huge, unmanned, and tethered steam-powered machine that was able to leave the ground but destroyed itself in the process. Gibbs-Smith calls it a "test rig" rather than an "aircraft" (Gibbs-Smith 1960, 25–26). Plate V(d) in that volume shows Maxim's machine.

26. There are detailed discussions of the work of the Esher committee in Walker 1974, chap. 12; Gollin 1984, chap. 13; and Driver 1997, 207–13.

27. On the RAF, see Hare 1990. With the advent of the Royal Air Force, the Royal Aircraft Factory was renamed the Royal Aircraft Establishment (RAE) to avoid confusion over their shared initials. For an authoritative picture of the RAE at the height of its powers in the 1960s, see Lighthill 1965.

28. Despite its name, the Factory never manufactured aircraft; it built and tested prototypes.

29. Glazebrook 1920. See also Glazebrook 1931. For Glazebrook's life and career, see Rayleigh and Selby 1936–38, 29–56. (This Rayleigh is the son of the Rayleigh who presided over the ACA.)

30. Schuster 1921; Strutt 1924, esp. 291–93.

31. Rayleigh 1883, 1891, 1900, 1902.

32. Pritchard 1958. Further biographical details and memories of O'Gorman are provided in this issue of the *Journal of the Royal Aeronautical Society* by F. M. Green, G. de Havilland, G. I. Taylor, B. Melvill Jones, and W. Farren (62:470–75).

33. On Greenhill, see H.F.B. 1928; on Shaw, see Gold 1945–48. Horace Darwin had been a student at Trinity, graduating in 1874. He knew both Rayleigh and Glazebrook and in 1881 col-

laborated with his brother George Darwin at the Cavendish in a study of the lunar disturbances to the force of gravity. In 1885 he founded the Cambridge Scientific Instrument Company, which supplied both the Cavendish and the NPL. William Napier Shaw had been one of the four original directors of the company but had relinquished the post in 1901. See Glazebrook 1937, and Cattermole and Wolfe 1987. Horace Darwin (1913) gave the first Wilbur Wright Memorial Lecture. On Mallock, see Boys 1935; on Petavel, see Robertson 1936; on Lanchester, see Ricardo 1948. In a letter to the *Times* of December 8, 1938, F. J. Selby said that Glazebrook "was largely responsible for the selection of the members of the committee" (RTG-TCC, M. 1872). J. L. Nayler gives a slightly different account, saying, "it is reputed that these two, and Shaw rather than Glazebrook, were mainly responsible for the action that was taken to appoint the ACA" (Nayler 1968, 1045). Lanchester says that he was appointed by Rayleigh. Driver accepts Lanchester's account but implies that Lanchester's appointment was made against Haldane's inclinations (Driver 1997, 220). This appears to be an inference from Haldane's supposed general attitudes.

34. The data come from an appreciation of Glazebrook that Selby wrote for the *Times* of December 8, 1938 (RTG-TCC, M. 1872.).

35. Nayler 1968.

36. Glazebrook and Shaw 1885. On the role of this book in establishing the Cavendish tradition of accurate measurement of so-called absolute units, see Schaffer 1992.

37. On the long-standing links between Rayleigh, Glazebrook, Shaw, and Mallock, and their role in developing the Cavendish laboratory, see Schaffer 1992, 1994, and 1995.

38. Warwick 2003. In this section my overall picture derives from Warwick's rich study.

39. Warwick's book contains a useful list of the top wranglers from 1865 to 1909, which identifies their coaches. See Warwick 2003, 512–25.

40. Lamb 1932, 35. The circumstances leading to the setting of the examination question, which arose out of a correspondence between Thomson and Stokes, are described in Cross 1985, 143–44.

41. Hardy 1906.

42. See the preface to G. H. Hardy's (1908) *Course of Pure Mathematics*: "It has been my good fortune during the last eight or nine years," said Hardy, "to have a share in the teaching of a good many of the ablest candidates for the mathematical Tripos; and it is rare indeed that I have encountered a pupil who could face the simplest problem involving the ideas of infinity, limit, or continuity, with a vestige of the confidence with which he would deal with questions of a different character and of far greater intrinsic difficulty" (1908, vi). His book was meant to put this right, and it was, he stressed, "a book for mathematicians" (i).

43. George Darwin (1845–1912), a fellow of Trinity College, has been described as "an applied mathematician in the school of Kelvin or Stokes" (Kopal 1980).

44. Moulton was also a fellow of the Royal Society and went on to perform important work presiding over the Explosives Supply Department during World War I. See Moulton 1922.

45. Historians have not seen the significance of Haldane's weekend in Cambridge. Walker (1974, 291) quotes the entirety of Haldane's introductory talk to the Esher committee, but only as an example of how to make a poor case for a good cause. Would the soldiers and politicians on the committee, he asks, really be interested in whether he was in Cambridge on Saturday or Monday or want to hear about his cozy chats with highbrow scientists? Gollin (1984, 406) quotes extensively from Haldane's speech but leaves out precisely those passages describing his visit to Cambridge and his discussions with Darwin.

46. PRO: DSIR (Department of Scientific and Industrial Research) 23/2. This document

constitutes the second of the technical reports of the committee, the first being, in effect, a reading list, that is, a list of publications relevant to aeronautics. It is an expanded version of minute 6 from the first meeting of the ACA; see PRO: DSIR 22/1.

47. The stress on models applies mainly to the NPL. At the Royal Aircraft Factory a tradition of full-scale experimentation and testing was to grow up, and this led to significant tensions between some of the leading figures.

48. For detailed documentation of the exploits of the pioneers, see Dallas Brett 1988.

49. Lanchester 1908. Lanchester was skeptical of the idea that the Wrights were scientific in their approach. Like Haldane, he suspected that they were simply empirics. In a letter of March 3, 1909, to Col. Fullerton, the honorary secretary of the Aeronautical Society, Lanchester said: "I think it was a mistake of the Aeronautical Society giving the Wrights a medal for their contribution to aeronautical *science*, I agree with their having the medal but it should have been for what they have done—i.e. for the advancement of aeronautical art" (FWL-RAeS, folder 11).

50. Gollin 1989, 39 (first two quotations) and 41 (third quotation).

51. As quoted in Driver 1997, 223.

52. Gollin 1989, 35.

53. *Flight* 1909a, 308.

54. Bryan 1909b.

55. Bairstow 1935; Brodetsky 1928.

56. On the baleful influence of these "practical men" on the British economy, see Barnett 2002, 94–99; and 2000, 41. See also Locke 1984.

57. *Aeronautics* 1910, 118.

58. Turner 1911, 298.

59. This view has its defenders today; see, e.g., Driver 1997.

60. Berriman 1913, 147. Berriman was the technical editor of *Flight*. The book contains abstracts from the official report on the trials on pp. 272–80.

61. *Aero* 1912b.

62. *Flight* 1912b.

63. *Aeronautics* 1917.

64. These facts are given by Driver, who, counter to the trend of his own argument, admits that the irate War Office dossier on Handley Page "amounts to a damning indictment of the firm's manufacturing capacity at this time, and cannot have made those in authority any more inclined to pursue procurement through the private sector" (Driver 1997, 92).

65. Grey 1913.

66. Horace Darwin had recommended that Busk be awarded the Winbolt Prize in engineering at Cambridge in 1908. See Busk 1925, 37. Busk was an assistant engineer-physicist at the Factory and was appointed by O'Gorman on the recommendation of Glazebrook (among others). He died on November 5, 1914, when the machine crashed on Laffan's Plain at Farnborough. See Hare 1990, 51 and 79.

67. Grey 1914b, 320.

68. Grey 1916a.

69. *Flight* 1916b. Cooper (1986, 30–31) supports the thesis that the threat was exaggerated. Driver (1979, 241) argues that the Fokker-scourge arguments were not really about tactics and armaments but were "more concerned with the crisis in the aviation industry, that is with the development of what amounted to a procurement-based monopoly and the suffocation of free, competitive, design initiative." In its own partisan way this corroborates the *Flight* editorial, that is, Driver supplies the ulterior motive.

70. *Flight* 1916a, 255.

71. For more detail regarding the two committees of inquiry, written from opposing points of view, see Hare 1990, 91–102, and Driver 1997, 239–48.

72. Like the other aeronautical journals, *Flight* carried lengthy, verbatim accounts of the reports of the official inquiries; for example, "Report on the Royal Aircraft Factory," *Flight* 8 (August 3, 1916): 650–54; and "R.F.C. Inquiry Committee: Interim Report," *Flight* 8 (August 17, 1916): 696–99.

73. *Flight* treated the reports of the Advisory Committee seriously and defended the Factory against the charge that it pirated ideas from the private sector. For example, "Editorial Comment: The Technical Report, 1912" contains a subtle defense of the role of fundamental research in aeronautics. The article "The Royal Aircraft Factory and Industry" (*Flight* 1914) effectively rebuts the charge of copying ideas from private enterprise.

74. *Flight* 1916d.

75. For details of the SE5, see Hare 1990, 263–96. Marshal of the Royal Air Force, Lord Douglas of Kirtleside, who flew the SE5 in 1917–18, is quoted as saying: "Many of us felt that it was the best fighter in Britain in the First World War" (Hare 1990, 291). Its rival for this title was the commercially manufactured Sopwith Camel. The Camel, however, was difficult to fly, and its stall characteristics claimed many lives. It was also unstable in a dive, and this made it difficult for the pilot to take aim and shoot accurately in a battle with another aircraft. For a test pilot's comparison of the two aircraft, which favors the SE5, see Hill 1920.

76. On Haldane's fate, see Koss 1969.

77. These words were used by O'Gorman many years later in a letter of January 8, 1943, to R. S. Whipple of Cambridge Instruments. They are quoted in Cattermole and Wolfe 1987, 127.

78. Billing also spread the rumor that the German Secret Service possessed a black book listing 47,000 British persons who had been sexually compromised and were working for the Germans. Haldane's name was said to be on the list. For an account of the trail of destruction left by this lunacy, see Hoare 1997. The Haldane accusation is also described (120).

79. See, for example, the references in *Flight* to "that most loathsome of all types . . . the 'Conscientious Objector'" (*Flight* 1917, 468).

80. *Aeroplane* 1916, 570.

81. Grey 1917a.

82. Churchill to Grey, July 4, 1917, CGG-RAeS, folder 6, Royal Aeronautical Society. An attached sheet gives the parliamentary reply to a question asked by Churchill: "In the special circumstances, however, the Military Representative has been instructed not to press for Mr. Grey to be made available for military service."

83. These words were quoted in the first annual report of the ACA. *Technical Report of the Advisory Committee for Aeronautics for the Year 1909–10*, 5.

84. See Technical Report 3, PRO: DSIR 23/3.

85. The details of the methods of measurement and the means by which the model was supported in the channel, as well as the various expedients to produce a steady flow (such as layers of gauze), are described in Stanton 1910.

86. The label "parameter variation" comes from Vincenti 1990, chap. 5.

87. Bairstow and Melvill Jones 1912b.

88. PRO: DSIR 22/1, p. 5.

89. Pilcher was killed in a flying accident in 1899 (Gibbs-Smith 1960, 30–31). Plate VI(c) shows Pilcher in flight.

90. Bryan and Williams 1904.

91. Bryan 1897, 1907.

92. See Bryan's response to a letter to *Nature* by Herbert Chatley (Bryan 1909c).

93. Bryan 1910a, 10.

94. On A. E. H. Love, see Milne 1939–41.

95. Routh 1884. Routh's Adams Prize essay was published as Routh 1877.

96. Bryan and Williams 1904, 115.

97. In an article on the aviation meeting and flying display at Rheims in 1909, Bryan noted the number of accidents and fatalities and concluded that problems of stability played a central role: "the obvious remedy is that aviators should wait until this subject has at least been thrashed out mathematically" (Bryan 1909a, 397). The same theme was taken up in a literature review when Bryan said, "it would have been better, cheaper, and probably quite as quick in the long run to have got everything done that could be done in studying the problems of aviation by the methods of exact science and to have developed the practical side subsequently" (Bryan 1910d, 229). In Bryan's view, it was wrong that "the skill of the aviator" was "being made to take the place of exact mathematical calculation, with uncertain results" (233).

98. Bryan 1911.

99. Strutt 1924, 27. For the context of Rayleigh's utterance, see the important chapter "Routh's Men," in Warwick 2003. Rayleigh's words are quoted on p. 246.

100. The reference to the projected book, which was never written, is in Bryan 1915, 49.

101. Harper 1913.

102. W.H.W. (probably Sir W. H. White) 1912.

103. Bryan 1911, 7.

104. Bryan 1913b.

105. *Aero* 1910, 216.

106. Hilton 1912, 344.

107. The work at the NPL on stability was summarized after the war in Bairstow 1920, chap. 10. On Bairstow, see Temple, Nayler, and Relf 1965. On Melvill Jones, see Hall and Morgan 1977.

108. North 1922, 409. North worked for the Boulton Paul Company and was later responsible for the Defiant night fighter that operated in World War II. The problems with the stability measurements to which he drew attention were overcome, after the Great War, by E. F. Relf, who suspended the model by wires. See Southwell 1925b, 148. The older, spindle-mount, method is illustrated in Bairstow 1920, 103.

109. See Bairstow's contribution on p. 27 to the discussion following Harper 1913. Using the pendulum to justify the application of mathematics to the real world has a long pedigree. Galileo claimed that the period of a simple pendulum was independent of the amplitude. As Kuhn (1962, 123) delicately put it, this was a discovery "that the normal science stemming from Galileo had to eradicate and that we are quite unable to document today."

110. According to Melvill Jones (1923, 477), "The classical stability calculations of Bryan and others are complicated enough, although they only deal with infinitesimal disturbances from steady straight flight. If we were to attempt to attack our problem in the same way, we should be involved in hundreds of times this complication before we could make a complete statement of the more involved motions."

111. Glauert 1920b, 339.

112. North 1922, 408.

113. Bairstow 1919, 853.

114. Jane 1990, 90, shows machines derived from the Gun Bus prototype.

115. Bairstow 1914. The article contains a drawing of one of the model gliders used in the talk. It possesses an unusual, vertical stabilizing fin midway along the fuselage. This disposition was recommended by Bryan (1911, 14) for its effectiveness in producing lateral stability.

116. For Low's comments, see p. 78 of the discussion included in Bairstow 1914.

117. On the modern relevance of Bryan's work, see Abzug and Larrabee 1997. Bryan's portrait provides the frontispiece of the book, and his equations are reproduced on p. 11. "Today's stability and control engineers," they say "are generally astonished when they see these equations. . . . Bryan's equations are identical to those used in analysis and simulation for the most advanced of today's aircraft" (Abzug and Larrabee 1997, 10).

118. The letter was published in the context of an exchange between Lanchester, Bryan, and Bairstow over Bryan's Wilbur Wright Memorial Lecture of 1916. Lanchester had objected that Bryan had ignored the fact that airplanes banked when they turned. Bairstow 1916, 17–19.

119. Greenhill 1914, 427. The same idea was repeated in a lecture at the Mathematical Association on January 9, 1915, and reported in *Nature*. Greenhill 1915, 573.

120. Cowley and Levy 1918, 264.

Chapter Two: The Air as an Ideal Fluid

1. Lamb 1879, 1. All later editions contain essentially, though not exactly, the same wording.

2. Lamb 1932. In the index there are over 100 references to Rayleigh and the lesser, but still significant, number of 17 references to Greenhill. In comparison, Helmholtz has 28 and Kirchhoff 31 entries. After Rayleigh the largest number of references is to Stokes (57) and Kelvin (55).

3. Greenhill 1911; Unwin 1911.

4. Newton 1966, 300–302.

5. Brush 1976, chap. 1.

6. Newton 1966 (1729 translation), book 2, proposition 34, 331–32. The discussion of water is empirical and comes later.

7. Detailed accounts of the history of fluid dynamics are to be found in Darrigol 2005 and Eckert 2006.

8. Some writers argue that the equations of fluid dynamics deal with volumes that are *large* enough to contain many molecules (justifying an averaging process) but *small* enough to act as an infinitesimal (justifying the use of the differential calculus). Others, doubting that these conflicting requirements can be reconciled, argue that the concept of density at a point only applies to an *idealized* fluid, and not to a *real* fluid. Cf. Anderson 1991, 90, and Rutherford 1959, 1.

9. The atmospheric pressure on a body at sea level is in the region of 2,000 lb. per sq. ft. If a body is moved through still air at around 200 miles per hour, the greatest augmentation of pressure on the body on the assumption of incompressibility is about 102 lb. per sq. ft. This figure compares with about 104 lb. per sq. ft. if compressibility is taken into account. So the error introduced by the assumption is about 2 percent for the pressure differences, which translates into an error of about 3 percent for local increases in density. The data are from Reid 1932, 10–11.

10. On the history and role of the calculus, see Boyer 1959 and Garber 1998.

11. Cowley and Levy 1918.

12. Wers-key 1988, 44–52, 115–31. Levy recalls that Glazebrook praised his work but warned the university department that he, Levy, was a troublemaker. Knowing Glazebrook from Trinity, the professor of mathematics promptly hired Levy.

13. Nayler 1966a, 84. Mitchell went on to design the Supermarine Spitfire. On Mitchell and the Spitfire, see Glancey 2007.

14. There are some slight notational inconsistencies between Cowley and Levy's diagrams and their text but these are not of any importance.

15. This volume would now be called a "control volume," although the term seems not to have been used in Britain until it was taken over from German usage after World War I. Control volume thinking appears to have originated in engineering practice and was not immediately taken up by physicists. See Vincenti 1990, chap. 4.

16. Different authors use different notations. Stokes wrote Du/Dt. I follow Cowley and Levy who simply use du/dt.

17. Almost any early work on aircraft construction would confirm this usage, e.g., Fage 1915, 8–10; Judge 1917, 184–89.

18. The concept of streamline appears to have arisen from the nautical interest in "water lines" or the path taken by water as it flows past a ship. It was developed mathematically by Rankine of Glasgow in the early 1860s (Rankine 1881, 495–521).

19. The intuitive argument is to be found in Glauert 1926, 41. Reid 1932, 32; Munk 1934, 238; Streeter 1948, 21. The shape of the fluid element actually plays no role in the argument.

20. Prandtl provides a good, intuitive account. See Tietjens 1957b, sec. 72.

21. Sneddon 1957, chap. 4, "Laplace's Equation." Lambe and Tranter 1964, 212–15.

22. The physical meaning of, say, $\nabla^2 Q$ at some point P is that it refers to the difference between the average value of the property Q throughout a small sphere centered on P and the value of Q at the point P itself. James Clerk-Maxwell called it the "concentration" of the property Q at the point P. Laplace's equation thus deals with phenomena having zero concentration in Maxwell's sense (Maxwell 1869–71a).

23. Hanlon 1869–71; Maxwell 1869–871b.

24. Relf 1924a. On Relf's influential role as an experimenter at the NPL and the author of many reports for the ACA and its successors, see Collar 1971.

25. On the broader analogies between hydrodynamic flows (both irrotational and rotational) and other physical phenomena, see Hesse 1961, chap. 4. Smith and Wise 1989, chap. 12.

26. Cowley and Levy did not identify Bernoulli's law by name but they derived the result by expressing the velocities in the Euler equations in terms of the stream function and adding the requirement that the flow is irrotational, that is, that the flow satisfies Laplace's equation (41). A simpler and more intuitive derivation was given in Bairstow 1920, 356–57. Bairstow's derivation is the one used in current text books such as Anderson 1989, 69–72.

27. Relf, Bramwell, and Fage 1912; McKinnon Wood 1923; Tietjens 1957a, 226–31.

28. For a history of the Pitot tube and the modifications needed for high-speed flight, see Anderson 1989, chap. 4.

29. There is nothing "imaginary" about the symbol i in the ordinary sense of the word, although it deviates from standard arithmetic in which negative numbers cannot have square roots. The expression $x + iy$ means go x units of distance parallel to the x-axis, turn through ninety degrees, and then go y units of distance parallel to the y-axis. The letter i designates the operation of turning through ninety degrees. The plus and minus signs signal direction, while $i \times i$, that is i^2, means "turn through ninety degrees and then turn through ninety degrees again." The result is a complete reversal of direction, that is, a change from a positive to a negative direction. This is why $i^2 = -1$ and hence why $i = \sqrt{-1}$. The symbol is explained in this way in Hardy 1908, chap. 3.

30. Modern textbooks would add the qualification that the function must be "analytic,"

which means having a single, finite value for each value of z and having a single-valued derivative. This is taken for granted by Cowley and Levy. I also take it as read in what follows.

31. Cowley and Levy do not go into this question for the case of the flow around the cylinder, but others did. Glauert said: "if the broken lines are taken to be the streamlines, it is necessary to assume a distribution of sources and sinks over the upper and lower halves of the circumference in order to satisfy the boundary condition, since the fluid has a definite velocity normal to the circumference of the circle" (Glauert 1926, 52).

32. The extended definition of a "polygon" is taken for granted by Cowley and Levy. The justification for the extended usage is discussed by later writers such as Streeter 1948, 156−57, and Milne-Thomson 1960, 264−66.

33. Some would suggest that it has distorted the field. "On account of the relative simplicity of such a procedure and on account of the existence of a very extensive mathematical theory of conjugate functions, this type of flow has in the past received perhaps undue emphasis" (Rutherford 1959, 45). Conjugate functions are related in the way that φ and ψ are related.

34. Hardy 1908, 105.

35. Burt 1932, 64−73; Shapin 1996, 57−64.

36. Wigner 1967; Colyvan 2001.

37. Taylor 1916b.

38. Lamb 1879, 244.

39. See, e.g., Rouse and Ince 1957, 193.

40. Helmholtz 1868. An English translation by Frederick Guthrie was also published in 1868. The translation given in the text is my own.

41. The possibility of such flows had been noted earlier by Stokes in a paper read to the Cambridge Philosophical Society in 1842 (Stokes 1880, 1:11).

42. The air on one side of the surface of discontinuity moves at a different speed from the air on the other side. For the "live," streaming, air, $v = V$; for the "dead" air, $v = 0$. Yet they have the same static pressure, namely, atmospheric pressure. Does this contradict Bernoulli's law, which states that the higher the speed, the lower the pressure? No, it is consistent with the law. The appearance of inconsistency derives from the truncated statement of the law, which leaves out the conditions under which the trade-off of speed and pressure take place. It is important to notice that the Bernoulli constant, the total head pressure, is different in the two bodies of fluid. In the dead air the total pressure is just the static pressure, that is, the atmospheric pressure. In the streaming air the total pressure is defined by the conditions at a distance from the wing. It is the sum of the static pressure (the atmospheric pressure) and the dynamic pressure ($\frac{1}{2}\rho V^2$), where V is the speed of the free stream. Bernoulli's law therefore does not preclude the two different bodies of air having the same static pressure behind the wing while moving at different speeds.

43. Kirchhoff 1869, 1876.

44. Lamb spells out the steps and transformations that are implicit in the original publications. See Lamb 1879, 100−109.

Chapter Three: Early British Work on Lift and Drag

1. Rayleigh 1876.

2. Rayleigh was not enthusiastic about relativity theory, quantum theory, or Bohr's theory of the atom. He was a "classical" rather than a "modern" theorist. See Lindsay 1981.

3. Kelvin's words are from a letter of December 8, 1896, declining an invitation to join the Aeronautical Society. They are quoted in Gibbs-Smith 1960, 35.

4. Rayleigh 1891.

5. For an analysis and commentary on Rayleigh's use of the empirical data, see Anderson 1997, 100–109.

6. Rayleigh repeated his diagnosis of the problem in Rayleigh 1900 and 1902. In the 2nd edition of *Hydrodynamics*, Lamb had repeated Rayleigh's original opinion that the agreement with the experiment was good (1895, 111), but in the 3rd edition he withdrew this judgment and referred the reader, without comment, to later publications (1906, 95).

7. Kelvin 1894.

8. Greenhill 1910.

9. Michell 1890. The information about Michell's Tripos comes from Warwick 2003.

10. Love 1891.

11. Greenhill's approach exemplifies Warwick's thesis that Tripos "students trained to solve mathematical problems took problem solving as a model of research" (2003, 229). Wranglers, Warwick argues, were "taught to formulate physical problems in terms of the mathematical tools to hand" (2003, 240). This is exactly what Greenhill was doing.

12. For a full discussion of this theme, see Warwick 2003, esp. chap 3, sec. 7. The characterization of the published solutions as an "archive" comes from Warwick.

13. Watson and Routh 1860.

14. Wolstenholme 1867. When the prestigious Fifth International Congress of Mathematicians was held in Cambridge in 1912, R. F. Scott, the vice chancellor, made reference to this book in his welcoming speech to the assembled delegates. The Senate House examinations, he said, had had a profound influence on mathematical studies at Cambridge. "I may illustrate what I mean by reference to the *Mathematical Problems* of the late Mr Joseph Wolstenholme, a form of work I believe without parallel in the mathematical literature of other nations" (Hobson and Love 1913, Introductory Address 37). See also the discussion in Warwick 2003, 156 and 261.

15. Greenhill 1876.

16. Lamb 1916, 98.

17. Ramsey 1913, 127. Ramsey's book was an update of an earlier textbook on hydrodynamics by Besant, who had been Ramsey's coach. On Besant, see Warwick 2003, 534.

18. Cooper 1911.

19. *Flight* 1911.

20. Greenhill nevertheless continued to stamp his personality on the reports he submitted. In a memorandum on bomb trajectories for the Advisory Committee, he invited his readers to test his mathematical analysis empirically. With Edwardian insouciance he declared: "A balloonist can experiment up in the air, throwing out the empty champagne bottle with a spin, and observing the helix described, with convolutions growing larger and larger" (1917, 4).

21. Greenhill 1912.

22. *Aeronautics* 1913.

23. Bryan 1912c.

24. Mallock 1909b.

25. Mallock was here calling upon ideas that he had already developed before joining the ACA; see Mallock 1907, 265.

26. Mallock 1909a. One other topic that Mallock mentioned was the resistance of ropes and stays. There has been much work on this, he said, but chiefly "by practical men."

27. Stanton and Bairstow 1910.

28. Eden 1912. The preliminary work is described in Eden 1911.

29. Bairstow and Jones 1912b.

30. Bairstow and Jones 1912a.

31. One of the first topics discussed in the ACA was the conditions for making such an inference. Rayleigh put the conclusion into the now standard form: that the ratio of the inertial and viscous forces in the two cases had to be the same for them to be truly similar. Such conditions are, however, very difficult to realize in practice. See Rayleigh 1909.

32. The photographic approach was extended by Relf using an improved photographic technique and a model aerofoil called the RAF 6. See Relf 1913.

33. Mises 1945, 259.

34. Kuhn 1962. Kuhn should not be given a purely psychological reading. For a group to treat something as business-as-usual only requires compliance; it does not need confidence at the individual level. Conversely, a fundamental change in practice, at the group level, does not necessarily imply a change in subjective convictions at the individual level. Kuhn is addressing the collective outcome and interaction of many varied individual states of belief, that is, he is describing sociological processes.

35. Bryan and Jones 1914.

36. Leathem 1915. (Leathem had helped Ramsey with his textbook.)

37. Levy 1915.

38. Greenhill 1916.

39. Levy 1916.

40. Selig Brodetsky, a collaborator of Bryan's and a senior wrangler at Trinity in 1908, was still defending the discontinuity account of lift in 1921. See Brodetsky 1921. To my knowledge, the last known sighting of a mathematician using discontinuity as an account of lift was in 1939. Walter Vincenti recalls that as a graduate student in aeronautical engineering at Stanford, he took a mathematics course on conformal mapping from the distinguished Hungarian mathematician Gabor Szego. Szego analyzed the lift of a wing in terms of Rayleigh flow. The young Vincenti felt compelled to tell Szego that this was not what was being taught in the Department of Aeronautics (Vincenti, pers. comm.).

41. Cowley and Levy 1918, 72–75; Levy 1919.

42. Bryan 1912b.

43. Bairstow 1913.

44. The text of the discussion follows that of the lecture. Greenhill's contribution to theological aerodynamics is on p. 124.

45. One version of this story is to be found in Fage 1966, 91, and another in Nayler 1966b, 79.

46. Taylor 1916a.

47. G. I. Taylor, "Turbulent Motion in Fluids," Adams Prize essay for 1913–14, GIT-TCC, file C. 2. Taylor describes some of his wartime adventures in Taylor 1971, where he suggests that Busk's death was due to an overflowing gasoline tank (527). The tank on the BE2 had to be filled manually but was not fitted with a fuel gauge.

48. Taylor's essay won the Adams Prize, and material for a number of his subsequent papers on turbulence was drawn from it. Taylor's career and work are described in Batchelor 1996. The Adams Prize work, though not the significance of the remarks in the preface, is described in Batchelor 1996, chap. 12.

49. Glazebrook 1914.

50. Grey usually used the word "scientific" in a negative way, but not always: "Lord Haldane was completely in error when he said the Wrights lacked science. As far as they went their science was as exact as that of the Government Advisory Committee, the N.P.L., and the R.A.F." (Grey 1916b, 833). This statement was more an attack on Haldane than a defense of the Wrights. Grey did not have a high opinion of their designs, writing on June 10, 1914, to Moore-Brabazon: "The Wright was a complete dead-end design. And it could never have been developed any further" (CGG-RAeS, folder 4).

51. For their part many "mathematicians" proudly cultivated accuracy. See Schaffer 1995.

52. Ledeboer 1916, 33.

53. Grey 1917b, 1284.

54. "The trouble is that these high-class scientists are accurate to many places of decimals in these mere calculations, but always forget or neglect the existence of some unknown or unknowable factor . . . and that is why it is always safer to trust a plane engineer than a slide-rule scientist with a string of letters after his name and no solid workshop experience" (Grey 1915, 153).

55. Thomson 1919; J.C. 1920, 471 (quote).

56. On Barnwell, see Anderson 2006.

57. Grey, in Barnwell 1917, 4.

58. Sayers 1922.

59. "It may be taken as an admitted fact that model experiments unsupported by full-sized tests, are at present of little value to anybody. There is some reason to believe that a great deal of harm has been done in the past by a failure in the part of the N.P.L. and the Advisory Committee for Aeronautics to recognise this" (Sayers 1919, 1310).

60. Grey 1914a, 423.

61. Aston 1911, 70. One feature of Eiffel's work that is singled out is that Eiffel proves "once and for all" that there is no difference between the case where an object moves through still air and the case where the air moves past a stationary object. This equivalence was still a matter of dispute among practical men. The equivalence was opposed by Rankine Kennedy (1910). It was, however defended vigorously in the correspondence columns of the *Aero* by, among others, Horatio Phillips (1910), who dismissed Kennedy as absurd. Kennedy insisted that there was empirical evidence for the lack of equivalence and cited the work of Du Buat. The modern account is that Du Buat's empirical findings were correct but that they were the result of differences in turbulence between the two cases of relative motion. The equivalence between a body moving uniformly through stationary air and air moving uniformly past a stationary body is now said to hold good provided turbulence is removed.

62. Ledeboer 1916, 33.

63. *Aero* 1912a.

64. Bryan 1912a, 266.

65. Newton 1966 (1729 translation), book 2, proposition 34, 331–32.

66. For some of the confusions about the status of Newton's fluid and the arguments that resulted, see Villamil 1913 and Giacomelli and Pistolesi 1934, 310–14.

67. Chatley 1910, vi.

68. H. S. Hele-Shaw (1854–1941) is not to be confused with Napier Shaw, the Cambridge-trained meteorologist on the ACA, but he was a significant figure in turn-of-the-century fluid dynamics. On the status of Hele-Shaw's photographs of fluid flow, see Bloor 2008b.

69. Berriman 1911, 1913.

70. Berriman 1913, 303.

71. Berriman 1911, 5.

72. On Langley's work, see Anderson 1997, 164–91. For Lanchester's response to Langley, see Lanchester 1907, 347–38. Lanchester was critical of Langley's reading of Newton and accused him of wrongly attributing to Newton the idea that the sine squared law applied to motion through atmospheric air.

73. Bryan 1912c. Bryan says that Berriman does not seem to make any distinction between the tangent of the difference between two angles and the difference of the tangents of the angles (1912c, 265).

74. Kuhn 1962, chap. 4.

75. Thurston 1911.

76. On the work of Phillips and Lilienthal, see Anderson 1997.

77. For a sympathetic account of Thurston's aeronautical work before, during, and after World War I, see Ackroyd, Bernstein, and Armstrong 2008, 358–65.

78. Handley Page 1911. See also Shaw 1919, 68–70 (once again, a different Shaw).

79. Handley Page 1921. Handley Page came to accept Prandtl's account of how the slot works. Prandtl said that the air coming through the slot gives kinetic energy to the air in the boundary layer and prevents separation of the flow from the upper surface. See Tietjens 1957a, 153–56. Though widely accepted, this account has been challenged. See Smith 1975, 518. Smith, the chief aerodynamic engineer in charge of research at Douglas Aircraft, argued that the slot is not a blowing device for controlling the boundary layer. I thank David Musker for drawing my attention to Smith's paper.

80. The significance of simultaneous discoveries and priority disputes as a tool for investigating the sociology of science was pointed out in the 1950s by Robert Merton. See Merton 1973.

81. On Lachmann, see Anderson 1997, 365–67. On Thurston and the long-standing patent controversy surrounding the leading-edge slot, see Musker (forthcoming). In his analysis of the patent disputes, Musker shows the negotiable character of the relevant concepts of "slot" and "aerofoil."

82. Low 1914.

83. "The lamps are going out all over Europe; we shall not see them lit again in our lifetime." These words were uttered by Sir Edward Grey, British foreign secretary, on Monday, August 3, 1914 (Grey 1925, 2:20).

84. Thomson 1919, 18.

85. For example, typical British aerofoils of the time were slender with sharp leading edges. Wouldn't this help them cut through the air? Intuitively, this may seem a desirable thing. If so, then it shows how questionable intuition can be. Later experts would come to see this form of aerofoil as a mistake for the flight speeds in question. Thin wings encourage leading-edge stalling. Thin wings with sharp leading edges only come into their own at supersonic speeds. See Mises 1945, 258–64.

Chapter Four: Lanchester's Cyclic Theory of Lift and Its Early Reception

1. Lanchester 1907, vi. Notice the difference between Lanchester's preference for "simple cases" and Taylor's focus on "simple bodies" (in the epigraph at the head of the previous chapter). The two subtly different appeals to simplicity are used to justify diametrically opposed approaches.

2. Elementary introductions to the circulatory theory can be found in Kármán 1963, Sutton 1965, and Wegener 1997.

3. Anderson and Eberhardt 2001, 16.

4. Inverted flight has its problems but not ones the popular theory can illuminate. In general terms a wing is less efficient in inverted flight. "The inefficiency of the wings in inverted flight causes the aeroplane to fly at an angle of incidence considerably greater than that for a corresponding speed in normal flight, and the pilot to receive the impression that the speed is lower than is actually the case. If the engine is off the gliding angle is relatively poor. . . . Added to this the aeroplane stalls at a higher speed, sometimes as much as 30 per cent in excess of its stalling speed in normal flight" (Hill 1923).

5. Russell 1903, 474.

6. Southwell 1931, 359–60. A similar idea was invoked by Albert Betz in his post–World War II review of incompressible flow, Betz 1948, 4.

7. The definition can be found in any standard textbook on hydrodynamics, for example, Lamb 1932, 33.

8. These experiments are discussed in chapter 9.

9. Thomson (Lord Kelvin) 1869, 247.

10. Ricardo 1948; Kingsford 1960.

11. Lanchester 1907.

12. "The hydrodynamic interpretation included in the present work has been added subsequently" (Lanchester 1907, ix). References to Lamb are found, for example, on pages 69, 79, 87, 101, 116.

13. For example, Lanchester 1907, 144, 145, and 161.

14. In his Lanchester Lecture, J. A. D. Ackroyd (1992) provides an analysis of Lanchester's *Aerodynamics* (including Lanchester's independent development of ideas in boundary-layer theory). Ackroyd shows convincingly that "in his calculation procedure for the prediction of wing performance," Lanchester "makes virtually no use" of his qualitative ideas about circulation and vorticity (130). Rather, Lanchester reverts to a version of the Newtonian theory of "sweep" and seeks to combine empirical results with the formulas of discontinuity theory as applied to the flow around flat plates. Ackroyd's important contribution in this lecture is to clarify Lanchester's thinking on these topics and to follow its mathematical development in detail. Ackroyd argues that it is these parts of the book that yield "the key results which can be obtained from the 'Aerodynamics' of 1907" (131). That is, unlike the theoretical picture of circulation, they yield the main quantitative results and mathematical formulas that can be extracted from the book.

15. Rayleigh 1877.

16. Apparently real tennis is so complicated that it takes two years to learn the rules, and it is "impossible to convey by means of the printed word a description of the game that would be understandable to those who have not seen it played" (*New Illustrated Universal Reference Book*, s.v. "Tennis," 678). Clearly this was a game for wranglers.

17. Strutt 1924, 25. Real tennis was also a great favorite of the pure mathematician G. H. Hardy (Snow 1967, 18, 32, 44, 50). I thank Steve Sturdy for drawing my attention to real tennis as the likely source of Rayleigh's question about the trajectory of a tennis ball.

18. On Magnus and the Magnus effect, see Hoffmann 1995.

19. In the original paper Rayleigh actually said the ball would go in the opposite direction to the circulation. This, along with what appears to be a typographical error in the integration, was corrected when Rayleigh published his collected papers.

20. Greenhill 1880.

21. In a chapter devoted to "special curves" in his textbook on the calculus, Lamb defines a

"trochoid" as follows: "The curve traced by *any* point fixed relative to a circle which rolls on a fixed straight line is called a 'trochoid'" (1919, 296).

22. In Lamb's contribution to the Royal Society obituary of Rayleigh, the attraction of the discontinuous-flow picture of the force on a body in a moving fluid was identified as its generality. It provided "the best *general* representation of the phenomena which we are yet able to get by *a priori* dynamical reasoning." See Schuster 1921, xlv.

23. Prandtl 1927a. The same volume of *Die Naturwissenschaften* also includes an appreciation of Runge as a mathematician by R. Courant (229–31), and as a spectroscopist by F. Paschen (231–33). I. Runge 1949, 135–37.

24. On the Parseval airships, see Hartcup 1974, 39–41.

25. F. W. Lanchester, *Notes concerning the Position of Aeronautics in 1908*, FWL-RAeS, 2–4.

26. Lanchester 1909.

27. Lanchester to C. C. Walker (of De Havilland), May 25, 1937, FWL-UC, 5–25. There is no specific reason to doubt Lanchester's recollection, but it is prudent to remember that the letter was written nearly thirty years after the event.

28. Kingsford (1960, 242) lists the fourteen reports for which Lanchester was responsible.

29. *Aeronautics* 1908, xiv.

30. *Aeronautical Journal* 1908.

31. *Engineering* 1908; *Engineer* 1908; *Times* 1908.

32. Judge 1917. Lanchester's figures are reproduced on pages 47 and 49.

33. Lanchester's biographer described the many talks and lectures given by Lanchester in the years before and during the Great War. See Kingsford 1960.

34. Judge 1917 also cited work by Joukowsky (spelled Schukowsky) but treated him as someone who had designed a single aerofoil section ("Dr. Schukowsky's wing"; 151) rather than a whole family of aerofoil-like shapes. A less than adequate response to Joukowsky was common ground between the "practical men" and the "mathematicians."

35. Ledeboer 1909.

36. North 1966.

37. Chatley 1914.

38. Grey 1917c, 722.

39. Grey 1916d.

40. Grey 1916c, 66.

41. *Flight* 1916c.

42. Lanchester 1917a.

43. The Air Board was the forerunner of the Air Ministry, and the two branches of the Air Service were the Royal Flying Corps and the Royal Naval Air Service.

44. Lanchester 1917c.

45. Lanchester to Balfour, March 10, 1916; and Lanchester to Steel, May 4, 1916; FWL-UC, 514.68 and 514.116, respectively. Murray Seuter was later to go into politics, and in the 1930s he became the Conservative MP for Hertford. He was a member of the Anglo-German Fellowship and accepted an invitation to attend the Nuremberg Rally. See Griffiths 1980, 185 and 225. Griffiths also gives information about C. G. Grey's political and racial opinions.

46. [Bryan?] 1908.

47. Kutta 1910, 1911.

48. Kutta 1902.

49. Finsterwalder 1901–8, 167n78a.

50. This point was made by J. D. Anderson (1997, 247–49 and 460–61).

51. Lamb 1879, "Irrotational Motion in Multiply-Connected Spaces," 49–51.

52. Lamb 1879, 37–38.

53. Many years later a version of this argument was put forward by the American mathematician Garrett Birkhoff (1950, 19) to illustrate one of the many "paradoxes" of wing theory.

54. Birkhoff's source was Mises and Friedrichs 1941, which are lecture notes from a wartime course on applied mathematics. The argument against the three-dimensional circulation theory is indeed made on page 108 of those lecture notes, but on page 109, Mises and Friedrichs give, and accept, Lanchester's answer.

55. Sometimes the identity of anonymous reviewers can be established from annotations on the old copies of the journal kept by the publishers, but this did not prove possible in the present case. I greatly appreciate the efforts made by Richard Webb, in the offices of *Nature*, to track down the author.

56. See Bryan 1905, 1908, 1909a, 1910c, 1910d, 1913a; Bryan and Harper 1910.

57. Bryan 1917. Joukowsky's name was spelled "Joukowski" throughout the review.

58. Joukowsky 1910/1912.

59. Joukowsky 1916.

60. G. I. Taylor, "Turbulent Motion in Fluids," Adams Prize essay for 1913–14, GIT-TCC.

61. "I have been unable to get any books of reference except Lamb's Hydrodynamics, the Reports of the Advisory Committee for Aeronautics and a paper of Dr Stanton's on friction in pipes" (Taylor, "Turbulent Motion in Fluids," 3).

62. Lanchester 1915.

63. The second part, which I do not discuss, was devoted to skin friction. Lanchester opposed the widespread idea, stemming from Langley, that skin friction was negligible and played no significant role in aerodynamics.

64. What causes what? Does circulation cause the pressure differences, or do the pressure differences cause the circulation? There is an evident logical circularity in the account as I have expressed it, though I do not think I have done any injustice to Lanchester's formulation. In Anderson 1991, the unequivocal answer is given that pressures are fundamental and circulation is merely a way of expressing this basic physical fact. "The Kutta-Joukowsky theorem is simply an alternative way of expressing the *consequences* of the surface pressure distribution; it is a mathematical expression that is consistent with the special tools we have developed for the analysis of inviscid incompressible flow. . . . Therefore, it is not quite proper to say that circulation 'causes' lift. Rather, lift is 'caused' by the net imbalance of the surface pressure distribution, and circulation is simply a defined quantity determined from the same pressures" (218).

65. *Flight* 1915.

66. Petavel 1913.

67. Bairstow 1913.

68. Nayler 1966a, 83.

69. On Campbell's scientific work, see Warwick 1992 and 1993.

70. Campbell 1920. Campbell's philosophical discussion is enmeshed with the debate with "practical men." "The desire of many half-educated persons to rely on 'practical conclusions' rather than the reasoning of the 'theorists' is founded merely on ignorance and on an inability to differentiate between the kinds of thought likely to lead to truth and those which may be associated with error. . . . The views of 'practical men' are usually derived from assumptions and arguments no less complex than those on which theory is based; they are more and not less liable to error because they are less openly expressed" (121). The same preoccupation is evident

in Campbell's small, introductory book on the nature of science originally published in 1921. See Campbell 1952, 181–83.

71. Pannell and Campbell 1916a. (The report was only made public in 1920.) In the report the biplane is referred to as the R.E. 5. This must have been a provisional designation because the model normally called the R.E. 5 was a biplane of about half the span of the airplane under test.

72. Pannell and Campbell 1916b, 138.

73. Pannell and Campbell 1916b. Despite the numbering of the R&M, this work was performed after that described in Pannell and Campbell 1916a.

74. Cowley and Levy 1918.

75. Levy had written to Lanchester to ask about "the physical conditions that determine the cyclic constant." Lanchester wrote back suggesting a collaboration, but nothing appears to have come of this. Levy and Jones to Lanchester, May 27, 1916; and Lanchester to Levy and Jones, May 30, 1916; WFL-UC, 514.126 and 514.127, respectively.

76. For an authoritative account of the Bayesian approach in the philosophy of science, see Hesse 1974. Experimental psychologists have also made use of this theory; see Broadbent 1973, 34. For a recent discussion of some of the problems of using Bayes' theorem as a model of human reasoning, see Gigerenzer 2000.

77. Although their degree of belief might have been enhanced, it does not follow that they were convinced, and it is clear that Pannell, for one, was far from convinced. He had a confrontation with Lanchester over the aerodynamic significance of formation flying. See Lanchester 1917b and Pannell 1917.

Chapter Five: Two Traditions

1. Frank 1951, 99. Frank's work is discussed in chapter 10.

2. Wimperis 1926, 668.

3. Lamb 1932, 84–85; Streeter 1948, 132. The boat-ellipse analogy can be used to explain why a drifting boat settles across the stream. See Milne-Thomson 1960, 170–71.

4. G. P. Thomson, "British Science in War Time," n.d., talk given in Canada during World War II, GPT-TCC: GPT H, file 92. In the talk Thomson drew on his extensive experience in both world wars. In World War II Thomson was in charge of the secret committee that oversaw the British atomic bomb project.

5. H. Levy and R. Jones to Lanchester, May 27, 1916, FWL-UC, 514.126.

6. Outside Cambridge, Rayleigh's paper was sometimes read differently. Prof. Friedrich Ahlborn, of Hamburg, saw it as a demonstration of the necessary role of friction in the creation of aerodynamic forces and used it to criticize those whom he saw as committed to an inviscid approach. Ahlborn 1927.

7. EP-CUL, vol. 39, 1910, p. 845.

8. "A rectangular aeroplane of considerable length and of breadth b is moving with velocity V in a direction perpendicular to its length, and inclined at an angle α to its plane. On the assumption that the problem may be treated as that of discontinuous motion in two dimensions of an incompressible fluid past a lamina, obtain a formula to determine the resultant thrust per unit length of the plane and the position of its centre of pressure." Question 4, Maths. Tripos pt. II, Friday, June 3, 1898, EP-CUL, vol. 17, 1898, p. 734.

9. CUA, examination lists, 24, Maths. 1910.

10. Ramsey 1913.

11. Lamb 1879.

12. In reviewing the book Rayleigh conveyed the impression that Lamb had been somewhat conservative. "During the last few years," said Rayleigh (1916), "much work has been done in connexion with artificial flight. We may hope that before long this may be coordinated and brought into closer relation with theoretical hydrodynamics. In the meantime one can hardly deny that much of the latter science is out of touch with reality."

13. Stokes 1899.

14. For the significant impact of meteorology on Cambridge physics, see Galison 1997, chap. 2.

15. The accuracy of Stokes' law became a factor in the disputes surrounding Millikan's measurements. See Holton 1978.

16. According to the philosopher of science Mary Hesse, this distinction could be identified as a "coherence condition" of the (British) conceptual network. See Hesse 1974, 51–57.

17. Fuhrmann 1910.

18. Prandtl 1914b.

19. Fuhrmann 1911–12.

20. Prandtl 1923.

21. Tietjens 1931.

22. Munk 1981.

23. On Navier, see Rouse and Ince 1957, chap. 12, and Tokaty 1994, 88–90.

24. Stokes in fact offered a nonrigorous justification for the assumption of linearity based on molecular speculation. Darrigol (2002, 139–43) points out there were a number of different approaches to the Navier-Stokes equations. These depended on different orientations to mathematical rigor and engineering practicality (the British mathematical physicists being least involved in engineering), but there was no simple divide between those who accepted and those who rejected molecular speculation.

25. In the 1950s the mathematician Clifford Truesdell made the following observation: "I could not accept the so-called derivations of the Navier-Stokes equations given in the textbooks. It seemed as unreasonable to suppose viscous stress a linear function of rate of deformation as to replace every curve by a straight line" (1955, 15). Truesdell (1952) also argued that the tests to which the Stokes equations had been subjected were, in reality, neither extensive nor conclusive and that many well-established facts indicated the falsity of the assumptions on which they were based. See also Truesdell 1968, 334–66.

26. Mises 1917, 1920. On the scope of von Mises' achievements and interests, see Frank's introduction to Birkhoff, Kuerti, and Szegö 1954, and Frank 1954b.

27. LP-MPGA, Abteilung III, Repositur 61, no. 1080. Some of this correspondence on the boundary layer was published; see Mises and Prandtl 1927–28.

28. Mises 1914.

29. On von Mises, see Gridgeman 1981 and Vogt 2007.

30. Mises 1918.

31. Prandtl to von Mises, January 8, 1919, LP-MPGA, Abteilung III, Repositur 61, no. 2386.

32. Mises 1909.

33. Mises 1945. In the preface, von Mises describes the evolution of this book from the *Flug-lehre* (1918).

34. Mises 1939b, see esp. 10 and 310. The analogy comes from Ryle 1949, 18.

35. "I am prepared to concede without further argument that all the theoretical constructs, including geometry, which are used in the various branches of physics, are only imperfect

instruments to enable the world of empirical fact to be reconstructed in our minds" (Mises 1939b, 10).

36. Mises 1922, 1930a, and 1930b.

37. Birkhoff 1960.

38. Hobson and Love 1913, vol. 1.

39. Larmor is reported as saying: "What we were suffering from was over specialisation. In his view mathematics included the whole of theoretical physics." This account is from the minutes of the discussion of Carl Runge's paper on the mathematical training of physicists, in Hobson and Love 1913, 2:598–607, quotation at 605.

40. "Prof Love . . . wished to associate himself with Sir J. Larmor's view. The ideal thing would be, every mathematician a physicist and every physicist a mathematician" (Hobson and Love 1913, 2:605).

41. Love 1906.

42. A picture of the emergence of what Lamb called the Cambridge school is given in Smith and Wise 1989, chap.6, titled "The Language of Mathematical Physics." See also the papers in Harman 1985.

43. Challis 1873. Challis aspired to a "union of mathematical reasoning with experimental research" (xxiii), but his idea of what counted as unity was clearly different from that of later workers. For a dismissive review, see Maxwell 1873.

44. The shift in emphasis is described in Garber 1998, chap. 7. Garber refers to the later, turn-of-the-century style as "theoretical physics."

45. Strutt 1924, 27.

46. Making reference to "schools" is not the prerogative of historical actors such as Horace Lamb, but it is the stock-in-trade of historians and other analysts. There is, however, a difference: actors use such terms to organize their lives, whereas analysts use them to organize their narratives. When Warwick refers to the Cambridge school in *Masters of Theory*, he is drawing particular attention to the group around Larmor. The Cambridge school, as self-identified by Lamb, surely encompassed this group but cannot be automatically equated with it. There is, however, no contradiction here. As scientists follow different professional trajectories and engage with different projects, for example, fluid dynamics rather than electrodynamics, their interactions and groupings will assume different forms, which will be expressed by the different foci of their labels and different boundaries in their classifications.

47. Föppl 1925.

48. On Föppl, see Holton 1973 and 1995, 275; Bromberg 1981. Darrigol (1993, 267–71) brings out Föppl's insistence that electrical charges are computational conveniences rather than realities. There is a detailed account of the relation between Einstein's 1905 relativity paper and Föppl's work, in Miller 1998, 142–46.

49. Schlichting 1975, 298.

50. Vogel-Prandtl 1993, 60.

51. Fricke 1905, 616; Schubring 1990, 274.

52. Föppl related his ideas on the role of mental pictures (*Gedankenbilder*) to Heinrich Hertz, who had been a student at the *technische Hochschule* in Munich. Föppl 1900a, 10.

53. A. Föppl 1910, sec. 67, 433. Examples from aerodynamics might include Lanchester's "sweep" from wing theory and the "rotary inflow factor" from propeller theory. Both were later assimilated into classical hydrodynamic theory. Prandtl eventually gave a rationalization of sweep; see Warner 1936, 91–92. The rotary inflow factor was a fudge that involved postulating a rotation in the air flowing into a propeller. It squared up theory and observation but was

unacceptable to British mathematicians because it violated Kelvin's theorem. The way to avoid this violation was explained by H. Glauert and G. I. Taylor. See Taylor 1971, 528.

54. Pyenson 1983; Manegold 1970.

55. Föppl's dismissive attitude was similar to that adopted by von Kármán in his exchange with Gustav Lilienthal (Otto Lilienthal's brother) when they clashed over the question of how to explain gliding and soaring flight. See Kármán 1922. In the private correspondence between von Kármán and Arnold Berliner, the editor, the exchange was treated as a joke. Berliner to von Kármán, February 21, 1922, and von Kármán to Berliner, March 3, 1922, TVK-CIT, 2.35.

56. Prandtl to von Mises, August 2, 1921, LP-MPGA, Abteilung II, Repositur 61, no. 1078.

57. Mises 1921, 3.

58. Warwick 2003, 252-53.

59. Lamb 1931.

60. Bryan 1910b, 243.

61. Prandtl to Taylor, November 30, 1935, GIT-TCC, D. 65, 1-22.

62. Why did Prandtl say it was "no longer" considered part of physics? I take this statement to be a reference to the long-past golden age of Gauss, much celebrated by Klein and Prandtl, when all the mathematical sciences formed a unified whole. See Kármán 1940; Klein 1979.

63. Lanchester was at the Fifth International Congress but I do not know whether he was present when Lamb made his comments about the Cambridge school. In any case Lanchester was using the label in a general (though not unreasonable) way. Given his positive view of Edward Busk, one wonders if he too was included in the Cambridge school. Busk was a product of the Engineering Tripos rather than the Mathematical Tripos.

64. F. W. Lanchester, "Memorandum by F. W. Lanchester concerning His Claim to Recognition by the Air Ministry," January 1936, FWL-RAeS, envelope 7.

65. Lanchester to Pritchard, May 16, 1931, FWL-RAeS, folder 11a.

66. PRO: DSIR 22/1-2, DSIR 22/38-40.

67. Nayler 1966a, 82.

68. Lanchester to Pritchard, August 19, 1937, p. 3, FWL-RAeS, folder 11a.

69. Ackroyd 1992, 132.

70. Sutton 1965, 92. Sutton's stress on isolation is difficult to reconcile with Lanchester's membership of the Advisory Committee for Aeronautics, though it is indisputable that Lanchester felt himself to be an outsider.

71. Levy 1945, chap. 1, and 1968.

72. On Levy, see Werskey 1988, 48-49. Werskey speaks of Levy going to Oxford to continue his "aerodynamic researches" and explains that "the country's leading theoretician in this area was based at Oxford" (48). I take this to be a reference to A. E. H. Love.

73. Bartlett 1932, chap. 16.

Chapter Six: *Technische Mechanik* in Action

1. Kutta 1910, 3.

2. Kutta 1901.

3. Kutta 1910.

4. The *Habilitationschrift* is a second thesis designed to go beyond the research required for the Ph.D. Successful completion gives one the right to present academic lectures in a German university.

5. Prandtl approached Finsterwalder's son in an effort to find a copy of Kutta's thesis but

apparently made no headway. Prandtl to Richard Finsterwalder, April 8, 1952, LP-MPGA, Abteilung III, Repositur 61, no. 447.

6. Kutta 1902.

7. Pfeiffer 1950.

8. Schubring 1981, 1994; Grattan-Guiness 1997; Archibald 2001.

9. Hashagen 1993.

10. Braun 1977.

11. Hunecke 1979; Gispen 1989, 154.

12. Predictably, not everything swung in unison. On the need to explain the differences between different parts of the TH system, see Harwood 2006.

13. Gispen 1989, 158.

14. Brämer 1941, 173.

15. Bendemann's remarks were published on pages 710–11 of the discussion following Riedler 1908, 702–7.

16. Parseval 1910.

17. For the Munich situation I am using the detailed research published by Ulf Hashagen. As well as the previously cited Hashagen 1993, see Hashagen 2000 and 2003.

18. Klein's complex role is discussed more fully in chapter 7.

19. Hashagen 2000, 270.

20. On Finsterwalder, see Hashagen 2003, 207–10, and for a tribute to Finsterwalder's influence, an indication of his professional concerns (with geometry and cartography and photographic surveying), and a bibliography, see Kneißl 1942. The quotation about Finsterwalder as the prototypical technological mathematician comes from Hashagen 1993, 78.

21. In the *Jahresbericht des Münchener Vereins für Luftschiffahrt* (1901, 20), S. Finsterwalder, of Leopoldstr. 51, is listed as having made three balloon flights. It is presumably no coincidence that the journal in which Kutta's 1902 report appeared, the *Illustrirte Aëronautische Mittheilungen*, was published by the Münchener und Oberrheinischen Verein für Luftschiffahrt.

22. Pfeiffer 1950.

23. Finsterwalder 1901–8.

24. Lilienthal 1889.

25. Kutta 1902. The reference to Lilienthal is on pages 133–34.

26. Kutta cites the picture, figure 33, that Lilienthal offers on page 88 of his 1889 book.

27. Kutta 1910.

28. I thank Dr. Zae-Young Ghim for his patience in guiding me through some of the reasoning in Kutta's paper when my recall of standard integrals and trigonometric identities proved unable to bridge a gap of forty years.

29. Finsterwalder 1909. A slightly revised version, which included diagrams, was published as Finsterwalder 1910 in the *Zeitschrift für Flugtechnik*.

30. Kutta referred to "ein allgemeiner, damals aufgestellter, seitdem von N. Joukowsky neu gefundener Satz" (1910, 4), that is, a general proposition put forward at that time and found afresh by Joukowsky.

31. Versions of the general proof can be found in any textbook, for example, Reid 1932, 62–67, and Tietjens 1957a, 163–66.

32. There has been a range of reactions to this discrepancy between the general lift theorem and the situation on the surface of the plate. Garrett Birkhoff was in no doubt that there is a "clear contradiction" here. The resolution proposed by Kutta was treated as an amusing evasion. The correct conclusion, according to Birkhoff, should be to acknowledge the "failure" of

the general theorem and accept that classical hydrodynamics is incorrect or incomplete. Ludwig Prandtl called the result a strange paradox ("ein eigenartigen Paradoxon") but immediately followed Kutta and suggested rounding off the leading edge. Prandtl's Göttingen colleague Albert Betz said it was only an "apparent contradiction" and cited Kutta's response as the correct one. See Birkhoff 1946, and 1950, 18; Prandtl 1949, 193; Betz 1935, 26–29.

33. Kutta 1911.

34. Mises 1945, 201. No source is given.

35. Hashagen 2003, 256–57.

36. Finsterwalder's enthusiasm for Kutta's work (both the aerodynamic result and the doctoral work in numerical methods for solving differential equations) contrasts with Ferdinand Lindemann's dismissive response to Kutta's Ph.D., which he described as "rather meager" ("etwas dürftig"). Lindemann was from the University of Munich, not the TH. Given the textbook status of the Kutta-Runge method, Finsterwalder's response seems to have been the more just. Hashagen 2003, 253.

37. Finsterwalder's respect is evinced in the obituary he wrote for Föppl (Finsterwalder 1924).

38. On the airship development, see Fritzsche 1992, 9–58.

39. It is sometimes called the Joukowsky condition, although Tani (1979) has argued that this is a misattribution.

40. The link is made by Mises 1945, 198–208.

41. Deimler 1912. Deimler was a *Privatdozent* at the THM. His doctorate of 1908, under Prandtl in Göttingen, had been on stability and involved an extension of the methods of Bryan and Williams. Deimler was killed on August 22, 1914, leading an infantry attack. See Finsterwalder 1913–14.

42. Kutta (1911, 67) mentions a 1910 dissertation at Jena by a certain Herr Sonnefeld. The dissertation was presumably written under Kutta's guidance. I have found no published material. Von Kármán and Burger say that there are French and Italian publications that follow Kutta closely, but I have found no English or German literature. See Kármán and Burger 1935, 91.

43. Ackroyd, Axcell, and Ruban 2001, 145–83. The book also contains translations of papers by Prandtl, Blasius, Joukowsky, and others.

44. The treatment of leading-edge singularities acquired a new urgency when flight speeds began to approach the speed of sound. See Jones 1950.

45. For the complex background of the *Verein Deutscher Flugtechniker*, and later the *Wissenschaftliche Gesellschaft für Flugtechnik*, see Trischler 1992, chap. 1.

46. The journal certainly did report research, and for a while the Aeronautical Society even had its own Laboratory Committee, which lasted from 1910 to 1913 and held eighteen meetings. Its members included A. R. Low, M. O'Gorman, B. Melvill Jones, and F. Handley Page. Some experiments were carried out, and literature was reviewed and reported in the journal. The ideas for research were similar to the program set out by the Advisory Committee for Aeronautics in 1909. The minute book of the committee is in the library of the Royal Aeronautical Society; I thank Brian Riddle for drawing it to my attention.

47. Joukowsky 1910/1912. Part 1 of Joukowsky's paper is translated in Ackroyd, Axcell, and Ruban 2001, 223–29, under the title "On the Contours of the Aerofoils of Hang Gliders." Rendering *Drachenflieger* as "hang glider" may be questioned. In 1910 the word covered powered aircraft in general; see Vorreiter 1909. In part 2 Joukowsky used the word *Drachenflieger* to describe the Antoinette powered monoplane (85).

48. There are minor notational differences between Kutta and Joukowsky. For example, in the 1910 paper Kutta used r, rather than a, for the radius and specified the lift for a wing of span b rather than citing the lift per unit length. In the 1902 paper, r was assumed to be of unit length.

49. Tschapligin's papers on wing theory have been translated. See Chaplygin 1956.

50. Joukowsky published two papers on this topic in 1906, one in French (Joukowsky 1906) and one in Russian. This latter one has now been translated as Joukowsky 2001.

51. Finsterwalder 1910, 9.

52. Joukowsky does not make this part of his argument clear. The steps that he took for granted are made explicit in Glauert 1926, 82–87. See also Batchelor 1967, 441–44.

53. Loukianoff 1912.

54. Mises 1945, 181.

55. Blumenthal was dismissed from his post at Aachen by the Hitler regime in 1933. Von Kármán tried, but failed, to get him a post in the United States. Their increasingly desperate correspondence is in the von Kármán papers, TVK-CIT, box 3, folder 3.10. Blumenthal and his wife died in the concentration camp at Theresienstadt in 1944. Annette Vogt kindly drew my attention to the moving account by Volkmar Felsch, "Der Aachener Mathematikprofessor Otto Blumenthal, Vortrag in der Volkschule Aachen," November 1, 2003, typescript.

56. Blumenthal 1913.

57. Trefftz 1913.

58. Joukowsky surely knew the formula. He made the analytical basis of both his 1910 constructions explicit in his book *Aérodynamique* (1916, 145–50). The continued use of two transformations suggests, however, that he may not have fully appreciated the generality of the "Joukowsky transformation."

59. The Joukowsky transformation involves the very formula that G. H. Hardy set as an exercise in 1908 when asking students to show that the formula transformed concentric circles into ellipses. When used to generate winglike shapes, however, it has to be applied to an off-center circle. It was also shown in chapter 2 that the formula provides a description of the flow around a circular cylinder.

60. Betz 1915. For some reason Betz's spelling of Joukowsky's name differs from that used on the title page of the journal.

61. Kármán and Trefftz 1918. On the first page the authors said that in using a sickle shape they were following an idea they had found in Kutta's 1911 paper, but were developing it in a mathematically simpler fashion.

62. Mises 1920, 72.

63. Betz 1924, 100.

64. Mises 1917, 1920.

65. Mises' approach was continued by Wilhelm Müller. See Müller 1923, 1924a, and 1924b. In 1939 Müller (1880–1968) was appointed to Arnold Sommerfeld's chair in Munich. The assumption among physicists had been that, as a theoretical physicist, Sommerfeld would be succeeded by Heisenberg. Heisenberg, however, became a target for Nazi zealots because of his association with Einstein. Müller, on the other hand, was a Nazi who wrote anti-Semitic tracts. See Litten 2000. Prandtl became involved in this affair when he wrote to Himmler on Heisenberg's behalf. The correspondence between Prandtl and Himmler is in Hentschel and Hentschel 1996, 172–78.

66. Pröll 1913.

67. The theme of the unity of theory and practice was not a new one. See Bendemann 1910.

68. On Ahlborn, see Eckert 2006, 38–42.

69. Ahlborn was right, but it is doubtful if anyone from Aachen or Göttingen thought otherwise. That a contour can be deduced from a circle by means of a mathematical transformation says nothing about its aerodynamic excellence. It is no guarantee that it will be superior to one created by other methods whether scientific or intuitive. Bairstow made the point in the second edition of his book. The Joukowsky transformation, he said, "has no special validity in the sense that it produces the correct form for an efficient wing" (Bairstow 1939, 325).

Chapter Seven: The Finite Wing

1. Ackeret 1925, 31. On Dirichlet, see Ore 1981.

2. In what follows I largely adopt the notation used by Prandtl in the published version of his lectures, that is, as in Tietjens 1931.

3. For a detailed account of the relevant history and politics of German aeronautical research, see Trischler 1992, pt. 1.

4. Prandtl 1904. In his original presentation Prandtl used vector notation.

5. Arnold Sommerfeld said Klein congratulated Prandtl on giving the best paper in the conference: "Ihr Vortrag war der schönste des ganzen Kongresses" (Sommerfeld 1935, 1).

6. Goldstein 1969.

7. Lighthill 1995, 796.

8. Eckert 2006, 32–38.

9. Boundary-layer removal became a significant topic of research for the Göttingen group. For the development of the suction idea and its application to aircraft wings, see the account of the work of Ackerat and Schrenk in Tietjens 1931, 95–96.

10. In his boundary-layer work Prandtl used what is, today, called the "method of matched asymptotic expansions" to address a "singular-perturbation problem." See Germain 2000, 31–40, and Lighthill 1995, 800–803, who gives an elementary example to illustrate the mathematical ideas involved. Viewed in terms of this mathematical technique, versions and anticipations of the boundary-layer idea can be found in earlier work by a number of other authors such as Rayleigh, Helmholtz, Love, Stokes, and Maxwell. See Van Dyke 1994. Prandtl's theory of the finite wing is now also recognized as a singular perturbation problem; see M. Van Dyke 1964, 602. I thank Walter Vincenti for drawing my attention to Van Dyke's work and Horst Nowacki for explaining some of the difficulties of attributing these new methods to earlier thinkers, including Prandtl himself.

11. Blasius 1908.

12. Boltze 1908; Hiemenz 1911; Töpfer 1912.

13. On the history and diffusion of the theory, see Dryden 1955 and Tani 1977.

14. Schlichting 1951.

15. Nickel 1973.

16. Finsterwalder 1910, 8; Kutta 1911, 72–73. There is also a brief, qualitative account of Prandtl's boundary-layer theory in the sixth volume of August Föppl's *Vorlesungen.* See Föppl 1910, 371–72.

17. O. Föppl 1912, 121. There were also other complications. The value of the coefficient is sensitive to the precise way in which the sphere is mounted in the wind channel. See Mises 1945, 100.

18. Prandtl 1914a and Wieselsberger 1914c.

19. Rowe 2001.

20. Prandtl 1925c.

21. Reid 1996; Dieudonné 1981a, 1981b; Pyenson 1979a, 1971b, 1982; Sigurdsson 1994, 1996. See also the papers in Gray 1999.

22. Mehrtens 1990, 394–400. Mehrtens uses the terms "modern" and "countermodern" in a technical rather than an everyday sense. To have a modern approach means treating the symbolic systems of mathematics as self-contained. A countermodern approach involves treating them as having an external reference and as deriving their truth and significance from something outside themselves.

23. Glas 1993, 2000; Rowe 1986, 1994.

24. Rowe 1989, 206.

25. Manegold 1968, 1970; Pyenson 1983; Schubring 1989; Tobies 1989, 2002.

26. Prandtl's own appreciation of Böttinger, with a brief sketch of his role, is given in an obituary. See Prandtl 1920a.

27. Runge and Prandtl 1907.

28. The document is quoted in Manegold 1970, 24.

29. "Daß in Göttingen Luftfahrtwissenschaft getrieben werden müsse, war allein seine Idee" (Prandtl 1925c).

30. For a detailed account of subsequent institutional changes (such as the role played by the Kaiser-Wilhelm-Gesellschaft), and all the negotiations surrounding finance and politics in which Klein, Böttinger, and Prandtl worked closely together, see Rotta 1990 and Trischler 1992.

31. Rowe 2004. The picture of these meetings as intimidating comes from Max Born.

32. Vogel-Prandtl 1993.

33. The commitment to this tradition is symbolized by the collection of scientific papers presented to August Föppl in 1924. As well as papers by Prandtl, and by Föppl's two sons, this collection also contained contributions from von Kármán and Timoschenko. See O. Föppl et al. 1924.

34. Runge and Prandtl 1907.

35. Runge 1949, 123.

36. Born and von Kármán shared a house with a lively group of friends. T. von Kármán, "Aus meiner Göttinger Studienzeit," unpublished talk, ca. 1957, TVK-CIT, 118.35. Prandtl seemed to have little awareness of von Kármán's interest in modern physics. This led to an embarrassing episode when Prandtl unwittingly sabotaged von Kármán's application for a physics chair at Göttingen. Prandtl to von Kármán, April 23, TVK-CIT, 23.41.

37. This picture is supported by the accounts given by Rotta 1990 and Eckert 2006. Prandtl's teaching commitments also point to an environment devoted to technical mechanics. For a year-by-year breakdown of Prandtl's teaching commitments covering the whole of Prandtl's career, see Wuest 2000.

38. Mises 1924, 88.

39. Klein 1900. See especially the four characteristics listed on p. 28.

40. Lorenz 1903. On the Lorenz episode, see Staley 1992, chap. 3.

41. The episode is discussed in Manegold 1970, 128–36. Pyenson points out that Nernst had used a similar metaphor the previous year in a letter to Althoff. See Pyenson 1983, 63.

42. Föppl 1901, vi.

43. The continued predominance of those with engineering training is supported by the full list of members of the Prandtl group at Göttingen. The list is given by Betz in the history of the institute that he wrote during World War II (Betz 1941, 63–66).

44. Dryden 1965; Kármán 1963, 67–73.

45. Kármán 1911–12; Kármán and Rubach 1912.

46. Von Kármán was a raconteur who could not resist telling stories at Prandtl's expense. He claimed that when Prandtl approached August Föppl for the hand of one of his daughters, he forgot to specify which one, so the family chose their elder daughter to be the required Frau Prandtl. Not surprisingly Prandtl's own daughter, Johanna, firmly denied this scandalous suggestion about her parents (Vogel-Prandtl 1993, 47).

47. Prandtl 1909.

48. Prandtl 1910a.

49. Prandtl 1910b.

50. Föppl 1900b, vii.

51. Prandtl 1918b, 451.

52. Prandtl 1912, 35.

53. Prandtl 1920b, 44.

54. Föppl 1897, 63–66; 1904, 106; and 1910, sec. 59, "Zusammenhang der Strömungsproblemen mit Problemen aus der Lehr vom Magnetismus."

55. The Biot-Savart law was formulated in 1820. On its history, see Whittacker 1951, 82–83. The electrical-hydrodynamic analogy was also to be found in British textbooks, for example, Campbell 1907, 20. Campbell was at pains to point out the limits of the analogy and to insist that the theory of perfect fluids does not provide a basis for understanding physical reality.

56. Föppl 1897, 66.

57. O. Föppl 1911a.

58. O. Föppl 1910, 1911b.

59. Prandtl 1912, 34–35.

60. A full derivation is given in Glauert 1926, 157–62.

61. Betz 1912.

62. Reid 1932, 166–76.

63. Betz 1913.

64. Betz produced a follow-up study of the centers of pressure of the biplane configurations used in this experiment: Betz 1914a.

65. Wieselsberger 1914a.

66. Betz 1914c.

67. Betz 1914b.

68. Wieselsberger 1914b.

69. Reißner 1912b.

70. Fromkin 2004. This work summarizes research that points to a war party in Berlin intent on engineering a war between Germany and Russia. For an account of the changing theories as to the causes of World War I, see Mombauer 2002.

71. Fuhrmann 1913; Prandtl 1913. Prandtl's article was reprinted as *Abriß der Lehre von der Flüssigkeits- und Gasbewegung* (Jena: Fischer, 1913).

72. Betz 1918.

73. Betz 1919.

74. The term "family resemblance" comes from Reid 1932, 90.

75. Tietjens 1931, 219.

76. Betz 1917b.

77. Munk 1917e.

78. Prandtl 1925a, 50–53.

79. The dating, and E. Pohlhausen's role, is taken from Prandtl 1918b, 474. Moritz Epple (2002, 180) points out that von Kármán dates this step about one year later.

80. Munk 1917c, 136.

81. Prandtl 1920b, 50.

82. Mises 1945, 243.

83. Siegmund-Schultze 2004, 364. The word "remarkable" occurs in Mises 1945, 235, 241, and 243.

84. Prandtl 1948, 92.

85. The relation between vortex lines and streamlines was emphasized when Prandtl produced a general summary of the theory in 1918. Thus Prandtl (1918b, 461) says, "die freien Wirbellinien unter den hier vorausgesetzen Bedingungen identisch sind mit Stromlinien. Diese Beziehung ist für die späteren Entwicklungen von größter Wichtigkeit." That is, "under the conditions that are being assumed, the free vortex lines are identical with the streamlines. This relationship is of the highest importance for the later development [of the theory]."

86. Rotta 1990, 190.

87. A picture of curved (bogenförmig) trailing vortices is found in Finsterwalder 1910, 6. Both the original Lanchester curved figure and Prandtl's later, straight-line figure, were given in Grammel 1917. Grammel described the curved figure as an early account of the vortex structure, while the newer version of the theory had the vortices going straight back, pulled by the free stream ("die von der Hauptströmung . . . sofort nach hinten mitgerissen wird"; 114). The source cited for the newer approach was Prandtl 1913. Grammel does not say whether Prandtl was merely correcting Lanchester or also correcting his, Prandtl's, own earlier views.

88. Lanchester 1907, 173. Lanchester gives a drawing of the apparatus on p. 174 (fig. 80). The text does not make it clear just how much Lanchester saw over and above a few dimples on the surface of the water.

89. Other physical processes might also have contributed to the visually detectable effect. Although the lift-giving circulation is zero at the tips, there will be a movement of air outward along the span on the underside which goes around the tip and inward along the span on the upper surface. This could contribute toward making the vortex sheet more visible at the tips. I thank Horst Novacki for this observation.

90. Parseval 1920, 63.

91. Munk 1917g, 199.

92. For an overview of the wartime work, see Rotta 1990, 115–97, and Eckert 2006, chap. 3.

93. Betz 1917a; Prandtl 1917, 1918a.

94. Undercarriages and machine-gun mountings: Wieselsberger 1918b; engine-cooling systems: Kumbruch 1917, Munk 1917h, Wieselsberger 1918c; and struts and bracing wires: Pohlhausen 1917, Munk 1917b, 1917k.

95. Lift and drag of the fuselage: Munk 1917d; effect of dividing the wing: Munk and Cario 1917; forces on fins and rudders: Munk 1917i, 1917j, 1918; and properties of the triplane configuration: Wieselsberger 1918a.

96. Munk 1917f.

97. Munk and Molthan 1918; Molthan 1918.

98. Munk and Cario 1918.

99. Munk 1917c; Munk and Pohlhausen 1917; Munk and Hückel 1917a, 1917b. Erich Hückel (1896–1980) was later to work on general relativity at Göttingen with David Hilbert. See Rowe 2004, 100.

100. Munk 1917a.

101. Munk and Hückel 1918.

102. Camm 1918, 489.

103. Eckert 2006, 76–77. Eckert's source is Weyl 1965, 225.

104. Munk 1917g.

105. Munk and Pohlhausen 1917, tables 149 and 151. See also Air Ministry 1925, 65–66, fig. 17.

106. The identification of the Göttingen 95 profile as the Einstein wing was made by Peter M. Grosz. I thank Dieter Hoffmann and Nelson Studart for drawing my attention to Grosz's discovery.

107. Einstein and Ehrhardt corresponded in August and September 1954. See *Interavia* 1955.

108. Einstein 1916. A reprint and notes on the incident are in Knox, Klein, and Schulmann 1996, 400–402.

109. Lamb 1932, 81–83.

110. Tietjens 1931, 174–75.

111. Rotta 1990, 194.

112. Rotta 1990, 195–96.

113. Prandtl 1920b.

114. Prandtl 1918b, 1919. The second report was presented on February 21, 1919. Here Prandtl summarized the work on multiplanes and used the theory to discuss wind-channel corrections.

115. Epple 2002.

116. Both Betz and Munk submitted parts of their wartime work for doctorates at Göttingen. See Betz 1919 and Munk 1919. Munk presented the mathematical side of his work for his Göttingen dissertation while also submitting the more empirical side to the TH at Hanover, from which he also received a doctorate. The authorities in Hanover, however, felt that Munk was treating them as second best. The old disparities between frontline officers and staff officers still rankled. See the account in Eckert 2006, 73–76.

Chapter Eight: "We Have Nothing to Learn from the Hun"

1. Taylor 1966, 113. Taylor's view was that the reconciliation called for a theory of turbulence because all motion in a real, viscous fluid, such as a wing moving through the air, would generate turbulence. As Taylor put it in his Wilbur Wright Lecture of 1921: "The ordinary hydrodynamical theory is therefore quite inapplicable to the case of bodies moving steadily through the air. One must seek for the explanation of the forces that are observed in these cases in the action of the eddying region on the flow." See Batchelor 1963, 3:53.

2. Wilde 1966, 1205 (orig. 1894).

3. Munk 1981, 2.

4. Buchan's most famous novel, *The Thirty-Nine Steps*, features foreign spies, a remote Scottish setting, and airplanes (Buchan 1915). Buchan's "shocker" was published later than the Dunne tests but in ample time to embellish the collective memory of the events. It is also possible that the stories surrounding the test and the events in the novel came from common sources. See Daniell 1992.

5. "The reports of the ACA dealing with aeroplane stability were not published during the war and the Germans were anxious to get these papers. It usually took about eighteen months before they saw copies" (Nayler 1966a, 83).

6. This would not have been difficult. There was little effective, covert intelligence activity in Britain on behalf of the Germans during World War I. See Boghardt 2004.

7. The wartime aeronautical press in Germany contained accounts of experimental work done at the NPL, but the data were prewar. There was also an account of the new aerodynamic research facilities at the Massachusetts Institute of Technology, but it must be remembered that the Unites States did not enter the war until 1917. See Munk 1915 and Wieselsberger 1916. The archives of Prandtl's institute in Göttingen contain a German translation, dated November 29, 1916, of the yearbook of the ACA for 1915–16, but this too could have arrived by open channels. LP-ADGLR, box 101–121, 1352. The same applies to the numerous extracts from technical journals (*Auszüge aus technischen Zeitschriften*) that were provided in the *Technische Berichte*.

8. Greenhill gave two papers: 1905a and 1905b.

9. Bryan 1904. A brief, qualitative account of Prandtl's boundary-layer work found its way into the fourth edition of Lamb's *Hydrodynamics* published in 1916, but it did not receive a mathematical discussion until the fifth edition of 1924.

10. In the stilted prose of a report in *Flight*: "One of the first things performed by the Committee had been the making of systematic arrangements whereby it should be kept in close touch with everything that was being done in connection with the study of flight all over the world" (*Flight* 1909c, 470).

11. Preliminary List of Periodicals and General Works on Aeronautics, PRO: DSIR 23/1.

12. Advisory Committee for Aeronautics, first meeting, minute 4, and second meeting, minute 16, PRO: DSIR 22/1.

13. Reports and Memoranda No. 1 (May 12, 1909) was reprinted as an appendix to the *Report of the Advisory Committee for Aeronautics for the Year 1909–10* (London: His Majesty's Stationary Office, 1910), 129–38.

14. "The report of the Advisory Committee for Aeronautics for the year 1913–14, was ready for issue at the outbreak of the war. It was immediately made a strictly confidential document but was communicated under terms of secrecy to certain British manufacturers." L. Bairstow, "War Type Designs of Heavier-Than-Air Craft," May 8, 1921, PRO: AIR 1/21 15/1/107.

15. Who scanned the foreign-language works and produced the abstracts? In one case, in a letter to Lanchester, Glazebrook said that F. J. Selby, secretary of the committee, was translating Prandtl's paper on the resistance of a sphere, but it is unclear whether this was a special case or an instance of the general procedure. Glazebrook to Lanchester, June 9, 1914, FWL-UC, F. 4, letter 4–7. Lanchester had drawn attention during a committee meeting to the importance of this paper. Meeting 54, July 7, 1914, minute 522, PRO: DSIR 22/1.

16. Advisory Committee for Aeronautics, meeting 17, December 6, 1910, minute 172, PRO: DSIR 22/1.

17. Selby wrote a belated acknowledgment to Prandtl on September 22, 1913, LP-ADGLR, Schriftwechsel, 1903–25, 3675. Selby also spoke of the forthcoming visit, but I have been unable to find any further account of it. From letters written after the war, however, it is clear that the visit took place and that Stanton was Prandtl's guide. See Prandtl to Fage, December 23, 1925, LP-MPGA, Abteilung III, Repositur 61, no. 425; and Prandtl to Glazebrook, March 3, 1926, LP-MPGA, Abteilung III, Repositur 61, no. 537.

18. These points were made explicitly in the letter to Lanchester of June 9, 1914 (FWL-UC, F. 4, letter 4–7.), which dealt with the forthcoming discussion of Prandtl's work on the resistance of spheres. As discussed previously, this work was prompted by discrepancies between wind-channel results in Paris and Göttingen. Lanchester had suggested that the discrepancy pointed to the total unreliability of wind-channel work. Glazebrook thought this was "alarmist"

and said, "I am a little afraid that if we circulated your note as it is it might land us in serious difficulties with the financial authorities who would be glad to cut down expenditure on experiments, and with the so-called 'practical man' who thinks nothing is to be gained from scientific investigation."

19. *Technical Report of the Advisory Committee for Aeronautics, 1911–12* (London: His Majesty's Stationary Office, 1912), 257–58.

20. On the pressure of the war years see *Technical Report of the Advisory Committee for Aeronautics for the Year 1915–16* (London: His Majesty's Stationary Office, 1916), 9.

21. Grey 1918, 381. Grey did, however, accept that there was much to learn from Germany in matters of quantity production.

22. For example, an editorial in *Aeronautics* of August 16, 1916, declared that "the Hun neither during the war nor before its outbreak ever originated a new principle in aeronautical design or construction."

23. Consider the remarks of Petavel, the chairman of the Aerodynamics Sub-Committee. On Tuesday, February 5, 1918, the committee (at a meeting attended by Glazebrook, Bairstow, O'Gorman, and Lanchester, among others) was discussing tests that had been made on a German Albatross wing section. Petavel said "that if as it appeared the German wing sections were less efficient than our own, it would seem there must be some additional merits in their aeroplanes which we had not yet discovered." Bairstow, however, was convinced of the aerodynamic inferiority of the German machine. Minutes of the Aerodynamics Sub-Committee, minute 53, PRO: DSIR 22/38.

24. Referring to the experimental work done at the behest of the Aerodynamics Sub-Committee, its chairman, Petavel, said it "had contributed in no small measure to the wonderful successes won recently by the allied arms." Minutes of the Aerodynamics Sub-Committee, meeting 12, October 1, 1918, minute 190, PRO: DSIR 22/38.

25. Dr. Audrey Glauert kindly gave me a photocopy of this rare document. For extracts, notes on its provenance, and a glossary of the "Book of Aeron," see *Journal of the Royal Aeronautical Society* 70 (1966): 85–88.

26. "I thought then that my messmates were rather a bright lot, but I never imagined that among them were three future peers of the realm, five Knights, three Nobel prize winners, professors and Fellows of the Royal Society galore" (Green 1958, 7). F. M. Green played an important role in the design of the engine for the SE5 and was, like O'Gorman, a victim of the political hate-campaign described earlier. See also McKinnon Wood 1960.

27. Technical Report T. 1252, Advisory Committee for Aeronautics, National Physical Laboratory, "Report as to the Position of Work, as the Programme for 1919–20," November 1918, PRO: AIR 2/2582. See also *Technical Report of the Aeronautical Research Committee for the Year 1920–21* (London: His Majesty's Stationary Office, 1921), 15.

28. Bairstow 1922 and Glazebrook 1923b. See also Glazebrook 1924.

29. An account of Knight's activities is given in Eckert 2006, 84–91.

30. A. Toussant, "Résumé of Theoretical Works in Aerodynamics at the Göttingen Laboratory. Published in the *Technische Berichte*" (extract from *Review of Aeronautical Works*, no. 2, U.S. National Advisory Committee of Aeronautics), Ae. Tech. 48, February 1920, PRO: DSIR 23/8805.

31. Col. Dorand, "Special Report on the New Aerodynamics Laboratory at Göttingen," T. 1516. Dorand's report and its provenance were noted in Aerodynamics Sub-Committee, meeting 33, November 2, 1920, minute 329, PRO: DSIR 22/39.

32. Aerodynamics Sub-Committee, meeting 33, minute 333 (c) (vii).

33. Technical Report T. 1516a, October 1920, PRO: DSIR 23/1530. As far back as 1912 lack of steadiness and uniformity in the flow had been an acknowledged problem for the National Physical Laboratory. See *Advisory Committee for Aeronautics: Report for the Year 1911–12* (London: His Majesty's Stationary Office, 1919), 7.

34. Minutes of the Aeronautical Research Committee, meeting 3, July 13, 1920, minute 28, PRO: DSIR 22/2.

35. This letter is quoted as document 2–19 in Hansen 2003, 537.

36. The letters exchanged between Knight and Prandtl describing these problems are found in LP-MPGA, Abteilung III, Repositur 61, no. 836–37.

37. Minutes of the Aeronautical Research Committee, meeting 9, March 8, 1921, minute 94, PRO: DSIR 22/2.

38. The British eventually published a two-volume translation of selections of the material from the technical reports. See Air Ministry 1925.

39. Bairstow, "War Type Designs of Heavier-Than-Air Craft," Appendix 3, p. 1. PRO: AIR 1/21 15/1/107.

40. Minutes of the Aeronautical Research Committee, meeting 1, PRO: DSIR 22/2.

41. The finances of the NPL had always been a source of difficulty, and the problems reached a new intensity under the DSIR. One can only admire the way in which Glazebrook battled against some of the devious bureaucracy with which he had to deal. See Hutchinson 1969. On the origins of the DSIR and the important role played by Haldane, Rayleigh, and Glazebrook, see Varcoe 1970 and MacLeod and Andrews 1970.

42. Minutes of the Aerodynamics Sub-Committee, meeting 9, July 2, 1918, PRO: DSIR 22/38.

43. Nayler mentions "differences" between Bairstow and Glazebrook but does not elaborate. Nayler 1966a, 83.

44. For a hostile, but no doubt well-deserved, account of Zaharoff's profit-making activities, see Brockway 1933, chap. 4. See also Allfrey 1989.

45. Hilken 1967, 158–63.

46. Minutes of the Aerodynamics Sub-Committee, meeting 14, December 3, 1918, minute 209, PRO: DSIR 22/38.

47. F. W. Lanchester, memorandum by F. W. Lanchester concerning his claim to recognition by the Air Ministry, January 1936, FWL-RAeS, envelope 7.

48. The exchange, in which Prandtl wrote in German and Southwell in English, is in LP-MPGA, box 21–40, 1365.

49. McKinnon Wood 1966 and *RAE News* 1968.

50. This information comes from the diaries of David Pinsent. I am grateful to Anne Pinsent, David Pinsent's niece, for permission to see the relevant diary entries. This was made possible through the good offices of Audrey Glauert. Pinsent was killed in a flying accident at Farnborough during the war. Wittgenstein, who was then in the Austrian army, dedicated the *Tractatus* to his memory.

51. Farren and Tizard 1935.

52. Glauert 1915a, 1915b. The two parts of 1915b provided an explanation of errors in the position of the Sun, Venus, and Mercury in terms of minute irregularities in the rotation of the earth. The problem was suggested to Glauert by A. S. Eddington. The 1915a paper was a generalization of work done by Sir George Darwin. Darwin had analyzed a rotating mass of viscous fluid in which all relative motion in the fluid had been damped by the viscosity. Glauert generalized the result by allowing for relative motion. To do this he assumed the fluid was inviscid and

satisfied Laplace's equation. (Glauert's later work in aerodynamics also involved moving from viscous to inviscid fluids.)

53. Hotel Hessler stood opposite the Zoo Bahnhoff, at Kantstraße 165–66. The area was heavily bombed during World War II, and the original building no longer exists.

54. Knight described himself in the letter to Prandtl as "accredited . . . to the British, French and Italian governments . . . through the U.S. N.A.C.A." It is unclear what, if anything, this means.

55. Bairstow 1920.

56. Keynes 1919. Keynes was a fellow of Kings College and had been twelfth wrangler in 1905. He was attached to the Treasury during the Great War and was an official representative at the Paris Peace Conference. For the Cambridge background to Keynes' prescient book and the economic circumstances in Europe that resulted from the war and the peace treaty, see Skidelsky 1983 and 1992. Keynes found himself engulfed in policy battles in which the confrontation was defined, by the actors themselves, in terms of "practical men" versus "Tripos mathematicians." Here "practical man" meant "banker." See, e.g., Skidelsky 1992, 326, 399.

57. Glauert to Prandtl, February 27, 1921, May 23, 1922, and November 8, 1923, LP-MPGA, Abteilung III, Repositur 61, no. 536.

58. Schroeder-Gudehus 1973. I return to this theme in chapter 10.

59. R. McKinnon Wood, "The Aerodynamics Laboratory at Göttingen," T. 1556, February 1921, PRO: DSIR 23/1580.

60. The Air Ministry vetoed the suggestion. Minutes of the Aeronautical Research Committee, meeting 28, December 12, 1922, minute 271, PRO: DSIR 22/3. The ban on formal relations and information exchange with Göttingen was not lifted until 1926. Minutes of the Aeronautical Research Committee, meeting 63, March 8, 1926, minute 583, PRO: DSIR 22/4.

61. Bairstow to Prandtl, February 9, 1926, LP-MPGA, Abteilung I, Repositur 44, no. 269.

62. Farren and Tizard 1935, 607.

63. For the pressures on British scientists of German extraction, see Badash 1979.

64. Prandtl's mysterious visit to the NPL in 1913 does not appear to have helped the cause of the circulation theory, but this may have been because of language problems. Before the war Hyman Levy and Robert Jones (a student of G. H. Bryan) visited Göttingen. Both were to work at the NPL during the Great War, but neither brought back from Germany news of, or enthusiasm for, the circulatory theory. On Jones' visit, see Hashimoto 1990, 37–38.

65. There may be a temptation to seize upon men like Glauert, who break ranks, and to interpret their behavior as evidence for some form of individualistic analysis. For example, Glauert's behavior may be used as an excuse to go back to "an older, more heroic history," that is, one in which individuality is set in fundamental opposition to society, and creativity is treated as the overcoming or avoidance of social influence. For this approach to innovation, see Constant 1980, 32. The stance that I should recommend is one that treats human agents as complex causal mechanisms that are in constant causal interaction with one another and are thus never truly outside the overall causal nexus of society. These interactions constantly change both the individual participants and the culture they sustain. As an antidote to individualism, see Barnes 2000. Barnes criticizes individualism as a stance in the analysis of moral action, but his arguments are readily transferred to the analysis of cognition in general.

66. Glauert 1920a.

67. H. Glauert, "Aerofoil Theory," T. 1563, PRO: DSIR 23/1577. This report later became R&M 723.

68. Farren and Tizard 1935, 609.

69. The figures to which North was referring must have been figures 85 and 86 on

pp. 177–78. These were the ones that had prompted doubts in Prandtl's mind and led him to the horseshoe model.

70. A wing providing lift must deflect air downward, thus creating a "downwash." By basic Newtonian principles the angle of downwash will be roughly proportional to lift.

71. The BE2E was a biplane, and for a biplane the downwash angle is the sum of the effects due to the two individual wings.

72. For a wing with an elliptical lift distribution, the induced normal velocity along (say) the trailing edge is constant. (Or, more precisely, the induced normal velocity is constant along the vortex line that represents the wing.) This may seem to conflict with Glauert's finding, but it does not mean that the downwash is constant, only that one of the factors contributing to the downwash is constant. Glauert was looking at the overall flow picture in the air surrounding the wing rather than the situation directly on the wing itself.

73. The approximations involved making allowance for wing stagger and taper. Glauert had discussed the Göttingen work on stagger in his previous report "Aerofoil Theory." He explored Trefftz's work on tapered wings in Technical Report T. 1756, which became Reports and Memoranda No. 824 in October 1922.

74. The quotation is from Simmons and Over 1924. Their aim was to see if the predicted vortices had "an actual existence." Their conclusion was that "the general agreement obtained may reasonably be taken as evidence in support of the theory." For an account of the problems involved in understanding downwash, see Reid 1932, 193–96.

75. Farren and Tizard 1935, 610.

76. Low 1923b. The discussion follows directly on pp. 60–72. Bairstow's introduction is on p. 37.

77. Low's remarks had been made in response to a talk by Glazebrook. See Glazebrook 1920, 451 (quotation).

78. Low joined the Fluid Motion Panel on June 10, 1931. He wrote report T. 3217, "Analogy between Taylor's Criterion for Instability of Steady Motion between Rotating Co-axial Cylinders and Rayleigh's Criterion for Instability of a Layer of Viscous Fluid Heated from Below." Taylor's comments are noted in the minutes of the Fluid Motion Panel, meeting 10, April 19, 1932, minute 61; and meeting 13, February 10, 1933, minute 75, PRO: DSIR 22/49. Southwell, from the chair, also congratulated Low on his results.

79. Citation practices became problematic during the Great War. Led by Wilhelm Wien, some German scientists campaigned against undue deference to foreign, and particularly English, scientists. See Wolf 2003. There was, however, little response to Wien's *Aufforderung*. It is noteworthy that Joukowsky's name remained on the title page of the *Zeitschrift für Flugtechnik* despite Germany and Russia being at war.

80. Low's talk was reported at length in *Engineering*. See Low 1922.

81. McKinnon Wood 1922. This was followed by a discussion reported on pp. 497–501.

82. Piercy 1923a, 1923b. Piercy played a significant role in British aeronautics in the interwar years and during World War II. See Ackroyd, Bernstein, and Armstrong 2008, 365–74.

83. M. Glauert 1923. The earlier report was T. 1675, "Theoretical Streamlines round a Joukowsky Aerofoil," which later became R&M 788. Southwell noted at a meeting of the Aerodynamics Sub-Committee that photographs taken by Relf of the flow of water round a Joukowsky aerofoil "did not agree with Miss Barker's calculations." Minutes of the Aerodynamics Sub-Committee, meeting 47, April 4, 1922, minute 450, PRO: DSIR 22/40.

84. Bieberbach was to gain notoriety a few years later when, during the Third Reich, he became an advocate of "Aryan mathematics." See Lindner 1980.

85. Glauert 1923d.

86. Glauert to Prandtl, November 8, 1923, LP-MPGA, Abteilung III, Repositur 61, no. 536.

87. Bairstow 1923.

88. Stanton and Pannell 1914.

89. As a model for such a deduction one may take Stokes' own explanation of Hele-Shaw's photographic representation of the streamlines of an inviscid fluid. See Stokes 1899. Bairstow gave a selection of Hele-Shaw's photographs in his *Applied Aerodynamics* (Bairstow 1920, 350).

90. Bairstow, Cave, and Lang 1922.

91. Bairstow, Cave, and Lang 1923.

92. On the background of Stokes' work on the sphere, see Darrigol 2005.

93. See Lamb 1916, 596–601.

94. Lamb 1911.

95. Prandtl to Bairstow, October 17, 1923, LP-MPGA, Abteilung III, Repositur 61, no. 77.

96. Glauert 1923c.

97. Boyer 1959, 216, 228.

98. In *The Analyst* of 1734 Bishop Berkeley famously, and effectively, made fun of Newton's obscure appeal to "evanescent" quantities in the development of the calculus. May we not, said Berkeley, "call them the ghosts of departed quantities?" (Fraser 1871, 283). Glauert was perhaps fortunate that Bairstow did not exploit the historical parallel and respond polemically by calling Glauert's evanescent viscosity "the ghost of a departed boundary layer."

99. Low 1923a.

100. For specifications and photographs of the various products of the Albatros Werke of Johannisthal, near Berlin, see Jane 1990.

101. AUE, "Graduates in Arts under the New Ordinances," 1901–3.

102. Geckeler 1922.

103. The information about the identity of the chairman is given on pp. 239 and 298 of the report of the congress.

104. Glazebrook 1923c. The discussion is on pp. 60–67.

105. Glazebrook's excuse has filtered into the literature. I mentioned in the introduction to this volume that it was accepted in Kingsford's 1960 biography of Lanchester. It was also accepted by the mathematician O. G. Sutton. Lanchester, said Sutton, "either could not, or would not, express his conclusions in the usual mathematical form." As a result they were, "at times, difficult to comprehend" (Sutton 1954, 156). The idea that Lanchester's ideas were not "readily accessible" because his "work was unclear and qualitative" is repeated in Constant 1980, 106.

Chapter Nine: The Laws of Prandtl and the Laws of Nature

1. Munk 1981, 2.

2. Christopherson 1972, 552.

3. Glauert to Prandtl, November 8, 1923, LP-MPGA, Abteilung III, Repositur 61, no. 536. A further positive sign was that Glauert had been invited by Glazebrook to contribute two articles to the multivolume *Dictionary of Applied Physics* that he was editing. The subject matter of the articles was, however, carefully circumscribed. They were presented, accurately but surely significantly, as the application of classical hydrodynamic theory to aeronautics. See Glauert 1923a, 1923b.

4. Muriel Barker, "Theoretical Streamlines round a Joukowsky Aerofoil," T. 1675, which, in December 1921, became R&M 788.

5. Minutes of the Aeronautical Research Committee, meeting 19, February 14, 1922, and meeting 21, April 11, 1922, PRO: DSIR 22/3. Minutes of the Aerodynamics Sub-Committee, meeting 45, February 9, 1922; meeting 46, March 7, 1922; and meeting 47, April 4, 1922, PRO: DSIR 22/40.

6. A. J. Sutton Pippard was a leading expert on aircraft structures and braced frameworks of the kind used in airships. See Skempton 1970.

7. R. V. Southwell, "On the Use of Soap-Films for Determining Theoretical Stream Lines round an Aerofoil in a Wind Tunnel," T. 1696, March 1922, PRO: DSIR 23/1710.

8. The method was originally used to solve equations in the theory of elasticity. See Griffith and Taylor 1917. On the work of Taylor and Griffith, see Hashimoto 1990, 79–84.

9. G. I. Taylor, "Note on T. 1696," T. 1696a, March 1922, PRO: DSIR 23/1710.

10. Relf 1924b.

11. H. Glauert, "Notes on the Flow Pattern round an Aerofoil," T. 1696b, May 1922, PRO: DSIR 23/1710.

12. G. I. Taylor, "Remarks on T. 1696b," T. 1696c, May 1922, PRO: DSIR 23/1710.

13. Minutes of the Aerodynamics Sub-Committee, meeting 54, July 3, 1923, PRO: DSIR 22/50; minutes of the Aerodynamics Research Committee, meeting 35, July 10, 1923, minute 327, PRO: DSIR 22/3.

14. H. Glauert, "Experimental Tests of the Vortex Theory of Aerofoils," T. 1850, November 1923, PRO: DSIR 23/1864.

15. G. I. Taylor, "Note on the Prandtl Theory," T. 1875, January 1924, PRO: DSIR 23/1889.

16. Bairstow 1939, 354.

17. L. W. Bryant and D. H. Williams, "An Investigation of the Flow of Air around an Aerofoil of Infinite Span," T. 1885, February 1924, PRO: DSIR 23/1889.

18. Bryant and Williams 1926, and Taylor 1926.

19. The report appears in an anonymous roundup of news and recent work titled "Societies and Academies," *Nature* 116 (July 4, 1925): 34.

20. Kármán and Burgers 1935.

21. In what appears to be the text (in English) of a lecture prepared in 1926 or 1927 at Aachen, von Kármán described Bryant and Williams' work as "marvellously accurate" and as providing a direct determination of the circulation. He went on: "Although . . . a wake behind the airfoil unavoidably arises, these two investigators have found the relation between circulation and lift to be very well in accordance with the theory, provided that the circulation integral intersects the wake approximately at right angles. G. I. Taylor has demonstrated theoretically that the vortex region actually has no effect on the relation between circulation and lift when the integral is taken over the path which intersects the vortex region at right angles to the line of flight. It can thus be understood why the circulation theory gives satisfactory results in spite of the fact that the conditions of absolute irrotational motion is not fulfilled for real fluids" (TVK-CIT, folder 117.16, pp. 11–12). But Taylor's point was that the "satisfactory" character of the results was an artifact.

22. Experiments were later carried out in the wind tunnel to test Taylor's analysis of the flow around an inclined plate. Although the flow could not be claimed to be Rayleigh flow, it approximated it, and the experiment confirmed that the measured lift and circulation were indeed related by the formula $L = \rho VT$, provided the circulation was taken along contours that cut the wake in the way Taylor specified. Fage and Johansen 1928.

23. Minutes of the Aerodynamic Sub-Committee, meeting 56, December 4, 1923, minute 534, PRO: DSIR 22/40.

24. Minutes of the Aerodynamic Sub-Committee, meeting 58, March 4, 1924, PRO: DSIR 22/40. The numbers indicating the circulation in Bryant and Williams' report T. 1885 differ from those in the later published version because in the latter absolute units are used.

25. Fage and Jones 1926.

26. Sears 1956, 491.

27. Preston 1954, 1.

28. See, for example, Howarth 1935a; 1935b, 563; Piercy, Preston, and Whitchead 1938, 811; Preston 1943, 2—3.

29. Fage 1915.

30. Collier 1978, 33.

31. Fage and Simmons 1926.

32. Lavender 1923. Fage and Simmons also used a hot-wire instrument but found that the two methods of measurement produced very similar results.

33. Fage and Simmons did not treat Taylor's criticisms of the earlier experiments as problematic. They presented their own work as "a continuation of that undertaken by L. W. Bryant and D. H. Williams, who confirmed experimentally the Kutta-Joukowsky relation connecting circulation and lift" (Fage and Simmons 1926, 303).

34. Clifford 1886. The quotation comes from the lecture "Atoms" (110—33). I thank Josipa Petrunic for drawing my attention to Clifford's metaphor.

35. Einstein 1905. For a translation of Einstein's paper, see Arons and Peppard 1965. On the background of Einstein's work, see Stuewer 1970 and Wheaton 1983. A simple account of Einstein's heuristic idea of photons is given in Frank 1948, chap. 3, sec. 10. Einstein's paper addressed the problematic interface between theories based on a continuum and theories based on particles. Such problems are never far beneath the surface when dealing with fluid dynamics.

36. Footprint analogies are discussed in Toulmin 1953. Toulmin analyses Robinson Crusoe's inference from a footprint to a fellow inhabitant of his island and contrasts inferences of this kind with the more significant forms of scientific inference. In running them together Clifford committed what Toulmin calls the "Man Friday Fallacy" (18). The difference in the character of the inductive inference between these two kinds of reasoning was first identified by Hume in the context of not science but natural theology. See Hume 1902, 143 (first published 1777).

37. Clifford's metaphor leads naturally to the metaphor of scientific knowledge as a "map." This fits well with Fage and Simmons' use of contour lines in their diagrams. The map metaphor brings out both the connection with an independent reality and the necessary involvement of conventions and interests in the construction of knowledge. See Toulmin 1953, chap. 4; Barnes 1977, chap. 1; Rudwick 1985, 454. One complication in applying the map metaphor should be noted. Sometimes scientists will treat their different "maps" as rivals and sometimes as nonrivals whose differences merely derive from different purposes, but *why* they are so treated need not be evident from the "maps" themselves. The historian must address this question from the point of view of the historical actors and not presume today's understanding of the status of the "map."

38. Christopherson 1972.

39. Lambe and Tranter 1961, chap. 10.

40. Southwell 1938, 173.

41. Southwell 1924; Southwell and Chitty 1930; Southwell and Squire 1932.

42. This remark was made on May 6, 1930, in a vote of thanks after Southwell had given the James Forrest Lecture. See Southwell 1931, 379.

43. Southwell exemplified his own negative insights about prediction when he said, "I do not believe that the helicopter is a form of aircraft which will prove to have a great military value" (Southwell 1925a, 401).

44. "Appendix to Minutes: The Aeroplane of 1930," in minutes of the Aeronautical Research Committee, meeting 8, minute 1, February 8, 1921, PRO: DSIR 22/2.

45. Southwell 1925b.

46. Barker 1922.

47. Southwell 1925a, 410.

48. See also Southwell 1931. Here the suppressed premise of the 1925 talk is made explicit: "whether the picture can be justified or not, it 'works'" (363).

49. On Wimperis, see Serby 1971 and Clark 1962, chaps. 1 and 2.

50. Wimperis 1926. See also Wimperis 1928 for a useful overview of the network of government bodies linked to aeronautical research in Britain.

51. Grammel 1917.

52. Pröll 1919.

53. Bader 1920; Fuchs and Hopf 1922. Bader 1920 mainly gives mathematical formulas derived from the circulation theory but little by way of theoretical explanation. It does, however, guide the reader to some of the relevant literature from Göttingen by Föppl and Betz.

54. Glauert 1926.

55. I base this statement on a remark recently made to me by a member of the aeronautics department at Stanford University.

56. Low 1927.

57. Martin 1924. Low indicates that the articles were available as a pamphlet.

58. Low 1923c.

59. Hobson 1891. Hobson became the Sadleirian Professor of Pure Mathematics at Cambridge. During the Great War he contributed the chapter on mathematics to Seward 1917, a volume designed to show the important contribution made by pure science to the war effort.

60. Southwell 1927.

61. This once again overlooks the role played by the discontinuity theory in the early years, but it is right to say that a purely empirical approach soon took hold. As a representative sample of work of this character, see, for example, Nayler et al. 1914. The report consists of over one hundred pages of dense numerical tables followed by nearly eighty graphs.

62. Southwell stressed the importance of the engineer's capacity to visualize in Southwell 1938, 172. This article was written ten years after the Glauert review and might represent a change in outlook, but I know of no evidence for any such change.

63. Southwell's warning against the pedagogic dangers of talk about "roller bearings" was taken up by Piercy 1944, 58.

64. See Fage and Nixon 1923, and Fage 1927. Some of Fage's important work at the NPL is summarized in Fage 1928.

65. Lanchester had been invited to give the Wright Lecture the previous year. See Lanchester 1926. Lanchester had not been actively engaged in aerodynamics for some time so the lecture was a recapitulation of earlier ideas.

66. Glauert to Prandtl, April 29, 1922; Prandtl to Glauert, May 9, 1922, both at LP-MPGA, Abteilung III, Repositur 61, no. 536.

67. Prandtl to Colonel the Master of Semple, March 19, 1927, LP-MPGA, Abteilung III, Repositur 61, no.1983.

68. Prandtl 1927c. The German text was published as Prandtl 1927b.

69. Von Kármán appears not to have fully accepted Prandtl's claim that he, Prandtl, and Lanchester had developed their ideas independently. Von Kármán did not say explicitly that Prandtl got his inspiration from reading Lanchester, but when he addressed the relation between the ideas of the two men in his book *Aerodynamics*, he launched into examples of absentminded scientists who mistook other peoples' suggestions for their own thoughts. Von Kármán (1963, 51–52) thus left the suggestion hanging in the air that Prandtl fell into this category. It is unclear how this matter could be resolved. Clearly von Kármán is talking about a real possibility, but there may be something questionable about the entire exercise of assigning ideas to individual owners, or attributing the origin of ideas to individuals. Perhaps "ideas," by their very nature, are collective constructs. This is the position, based on both commonsense observation and laboratory experiment, taken in Bartlett's work in social psychology that informs my own approach. See Bartlett 1932, chap. 16.

70. Was Prandtl himself satisfied with it? The 1927 argument was not repeated in the 1929 volume of Prandtl's Göttingen lectures. The diagrams of the struts surrounded by fluid lines were reproduced from the Wright Lecture, but the accompanying argument is truncated. Tietjens 1929, 179–80.

71. See, for example, the critical comments by Larmor on Helmholtz's proofs in Ramsey 1913, 216, and by Goldstein in Ramsey 1935, 217. There are also historical remarks and comparisons on various associated proofs in Lamb 1932, 203–7.

72. Klein 1909–10. For a discussion of Klein's coffee-spoon experiment and the central idea of Prandtl's argument, see Saffman 1992, 97 and 109.

73. Lagally 1915; Jaffé 1920; Prandtl 1924; Betz 1930, 1950; Ackeret 1936. Ackeret was the only one who took issue with Klein's argument. He said that Klein wrongly assumed that the removal of the "spoon" had no effect on the fluid, for example, by altering its kinetic energy.

74. Cook 1990.

75. Jeffreys 1925.

76. Prandtl to Jeffreys, January 6, 1926, LP-MPGA, Abteilung III, Repositur 61, no. 763.

77. Jeffreys to Prandtl, April 22, 1926, LP-MPGA, Abteilung III, Repositur 61, no. 763.

78. Jeffreys 1930.

79. Lamb 1924b, 548.

80. Jeffreys 1928a, 479. See also Jeffreys 1928b.

81. Southwell 1931.

82. Mallock 1912, 213–15. See also Mallock 1907 and 1911.

83. Lamb 1924a, 22.

84. Taylor had not been able to attend Prandtl's Wright Lecture but, like Lamb and Southwell, he had been impressed by the photographs and films. Taylor to Prandtl, April 24, 1927, LP-MPGA, Abteilung III, Repositur 61, no. 1653.

85. Glauert 1929; Lamb 1929. The theme of the artificiality of bound vortices was taken up in Morris 1937.

86. Lamb 1925.

87. In the final, 1932 edition of his *Hydrodynamics*, Lamb expressed similar reservations about the boundary-layer equations. The derivations from the Stokes equations offered by Prandtl and Blasius were based on approximations that were not properly justified mathematically. "The approximations are explained in greater detail by Blasius," said Lamb, "they may be justified in the last resort by comparison with the results deduced" (685). This was a last resort because it was an empirical not a mathematical justification. As one mathematician remarked,

"Lamb expressed very well the exasperation one must feel with these arguments by the faint praise he can give them" (Shinbrot 1973, 134).

88. Minutes of the Fluid Motion Panel of the Aerodynamics Sub-Committee, meeting 15, June 30, 1933, minute 92, PRO: DSIR 22/49.

89. Goldstein 1938. The book is dedicated to Lamb's memory. On Goldstein, see Lighthill 1990.

90. Goldstein, unlike Lamb, said that Prandtl gave a "clear explanation" of the origin of circulation. Goldstein 1938, 46.

91. Low to Prandtl, undated [reply to Prandtl's letter of June 4, 1927], LP-MPGA, Abteilung III, Repositur 61, no. 1983.

92. On the subsequent history of turbulence research, see Darrigol 2005, chap. 5, and Eckert 2008, 39–71.

93. Maccoll 1930; for discussion, see p. 697.

94. Maccoll was to do distinguished work in collaboration with G. I. Taylor and was the coauthor with Taylor of the chapter on compressible flow in Durand's multivolume review. Taylor and Maccoll 1935.

95. Bairstow also published collaborative work with the Cambridge mathematician Arthur Berry; see Bairstow and Berry 1919. Berry was a fellow of King's College and had been senior wrangler in 1885 (ahead of A. E. H. Love). He was coached by Routh. This information comes from Warwick 2003, 517.

96. Bairstow 1939. "From a consideration of all available experimental results it may be concluded that the main effects of finite span can be reproduced by potential flow theory involving trailing vortices" (435). Notice the word "reproduce." The suggestion is that perfect fluid theory can simulate certain aspects of real fluids but not provide a true description of underlying processes.

97. Bairstow 1930. Bairstow may have been alerted to this comparison because the leading British expert on the wave-particle duality was none other than G. P. Thomson, the friend of Glauert and Pinsent from the Chudley days, author of *Applied Aerodynamics*, and, of course, the son of J. J. Thomson. On G. P Thomson's work in physics in the years after his aeronautical experiences in the Great War, see Navarro 2010.

98. "Potential flow calculations are useful in that they serve to provide orders of magnitude for some characteristics, which can be regarded as limiting values when the Reynolds number is infinitely large. Thus the departure of a measured characteristic from its theoretical potential flow is a measure of the influence of the boundary layer and some idea of the probable scale effects can be obtained" (Preston 1949, 1).

99. Milne-Thomson 1958.

100. This was exactly the fate of Warwick's early papers on the Cambridge response to Einstein's work when they were assimilated into Brush's survey of reasons why relativity was accepted. See Warwick 1992 and 1993, and Brush 1999.

101. I am closely paraphrasing the account given in O'Hear 1989, 210–16. Many other sources, both earlier and later, could have been used, for example, Shapere 1986, McMullin 1987, and Haack 1996.

102. If there is a divergence between the historical case studied by Warwick in *Masters of Theory* and the picture that has emerged from my study, it concerns the relative priority the actors afforded to mathematical and physical considerations. Warwick found that, in the area of electromagnetic theory, Cambridge mathematics did not simply articulate a prior "ontologi-

cal" commitment to the physics of the ether. The deepest commitment appeared to be to the mathematical methods themselves as cultivated in the Tripos—hence the willingness, in some quarters, to postulate, say, multiple ethers if the mathematics seemed to demand it. As Warwick put it, a form of pedagogy became a form of knowing (2003, 3–4). As I have said previously, in the case of aerodynamics, this priority accorded to mathematical method seems to be true of the work of Bryan and Greenhill, but it seems less true of the later phases of my story where the concern was to do justice to the physical realities of viscous flow. This was made explicit by G. I. Taylor but was central to British work in general. The commitment was to physical realism rather than mathematical method. But this difference between the two episodes should not be overstated. The conception of physical reality that was operative in the British aerodynamic work was embodied in the Stokes equations, whose truth was taken for granted. In this sense one could still say that the mathematics defined the physics.

Chapter Ten: Pessimism, Positivism, and Relativism

1. Wittgenstein 1969, para. 132. It is not known who Wittgenstein had in mind, but such sentiments were widespread among enthusiasts for aviation. Claude Grahame-White, whose aero-charabanc is shown in figure 1.7 of this volume, wrote lyrically of the "friendly intercourse between nations" promoted by commercial aviation (1930, 257). Another example arose in the 1930s when a group of left-wing scientists, including Bernal, Needham, and Waddington, objected to the financial connections between the Cambridge Department of Aeronautics and the armaments industry. Melvill Jones replied: "Those of us who work on the science of aeronautics at Cambridge believe that the ultimate influence of the aeroplane on civilisation will be pacific rather than warlike; we strongly resent the suggestion that our main interest is the development of armaments" (BMJ-RAFH, box AC 76/6/81).

2. On the Strong Program, see Bloor 1973, 1976, 1997; Barnes 1974, 1977; and Barnes, Bloor, and Henry 1996. For the general principles behind the application of this approach to technology, see Barnes 1982 and Pinch 1988. For a concrete example of this approach to the sociology of technological knowledge see, for example, MacKenzie 1990.

3. Frank 1914, 1918, 1921, 1933; Frank and Löwner 1919.

4. Philosophers of science frequently seek to turn the asymmetrical structure of commonsense curiosity into a metaphysical and methodological stance. For examples, see Lakatos 1971 and Laudan 1977. These authors postulate an internal rationality for science and treat rational processes as fundamentally distinct from the nexus of sociopsychological causes. The latter provide general conditions for, or explain deviations from, rationality, not the operation of rationality itself, which (allegedly) has its own internal drive and is not susceptible to sociopsychological causal analysis. Such an approach, with minor variations, is still widely taken for granted, for example, Newton-Smith 1981, 264–65; Haack 1996; Okasha 2000; Brown 2001. Further arguments of this kind are mentioned shortly.

5. Two exemplary studies of this kind are Perrow 1984 and Vaughan 1996.

6. Recent examples of the false imputation that, according to a Strong Program analysis, scientists respond to society *rather than* to the natural world are to be found in Nola and Sankey 2000, 41, 71. The truth, surely, is that scientists can and must respond collectively to the natural world.

7. Sturm and Gigerenzer 2006.

8. For an angry expression of this view, see Laudan 1996, 209. The anger is directed against the present author.

9. Bloor 1973. The argument is developed in more detail in Bloor 1997.

10. Wittgenstein 1967.

11. Wittgenstein is normally read as an enemy of causal explanation in sociology. Such a reading can certainly be supported by quotations. The crucial facts to keep in mind, however, are (1) that this is possible because Wittgenstein was not a wholly consistent thinker, and (2) that the causal analysis to which I refer is clearly there to be seen in the text. My policy is to look at Wittgenstein's practice rather than his comments on that practice, that is, his arguments rather than his opinions. Others prefer to trivialize Wittgenstein so that there is nothing of naturalistic significance to be learned from him, for example, see Friedman 1998, 251–56.

12. "But what would refute . . . the complete scepticism of the strong programme . . . is even one case of a scientific theory with proper evidential warrant. There are many such cases" (Norton 2000, 72).

13. Wittgenstein's naturalistic argument about rule following can be evaded by a sufficiently determined resort to a nonnaturalistic metaphysics, for example, by postulating a Platonist theory of logical and mathematical reality and a teleological view of human reasoning. The price to be paid is that such claims operate wholly outside the framework of natural science as it is currently understood. I explored these points in Bloor 1973, secs. (iii) and (iv).

14. The teacher–pupil interaction of Wittgenstein's example can, of course, also be seen as a microcosm of Cambridge coaching, so the example is relevant to Warwick's *Masters of Theory.* Even senior wranglers had to begin their mathematical careers by learning how to follow the rule "add 2." The fact that the mathematics in the example is trivial makes no difference to the fundamental question that is at stake, namely, the ultimate source of logical compulsion and mathematical necessity.

15. Mayr 1976.

16. Such a view is confirmed by the analysis of citation practices. See Price 1969.

17. Barnes 1982.

18. The assumption that there is a strong dependency of technology on science is sometimes called the "liner model," but see Edgerton 2004.

19. For warnings against an overemphasis on novelty, see Edgerton 2006a. The phrase "the shock of the old" comes from the title of Edgerton's book.

20. Barnes 2005.

21. Zilsel 2000.

22. Werskey 1988.

23. Levy 1939, chap. 26. See also Levy 1938a and 1938b.

24. Part of that difference is related to the relative status of the two fields and their roles in the generation of novel technology. For Glazebrook, applied science was simply the application of pure science to particular problems. Everything flowed from pure science. Glazebrook 1917, 15. See also Seward 1917.

25. Hilken 1967, chap. 5.

26. Ewing 1918–19.

27. Goody 2005.

28. Hopkinson 1898.

29. McKinnon Wood 1966.

30. Farren 1956, 431.

31. Handwritten and typed notes for a talk given at the Royal Aircraft Establishment, December 3, 1957, BMJ-RAFH, box AC/76/6/139.

32. What if Haldane had recruited from non-Cambridge engineers? To some extent he

did. Petavel was trained in London and Manchester but provided no significant support for Lanchester. Even Mervin O'Gorman, trained in engineering at University College, Dublin, and a friend of Lanchester's, was unsympathetic to Lanchester's theoretical ideas. He took his cue from Lamb and Taylor. O'Gorman to Melvill Jones, January 10, 1957, and February 16, 1953, BMJ-RAFH, box AC/17/6/162−167.

33. A corresponding counterfactual question can be asked about the German side of the story. What would have happened if German physicists, rather than German engineers, had been put in charge of aerodynamic research? The Einstein episode suggests that they, like their British counterparts, might have been defeated by the task of developing a workable theory of the wing.

34. Wiener 1992.

35. Barnett 2002. The reference to men who are hard of mind and will comes from p. 20.

36. Barnett 1986, 219.

37. Pollard 1994, 80.

38. Pollard 2000, 87.

39. Pollard 1990, chap. 3.

40. Sanderson 1972, 1988.

41. Edgerton 1991c, 1991a, 1996, and 2006b.

42. Crowther 1968, 89.

43. Declinists have argued that the country houses favored by the wealthy symbolized a retreat from the world of science and technology. This notion is challenged in Schaffer 1998.

44. On the army and navy as a source of funding for British science, see Brock 1976; Turner 1980; Edgerton 1991b, 2006.

45. At the beginning of the Great War, G. H. Bryan had warned that the Germans had some of their best mathematicians at work on stability. His comments were grounded on fact because, before the war, he had been in contact with Reißner and had a sense of who was involved in the German work. See, for example, Deimler 1910; Reißner 1910, 1912a; Runge 1911, 1912. (Reißner mentions recent, friendly contact with Bryan in Reißner 1912a, 40.) Kármán and Trefftz 1914−15. Although the German work on stability was mathematically sophisticated, it represented a much smaller proportion of their wartime aerodynamic effort than did the respective British work on stability. Thus, of the 340 pages of text in volume 1 of the *Technische Berichte*, only 22 were devoted to stability, and these dealt with static rather than dynamic stability; see Fuchs and Hopf 1917, and Gumbel and Hopf 1917.

46. See James 1990, 123.

47. Barnett 2002, 105.

48. I would treat *Masters of Theory* as a member of this category. It is usefully read in conjunction with Mackenzie 1981 and 2001. MacKenzie also makes some important observations about the limitations of appeals to training in the explanation of scientific and mathematical controversies; see MacKenzie 1981, 27.

49. The generalized form of Lanchester's metaphor of science as a "game," with its identification of the radical contingency of concept application, is to be found in the position that has come to be known as "meaning finitism." See Hesse 1974 and Barnes, Bloor, and Henry 1996, chap. 3.

50. Frank 1954a.

51. Holton and Cohen 1981.

52. Stadler 2001.

53. Frank and Mises 1925–27.

54. Von Kármán was also commissioned to write the section on viscous flow but failed to produce the required chapter. Frank then commissioned Oseen for this task. Neither von Kármán nor Prandtl were impressed by Oseen's treatment. TVK-CIT, folder nos. 9.37 and 23.42.

55. Siegmund-Schultze 2007.

56. Frank 1917; translated in Frank 1961, 69–85.

57. Frank 1928, 1929, 1932 (translated as Frank 1998).

58. The concept of the "unity of science" changed in the move from Vienna to Harvard. See Galison 2001.

59. Frank 1948, 1957, 1961.

60. Rudolf Carnap, a friend and colleague of Frank's in Prague, built a policy-parameter into his (otherwise) wholly formal account of inductive inference. It measured the degree of risk or caution informing the inference. See Carnap 1952. Carnap treated λ as a psychological variable, but it is better seen as a sociological phenomenon.

61. Fritzsche 1992.

62. Lamb 1904.

63. On Stokes' deep religious commitment, see Wilson 1984.

64. Klein 1900, 39.

65. "So too the fact that it can be described by Newtonian mechanics asserts nothing about the world; but *this* asserts something, namely, that it can be described in that particular way in which as a matter of fact it is described" (Wittgenstein 1922, 177, para. 6.342). This passage is quoted in Frank 1957, 107.

66. Lagally 1921. I am following the treatment given in Ramsey 1935, 241–42.

67. Blasius 1910. Blasius' theorem gives the forces on the body and also the moments of the forces.

68. Both the Kutta formula and the Lagally formula belong to what is now called the class of "reciprocity theorems." These began to receive a unified treatment as a branch of potential theory in the 1960s. I thank Horst Nowacki for drawing my attention to this development.

69. "The concept 'ellipse' that was, in the simple planetary motion, only an instrument to make the description of facts simple and convenient becomes, in the theory of perturbation, the indispensable instrument for the solution of the problem of 'perturbed motion' that is of an immensely higher complication than the simple planetary motion covered by Kepler's law" (Frank 1957, 308).

70. Durkheim 1915.

71. Uebel 2000. A similar claim is made alongside the otherwise sympathetic treatment of the Strong Program in Uebel 1996, 339.

72. Frank 1951. This work is not mentioned in Uebel's paper though it is discussed in Nemeth 2003. Nemeth seeks to put distance between Frank and current work in the sociology of knowledge (134). Such work is described and criticized as relativist (without quotation marks), whereas Frank is always called a "relativist" (with quotation marks). The quotation marks are used by Nemeth as a device to distance Frank from relativism. Nemeth never brings Frank's relativism into clear focus and merely uses his work to support the idea that "philosophers" have important ideological, educational, and justificatory functions to perform.

73. Frank 1952.

74. "My propositions are elucidatory in this way: he who understands me finally recognises

them as senseless, when he has climbed out through them, on them, over them" (Wittgenstein 1922, para. 6.54).

75. Anticommunist hysteria played a role both outside and inside the philosophical profession. For a fascinating, but depressing, account of the pressures on Frank and his colleagues, see Reisch 2005.

76. Holton et al. 1968. A similar thing had happened in 1965. The second volume of *Boston Studies in the Philosophy of Science* was dedicated to Frank and prefaced by no less than ten expressions of appreciation of his work, covering some twenty pages. Only one person mentioned Frank's relativism. See Cohen and Wartofsky 1965, ix–xxx. The exception was the mathematician Hilda Pollaczek-Geiringer, the widow of Richard von Mises. On Pollaczek-Geiringer, see Siegmund-Schultze 1993.

77. Mises 1939a, 159 (translated as Mises 1956; the quoted phrase is on p. 147).

78. I have analyzed and documented these arguments in Bloor 2007.

79. Even if it were true that all subjectivists were relativists, it would not follow, and would not be true, that all relativists are subjectivists. In any case the premise is wrong. Some subjectivists, for example, religious mystics, are absolutists.

80. A vivid contemporary account of the V-1 attacks was provided by the novelist H. E. Bates, who had the rank of squadron leader in the RAF and who published a number of works of fiction under the name Pilot Officer X. Bates' (nonfiction) account of the V-1 attacks was written at the request of the Air Ministry but suppressed on the grounds that it revealed too much sensitive information. Many years later the text was found in the Public Record Office and has now been published. See Bates 1994. The technique of using the slipstream to topple the bombs is described on pp. 116–18. There exists a remarkable ground-to-air photograph of a Spitfire at the moment of encounter with a V-1 in Lyall 1994, opposite p. 209. See also the account on pp. 346–47.

81. Spalart 1998, 107. Spalart states: "Two radically different concepts of the long-term behaviour of trailing vortices have been held for years. One [the predictable decay theory DB] is more empirical and rooted in the government/industry community; the other [the stochastic collapse theory DB] is more theoretical and rooted in the academic community" (119). Some of the continuing and unresolved problems in this area are identified in Spalart 2008.

82. Vincenti 1990, chap. 3.

83. Hume 1960, book 1, pt. 3 (first published 1739–40). Hume did not deny that there were causal relations. His skepticism was directed at the idea that such relations can be known a priori or that experience reveals any necessary connection between cause and effect. Early in his career (1907) Frank pointed out that a general principle of causality, such as "like causes will always be followed by like effects," can always be rendered immune to refutation by postulating hidden processes to explain exceptions. This is true but in no way reduces the force of Hume's argument. See Frank 1961, 62–68.

84. Maskell and Weber 1959; Küchemann 1967, 14–16; 1970, 28.

85. Thwaites 1960, 511–12.

86. Atkins 2007. The author is an Oxford professor of chemistry.

87. I am grateful to Dr. Audrey Glauert, Hermann Glauert's daughter, for giving me her personal account. The date of the accident is given incorrectly as August 3 in Glauert's Royal Society obituary.

88. Muriel Glauert to von Kármán, October 10, 1934, TVK-CIT, folder 11.16.

89. Von Kármán to Muriel Glauert, November 18, 1934, TVK-CIT, folder 11.16.

90. See "The Internationalization of Fluid Mechanics in the 1920s," in Eckert 2006, 83–105.

91. The predictable answer is that, before long, von Kármán was working closely with the military. In later years he was also involved in space research, which did not please his old friend Max Born. "This is all very clever," wrote Born in 1958, "but at the same time immensely stupid. Not a penny would be given for all this if the military would not think it useful for their purposes. When my memory does not deceive me, you would have shared my views when we were young. I am sorry that you have been enveloped in all kind of activities which I think are very dangerous for the human race." Born to von Kármán, March 7, 1958, TVK-CIT, folder 3.27. For more of Born's views on this theme, see Born 1968, 150–59. For von Kármán's work in the United States, see Hanle 1982, esp. 135–39, and Goodstein 1991, chap. 8.

92. The empirical work on the transnational character of science is impressive. See Kevles 1971; Forman 1973; Crawford, Shinn, and Sörlin 1993; Ash and Söllner 1996; Siegmund-Schultze 2001; Wolf 2003; Eckert 2005.

93. Owen and Maskell 1980.

94. Combined Intelligence Objectives Sub-Committee (CIOS) file XXV-22. See also CIOS file XXII-5. Copies are held in the library of the Max-Planck-Institut für Wissenschaftsgeschichte, Berlin.

95. Lasby 1971, Bower 1987, and Hunt 1991.

96. Küchemann 1978.

97. Truesdell 1952, 80.

98. There are similar sentiments in the prologue of Thwaites 1960, 1.

99. Fefferman 2008. Fefferman points out that similar mathematical obscurity surrounds the Euler equations as well.

100. Anderson 1995.

101. Phillips 2004, 2005; Phillips and Snyder 2000. I thank Brian Riddle for drawing my attention to this body of work.

102. Can you get "closer" to "absolute" truth? No. You can no more get closer to absolute truth than you can get closer to infinity by counting for longer. On relativist accounts of progress, see Bloor 2007, 263–67.

103. Unlike Frank, Küchemann did not use the label "relativist" to describe himself. In the opening pages of his book he says he is a follower of Karl Popper, but there is no necessary inconsistency here. If, as Popper insists, all knowledge is conjectural (which is the essential point taken up by Küchemann), then it follows that no knowledge claim can have an absolute character. On Frank's definition this makes both Küchemann and Popper relativists. Popper denies that he is a "relativist," but he is able to do this because he gives the word a special meaning. Popper defines relativism as the view that "the choice between competing theories is arbitrary." See Popper 1962, 2:369. This stipulation lacks the clarity of Frank's definition, and it may be doubted if there has ever been anyone who conforms to it.

104. The polemical literature denouncing "relativism" is extensive. See Gross and Levitt 1994; Gross, Levitt, and Lewis 1996; Koertge 1998; Sokal and Bricmont 1998; Boghossian 2006. Like Popper's rejection of "relativism," this literature scores some easy points against some easy targets, but when serious work in the history and sociology of science is attacked, the weaknesses of the antirelativist case become apparent. For some effective replies that expose these weaknesses, see MacKenzie 1999, and Shapin and Schaffer 1999.

105. Dawkins 1995. I have discussed Dawkins' argument further in Bloor 2008a.

106. See, for example, Norris 1997, 314.

107. For an analysis of this thought-provoking episode see Vincenti 1990, chap. 2. Vincenti explores the question of why the Davis wing performed so well in the wind tunnel. He concludes

that, by chance, Davis had hit on a profile that generated a laminar flow. Low-drag, laminar-flow aerofoils were later developed on scientific principles. Vincenti also points out that under operational, as distinct from laboratory, conditions laminar-flow wings usually turn out to be no better than conventional aerofoils.

108. I am here using the word "tribe" in a metaphorical way as, for example, in Becker 1989, 22–24. There is more apposite terminology available to the sociologist, and it would be better to characterize my two groups as "status groups." See Barnes 2003 and 1995, chap. 5. Warwick (2003) also draws a comparison between his work and the studies of the anthropologist. He talks of Cambridge academics as "natives" with their "rituals, ceremonies, and forms of sociability" (46–47). Importantly and correctly, Warwick insists that what these "natives" count as knowledge must be "described and explained as an integral part of their local cultural life" (47) in exactly the same way that it is described in the ethnography of, say, a simpler culture. In other words, Warwick insists on methodological "symmetry."

109. Many philosophers claim to be neither relativists nor absolutists. They say these are "extreme" positions, and they lay claim to a moderate middle ground. Like Popper, they justify this position by arbitrary and confused definitions of "relativism" that systematically conflate the relativist–absolutist dichotomy with other quite different dichotomies. On these confusions see Bloor 2011.

110. The wing used the principle of "controlled separation." Küchemann 1978, 69–70 and 339–40. See also Thwaites 1960, 511–16.

111. Davis 1969; Wiggs 1971; Wilson 1973; Costello and Hughes 1976; Feldman 1985; Arkes and Ayton 1999.

112. Frank 1951, 20.

113. The phrase comes from Friedman 1998, 263. Friedman's aim is to distinguish between the role of the philosopher and that of the sociologist of knowledge in order to identify some especially important, but essentially non-empirical, task that philosophers alone can perform as guardians of our values. Exactly what taking "reflective responsibility" for the normativity of our cognitive categories involves remains unclear.

114. Consider the celebrated intervention by the physicist Richard Feynman in the inquiry into the causes of the loss of the Challenger space shuttle and the death of its crew. The shuttle was launched after standing all night in icy conditions, and the rubber O-ring seals used on the booster rocket failed with disastrous consequences. Feynman is widely thought to have demonstrated the cause of the disaster when he appeared on television and dipped a piece of rubber into a cold liquid to show how hard it became. Diane Vaughan's detailed account of the subsequent inquiry suggests that Feynman's actions obscured more than they illuminated. The engineers involved in the launch decision knew that rubber compounds go brittle when they get cold. That was not what the difficult decision to rely on the seals was all about. The problem was to interpret a complex pattern of partial failures in past and successful launches. All the technology involved in the shuttle was being pushed to its limits, and vast numbers of components had acknowledged patterns of failure. If Feynman's intervention proved anything it was that physicists are not well equipped to comment on engineering decisions. Feynman's failure to take into account the complex relativities of the actual decision process did an injustice to the persons who had to bear the terrible responsibility for the launch and the death of the crew. The over simplification contributed more to the public misunderstanding of science than to its understanding. Vaughan 1996, 39. See also Gieryn and Figert 1990.

115. Hume 1960, xix (first published 1739–40).

116. Hume 1902, 160 (first published 1777).

117. Philosophers do not routinely treat fluid dynamics as a special case when they offer their general accounts of science. Where fluid dynamics is treated at all, it is taken as a representative piece of science, though its characteristics are identified differently by different writers: for example, Böhme 1980; Cartwright 1983, 14; Steiner 1998, 36–37; Morrison 1999; Heidelberger 2006.

Bibliography

ARCHIVAL SOURCES

AUE: Archives of the University of Edinburgh
BMJ-RAFH: Papers of Bennett Melvill Jones, Royal Air Force Museum, Hendon
CGG-RAES: Papers of C. G. Grey, Royal Aeronautical Society
CUA: Cambridge University Archives
EP-CUL: Examination Papers, Cambridge University Library
FWL-RAES: Papers of F. W. Lanchester, Royal Aeronautical Society
FWL-UC: Papers of F. W. Lanchester, University of Coventry
GIT-TCC: Papers of G. I. Taylor, Trinity College, Cambridge
GPT-TCC: Papers of G. P. Thomson, Trinity College, Cambridge
LP-ADGLR: Papers of Ludwig Prandtl, Archive of the Deutsche Gesellschaft für Luft-und Raumfahrt, Göttingen
LP-MPGA: Papers of Ludwig Prandtl, Archiv zur Geschichte der Max-Planck-Gesellschaft, Berlin
PRO: Public Record Office, National Archive, Kew
RBH-NLS: Papers of R. B. Haldane, National Library of Scotland, Edinburgh
RTG-TCC: Papers of R. T. Glazebrook, Trinity College, Cambridge
TVK-CIT: Papers of Theodore von Kármán, Archive of the California Institute of Technology

BOOKS AND ARTICLES

This bibliography combines primary and secondary sources, including the Reports and Memoranda of the Advisory Committee for Aeronautics. Unpublished technical reports and archival sources are not included but are cited in full in the endnotes.

Abzug, M. J., and E. E. Larrabee. 1997. *Airplane Stability and Control: A History of the Technologies That Made Aviation Possible.* Cambridge: Cambridge University Press.

Ackeret, J. 1925. *Das Rotorschiff und seine physikalischen Grundlagen.* Göttingen: Vandenhoeck und Ruprecht.

———. 1936. "Über die Bildung von Wirbeln in reibungslosen Flüssigkeiten." *Zeitschrift für angewandte Mathematik und Mechanik* 15:3–4.

Ackroyd, J. A. D. 1992. "Lanchester—The Man." *Aeronautical Journal* 88:119–40.

Ackroyd, J. A. D., B. P. Axcell, and A. I. Ruban, eds. 2001. *Early Developments of Modern Aerody-namics*. Oxford: Butterworth and Heinemann.

Ackroyd, J. A. D., L. Bernstein, and F. W. Armstrong. 2008. "One Hundred Years of Aeronautics in East London." *Aeronautical Journal* 112:357–80, 358–65.

Aero. 1910. "The Principles of Mechanical Flight: A Discussion at the British Association." 3:216–17.

———. 1912a. "The Editorial View." 6:350.

———. 1912b. "A Lesson for British Makers." 6:250.

Aeronautical Journal. 1908. Review of *Aerodynamics*, by F. W. Lanchester. 12:40–42.

Aeronautics. 1908. "Review of Books." Review of *Aerodynamics*, by F. W. Lanchester. 1:xiv.

———. 1910. "Notes." 3:118.

———. 1913. Review of *The Dynamics of Mechanical Flight*, by A. G. Greenhill. 6:64.

———. 1916. Editorial. 11:97.

———. 1917. "Physician Heal Thyself." 13:185.

———. 1919. "John William Dunne." 17:208–10.

Aeroplane. 1912a. "Aviation Science." 2:188.

———. 1912b. "The German Airship over Sheerness." 3:497.

———. 1916. "The Farnborough 'Funk Hole.'" 11:570.

Ahlborn, F. 1927. "Die Ablösungstheorie der Grenzschichten und die Wirbelbildung." *Jahrbuch der wissenschaftlichen Gesellschaft für Luftfahrt*, 171–88.

Air Ministry. 1925. *Translated Abstracts of Technische Berichte, 1917.* Vol. 1 (Air Publication 1120) and vol. 2 (Air Publication 1121). London: Air Ministry.

Allen, J. E., and Joan Bruce, eds. 1970. *The Future of Aerodynamics.* London: Hutchinson.

Allfrey, A. 1989. *Man of Arms: The Life and Legend of Sir Basil Zaharoff.* London: Weidenfeld and Nicolson.

Anderson, David F., and S. Eberhardt. 2001. *Understanding Flight.* New York: McGraw-Hill.

Anderson, J. D. 1989. *Introduction to Flight: Its Engineering and History.* New York: McGraw-Hill.

———. 1991. *Fundamentals of Aerodynamics.* New York: McGraw-Hill.

———. 1995. *Computational Fluid Dynamics: The Basics with Applications.* New York: McGraw-Hill.

———. 1997. *A History of Aerodynamics and Its Impact on Flying Machines.* Cambridge: Cambridge University Press.

———. 2006. "Airplaner Design Methodology: Setting the Gold Standard." *American Institute of Aeronautics and Astronautics* 44:2817–19.

Archibald, T. 2001. "Images of Mathematics in the German Mathematical Community." In Bottazzini and Dalmedico, *Changing Images in Mathematics*, 49–67.

Arkes, H. R., and P. Ayton. 1999. "The Sunk Cost Fallacy and Concorde Effects: Are Humans Less Rational Than Lower Animals?" *Psychological Bulletin* 125:591–600.

Arons, A. B., and M. B. Peppard. 1965. "Einstein's Proposal of the Photon Concept: A Translation of the 'Annalen der Physik' Paper of 1905." *American Journal of Physics* 33:367–74.

Ash, M. G., and A. Söllner, eds. 1996. *Forced Migration and Scientific Change: Émigré German-Speaking Scientists and Scholars after 1933.* New York: Cambridge University Press.

Ashby, E., and Mary Anderson. 1974. *Portrait of Haldane at Work on Education.* London: Macmillan.

Aston, W. G. 1911. "M. G. Eiffel's Research: Extracts from *La resistance de l'air et l'aviation.*" *Aero* 6:70–73.

Atkins, Peter. 2007. "Science's Future is an Interversity Challenge." *Times Higher*, May 25, 12.

Babbington Smith, C. 1961. *Testing Time: A Study of Man and Machine in the Test-Flying Era.* London: Cassell.

Badash, L. 1979. "British and American Views of the German Menace in World War I." *Notes and Records of the Royal Society of London* 34:91–121.

Bader, F. G. 1920. *Grundlagen der Flugtechnik: Entwerfen und Berechnen von Flugzeugen.* Leipzig: Teubner.

Bairstow, L. 1913. "The Laws of Similitude." *Aeronautical Journal* 17:17–126.

———. 1914. "The Stability of Aeroplanes." *Aeronautical Journal* 18:68–85.

———. 1916. "Inherent Controllability of Aeroplanes: Notes Arising from Prof. Bryan's Wilbur Wright Memorial Lecture." *Aeronautical Journal* 20:10–19.

———. 1919. "Progress of Aviation in the War Period." *Flight* 11:853–57.

———. 1920. *Applied Aerodynamics.* London: Longmans, Green.

———. 1922. "Future of Air Power: Expenditure on Research. Ministry's Unscientific Methods." *Times*, July 22, 16.

———. 1923. "The Fundamentals of Fluid Motion in Relation to Aeronautics." In Lockwood Marsh, *International Air Congress, London, 1923*, 239–45.

———. 1930. "Twenty-One Years' Progress in Aerodynamic Science." *Flight* (January 3): 28–30.

———. 1935. "George Hartley Bryan, 1864–1928." *Obituary Notices of Fellows of the Royal Society* 1:139–42.

———. 1939. *Applied Aerodynamics.* 2nd ed. London: Longmans, Green.

Bairstow, L., and A. Berry. 1919. "Two-Dimensional Solutions of Poisson's and Laplace's Equations." *Proceedings of the Royal Society of London* A 95:457–75.

Bairstow, L., and B. Melvill Jones. 1912a. "Experiments on Models of Aeroplane Wings." Reports and Memoranda No. 60.

———. 1912b. "Notes on the Properties of Aerofoils as Deduced from the Results of the Various Aeronautical Laboratories." Reports and Memoranda No. 53.

Bairstow, L., Miss B. M. Cave, and Miss E. D. Lang. 1922. "The Two-Dimensional Slow Motion of Viscous Fluids." *Proceedings of the Royal Society of London* A 100:394–413.

———. 1923. "The Resistance of a Cylinder Moving in a Viscous Fluid." *Philosophical Transactions of the Royal Society of London* A 223:383–432.

Barker, M. 1922. "On the Use of Very Small Pitot-Tubes for Measuring Wind Velocity." *Proceedings of the Royal Society of London* A 101:435–45.

Barnes, B. 1974. *Scientific Knowledge and Sociological Theory.* London: Routledge and Kegan Paul.

———. 1977. *Interests and the Growth of Knowledge.* London: Routledge.

———. 1982. "The Science–Technology Relationship: A Model and a Query." *Social Studies of Science* 12:166–71.

———. 1995. *The Elements of Social Theory.* London: UCL Press.

———. 2000. *Understanding Agency: Social Theory and Responsible Action.* London: Sage.

———. 2003. "Thomas Kuhn and the Problem of Social Order in Science." In Nickles, *Thomas Kuhn*, 122–41.

———. 2005. "Elusive Memories of Technoscience." *Perspectives on Science* 13:42–165.

Barnes, B., D. Bloor, and J. Henry. 1996. *Scientific Knowledge: A Sociological Analysis.* Chicago: University of Chicago Press.

Barnett, C. 1986. *The Audit of War: The Illusion and Reality of Britain as a Great Nation.* London: Macmillan.

———. 2000. "Interview." In English and Kenny, *Rethinking British Decline*, 40–49.

———. 2002. *The Collapse of British Power*. London: Pan Macmillan.

Barnwell, F. S. 1917. *Aeroplane Design*. London: McBride, Nast.

Bartlett, F. C. 1932. *Remembering: A Study in Experimental and Social Psychology*. Cambridge: Cambridge University Press.

Batchelor, G. K., ed. 1963. *The Scientific Papers of Sir Geoffrey Ingram Taylor*. Cambridge: Cambridge University Press.

———. 1966. *The Life and Legacy of G. I. Taylor*. Cambridge: Cambridge University Press.

———. 1967. *An Introduction to Fluid Dynamics*. Cambridge: Cambridge University Press.

Bates, H. E. 1994. *Flying Bombs over England*. Ed. Bob Ogley. Westerham, Kent: Froglets Publications.

Becker, T. 1989. *Academic Tribes and Territories: Intellectual Enquiry and the Cultures of Disciplines*. Milton Keynes: Open University Press.

Bendemann, F. 1910. "Der heutige Stand der Flugtechnik in Theorie und Praxis." *Zeitschrift des Vereines deutscher Ingenieure* 54:786–91, 888–96, 933–37.

Berkeley, G. 1734. "The Analyst; or, a Discourse Addressed to an Infidel Mathematician." In Fraser, *The Works of George Berkeley*, 253–98.

Berriman, A. E. 1911. "The Mathematics of the Cambered Plane." *Flight* 3:58–60; *Aeronautics* 4:5–8.

———. 1913. *Aviation: An Introduction to the Elements of Flight*. London: Methuen.

Betz, A. 1912. "Auftrieb und Widerstand einer Tragfläche in der Nähe einer horizontalen Ebene (Erdboden)." *Zeitschrift für Flugtechnik und Motorluftschiffahrt* 3:217–20.

———. 1913. "Auftrieb und Widerstand eines Doppeldeckers." *Zeitschrift für Flugtechnik und Motorluftschiffahrt* 4:1–3.

———. 1914a. "Angriffspunkte der Windkräfte bei Doppeldeckern." *Zeitschrift für Flugtechnik und Motorluftschiffahrt* 5:162–64.

———. 1914b. "Die gegenseitige Beeinflussung zweier Tragflächen." *Zeitschrift für Flugtechnik und Motorluftschiffahrt* 5:253–58.

———. 1914c. "Untersuchungen von Traflächen mit verwundenen und nach rückwärts gerichteten Enden." *Zeitschrift für Flugtechnik und Motorluftschiffahrt* 5:237–39.

———. 1915. "Untersuchung einer Schukowskyschen Tragfläche." *Zeitschrift für Flugtechnik und Motorluftschiffahrt* 6:173–79.

———. 1917a. "Berechnung der Luftkräfte auf ein Doppeldeckerzelle aus den entsprechenden Werten für Eindeckertragflächen." *Technische Berichte* 1:103–7.

———. 1917b. "Einfluß der Spannweite und Flächenbelastung auf die Luftkräfte von Tragflächen." *Technische Berichte* 1:98–102.

———. 1918. "Einführung in die Theorie der Flugzeug-Tragflügel." *Die Naturwissenschaften* 6:556–62, 573–78.

———. 1919. *Beiträge zur Tragflügeltheorie mit besonderer Berücksichtigung des einfachen rechteckigen Flügel*. Munich: Oldenbourg.

———. 1924. "Eine Verallgemeinerung der Schukowskyschen Flügelabbildung." *Zeitschrift für Flugtechnik und Motorluftschiffahrt* 15:100.

———. 1930. "Wirbelbildung in idealen Flüssigkeiten und Helmholtzscher Wirbelsatz." *Zeitschrift für Mathematik und Mechanik* 10:413–15.

———. 1935. "Applied Airfoil Theory." In Durand, *Aerodynamic Theory*, 4:1–129.

———. 1941. "Die Aerodynamische Versuchsanstalt Göttingen. Ein Beitrag zur Geschichte."

In *Beiträge zur Geschichte der Deutschen Luftfahrtwissenschaft und –technik*, 63–66. Berlin: Deutsche Akademie der Luftfahrtforschung.

———. 1948. "Inkompressible Strömmungen." In *FIAT Review of German Science, 1939–1946*, 1–19. Wiesbaden: Office of Military Government for Germany.

———. 1950. "Wie entsteht ein Wirbel in einer wenig zähen Flüssigkeit? *Die Naturwissenschaften* 37:193–96.

Birkenhead, Earl of. 1961. *The Prof in Two Worlds: The Official Life of Professor F. A. Lindemann, Viscount Cherwell*. London: Collins.

Birkhoff, G. 1946. "Reversibility and Two-Dimensional Airfoil Theory." *American Journal of Mathematics* 68:247–56.

———. 1950. *Hydrodynamics: A Study in Logic, Fact and Similitude*. Princeton: Princeton University Press.

———. 1960. *Hydrodynamics: A Study in Logic, Fact and Similitude*. 2nd ed. Princeton: Princeton University Press.

Birkhoff, G., G. Kuerti, and G. Szegö, eds. 1954. *Studies in Mathematics and Mechanics (Presented to Richard von Mises by Friends, Colleagues, and Pupils)*. New York: Academic Press.

Blasius, H. 1908. "Grenzschichten in Flüssigkeiten mit kleiner Reibung." *Zeitschrift für Mathematik und Physik* 56:1–37.

———. 1910. "Funktionentheoretische Methoden in der Hydrodynamik." *Zeitschrift für Mathematik und Physik* 58:90–110.

Bloor, D. 1973. "Wittgenstein and Mannheim on the Sociology of Mathematics." *Studies in the History and Philosophy of Science* 4:173–91.

———. 1976. *Knowledge and Social Imagery*. London: Routledge. (2nd ed., 1997, University of Chicago Press.)

———. 1997. *Wittgenstein on Rules and Institutions*. London: Routledge.

———. 2007. "Epistemic Grace: Anti-Relativism as Theology in Disguise." *Common Knowledge* 13:250–80.

———. 2008a. "Relativism at 30,000 Feet." In Mazzotti, *Knowledge as Social Order*, 13–33.

———. 2008b. "*Sichtbarmachung*, Construction and Common Sense in Fluid Mechanics." *Studies in the History and Philosophy of Science* 39:349–58.

———. 2011. "Relativism and the Sociology of Scientific Knowledge." In Hales, *Blackwell's Companion to Relativism*, 433–55.

Blumenthal, O. 1913. "Über die Druckverteilung längs Joukowskischer Tragflächen." *Zeitschrift für Flugtechnik und Motorluftschiffahrt* 4:123–30.

Boghardt, T. 2004. *Spies for the Kaiser: German Covert Operations in Great Britain during the First World War Era*. New York: Palgrave Macmillan.

Boghossian, P. A. 2006. *Fear of Knowledge: Against Relativism and Constructivism*. Oxford: Clarendon Press.

Böhme, G. 1980. "On the Possibility of 'Closed Theories.'" *Studies in History and Philosophy of Science* 11:163–72.

Boltze, B. 1908. "Grenzschichten an Rotationskörpern." Doctoral dissertation. Göttingen.

Born, M. 1968. *My Life and Views*. New York: Charles Scribner's.

Bottazzini, U., and A. D. Dalmedico, eds. 2001. *Changing Images in Mathematics from the French Revolution to the New Millennium*. London: Routledge.

Bower, T. 1987. *The Paperclip Conspiracy: The Battle for the Spoils and Secrets of Nazi Germany*. London: Michael Joseph.

Boyer, Carl B. 1959. *The History of the Calculus and Its Conceptual Development*. New York: Dover.

Boys, C. V. 1935. "Henry Reginald Arnulph Mallock, 1851–1933." *Obituary Notices of Fellows of the Royal Society* 1:95–100.

Brämer, J. 1941. "Die Deutsche Versuchsanstalt für Luftfahrt." In *Beiträge zur Geschichte der Deutschen Luftfahrtwissenschaft und -technik*, 169–362. Berlin: Deutsche Akademie der Luftfahrtforschung.

Braun, H.-J. 1977. "Methodenprobleme der Ingenieurwissenschaft, 1850 bis 1900." *Technikgeschichte* 44:1–18.

Broadbent, D. 1973. *In Defence of Empirical Psychology*. London: Methuen.

Brock, W. F. 1976. "The Spectrum of Scientific Patronage." In Turner, *Patronage of Science in the Nineteenth Century*, 173–206.

Brockway, F. 1933. *The Bloody Traffic*. London: Gollancz.

Brodetsky, S. 1921. *Mechanical Principles of the Aeroplane*. London: Churchill.

———. 1928. "Obituary: Prof. G. H. Bryan." *Nature* 122:849–50.

Bromberg, J. 1981. "Föppl, August, 1854–1924." In Gillispie, *Dictionary of Scientific Biography*, 5:63–64.

Brown, J. R. 2001. *Who Rules in Science? An Opinionated Guide to the Wars*. Cambridge: Harvard University Press.

Brown, L. M., A. Pais, and B. Pippard, eds. 1995. *Twentieth Century Physics*. Philadelphia: American Institute of Physics.

Brush, S. G. 1976. *The Kind of Motion We Call Heat: A History of the Kinetic Theory of Gases in the 19th Century*. Amsterdam: North-Holland Publishing.

———. 1999. "Why Was Relativity Accepted?" *Physics in Perspective* 1:184–214.

Bryan, G. H. 1897. "Artificial Flight." *Science Progress* 6:531–53.

———. 1904. "The Third International Congress of Mathematicians." *Nature* 70:417.

———. 1905. "Progress in Aërial Navigation." *Nature* 71:463–65.

———. 1907. "The Problem of the Flying Machine." *Cornhill Magazine* 22 (n. s.): 605–19.

———. 1908. "Progress in Aviation." *Nature* 78:668–72.

———. 1909a. "Aviation." *Nature* 81:397–99.

———. 1909b. "The Government and Aeronautical Research." *Nature* 80:313–14.

———. 1909c. "Stability in Aeroplanes." *Nature* 81:306.

———. 1910a. "Aeroplane Stability." *Nature* 83:10–13.

———. 1910b. "Dynamics in England, France and Germany." *Nature* 83:243.

———. 1910c. "The London to Manchester Flight." *Nature* 83:278–79.

———. 1910d. "Recent Aeronautical Publications." *Nature* 84:229–33.

———. 1911. *Stability in Aviation: An Introduction to Dynamical Stability as Applied to the Motions of Aeroplanes*. London: Macmillan.

———. 1912a. "The Dynamics of Mechanical Flight." *Aeronautical Journal* 16:264–67.

———. 1912b. Review of *ABC of Hydrodynamics*, by R. de Villamil. *Mathematical Gazette* 6:379–80.

———. 1912c. Review of *The Dynamics of Mechanical Flight*, by A. G. Greenhill. *Aeronautical Journal* 16:264–67.

———. 1913a. "A Mathematician's Lectures on Aeronautics." *Nature* 90:535–36.

———. 1913b. "Remarks on the Discussion." *Aeronautical Journal* 17:56–48.

———. 1915. "The Rigid Dynamics of Circling Flight." *Aeronautical Journal* 19:46–111.

————. 1916. "Researches in Aeronautical Mathematics." *Nature* 96:509–10.

————. 1917. "Aeronautical Theories." *Nature* 98:465–67.

[Bryan, G. H.?] 1908. "A Treatise on Aërial Flight." *Nature* 78:337–38.

Bryan, G. H., and E. H. Harper. 1910. "Modern Aëronautics." *Nature* 83:132–34.

Bryan, G. H., and R. Jones. 1914. "Discontinuous Fluid Motion Past a Bent Plane, with Special Reference to Aeroplane Problems." *Proceedings of the Royal Society* A 91:354–70.

Bryan, G. H., and W. E. Williams. 1904. "The Longitudinal Stability of Aerial Gliders." *Proceedings of the Royal Society of London* 73:100–116.

Bryant, L.W., and D. H. Williams. 1924. "An Investigation of the Flow of Air around an Aerofoil of Infinite Span." T. 1885, February 1924. PRO: DSIR 23/1889.

————. 1926. "An Investigation of the Flow of Air around an Aerofoil of Infinite Span." *Philosophical Transactions of the Royal Society* A 225:199–237.

Buchan, J. 1915. *The Thirty-Nine Steps.* London: Blackwood.

Buck, R. C., and R. S. Cohen, eds. 1971. *Boston Studies in the Philosophy of Science.* Vol. 8. Dordrecht: Reidel.

Bud, R., and S. E. Cozzens, eds. 1992. *Invisible Connections: Instruments, Institutions and Science.* Bellingham, WA: SPIE Optical Engineering Press.

Burt, E. A. 1932. *The Metaphysical Foundations of Modern Physical Science.* London: Routledge and Kegan Paul.

Busk, M. 1925. *E. T. Busk: A Pioneer in Flight.* London: John Murray.

Cahan, D. 1982. "Werner Siemens and the Origin of the Physikalisch-Technische Reichsanstalt, 1872–1887." *Historical Studies in the Physical Sciences* 12:253–83.

————. 1985. "The Institutional Revolution in German Physics, 1865–1914." *Historical Studies in the Physical Sciences* 15:1–65.

Callebaut, W., ed. 1993. *Taking the Naturalistic Turn or How Real Philosophy of Science Is Done.* Chicago: Chicago University Press.

Camm, S. 1918. "Enemy Aerofoil Sections." *Flying* 4:488–90.

Campbell, N. R. 1907. *Modern Electrical Theory.* Cambridge: Cambridge University Press.

————. 1920. *Physics: The Elements.* Cambridge: Cambridge University Press. Reprinted in 1957 as *Foundations of Science: The Philosophy of Theory and Experiment* (New York: Dover).

————. 1952. *What Is Science?* New York: Dover. (Orig. pub. 1921.)

Carnap, R. 1952. *The Continuum of Inductive Methods.* Chicago: Chicago University Press.

Cartwright, N. 1983. *How the Laws of Physics Lie.* Oxford: Clarendon Press.

Cattermole, M. J. G., and A. F. Wolfe. 1987. *Horace Darwin's Shop: A History of the Cambridge Scientific Instrument Company, 1878–1968.* Boston: Adam Hilger.

Challis, J. 1873. *An Essay on the Mathematical Principles of Physics with Reference to the Study of Physical Science by Candidates for Mathematical Honours in the University of Cambridge.* Cambridge: Deighton Bell.

Chaplygin, S. 1956. *The Selected Works on Wing Theory of Sergei A. Chaplygin.* Ed. M. A. Garbell. San Francisco: Garbell Research Foundation.

Chatley, H. 1907/1910/1921. *The Problem of Flight: A Text-Book of Aerial Engineering.* 1st, 2nd, and 3rd eds. London: Charles Griffin.

————. 1914. "'X Chasing' v. 'Rule-of-Thumb.' Or a Rational Application of Mathematics to Aeronautics." *Aeronautics* 7:46–47.

Chikara, S., S. Mitsuo, and J. W. Dauben, eds. 1994. *The Intersection of History and Mathematics.* Basel, Boston, and Berlin: Birkhäuser.

Christopherson, D. G. 1972. "Richard Vynne Southwell, 1888–1970." *Biographical Memoirs of Fellows of the Royal Society* 18:549–65.

Clark, R. W. 1962. *The Rise of the Boffin.* London: Phoenix House.

Clarke, I. F. 1966. *Voices Prophesying War, 1763–1984.* London: Oxford University Press.

Clifford, W. K. 1886. *Lectures and Essays.* Ed. L. Stephen and F. Pollock. 2nd ed. London: Macmillan.

Cockcroft, J., ed. 1965. *The Organization of Research Establishments.* Cambridge: Cambridge University Press.

Cohen, R. S., and M. W. Wartofsky, eds. 1965. *Boston Studies in the Philosophy of Science.* Vol. 2. New York: Humanities Press.

Collar, A. R. 1971. " Ernest Frederick Relf, 1880–1970." *Biographical Memoirs of Fellows of the Royal Society* 17:593–616.

Collier, A. E. 1978. "Arthur Fage, 1890–1977." *Biographical Memoirs of Fellows of the Royal Society* 24:33–53.

Collins, B., and K. Robbins, eds. *British Culture and Economic Decline.* London: Weidenfeld and Nicolson.

Collins, H. 2004. *Gravity's Shadow: The Search for Gravitational Waves.* Chicago: University of Chicago Press.

Colyvan, M. 2001. "The Miracle of Applied Mathematics." *Synthese* 127:265–77.

Constant, E. W. 1980. *The Origins of the Turbojet Revolution.* Baltimore: Johns Hopkins University Press.

Cook, A. 1990. "Sir Harold Jeffreys, 1891–1989." *Biographical Memoirs of Fellows of the Royal Society* 36:303–33.

Cooper, Bertram G. [B.G.C.]. 1911. "Review of Greenhill, R&M No. 19." *Aeronautical Journal* 15:94.

Cooper, Malcolm. 1986. *The Birth of Independent Air Power.* London: Allen and Unwin.

Costello, J., and T. Hughes. 1976. *Concorde Conspiracy.* New York: Scribner's.

Cowley, W. L., and H. Levy. 1918. *Aeronautics in Theory and Experiment.* London: Edward Arnold.

Cozzens, S., and T. Gieryn, eds. 1990. *Theories of Science in Society.* Bloomington: Indiana University Press.

Crawford, E., T. Shinn., and S. Sörlin, eds. 1993. *Denationalizing Science: The Contexts of International Scientific Practice.* Dordrecht: Kluwer.

Cross, J. J. 1985. "Integral Theorems in Cambridge Mathematical Physics, 1830–55." In Harman, *Wranglers and Physicists*, 112–46.

Crouch, T. D. 1989. *The Bishop's Boys: A Life of Wilbur and Orville Wright.* New York: Norton.

Crowther, J. G. 1968. *Scientific Types.* London: Barrie and Rockliff.

Dallas Brett, R. 1988. *History of British Aviation 1908–1914.* Surbiton: Air Research Publications. (Orig. pub. 1933.)

Daniell, D. 1992. "That Infernal Aeroplane." *John Buchan Journal,* no. 11, 10–14.

Darrigol, O. 1993. "The Electrodynamic Revolution in Germany as Documented by Early German Expositions of Maxwell's Theory." *Archives for the History of the Exact Sciences* 45:89–280.

———. 2002. "Between Hydrodynamics and Elasticity Theory: The First Five Births of the Navier-Stokes Equation." *Archives for the History of the Exact Sciences* 56:95–150.

———. 2005. *Worlds of Flow: A History of Hydrodynamics from the Bernoullis to Prandt.* New York: Oxford University Press.

Darwin, H. 1913. "Scientific Instruments, Their Design and Use in Aeronautics." *Aeronautical Journal* 17:170–90.

Davis, J. 1969. *The Concorde Affair*. London: Frewin.

Dawkins, R. 1995. *River Out of Eden*. London: Weidenfeld and Nicolson.

Deimler, W. 1910. "Stabilitätesuntersuchungen über symmetrische Gleitflieger." *Zeitschrift für Flugtechnik und Motorluftschiffahrt* 1:49–53, 64–66, 91–96, 106–8.

———. 1912. "Zeichnungen zur Kuttaströmung." *Zeitschrift für Flugtechnik und Motorluftschiffahrt* 3:93–96, 107–8.

Dieudonné, J. 1981a. "Minkowski, Hermann." In Gillispie, *Dictionary of Scientific Biography*, 9:411–14.

———. 1981b. "Weyl, Hermann." In Gillispie, *Dictionary of Scientific Biography*, 14:281–85.

Dixon, B., ed. 1968. *Journeys in Belief*. London: Allen and Unwin.

Driver, H. 1997. *The Birth of Military Aviation, Britain, 1903–1914*. London: Boydell Press.

Dryden, H. L. 1955. "Fifty Years of Boundary-Layer Theory and Experiment." *Science* 121: 375–78.

———. 1965. "Theodore von Kármán, 1881–1963." *Biographical Memoirs of the National Academy of Sciences* 35:345–84.

Duhem, P. 1954. *The Aim and Structure of Physical Theory*. Trans. P. P. Wiener. Princeton: Princeton University Press.

Dunne, J. W. 1913. "The Dunne Aeroplane." *Aeronautical Journal* 17:83–102.

Durand, W. F., ed. 1934–36. *Aerodynamic Theory: A General Review of Progress*. 6 vols. Berlin: Springer.

Durkheim, E. 1915. *The Elementary Forms of the Religious Life*. Trans. J. W. Swain. London: Allen and Unwin.

———. 1956. *Education and Sociology*. London: Collier-Macmillan.

Eckert, M. 2005. "Strategic Internationalism and the Transfer of Technical Knowledge: The United States, Germany, and Aerodynamics after World War I." *Technology and Culture* 46:104–31.

———. 2006. *The Dawn of Fluid Dynamics: A Discipline between Science and Technology*. Weinheim: Wiley-VCH.

———. 2008. "Turbulenz—Ein problemhistorischer Abriss." *Naturwissenschaften, Technik und Medizin* 16:39–71.

Eden, C. G. 1911. "Apparatus for the Visual and Photographic Study of the Distribution of the Flow Round Plates and Models in a Current of Water." Reports and Memoranda No. 31.

———. 1912. "Investigation by Visual and Photographic Methods of the Flow Past Plates and Models." Reports and Memoranda No. 58.

Edgerton, D. 1991a. *England and the Aeroplane: An Essay on a Militant and Technological Nation*. London: Macmillan.

———.1991b. "Liberal Militarism and the British State." *New Left Review*, no. 185, 58–169.

———. 1991c. "The Prophet Militant and Industrial: The Peculiarities of Correlli Barnett." *Twentieth Century British History* 2:360–79.

———. 1996. *Science, Technology and British Industrial Decline, 1870–1970*. Cambridge: Cambridge University Press.

———. 2004. "The 'Linear Model' Did Not Exist." In Grandin, Wormbs, and Widmalm, *Science–Industry Nexus*, 31–57.

———. 2006a. *The Shock of the Old: Technology and Global History since 1900*. London: Profile Books.

———. 2006b. *Warfare State Britain, 1920–1970.* Cambridge: Cambridge University Press.

Einstein, A. 1905. "Über einen Erzeugung und Verwandlung des Lichtes betreffenden heuristischen Gesichtspunkt." *Annalen der Physik* 17:132–48.

———. 1916. "Elementare Theorie der Wasserwellen und des Fluges." *Die Naturwissenschaften* 4:509–10.

Elliott, B., ed. 1988. *Technology and Social Process.* Edinburgh: Edinburgh University Press.

Engelhardt, H. T., and A. I. Caplan, eds. 1987. *Scientific Controversies: Case Studies in the Resolution and Closure of Disputes in Science and Technology.* Cambridge: Cambridge University Press.

Engineer. 1908. "Literature." Review of *Aerodynamics,* by F. W. Lanchester. 105:616–17.

Engineering. 1908. "Literature." Review of *Aerodynamics,* by F. W. Lanchester. 85:460–61.

English, R., and M. Kenny, eds. 2000. *Rethinking British Decline.* London: Macmillan.

Epple, M. 2002. "Präzision *versus* Exaktheit: Konfligierende Ideale der angewandten mathematischen Forschung. Das Beispiel der Tragflügeltheorie." *Berichte zur Wissenschaftsgeschichte* 25:171–93.

Ewing, J. A. 1918–19. "Bertram Hopkinson, 1874–1918." *Proceedings of the Royal Society* A 95:xxvi–xxxvi.

Fage, A. 1915. *The Aeroplane: A Concise Scientific Study.* London: Charles Griffin.

———. 1927. "The Flow of Air and of an Inviscid Fluid around an Elliptic Cylinder and an Aerofoil of Infinite Span, especially in the Region of the Forward Stagnation Point." *Philosophical Transactions of the Royal Society of London* 227:1–19.

———. 1928. "Some Recent Experiments on Fluid Motion." *Journal of the Royal Aeronautical Society* 32:296–321; subsequent discussion, 322–30.

———. 1966. "Early Days. Memories of People and Places." *Journal of the Royal Aeronautical Society* 70:91–92.

Fage, A., and F. C. Johansen. 1928. "The Connection between Lift and Circulation for an Inclined Flat Plate." Reports and Memoranda No. 1139.

Fage, A., and L. J. Jones. 1926. "On the Drag of an Aerofoil for Two-Dimensional Flow." *Proceedings of the Royal Society of London* A 111:592–603.

Fage, A., and H. L. Nixon. 1923. "The Prediction on the Prandtl Theory of the Lift and Drag for Infinite Span from Measurements on Aerofoils of Finite Span." Reports and Memoranda No. 903.

Fage, A., and L. F. G. Simmons. 1926. "An Investigation of the Air-Flow Pattern in the Wake of an Aerofoil of Finite Span." *Philosophical Transactions of the Royal Society* 225:303–30.

Farren, W. S. 1956. "The Aerodynamic Art." *Journal of the Royal Aeronautical Society* 60: 431–49.

Farren, W. S., and H. T. Tizard. 1935. "Hermann Glauert, 1892–1934." *Obituary Notices of the Royal Society of London* 4:607–10.

Fauval, J., R. Flood, and R. Wilson, eds. 1994. *Möbius und sein Band. Der Aufstieg von Mathematik und Astronomie im Deutschland des 19. Jahrhunderts.* Basel: Birkhäuser.

Fefferman, C. 2008. "The Euler and Navier-Stokes Equations." In Gowers, *Princeton Companion to Mathematics,* 194–96.

Feldman, E. J. 1985. *Concorde and Dissent: Explaining High Technology Project Failures in Britain and France.* Cambridge: Cambridge University Press.

Felsch, V. 2003. "Der Aachener Mathematikprofessor Otto Blumenthal. Vortrag in der Volkschule Aachen, 1.10.2003." Unpublished typescript.

Finsterwalder, S. 1901–8. "Aërodynamik." In Klein and Müller, *Encyklopädie der mathematischen Wissenschaften* 4:149–84.

———. 1909. "Die Aerodynamik als Grundlage der Luftschiffahrt." In *Verhandlungen der Schweizer Naturforschenden Gesellschaft*, 92 Jahrversammlung, Lausanne, 1:69–90.

———. 1910. "Die Aerodynamik als Grundlage der Luftschiffahrt." *Zeitschrift für Flugtechnik und Motorluftschiffahrt* 1:6–10, 30–31.

———. 1913–14. "Wilhelm Deimler." *Jahresbuch: Königliche Bayerische Technische Hochschule in München* (suppl.) 6:3–4.

———. 1924. "August Föppl." *Jahrbuch der Bayerischen Akademie der Wissenschaft*, 33–37.

Flight. 1909a. "The Advisory Committee for Aeronautics." 1:308.

———. 1909b. "An Epoch Making Week for Britain." 1:256–57.

———. 1909c. "Government Attitude on Airships and Aeroplanes: Mr. Haldane's Statement as to the Expenditure during the Coming Years." 1:470–72.

———. 1910. "The Blair Atholl Experiments." 2:709–10.

———. 1911. "Advisory Committee for Aeronautics: Reports and Memoranda, No. 19." 3:335.

———. 1912a. "The Technical Report." Editorial comment. 4:1041–44.

———. 1912b. "What Is Military Aviation For?" Editorial comment. 4:955–58.

———. 1914. "The Royal Aircraft Factory and Industry." 6:112–13.

———. 1915. "The Vortex Theory of Sustentation." 7:187–88.

———. 1916a. "The Air War at Home." Editorial comment. 8:255–57.

———. 1916b. "The Deadly Fokker." Editorial comment. 8:69–72.

———. 1916c. "Politics and Personalities." 8:347–48.

———. 1916d. "The R.A.F. Committee Report." Editorial comment. 8:637–40.

———. 1917. Editorial comment. 9:467–70.

Floud, R., and D. McCloskey, eds. 1994. *The Economic History of Britain since 1700*. Vol. 2, *1860–1939*. Cambridge: Cambridge University Press.

Föppl, A. 1897. *Die Geometrie der Wirbelfelder*. Leipzig: Teubner.

———. 1900a. *Vorlesungen über technische Mechanik*. 2nd ed. Vol. 1. Leipzig: Teubner.

———. 1900b. *Vorlesungen über technische Mechanik*. Vol. 3. Leipzig: Teubner.

———. 1901. *Vorlesungen über technische Mechanik*. Vol. 4. Leipzig: Teubner.

———. 1904. *Einführung in die Maxwellsche Theorie der Elekrizität*. Leipzig: Teubner. (Orig. pub. 1894; subsequent editions were edited by Max Abraham.)

———. 1910. *Vorlesungen über technische Mechanik*. Vol. 6. Leipzig: Teubner.

———. 1925. *Lebenserinnerungen. Rückblick aus meine Lehr-und Aufstiegjahren*. Munich: Oldenbourg.

Föppl, O. 1910. "Einfluß des Seitenverhältnisses auf die Windkräfte bei gewölbten Platten. Widerstand von Drähten." *Zeitschrift für Flugtechnik und Motorluftschiffahrt* 1:193–95.

———. 1911a. "Mitteilungen aus dem Göttinger Modellversuchsanstalt. 9. Auftrieb und Widerstand eines Höhensteuers, das hinter der Tragfläche angeordnet ist." *Zeitschrift für Flugtechnik und Motorluftschiffahrt* 2:182–84.

———. 1911b. "Windkäfte an ebenen und gewölbten Platten." *Jahrbuch der Motorluftschiff-Studiengesellschaft* 4:51–119.

———. 1912. "Ergebnisse der aerodynamischen Versuchanstalt von Eiffel, verglichen mit den Göttinger Resultaten." *Zeitschrift für Flugtechnik und Motorluftschiffahrt* 3:18–121.

Föppl, O., L. Prandtl, L. Föppl, and H. Thoma, eds. 1924. *Beiträge zur technischen Mechanik und technischen Physik*. Berlin: Springer.

Forman, P. 1973. "Scientific Internationalism and the Weimar Physicists: The Ideology and Its Manipulation in Germany after World War I." *Isis* 64:50–180.

Frank, P. 1914. "Zur Differentialgeometrie der Brachistochronen (mit Anwendungen auf Hydrodynamik und Variationsrechnung)." *Sitzungberichte der mathematisch-naturwissenschaftlichen Klasse der kaiserlichen Akademie der Wissenschaften* (Vienna), 665–77.

———. 1917. "Die Bedeutung der physikalischen Erkenntnistheorie Machs für das Geistesleben der Gegenwart." *Die Naturwissenschaften* 5:65–72.

———. 1918. "Mathematiche Analogie zwischen einem Problem aus der Optik bewegter Medien und einer Aufgabe aus dem Gebiete der Flugzeugbewegung." *Physikalische Zeitschrift* 19:23–24.

———. 1921. "Ein Satz über Potentialströmungen." *Mathematische Zeitschrift* 11:105–7.

———. 1928. "Über die 'Anschaulichkeit' physikalischer Theorien." *Die Naturwissenschaften* 16:121–28.

———. 1929. "Was bedeuten die gegenwärtigen physikalischen Theorien für die allgemeine Erkenntnislehre?" *Die Naturwissenschaften* 17:971–94.

———. 1932. *Das Kausalgesetz und seine Grenzen.* Vienna: Springer.

———. 1933. "Die schnellste Flugverbindung zwischen zwei Punkten." *Zeitschrift für angewandte Mathematik und Mechanik* 13:88–91.

———. 1948. *Einstein: His Life and Times.* London: Jonathan Cape.

———. 1951. *Relativity: A Richer Truth.* London: Jonathan Cape.

———. 1952. *Wahrheit—Relativ oder absolut?* Zurich: Pan-Verlag.

———. 1954a. "On the Variety of Reasons for the Acceptance of Scientific Theories." *Scientific Monthly* 79:139–45.

———. 1954b. "The Work of Richard von Mises: 1883–1953." *Science* 119:823–24.

———. 1957. *Philosophy of Science: The Link between Science and Philosophy.* Englewood Cliffs, NJ: Prentice-Hall.

———. 1961. *Modern Science and Its Philosophy.* New York: Collier.

———. 1998. *The Law of Causality and Its Limits.* Trans. Marie Neurath and Robert Cohen. Dordrecht: Kluwer.

Frank, P., and K. Löwner. 1919. "Eine Anwendung des Koebeschen Verzerrungsatzes auf ein Problem der Hydrodynamik." *Mathematische Zeitschrift* 3:78–86.

Frank, P., and R. von Mises, eds. 1925–27. *Die Differential- und Integralgleichungen der Mechanik und Physik.* 2 vols. Braunschweig: Vieweg.

Fraser, A. C., ed. 1871. *The Works of George Berkeley, D.D.* Vol. 3. Oxford: Clarendon Press.

Fricke, R. 1905. "Bemerkungen über den mathematischen Unterricht an den technischen Hochschulen in Deutschland." In Krazer, *Verhandlungen des dritten Internationalen Mathematiker-Kongresses,* 615–21.

Friedman, M. 1998. "On the Sociology of Scientific Knowledge and Its Philosophical Agenda." *Studies in the History and Philosophy of Science* 29:239–71.

Fritzsche, P. 1992. *A Nation of Flyers: German Aviation and the Popular Imagination.* Cambridge: Harvard University Press.

Fromkin, D. 2004. *Europe's Last Summer: Why the World Went to War in 1914.* London: Heinemann.

Fuchs, R., and L. Hopf. 1917. "Momentenausgleich und statische Langenstabilität." *Technische Berichte* 1:15–30.

———. 1922. *Aerodynamik.* Berlin: Richard Carl Schmidt.

Fuhrmann, G. 1910. "Bücher-Besprechungen, *Aerodynamik. Ein Gesamtwerk über das Fliegen,*

I Band. (übersetzt von C. und A. Runge) Leipzig und Berlin, Teubner, 1909." *Zeitschrift für Flugtechnik und Motorluftschiffahrt* 1:231.

———. 1911–12. "Theoretische und experimentelle Untersuchungen an Ballonmodellen." *Jahrbuch der Motorluftschiff-Studiengesellschaft* 5:64–123.

———. 1913. "Flüssigkeit." In Korschelt et al., *Handwörterbuch der Naturwissenschaften,* 73–84.

Galison, P. 1997. *Image and Logic: The Material Culture of Microphysics.* Chicago: University of Chicago Press.

———. 2001. "The Americanization of Unity." In Galison, Graubard, and Mendelsohn, *Science in Culture,* 45–71.

Galison, P., S. Graubard, and E. Mendelsohn, eds. 2001. *Science in Culture.* New Brunswick, NJ: Transaction Publishers.

Galison, P., and A. Rowland, eds. 2000. *Atmospheric Flight in the Twentieth Century.* Dordrecht: Kluwer.

Garber, E. 1998. *The Language of Physics: The Calculus and the Development of Theoretical Physics in Europe, 1750–1914.* Boston: Birkhäuser.

Gavroglu, K., J. Christianidis, and E. Nicolaidis, eds. 1994. *Trends in the Historiography of Science.* Dordrecht: Kluwer.

Geckeler, J. 1922. "Über Auftrieb und statische Längstabilität von Flugzeugtragflügeln in ihrer Abhängigkeit von Profilform." *Zeitschrift für Flugtechnik and Motorluftschiffahrt* 13:137–45, 191–95.

Germain, P. 2000. "The 'New' Mechanics of Fluids of Ludwig Prandtl." In Meier, *Ludwig Prandtl,* 31–40.

Giacomelli, R., and E. Pistolesi. 1934. "Historical Sketch." In Durand, *Aerodynamic Theory,* 1:305–94.

Gibbs-Smith, C. 1960. *The Aeroplane: An Historical Survey of Its Origins and Development.* London: Her Majesty's Stationary Office.

Gieryn, T. F., and Anne E. Figert. 1990. "Ingredients for a Theory of Science in Society: O-Rings, Ice Water, C-Clamps, Richard Feynman, and the *New York Times.*" In Cozzens and Gieryn, *Theories of Science in Society,* 67–97.

Gigerenzer, G. 2000. *Adaptive Thinking: Rationality in the Real World.* Oxford: Oxford University Press.

Gillispie, C. C., ed. 1981. *Dictionary of Scientific Biography.* 18 vols. New York: Charles Scribner's.

Gispen, K. 1989. *New Profession, Old Order: Engineers and German Society, 1815–1914.* Cambridge: Cambridge University Press.

Glancey, J. 2007. *Spitfire: The Biography.* London: Atlantic Books.

Glas, E. 1993. "From Form to Function: A Reassessment of Felix Klein's Unified Programme of Mathematical Research, Education and Development." *Studies in History and Philosophy of Science* 24:611–31.

———. 2000. "Model-Based Reasoning and Mathematical Discovery." *Studies in History and Philosophy of Science* 31:71–86.

Glauert, H. 1915a. "The Form of a Rotating Fluid Mass, as Disturbed by a Satellite." *Monthly Notices of the Royal Astronomical Society* 75:629–48.

———. 1915b. "The Rotation of the Earth." *Monthly Notices of the Royal Astronomical Society* 75 (pt. 1): 489–95; (pt. 2): 685–87.

———. 1920a. "Prandtl's Aerofoil Theory and Its Application to Wing Structures." *Aircraft Engineering* 1:160–61.

———. 1920b. "Summary of the Present State of Knowledge with Regard to Stability and Control of Aeroplanes." Reports and Memoranda No. 701.

———. 1923a. "Art: Hydrodynamics in Its Application to Aeronautics." In Glazebrook, *A Dictionary of Applied Physics*, 5:182–88.

———. 1923b. "Art: Wing Surfaces, the Hydrodynamical Theory of." In Glazebrook, *A Dictionary of Applied Physics*, 5:217–20.

———. 1923c. "Some Aspects of Modern Aerofoil Theory." In Lockwood Marsh, *International Air Congress, London, 1923*, 245–55.

———. 1923d. "Theoretical Relationships for the Lift and Drag of an Aerofoil Structure." *Journal of the Royal Aeronautical Society* 27:512–18.

———. 1926. *The Elements of Aerofoil and Airscrew Theory*. Cambridge: Cambridge University Press.

———. 1929. "The Accelerated Motion of a Cylindrical Body through a Fluid." Reports and Memoranda No. 1215.

Glauert, M. 1921. "Theoretical Stream-Lines round a Joukowsky Aerofoil." Reports and Memoranda No. 788.

———. 1923. "Two-Dimensional Aerofoil Theory." *Journal of the Royal Aeronautical Society* 27:348–66.

Glazebrook, R. T. 1902. "The Aims of the National Physical Laboratory." *Proceedings of the Royal Institution of Great Britain* 16:656–67.

———. 1914. "The Development of the Aeroplane." *Aeronautics* 7:169–76.

———. 1917. *Science and Industry: The Place of Cambridge in Any Scheme for Their Combination.* Cambridge: Cambridge University Press.

———. 1920. "Some Points of Importance in the Work of the Advisory Committee for Aeronautics." *Aeronautics* 18:435–37, 450–52.

———, ed. 1923a. *A Dictionary of Applied Physics*. Vol. 5. London: Macmillan.

———. 1923b. "Scientific Research." *Times*, February 28, 8.

———. 1923c. "Standardisation of Methods of Research." In Lockwood Marsh, *International Air Congress, London, 1923*, 21–37; subsequent discussion, 60–67.

———. 1924. "Science and Service." In Humberstone, . *Science and Labour*, 18–23.

———. 1931. "Aeronautical Research in England: The Aeronautical Research Committee, 1920–30." In *Cinquième Congrès International de la Navigation Aérienne*, 691–705. The Hague: Martinus Nijhoff.

———. 1937. "Darwin, Sir Horace, 1851–1928." In Weaver, *Dictionary of National Biography*, 239.

Glazebrook, R. T., and W. N. Shaw. 1885. *Practical Physics*. London: Longmans, Green.

Gold, E. 1945–48. "Sir William Napier Shaw, 1854–1945." *Obituary Notices of Fellows of the Royal Society* 5:203–30.

Goldberg, S. 1984. *Understanding Relativity: Origin and Impact of a Scientific Revolution.* Boston: Birkhäuser.

Goldstein, S., ed. 1938. *Modern Developments in Fluid Dynamics.* Oxford: Clarendon Press.

———. 1969. "Fluid Mechanics in the First Half of This Century." *Annual Review of Fluid Mechanics* 1:1–28.

Gollin, A. 1984. *No Longer an Island: Britain and the Wright Brothers, 1902–1909.* London: Heinemann.

———. 1989. *The Impact of Air Power on the British People and Their Government, 1909–1914.* Stanford: Stanford University Press.

Goodstein, J. R. 1991. *Millikan's School: A History of the California Institute of Technology*. New York: Norton.

Goody, G. 2005. "Fear, Shunning, and Valuelessness: Controversy over the Use of 'Cambridge' Mathematics in Late Victorian Electro-Technology." In Kaiser, *Pedagogy and the Practice of Science*, 111–49.

Gowers, T., ed. 2008. *The Princeton Companion to Mathematics*. Princeton: Princeton University Press.

Grahame-White, C. 1930. *Flying: An Epitome and a Forecast*. London: Chatto and Windus.

Grammel, R. 1917. *Die hydrodynamischen Grundlagen des Fluges*. Braunschweig: Vieweg und Sohn.

Grandin, K., N. Wormbs., and S. Widmalm, eds. 2004. *The Science–Industry Nexus: History, Policy, Implications*. New York: Watson.

Grattan-Guiness, I. 1997. *The Fontana History of the Mathematical Sciences*. London: Fontana Press.

Gray, J. 1999. *The Symbolic Universe: Geometry and Physics, 1890–1930*. Oxford: Oxford University Press.

Green, F. M. 1958. "The Chudleigh Mess." *RAE News* (January): 5–7.

Greenhill, G. 1876. *Solutions of the Cambridge Senate-House Problems and Riders for the Year 1875*. London: Macmillan.

———. 1880. "Notes on Hydrodynamics, I." *Messenger of Mathematics* 9:113–17.

———. 1905a. "The Mathematical Theory of the Top Considered Historically." In Krazer, *Verhandlungen des dritten Internationalen Mathematiker-Kongresses*, 100–108.

———. 1905b. "Teaching of Mechanics by Familiar Applications on a Large Scale." In Krazer, *Verhandlungen des dritten Internationalen Mathematiker-Kongresses*, 582–85.

———. 1910. "Report on the Theory of a Stream Line Past a Plane Barrier and of the Discontinuity Arising at the Edge, with an Application of the Theory to an Aeroplane." Reports and Memoranda No. 19.

———. 1911. "Hydromechanics." In *The Encyclopaedia Britannica*, 11th ed., 14:115–35. New York: Encyclopaedia Britannica.

———. 1912. *The Dynamics of Mechanical Flight*. London: Constable.

———. 1914. "Obituary of the Rev. Sir John Twisden." *Nature* 94:427.

———. 1915. "Mathematics in Artillery Science" *Nature* 94:573–74.

———. 1916. "Theory of a Stream Line Past a Curved Wing." Appendix to Report No. 19 (1910). London: His Majesty's Stationary Office.

———. 1917. "Motion of a Bomb Released in Flight." Reports and Memoranda No. 369.

Grey, C. G. 1913. "The Best in the World." *Aeroplane* 4:635–38.

———. 1914a. "The Improvement of Aeroplanes." *Aeroplane* 6:423–24.

———. 1914b. "The Victims of Science." *Aeroplane* 6:320–21.

———. 1915. "Scientific Accuracy versus Practical Truth." *Aeroplane* 9:153–54.

———. 1916a. "The Aeronautical Society." *Aeroplane* 8:562.

———. 1916b. "The Constitution of the Air Board." *Aeroplane* 10:833–34.

———. 1916c. "On a Professor." *Aeroplane* 10:166.

———. 1916d. "On Professors and a Book." *Aeroplane* 10:153–58.

———. 1917a. "The Advisory Committee's Report." *Aeroplane* 12:310–15.

———. 1917b. "Editorial Comment." *Aeroplane* 12:1284.

———. 1917c. "On Mr. Lanchester and his Friends." *Aeroplane* 13:711–22.

———. 1918. "On Expectations and Anticipations II." *Aeroplane* 14:381–84.

Grey, Viscount [Edward]. 1925. *Twenty Five Years, 1892–1916*. 2 vols. New York: F. A. Stokes.

Gridgeman, N. T. 1981. "Mises, Richard von." In Gillispie, *Dictionary of Scientific Biography*, 9:419–20.

Griffith, A. A., and G. I. Taylor. 1917. "The Use of Soap Films in Solving Torsion Problems." In *The Institution of Mechanical Engineers: Proceedings*, 755–809.

Griffiths, R. 1980. *Fellow Travellers of the Right: British Enthusiasts for Nazi Germany*. London: Constable.

Gross, P., and N. Levitt. 1994. *Higher Superstition: The Academic Left and Its Quarrels with Science*. Baltimore: Johns Hopkins University Press.

Gross, P., N. Levitt, and M. Lewis, eds. 1996. *Flight from Science and Reason*. New York: New York Academy of Sciences.

Gruber, W. H., and D. G. Marquis, eds. 1969. *Factors in the Transfer of Technology*. Cambridge: MIT Press.

Gumbel, E. J., and L. Hopf. 1917. "Die Druckpunktwanderung und ihr Einfluß auf die Stabilität." *Technische Berichte* 1:108–15.

Guthrie, F. 1868. "On Discontinuous Movements of Fluids." *London, Edinburgh and Dublin Philosophical Magazine* 36:337–46.

Haack, S. 1996. "Towards a Sober Sociology of Science." In Gross, Levitt, and Lewis, *Flight from Science and Reason*, 259–65.

Haldane, R. B. 1903–4. *The Pathway to Reality*. 2 vols. London: John Murray.

———. 1921. *The Reign of Relativity*. London: John Murray.

———. 1929. *Richard Burdon Haldane: An Autobiography*. London: Hodder and Stoughton.

Hales, S. D., ed. 2011. *Blackwell's Companion to Relativism*. Oxford: Blackwell.

Hall, A,. and M. Morgan. 1977. "Bennett Melvill Jones." *Biographical Memoirs of Fellows of the Royal Society* 23:253–82.

Handley Page, F. 1911. "The Pressure on Plane and Curved Surfaces Moving through the Air." *Aeronautical Journal* 15:47–64.

———. 1921. "The Handley Page Wing." *Aeronautical Journal* 25:263–89.

Hanle, P. A. 1982. *Bringing Aerodynamics to America*. Cambridge: MIT Press.

Hanlon, G. O. 1869–71. "The Vena Contracta." *Proceedings of the London Mathematical Society* 3:4–5.

Hansen, James R., ed. 2003. *The Ascent of the Airplane*. Vol. 1 of *The Wind and Beyond: A Documentary Journey into the History of Aerodynamics in America*. Washington, DC: National Aeronautics and Space Administration.

Hardy, G. H. 1906. "The Reform of the Mathematical Tripos." *Cambridge Review* 28:8–10.

———. 1908. *A Course of Pure Mathematics*. Cambridge: Cambridge University Press.

Hare, P. A. 1990. *The Royal Aircraft Factory*. London: Putnam.

Harman, P. M., ed. 1985. *Wranglers and Physicists: Studies on Cambridge Physics in the Nineteenth Century*. Manchester: Manchester University Press.

Harper, E. H. 1913. "The Mathematical Theory of Aeroplane Stability." *Aeronautical Journal* 17:11–23; subsequent discussion, 23–31.

Hartcup, G. 1974. *The Achievement of the Airship: A History of the Development of Rigid, Semi-Rigid and Non-Rigid Airships*. Newton Abbott: David and Charles.

Harwood, J. 2006. "Engineering Education between Science and Practice: Rethinking the Historiography." *History and Technology* 22:53–79.

Hashagen, U. 1993. "Mathematik für Ingenieure oder Stellenmarkt für Mathematiker. Die ersten

50 Jahre Mathematikunterricht an der TH München (1868–1918)." In Wengenroth, *Die TU München*, 39–86.

———. 2000. "Der Mathematiker Walther von Dyck und die 'wissenschaftliche' Technischer Hochschule." In Schneider, Trischler, and Wengenroth, *Oszillation*, 267–96.

———. 2003. *Walther von Dyck (1856–1934). Mathematik, Technik und Wissenschaftorganisation an der TH München.* Stuttgart: Franz Steiner Verlag.

Hashimoto, T. 1990. "Theory, Experiment, and Design Practice: The Formation of Aeronautical Research, 1909–1930." Ph.D. dissertation, Johns Hopkins University.

———. 2000. "The Wind Tunnel and the Emergence of Aeronautical Research in Britain." In Galison and Rowland, *Atmospheric Flight in the Twentieth Century*, 223–39.

———. 2007. "Leonard Bairstow as a Scientific Middleman: Early Aerodynamic Research on Airplane Stability in Britain, 1909–1920." *Historia Scientiarum* 17:103–20.

Hattersley, R. 2006. *The Edwardians*. London: Abacus, 2006.

Heidelberger, M. 2006. "Applying Models in Fluid Dynamics." *International Studies in the Philosophy of Science* 20:49–67.

Heidelberger, M., and F. Stadler, eds. 2003. *Philosophy of Science and Politics*. Vienna: Springer.

Helmholtz, H. von. 1868. "Über discontinuirliche Flüssigkeits-Bewegungen." *Monatsbericht der königlich preussischen Akademie der Wissenschaften zu Berlin*, 215–28.

Hentschel, K., and Ann Hentschel, eds. 1996. *Physics and National Socialism: An Anthology of Primary Sources*. Basel: Birkhäuser.

Hesse, M. 1961. *Forces and Fields: The Concept of Action at a Distance in the History of Physics*. London: Thomas Nelson.

———. 1974. *The Structure of Scientific Inference*. London: Macmillan.

H.F.B. 1928. "Alfred George Greenhill, 1847–1927." *Proceedings of the Royal Society* A 119:i–iv.

Hiemenz, K. 1911. "Die Grenzschicht an einem in den gleichförmigen Flüssigkeitsstrom eingetauchten geraden Kreiszylinder." *Dinglers Polytechnisches Journal* 326:321–24, 344–48, 357–62, 372–76, 391–93, 407–10.

Hilken, T. J. N. 1967. *Engineering at Cambridge University, 1783–1965*. Cambridge: Cambridge University Press.

Hill, R. M. 1920. "The Influence of Military and Civil Requirements on the Flying Qualities of Aeroplanes." Reports and Memoranda No. 678.

———. 1923. "The Manoeuvres of Inverted Flight." *Journal of the Royal Aeronautical Society* 27:569–605.

Hilton, H. 1912. Review of *Stability in Aviation*. *Mathematical Gazette* 6:343–44.

Hoare, P. 1997. *Wilde's Last Stand: Decadence, Conspiracy and the First World War*. London: Duckworth.

Hobson, E. W. 1891. *Treatise on Plane Trigonometry*. Cambridge: Cambridge University Press.

Hobson, E. W., and A. E. H. Love, eds. 1913. *Proceedings of the Fifth International Congress of Mathematicians (Cambridge, 22–28 August, 1912)*. 2 vols. Cambridge: Cambridge University Press.

Hoffmann, D., ed. 1995. *Gustav Magnus und sein Haus*. Stuttgart: Verlag für Geschichte der Naturwissenschaften und der Technik.

Holton, G. 1973. "Influences on Einstein's Early Work." In *Thematic Origins of Scientific Thought: Kepler to Einstein*. Cambridge: Harvard University Press.

———. 1978. *The Scientific Imagination*. Cambridge: Cambridge University Press.

———. 1995. "Einstein and Books." In Knox and Siegel, *No Truth Except in the Details*, 273–79.

Holton, G., and R. S. Cohen. 1981. "Frank, Philipp." In Gillispie, *Dictionary of Scientific Biography*, 5:122–23.

Holton, G., E. C. Kemble, W. V. Quine, S. S. Stevens., and M. G. White. 1968. "In Memory of Philipp Frank." *Philosophy of Science* 35:1–5.

Hopkinson, B. 1898. "On Discontinuous Fluid Motions involving Sources and Vortices." *Proceedings of the London Mathematical Society* 29:142–64.

Howarth, L. 1935a. "Note on the Development of the Circulation around a Thin Elliptic Cylinder." *Proceedings of the Cambridge Philosophical Society* 31:582–94.

———. 1935b. "The Theoretical Determination of the Lift Coefficient for a Thin Elliptic Cylinder." *Proceedings of the Royal Society of London* A 149:558–86.

Humberstone, T. L., ed. 1924. *Science and Labour*. London: Ernest Benn.

Hume, D. 1902. *An Enquiry concerning Human Understanding*. Ed. L. A. Selby-Bigge. Oxford: Clarendon Press. (Orig. pub. 1777.)

———. 1960. *A Treatise of Human Nature*. Oxford: Clarendon Press. (Orig. pub. 1739–40.)

Hunecke, V. 1979. "Der 'Kampf ums Dasein' und die Reform der technischen Erziehung im Denken Alois Riedlers." In Ruerup, *Wissenschaft und Gesellschaft*, 301–13.

Hunt, L. 1991. *Secret Agenda: The United States Government, Nazi Scientists, and Project Paperclip, 1945–1990*. New York: St. Martin's Press.

Hutchinson, E. 1969. "Scientists and Civil Servants: The Struggle over the National Physical Laboratory in 1918." *Minerva* 7:373–98.

Interavia. 1955. "Professor Einsteins 'Leichtsinn': Ein unbekanntes Dokument: Der Schöpfer der Relativitätstheorie als Aerodynamiker." 10:684–85.

Jaffé, G. 1920. "Bemerkung über die Entstehung von Wirbeln in Flüssigkeiten." *Physikalische Zeitschrift* 21:541–43.

Jakab, P. L. 1990. *Visions of a Flying Machine: The Wright Brothers and the Process of Invention*. Washington, DC: Smithsonian Institution Press.

James, H. 1990. "The German Experience and the Myth of British Cultural Exceptionalism." In Collins and Robbins, *British Culture and Economic Decline*, 91–128.

Jane, F. 1990. *Jane's Fighting Aircraft of World War I*. London: Studio Editions. (Orig. pub. 1919.)

J.C. 1920. "A Book for Designers." *Aeronautics* 18:471.

Jeffreys, H. 1925. "On the Circulation Theory of Aeroplane Lift." *Philosophical Magazine* 50:815–19.

———. 1928a. "The Equations of Viscous Motion and the Circulation Theorem." *Proceedings of the Cambridge Philosophical Society* 24:477–79.

———. 1928b. "On Aerofoils of Small Thickness." *Proceedings of the Royal Society of London* A 121:22–28.

———. 1930. "The Wake in Fluid Flow Past a Solid." *Proceedings of the Royal Society of London* A 128:376–93.

Jex, H. R., and F. E. C. Culick. 1985. "Flight Control Dynamics of the 1903 Wright Flyer." In *Proceedings of the 12th Atmospheric Flight Control Conference*, 534–48. New York: American Institute of Aeronautics and Astronautics.

Johannessen, K. S., and T. Nordenstam, eds. 1996. *Wittgenstein and the Philosophy of Culture*. Vienna: Hölder-Pichler-Tempsky.

Jones, B. M. 1923. "Control of Aeroplanes at Low Speeds." *Journal of the Royal Aeronautical Society* 27:473–78.

Jones, R. Y. 1950. "Leading-Edge Singularities in Thin-Airfoil Theory." *Journal of the Aeronautical Sciences* (May): 307–10.

Joukowsky, N. 1906. "De la chute dans l'air de corps legers de forme allongée, animés d'un movement rotatoire." *Bulletin de l'Institut Aërodynamique de Koutschino* 1:51–65.

———. 1910/1912. "Über die Konturen der Drachenflieger." *Zeitchrift für Flugtechnik und Motorluftschiffahrt* 1:281–84; 3:81–86.

———. 1916. *Bases théoriques de l'aéronautique: Aérodynamique*. Paris: Gauthier-Villars.

———. 2001. "On Annexed Vortices" (1906). In Ackroyd, Axcell, and Ruban, *Early Developments of Modern Aerodynamics*, 89–104.

Judge, A. W. 1917. *The Properties of Aerofoils and Aerodynamic Bodies: A Text-Book for Aeronautical Engineers, Draughtsmen and Students*. London: Whittaker.

Kaiser, D., ed. 2005. *Pedagogy and the Practice of Science*. Cambridge: MIT Press.

Kármán, T. von. 1911–12. "Über den Mechanismus des Widerstandes, den ein bewegter Körper in einer Flüssigkeit erfährt." *Nachrichten der Königl Gesellschaft der Wissenschaften zu Göttingen. Math.-phys. Klasse* (1911): 509–17; (1912): 547–56.

———. 1922. "Von Kármáns Erklärungen des Segelflugs." *Die Naturwissenschaften* 10:432–34.

———. 1940. "Some Remarks on Mathematics from the Engineer's Viewpoint." *Mechanical Engineering* 62:308–10.

———. 1963. *Aerodynamics: Selected Topics in the Light of Their Historical Development*. New York: McGraw-Hill. (Orig. pub. 1954.)

Kármán, T. von, and J. M. Burgers. 1935. "General Aerodynamic: Theory. Perfect Fluids." In Durand, *Aerodynamic Theory*, vol. 2. (This work comprises the whole of vol. 2.)

Kármán, T. von, and T. Levi-Civita, eds. 1924. *Vorträge aus dem Gebeite der Hydro-und Aerodynamik*. Berlin: Springer.

Kármán, T. von, and H. Rubach. 1912. "Über den Mechanismus des Flüssigkeits- und Luftwiderstandes." *Physikalische Zeitschrift* 13:49–59.

Kármán, T. von, and E. Trefftz. 1914–15. "Über Längsstabilität und Längsschwingungen von Flugzeugen." *Jahrbuch der Wissenschaftlichen Gesellschaft für Luftfahrt* 3:116–38.

———. 1918. "Potentialströmung um gegebene Tragflächenquerschnitte." *Zeitschrift für Flugtechnik und Motorluftschiffahrt* 9:111–16

Kelvin, Lord. 1894. "On the Doctrine of Discontinuity of Fluid Motion, in Connection with the Resistance against a Solid Moving through a Fluid." *Nature* 50:524–25, 549, 573–75, 597–98.

Kennedy, R. 1910. "The National Laboratory Experiments." *Aero* 2:339–40.

Kevles, D. J. 1971. "'Into Hostile Political Camps': The Reorganisation of International Science in World War I." *Isis* 62:47–60.

Keynes, J. M. 1919. *The Economic Consequences of the Peace*. London: Macmillan.

Kingsford, P. W. 1960. *F. W. Lanchester: The Life of an Engineer*. London: Arnold.

Kirchhoff, G. 1869. "Zur Theorie freier Flüssigkeitsstrahlen." *Crelle's Journal für reine und angewandte Mathematik* 70:289–98.

———. 1876. *Vorlesungen über Mathematische Physik (Mechanik)*. Leipzig: Teubner.

Klein, F. 1900. "Ueber technischer Mechanik." In Klein and Reicke, *Über angewandte Mathematik und Physik*, 26–41.

———. 1909–10. "Über die Bildung von Wirbeln in reibungslosen Flüssigkeiten." *Zeitschrift für Mathematik und Physik* 58:259–62.

————. 1979. *Development of Mathematics in the 19th Century*. Trans. M. Ackerman. Brookline, MA: Math Sci Press.

Klein, F., and C. Müller, eds. 1901–8. *Encyklopädie der mathematischen Wissenschaften*. Leipzig: Teubner.

Klein, F., and E. Reicke, eds. 1900. *Über angewandte Mathematik und Physik in ihrer Bedeutung für den Unterricht an den höheren Schulen nebst Erläuterung der bezüglichen Göttinger Universitätseinrichtungen*. Leipzig: Teubner.

Kneißl, M. 1942. "Sebastian Finsterwalder zum 80. Geburtstag." *Bildmessung und Luftbildwesen* 17:53–64.

Knox, A. J., M. J. Klein, and R. Schulmann, eds. 1966. *Collected Papers of Albert Einstein*. Vol. 6, *The Berlin Years, 1914–1917*. Princeton: Princeton University Press.

Knox, A. J., and D. M. Siegel, eds. 1995. *No Truth Except in the Details: Essays in Honor of Martin J. Klein*. Dordrecht: Kluwer.

Koertge, N. ed. 1998. *A House Built on Sand: Exposing Postmodernist Myths about Science*. Oxford: Oxford University Press.

Kopal, Z. 1980. "Darwin, George Howard." In Gillispie, *Dictionary of Scientific Biography*, 3:582–84.

Korschelt, E., et al., eds. 1913. *Handwörterbuch der Naturwissenschaften*. Jena: G. Fischer.

Koss, S. E. 1969. *Lord Haldane: Scapegoat for Liberalism*. New York: Columbia University Press.

Krazer, A., ed. 1905. *Verhandlungen des dritten Internationalen Mathematiker-Kongresses*. Leipzig: Teubner.

Küchemann, D. 1967. "Some Developments in Aerofoil Theory." Technical Report 67052. Farnborough: Royal Aircraft Establishment.

————. 1970. "An Aerodynamicist's Prospect of the Second Century." In Allen and Bruce, *Future of Aerodynamics*, 24–38.

————. 1978. *The Aerodynamic Design of Aircraft*. Oxford: Pergamon.

Kuhn, T. S. 1962. *The Structure of Scientific Revolutions*. Chicago: University of Chicago Press.

Kumbruch, H. 1917. "Der Luftwiderstand von Stirnkühlern." *Technische Berichte* 2:1–12.

Kutta, W. M. 1901. "Beitrag zur näherungsweisen Integration totaler Differentialgleichungen." *Zeitschrift für Mathematik und Physik* 46:435–53.

————. 1902. "Auftriebskraft in strömenden Flüssigkeiten." *Illustrirte Aëronautische Mittheilungen* 6:133–35.

————. 1910. "Über eine mit den Grundlagen des Flugproblems in Beziehung stehende zweidimensionale Strömung." *Sitzungsberichte der Königlich Bayerischen Akademie der Wissenschaften Mathematisch-physikalische Klass*, 3–58.

————. 1911. "Über ebene Zirkulationsströmungen nebst flugtechnischen Anwendungen." *Sitzungsberichte der mathematisch-physikalischen Klasse der K.B. Akademie der Wissenschaften zu München*, 65–125.

Lagally, M. 1915. "Zur Theorie der Wirbelschichten." *Sitzungsberichte der mathematisch-physikalischen Klasse der K.B. Akademie der Wissenschaften zu München*, 79–107.

————. 1921. "Über den Druck einer strömenden Flüssigkeit auf eine geschlossene Fläche." *Sitzungsberichte der mathematisch-physikalischen Klasse der Bayerischen Akademie der Wissenschaften zu München*, 209–26.

Lakatos, I. 1971. "The History of Science and Its Rational Reconstructions." In Buck and Cohen, *Boston Studies in the Philosophy of Science*, 8:91–136.

Lamb, H. 1879. *A Treatise on the Mathematical Theory of the Motion of Fluids*. Cambridge: Cambridge University Press.

———. 1895. *Hydrodynamics.* 2nd ed. Cambridge; Cambridge University Press.

———. 1904. "Presidential Address, Section A." *Report of the British Association for the Advancement of Science, Cambridge,* 421–30.

———. 1906. *Hydrodynamics.* 3rd ed. Cambridge; Cambridge University Press.

———. 1911. "On the Motion of a Sphere through a Viscous Fluid." *Philosophical Magazine* 21:112–20.

———. 1916. *Hydrodynamics.* 4th ed. Cambridge: Cambridge University Press.

———. 1919. *An Elementary Course of Infinitesimal Calculus.* 3rd ed. Cambridge: Cambridge University Press.

———. 1924a. *The Evolution of Mathematical Physics: Rouse Ball Lecture (1924).* Cambridge: Cambridge University Press.

———. 1924b. *Hydrodynamics.* 5th ed. Cambridge: Cambridge University Press.

———. 1925. "The Presidential Address." *Report of the British Association for the Advancement of Science, Southampton,* 1–14.

———. 1929. "The Hydrodynamic Forces on a Cylinder Moving in Two Dimensions." Reports and Memoranda No. 1218.

———. 1931. *Lehrbuch der Hydrodynamik.* 2nd ed. Ed. Elise Helly, with a foreword and additions by R. von Mises. Leipzig: Teubner.

———. 1932. *Hydrodynamics.* 6th ed. Cambridge: Cambridge University Press.

Lambe, C. G., and C. J. Tranter. 1964. *Differential Equations for Engineers and Scientist.* London: English Universities Press.

Lanchester, F. W. 1907. *Aerodynamics, constituting the First Volume of a Complete Work on Aerial Flight.* London: Constable.

———. 1908. "The Voisin and the Wright Aeroplanes." *Engineering* 86:828.

———. 1909. *Aerodynamik. Ein Gesamptwerk über das Fliegen.* Trans. C. Runge and A. Runge. Leipzig: Teubner.

———. 1915. "The Flying Machine: The Aerofoil in the Light of Theory and Experiment." *Proceedings of the Institution of Automobile Engineers* 9:171–259.

———. 1917a. "A Campaign of Slander: A Few Facts." *Flying* 1:51–54.

———. 1917b. "Formation Flying as Taught by Nature." *Flying* 1:35–37.

———. 1917c. "The Foundation Stones." *Flying* 2:354–56.

———. 1926. "Sustentation in Flight: Wilbur Wright Memorial Lecture." *Journal of the Royal Aeronautical Society* 30:587–606.

Lasby, C. G. 1971. *Project Paperclip: German Scientists and the Cold War.* New York: Atheneum.

Laudan, L. 1977. *Progress and Its Problems.* Berkeley: University of California Press.

———. 1996. *Beyond Positivism and Relativism: Theory, Method, and Evidence.* Oxford: Westview Press.

Lavender, T. 1923. "A Direction and Velocity Meter for Use in Wind-Tunnel Work." *Report and Memoranda No. 844.*

Leathem, J. G. 1915. "Some Applications of Conformal Transformations to Problems of Hydrodynamics." *Philosophical Transactions of the Royal Society* A 215:439–89.

Ledeboer, J. H. [J. H. L.]. 1909. "Review of Lanchester's Aerodonetics." *Aeronautics* 2:6–7.

———. [J. H. L.]. 1916. "The Function of Literature." *Aeronautics* 11:33.

Levy, H. 1915. "On the Resistance Experienced by a Body Moving in a Fluid." *Proceedings of the Royal Society of Edinburgh* 35:95–109.

———. 1916. "Discontinuous Fluid Motion Past a Curved Boundary." *Proceedings of the Royal Society of London* A 92:285–304.

———. 1919. "From Model to Full Scale in Aeronautics." *Aeronautical Journal* 23:326–56.

———. 1938a. *A Philosophy for a Modern Man.* London: Gollancz.

———. 1938b. *The Universe of Science.* London: Watts.

———. 1939. *Modern Science: A Study of Physical Science in the World Today.* London: Hamish Hamilton.

———. 1945. *Social Thinking.* London: Cobbett Press.

———. 1968. "From Myth to Rational Action." In Dixon, *Journeys in Belief,* 161–76.

Lighthill, J. 1965. "The Royal Aircraft Establishment." In Cockcroft, *Organization of Research Establishments,* 28–54.

———. 1990. "Sidney Goldstein, 1903–1989." *Biographical Memoirs of Fellows of the Royal Society* 36:175–97.

———. 1995. "Fluid Mechanics." In Brown, Pais, and Pippard, *Twentieth Century Physics,* 795–912.

Lilienthal, O. 1889. *Der Vogelflug als Grundlage der Fliegerkunst. Ein Beitrag zur Systematik der Flugtechnik.* Berlin: R. Gaertners Verlagsbuchhandlung.

Lindner, H. 1980. "'Deutsch' und 'gegentypische' Mathematik. Zur Begründung einer 'arteigenen' Mathematik im 'Dritten Reich' durch Ludwig Bieberbach." In Mehrtens and Richter, *Naturwissenschaft Technik und N-S Ideologie,* 88–115.

Lindsay, R. B. 1981. "Strutt, John William, Third Baron Rayleigh." In Gillispie, *Dictionary of Scientific Biography,* 13:100–107.

Litten, F. 2000. *Mechanik und Antisemitismus. Wilhelm Müller (1880–1968).* Munich: Institut für Geschichte der Naturwissenschaften.

Locke, R. R. 1984. *The End of the Practical Man: Entrepreneurship and Higher Education in Germany, France, and Great Britain, 1880–1940.* London: JAI Press.

Lockwood Marsh, W., ed. *International Air Congress, London, 1923: Report.* London: Royal Aeronautical Society.

Lorenz, H. 1903. "Der Unterricht in angewandter Mathematik und Physik an den deutschen Universitäten." *Jahresbericht der Deutschen Mathematiker-Vereinigung* 12:565–72.

Loukianoff, G. S. 1912. "Tragflächen des aerodynamischen Laboratoriums der Technischen Hochschule Moskau." *Zeitschrift für Flugtechnik und Motorluftschiffahrt* 3:153–59.

Love, A. E. H. 1891. "On the Theory of Discontinuous Fluid Motions in Two Dimensions." *Proceedings of the Cambridge Philosophical Society* 7:75–201.

———. 1906. *Treatise on the Mathematical Theory of Elasticity.* 2nd ed. Cambridge: Cambridge University Press.

Low, A. R. 1914. "The Rational Design of Aeroplanes." *Aeronautical Journal* 18:135–49.

———. 1922. "The Theory of the Airscrew." *Engineering* 114:593, 739.

———. 1923a. "The Circulation Theory of Lift, with an Example Worked out for an Albatross Wing-Profile." In Lockwood Marsh, *International Air Congress, London, 1923,* 255–79.

———. 1923b. "Review of Airscrew Theories." *Journal of the Royal Aeronautical Society* 27:38–59; subsequent discussion, 60–72.

———. 1923c. "A Treatise on Modern Aerodynamics." Review of *Aerodynamik,* by R. Fuchs and L. Hopf. *Aeroplane* 17:50–52.

———. 1927. "Aerofoil and Airscrew Theory." *Journal of the Royal Aeronautical Society* 31:167–68.

Lyall, G., ed. 1994. *The War in the Air, 1939–1945.* London: Pimlico.

Maccoll, J. W. 1930. "Modern Aerodynamical Research in Germany." *Journal of the Royal Aeronautical Society* 35:649–89; subsequent discussion, 697.

MacKenzie, D. 1981. *Statistics in Britain, 1865–1930: The Social Construction of Scientific Knowledge*. Edinburgh: Edinburgh University Press.

———. 1990. *Inventing Accuracy: A Historical Sociology of Nuclear Missile Guidance*. Cambridge: MIT Press.

———. 1999. "The Science Wars and the Past's Quiet Voice." *Social Studies of Science* 29:199–213; subsequent discussion, 215–34.

———. 2001. *Mechanizing Proof: Computing, Risk and Trust*. Cambridge: MIT Press.

MacLeod, R. M., and E. K. Andrews. 1970. "The Origins of the D.S.I.R.: Reflections on Ideas and Men, 1915–1916." *Public Administration* 48:23–48.

Mallock, A. 1907. "On the Resistance of Air." *Proceedings of the Royal Society of London* A 79:262–73.

———. 1909a. "Memorandum on General Questions to be Studied." Reports and Memoranda No. 2.

———. 1909b. "Note on Experiments made by Mr. R. E. Froude FRS: On the Forces Operating on Plane and Curved Surfaces when Travelling at Various Speeds in Water." Reports and Memoranda No. 16.

———. 1911. "Influence of Viscosity on the Stability of the Flow of Fluids." *Proceedings of the Royal Society of London* A 84:482–91.

———. 1912. "Aerial Flight: The James Forrest Lecture." *Minutes of the Proceedings of the Institution of Civil Engineers* 190:200–232.

Manegold, K.-H. 1968. "Felix Klein als Wissenschaftsorganisator. Ein Beitrag zum Verhältnis von Naturwissenschaft und Technik im 19. Jahrhundert." *Technikgeschichte* 35:177–204.

———. 1970. *Universität, Technische Hochschule und Industrie. Ein Beitrag zur Emanzipation der Technik im 19. Jahrhundert unter besonderer Berücksichtigung der Bestrebungen Felix Kleins*. Berlin: Duncker und Humblot.

Martin, H. M. 1924. "The Elements of the Lanchester Prandtl Theory of Aeroplane Lift and Drag." *Engineering* 117:1–3, 35–37, 100–102, 169–71, 258–60.

Maskell, E. C., and J. Weber. 1959. "On the Aerodynamic Design of Slender Wings." *Journal of the Royal Aeronautical Society* 63:709–21.

Maxwell, J. C. 1869–71a. "Remarks on the Mathematical Classification of Physical Quantities." *Proceedings of the London Mathematical Society* 3:224–32.

———. 1869–71b. "Remarks on the Preceding Paper." *Proceedings of the London Mathematical Society* 3:6–8.

———. 1873. Review of *Essay on the Mathematical Principles of Physics*, by J. Challis. *Nature* 8:279–80.

Mayr, O. 1976. "The Science–Technology Relationship as a Historiographic Problem." *Technology and Culture* 17:663–72.

Mazzotti, M., ed. 2008. *Knowledge as Social Order: Rethinking the Sociology of Barry Barnes*. Aldershot: Ashgate.

McKinnon Wood, R. 1922. "The Co-Relation of Model and Full Scale Work." *Journal of the Royal Aeronautical Society* 26:480–97; subsequent discussion, 497–501.

———. 1923. "Aerodynamic Research, Full Scale." In Glazebrook, *Dictionary of Applied Physics*, 5:1–19.

———. 1960. "Memories of the Chudleigh Mess." *New Scientist* 7 (June 16): 1532–34.

———. 1966. "Recollections 1914–1934." *Journal of the Royal Aeronautical Society* 70:89–90.

McMullin, E. 1987. "Scientific Controversy and Its Termination." In Engelhardt and Caplan, *Scientific Controversies*, 49–91.

Mehrtens, H. 1990. *Moderne-Sprache-Mathematik*. Frankfurt am Main: Suhrkamp.

Mehrtens, H., H. Bos, and T. Schneider, eds. 1981. *Social History of Nineteenth Century Mathematics*. Basel: Birkhäuser.

Mehrtens, H., and S. Richter, eds. 1980. *Naturwissenschaft Technik und N-S Ideologie, Beiträge zur Wissenschaftsgeschichte des Dritten Reichs*. Frankfurt am Main: Suhrkamp.

Meier, G. E. A., ed. 2000. *Ludwig Prandtl, ein Führer in Strömungslehre. Biographische Artikel zum Werk Ludwig Prandtls*. Braunschweig: Vieweg.

Merton, R. K. 1973. *The Sociology of Science: Theoretical and Empirical Investigations*. Chicago: University of Chicago Press.

Michell, J. H. 1890. "On the Theory of Free Stream Lines." *Philosophical Transactions of the Royal Society* A 18:389–431.

Miller, A. I. 1998. *Albert Einstein's Special Theory of Relativity: Emergence (1905) and Early Interpretation (1905–1911)*. New York: Springer-Verlag.

Milne, E. A. 1939–41. "Augustus Edward Hough Love, 1863–1940." *Obituary Notices of Fellows of the Royal Society of London* 3:466–82.

Milne-Thomson, L. M. 1958. *Theoretical Aerodynamics*. London: Macmillan.

———. 1960. *Theoretical Hydrodynamics*. London: Macmillan.

Mises, R. von. 1909. "Über die Probleme der technischen Hydrodynamik." *Zeitschrift für das gesamte Turbinenwesen* 6:165–68.

———. 1914. *Elemente der technischen Hydromechanik*. Leipzig: Teubner.

———. 1917. "Zur Theorie des Tragflächenauftriebes I." *Zeitschrift für Flugtechnik und Motorluftschiffahrt* 8:157–63.

———. 1918. *Fluglehre. Vorträge über Theorie und Berechnung der Flugzeuge in Elementarer Darstellung*. Berlin: Springer.

———. 1920. "Zur Theorie des Tragflächenauftriebes II." *Zeitschrift für Flugtechnik und Motorluftschiffahrt* 11:68–73, 87–89.

———. 1921. "Zur Einführung: Über die Aufgaben und Ziele der angewandten Mathematik." *Zeitschrift für angewandte Mathematik und Mechanik* 1:1–15.

———. 1922. "Über die gegenwärtige Krise der Mechanik." *Die Naturwissenschaften* 10:25–29.

———. 1924. "Felix Klein. Zu seinem 75. Geburtstag am 25. April 1924." *Zeitschrift für angewandte Mathematik und Mechanik* 4:86–92.

———. 1930a. "Über das naturwissenschaftliche Weltbild der Gegenwart." *Die Naturwissenschaften* 18:885–93.

———. 1930b. "Über kausale und statistische Gesetzmäßigkeit in der Physik." *Die Naturwissenschaften* 18:145–53.

———. 1939a. *Kleines Lehrbuch des Positivismus. Einführung in die empiristische Wissenschaftsauffassung*. The Hague: Van Stockum & Zoon.

———. 1939b. *Probability, Statistics and Truth*. Trans. J. Neyman, D. Scholl, and E. Rabinowitsch. London: William Hodge.

———. 1945. *Theory of Flight*. New York: McGraw-Hill.

———. 1956. *Positivism: A Study in Human Understanding*. New York: Braziller.

Mises, R. von, and Kurt Friedrichs. 1941. "Fluid Dynamics." Brown University Lecture Notes.

Mises, R. von, and L. Prandtl. 1927–28. "Bemerkungen zur Hydrodynamik." *Zeitschrift für angewandte Mathematik und Mechanik* 7:425–31; 8:249–51.

Molthan, W. 1918. "Messungen an einem Modell des D-Flugzeugs T. 29 der Deutschen Flugzeug-Werke." *Technische Berichte* 3:253–60.

Mombauer, A. 2002. *The Origins of the First World War: Controversies and Consensus.* London: Longman.

Morgan, M. S., and M. Morrison, eds. *Models as Mediators: Perspectives on Natural and Social Science.* Cambridge: Cambridge University Press.

Morris, R. 1937. "The Two-Dimensional Hydrodynamic Theory of Moving Aerofoils I." *Proceedings of the Royal Society* A 161:406–19.

Morrison, M. 1999. "Models as Autonomous Agents." In Morgan and Morrison, *Models as Mediators,* 38–65.

Mosely, R. 1978. "The Origins and Early Years of the National Physical Laboratory: A Chapter in the Pre-history of British Science Policy." *Minerva* 16:222–50.

Moulton, H. F. 1922. *The Life of Lord Moulton.* London: Nisbet.

Müler, U., J. K. Roesner, and B. Schmidt, eds. 1979. *Recent Developments in Theoretical and Experimental Fluid Mechanics.* Berlin: Springer.

Müller, W. 1923. "Über ebene Profilströmung und Zirkulation." *Zeitschrift für angewandte Mathematik und Mechanik* 3:117–28.

———. 1924a. "Zur Konstruktion von Tragflächenprofilen." *Zeitschrift für angewandte Mathematik und Mechanik* 4:213–31.

———. 1924b. "Zur Theorie der Misesschen Profilachsen." *Zeitschrift für angewandte Mathematik und Mechanik* 4:186–87.

Munk, M. 1915. "Die Modellversuchsanstalt des Massachusetts Institute of Technology." *Zeitschrift für Flugtechnik und Motorluftschiffahrt* 6:103–5.

———. 1917a. "Beitrag zur Aerodynamik der Flugzeugtragorgane." *Technische Berichte* 2:187–273.

———. 1917b. "Bericht über Luftwiderstandsmessungen von Streben." *Technische Berichte* 1:85–97.

———. 1917c. "Die Messungen an Flügelmodellen in der Göttinger Anstalt." *Technische Berichte* 1:135–47.

———. 1917d. "Messungen an Rumpfmodellen." *Technische Berichte* 2:23–24.

———. 1917e. "Modellmessungen an drei Tragflächen von verschiedener Spannweite." *Technische Berichte* 1:203.

———. 1917f. "Rumpf und Schraube." *Technische Berichte* 2:25–29.

———. 1917g. "Spannweite und Luftwiderstand." *Technische Berichte* 1:199–202.

———. 1917h. "Stirnkühler und Tragflächernkühler." *Technische Berichte* 2:19–21.

———. 1917i. "Systematische Versuche an Leitwerkmodellen." *Technische Berichte* 1:168–89.

———. 1917j. "Untersuchung eines Leitwerkes mit verschobener Ruderachse." *Technische Berichte* 1:223–30.

———. 1917k. "Weitere Widerstandsmessungen an Streben." *Technische Berichte* 2:15–17.

———. 1918. "Anblasversuche mit Leitwerken." *Technische Berichte* 2:401–505.

———. 1919. *Isoperimetrische Aufgaben aus der Theorie des Flüge.* Göttingen: Universitäts Buchdruckerei.

———. 1934. "Fluid Mechanics." In Durand, *Aerodynamic Theory,* 1:224–304.

———. 1981. "My Early Aerodynamic Research—Thoughts and Memories." *Annual Review of Fluid Mechanics* 13:1–7.

Munk, M., and G. Cario. 1917. "Flügel mit Spalt in Fahrtrichtung." *Technische Berichte* 1:219–22.

———. 1918. "Luftstromneigung hinter Flügeln." *Technische Berichte* 3:10–15.

Munk, M., and E. Hückel. 1917a. "Systematische Messungen an Flügelprofilen." *Technische Be-richte* 1:148–63.

———. 1917b. "Weiterer Untersuchungen von Flügelprofilen." *Technische Berichte* 1:204–18.

———. 1918. "Der Profilwiderstand von Tragflügeln. Eine Zusammenfassung der bisherigen Göttinger Flügelmessungen." *Technische Berichte* 3:451–461.

Munk, M., and W. Molthan. 1918. "Messungen an einem Flugzeugmodell Aeg D 1 der Allge-meinen Elektricitäts-Gesellschaft, A.-G., Abteilung Flugzeugbau." *Technische Berichte* 3:30–38.

Munk, M., and C. Pohlhausen. 1917. "Messungen an einfachen Flügelprofilen." *Technische Be-richte* 1:164–67.

Musker, D. Forthcoming. "Pride and Priority: Knowledge, Power and Control in the Early Aerospace Industry." Typescript.

Navarro, J. 2010. "Electron Diffraction chez Thomson: Early Responses to Quantum Physics in Britain." *British Journal for the History of Science* 43:245–75.

Nayler, J. L. 1966a. "Aeronautical Research at the NPL." *Journal of the Royal Aeronautical Society* 70:82–84.

———. 1966b. "The Aeronautical Research Council." *Journal of the Royal Aeronautical Society* 70:79–82.

———. 1968. "Early Days of British Aeronautical Research." *Journal of the Royal Aeronautical Society* 72:1045–54.

Nayler, J. L., et al. 1914. "Experiments on Models of Aeroplane Wings." Reports and Memoranda No. 152.

Nemeth, E. 2003. "Philosophy of Science and Democracy: Some Reflections on Philipp Frank's *Relativity a Richer Truth.*" In Heidelberger and Stadler, *Philosophy of Science and Politics*, 119–38.

New Illustrated Universal Reference Book. N.d. London: Odhams Press.

Newton, I. 1966. *Mathematical Principles of Natural Philosophy and His System of the World.* Trans. Andrew Motte (1729). Rev. Florian Cajori. Berkeley: University of California Press.

Newton-Smith, W. H. 1981. *The Rationality of Science.* London: Routledge and Kegan Paul.

Nickel, K. 1973. "Prandtl's Boundary-Layer Theory from the Viewpoint of a Mathematician." *Annual Review of Fluid Mechanics* 5:405–28.

Nickles, T. ed. 2003. *Thomas Kuhn.* Cambridge: Cambridge University Press.

Nola, R., and H. Sankey, eds. 2000. *After Popper, Kuhn and Feyerabend: Recent Issues in Theories of Scientific Method.* Dordrecht: Kluwer.

Norris, C. 1997. *Against Relativism: Philosophy of Science, Deconstruction and Critical Theory.* Oxford: Blackwell.

North, J. D. 1922. "Stability Calculations in the Process of Design." *Aeronautical Journal* 26:408–12.

———. 1966. "Fifty-Five Years in Aviation." *Journal of the Royal Aeronautical Society* 70:146–47.

Norton, J. D. 2000. "How Do We Know about Electrons?" In Nola and Sankey, *After Popper, Kuhn and Feyerabend*, 67–97.

O'Hear, A. 1989. *Introduction to the Philosophy of Science.* Oxford: Clarendon Press.

Okasha, S. 2000. "The Underdetermination of Theory by Data and the 'Strong Programme' in the Sociology of Knowledge." *International Studies in the Philosophy of Science* 14:283–97.

Ore, O. 1981. "Dirichlet, Gustav Peter Lejeune, 1805–1859." In Gillispie, *Dictionary of Scientific Biography*, 4:123–27.

Owen, P. R., and E. C. Maskell. 1980. "Dietrich Küchemann, 1911–1976." *Biographical Memoirs of Fellows of the Royal Society* 26:305–26.

Pannell, J. R. 1917. "The Aerodynamics of Formation Flying." *Flying* 2:131–32.

Pannell, J. R., and N. R. Campbell. 1916a. "The Balancing of Wing Flaps." Reports and Memoranda No. 200.

———. 1916b. "The Flow of Air Round a Wing Tip." Reports and Memoranda No. 197.

Parseval, A. von. 1910. "Der Akademisch Gebildete Ingenieure und die Aviatik." *Zeitschrift des Verbandes deutscher Diplom-Ingenieure* 1:108–9.

———. 1920. "Über Wirbelbildung an Tragflachen." *Berichte und Abhandlungen der Wissenschaftlichen Gesellschaft für Luftfahrt* 1:61–64.

Perrow, C. 1984. *Normal Accidents: Living with High-Risk Technologies.* New York: Basic Books.

Petavel, J. 1913. "Aeronautics: Three Lectures." *Journal of the Royal Society of Arts* 61:1065–75, 1082–90, 1100–1108.

Pfeiffer, F. 1950. "Wilhelm Kutta zum Gedenken." In *Technische Hochschule Stuttgart, Reden und Aufsätze* 16:46–57. Stuttgart: W. Kohlhammer.

Pfetsch, F. 1970. "Scientific Organisation and Science Policy in Imperial Germany, 1871–1914: The Foundation of the Imperial Institute of Physics and Technology." *Minerva* 8:557–80.

Phillips, H. 1910. *Aero* 2:380.

Phillips, W. F. 2004. "Lifting-Line Analysis for Twisted Wings and Washout-Optimized Wings." *Journal of Aircraft* 41:128–36.

———. 2005. "New Twist on an Old Wing Theory." *Aerospace America* (January): 27–30.

Phillips, W. F., and D. O. Snyder. 2000. "Modern Adaptation of Prandtl's Classic Lifting-Line Theory." *Journal of Aircraft* 37:662–70.

Piercy, N. A. V. 1923a. "Note on the Experimental Aspect of One of the Assumptions of Prandtl's Aerofoil Theory." *Journal of the Royal Aeronautical Society* 27:501–11.

———. 1923b. "On the Vortex Pair Quickly Formed by Some Aerofoils." *Journal of the Royal Aeronautical Society* 27:488–500.

———. 1944. *A Complete Course in Elementary Aerodynamics.* London: English Universities Press.

Piercy, N. A. V., J. H. Preston., and L. G. Whitehead. 1938. "The Approximate Prediction of Skin Friction and Lift." *Philosophical Magazine* 26:791–815.

Pinch, T. 1988. "Understanding Technology; Some Possible Implications of Work in the Sociology of Science." In Elliott, *Technology and Social Process,* 70–83.

Pohlhausen, K. 1917. "Widerstandmessungen an Seilen und Profildrähten." *Technische Berichte* 2:13.

Pollard, S. 1990. *Britain's Prime and Britain's Decline.* London: Edward Arnold.

———. 1994. "Entrepreneurship, 1870–1914." In Floud and McCloskey, *Economic History of Britain since 1700,* vol. 2: *1860–1939,* 62–89.

———. 2000. "Interview." In English and Kenny, *Rethinking British Decline,* 76–91.

Popper, K. R. 1962. *The Open Society and Its Enemies.* London: Routledge and Kegan Paul.

Prandtl, L. 1904. "Über Flüssigkeitsbewegung bei sehr kleiner Reibung." In Krazer, *Verhandlungen des dritten Internationalen Mathematiker-Kongresses,* 484–91. (Although the publication in which this article appeared bears the date 1905, it is conventional to cite the paper as 1904, i.e., the date of the congress at which it was presented.)

———. 1909. "Die Bedeutung von Modellversuchen für die Luftschiffahrt und Flugtechnik und die Einrichtungen für solche Versuche in Göttingen." *Zeitschrift des Vereins deutscher Ingenieure* 53:1711–19.

————. 1910a. "Betrachtungen über das Flugproblem." *Zeitschrift des Vereins deutscher Ingenieure* 54:698–702.

————. 1910b. "Einige für die Flugtechnik wichtige Beziehungen aus der Mechanik." *Zeitschrift für Flugtechnik und Motorluftschiffahrt* 1:3–6, 25–30, 61–64, 73–76.

————. 1912. "Ergebnisse und Ziele der Göttinger Modellversuchsanstalt." *Zeitschrift für Flugtechnik und Motorluftschiffahrt* 3:33–36.

————. 1913. "Flüssigkeitsbewegung." In Korschelt et al., *Handwörterbuch der Naturwissenschaften*, 101–40. Reprinted in 1913 as *Abriß der Lehre von der Flüssigkeits- und Gasbewegung* (Jena: Fischer).

————. 1914a. "Der Luftwiderstand von Kugeln." *Nachrichten der Gesellschaft der Wissenschaften zu Göttingen. Math.-phys. Klasse,* 177–90.

————. 1914b. "Georg Fuhrmann." *Zeitschrift für Flugtechnik und Motorluftschiffahrt* 5:267.

————. 1917. "Näherungsformel für den Widerstand von Tragwerken." *Technische Berichte* 2:275–78.

————. 1918a. "Der induzierte Widerstand von Mehrdeckern." *Technische Berichte* 3:309–15.

————. 1918b. "Tragflügeltheorie. I Mitteilung." *Nachrichten der Gesellschaft der Wissenschaften zu Göttingen. Math.-phys. Klasse,* 451–77.

————. 1919. "Tragflügeltheorie. II Mitteilung." *Nachrichten der Gesellschaft der Wissenschaften zu Göttingen. Math.-phys. Klasse,* 107–87.

————. 1920a. "H. Th. v. Böttinger." *Zeitschrift für Flugtechnik u. Motorluftschiffahrt* 11:169.

————. 1920b. "Tragflächen-Auftrieb und -Widerstand in der Theorie." *Jahrbuch der Wissenschaften Gesellschaft für Luftfahrt* 5:37–65.

————. 1923. "Applications of Modern Hydrodynamics to Aeronautics." NACA Report No. 116. Washington, DC: National Advisory Committee for Aeronautics.

————. 1924. "Über die Entstehung von Wirbeln in der idealen Flüssigkeit, mit Anwendung auf die Traflügeltheorie und anderen Aufgaben." In von Kármán and Levi-Civita, *Vorträge aus dem Gebeite der Hydro-und Aerodynamik,* 18–33.

————. 1925a. *Ergebnisse der Aerodynamischen Versuchsanstalt zu Göttingen.* Vol. 1, *Lieferung.* Munich: Oldenbourg.

————. 1925b. "Experimentelle Prüfung der Umrechnungsformeln." In Prandtl, *Ergebnisse der Aerodynamischen Versuchsanstalt zu Göttingen,* 1:50–53.

————. 1925c. "Felix Klein zum Gedächnis." *Zeitschrift für Flugtechnik und Motorluftschiffahrt* 16:1.

————. 1927a. "Carl Runge." *Die Naturwissenschaften* 15:227–29.

————. 1927b. "Die Entstehung von Wirbeln in einer Flüssigkeit mit kleiner Reibung." *Zeitschrift für Flugtechnik und Motorluftschiffahrt* 18:389–496.

————. 1927c. "The Generation of Vortices in Fluids of Small Viscosity." *Journal of the Royal Aeronautical Society* 31:720–41.

————. 1948. "Mein Weg zu hydrodynamischen Theorien." *Physikalische Blätter* 4:89–92.

————. 1949. *Führer durch die Strömunglehre.* Braunschweig: Vieweg und Sohn.

Preston, J. H. 1943. "The Approximate Calculation of the Lift of Symmetrical Aerofoils taking Account of the Boundary Layer, with Application to Control Problems." Reports and Memoranda No. 1996.

————. 1949. "The Calculation of Lift taking Account of the Boundary Layer." Reports and Memoranda No. 2725.

————. 1954. "Note on the Circulation in Circuits Which Cut the Streamlines in the Wake of an Aerofoil at Right Angles." Reports and Memoranda No. 2957.

Price, D. J. de S. 1969. "The Structures of Publication in Science and Technology." In Gruber and Marquis, *Factors in the Transfer of Technology*, 91–104.

Pritchard, J. L. 1958. "Mervyn O'Gorman, 1871–1958." *Journal of the Royal Aeronautical Society* 62:469–70.

Pröll, A. 1913. "Luftfahrt und Mechanik." *Jahrbuch der Wissenschaftlichen Gesellschaft für Flugtechnik* 2:94–117.

———. 1919. *Flugtechnik: Grundlagen des Kunstfluges*. Munich: Oldenbourg.

Pyenson, L. 1979a. "Mathematics, Education, and the Göttingen Approach to Physical Reality, 1890–1914." *Europa: A Journal of Interdisciplinary Studies* 2:91–126.

———. 1979b. "Physics in the Shadow of Mathematics: The Göttingen Electron-Theory Seminar of 1905." *Archive for History of Exact Sciences* 21:55–89.

———. 1982. "Relativity in Late Wilhelmian Germany: The Appeal to a Preestablished Harmony between Mathematics and Physics." *Archive for History of Exact Sciences* 27:137–55.

———. 1983. *Neohumanism and the Persistence of Pure Mathematics in Wilhelmian German*. Philadelphia: American Philosophical Society.

RAE News. 1968. "Farnborough Pioneer Dies." Pp. 22, 28.

Ramsey, A. S. 1913. *A Treatise on Hydromechanics: Part II, Hydrodynamics*. London: G. Bell.

———. 1935. *A Treatise on Hydromechanics: Part II, Hydrodynamics*. 4th ed. London: G. Bell.

Rankine, W. J. Macquorn. 1881. "On Plane Water-Lines in Two Dimensions." In *Miscellaneous Scientific Papers*, 495–521. London: Charles Griffin.

Rayleigh, Lord. 1876. "On the Resistance of Fluids." *Philosophical Magazine* 2:430–41.

———. 1877. "On the Irregular Flight of a Tennis-Ball." *Messenger of Mathematics* 7:14–16. Reprinted in 1899, in *Scientific Papers*, 1:344–46 (Cambridge: Cambridge University Press).

———. 1883. "The Soaring of Birds." *Nature* 27:534–35.

———. 1891. "Experiments in Aerodynamics." *Nature* 45:108–9. Reprinted in 1902, in *Scientific Papers*, 3:491–95 (Cambridge: Cambridge University Press).

———. 1900. "The Mechanical Principles of Flight." *Manchester Memoirs* 44:1–26.

———. 1902. "Flight." *Proceedings of the Royal Institution* 16:233–34.

———. 1909. "Note as to the Application of the Principle of Dynamic Similarity." Reports and Memoranda No. 15.

———. 1916. "Lamb's *Hydrodynamics*." *Nature* 97:318.

Rayleigh, Lord (Fourth Baron Rayleigh), and F. J. Selby. 1936–38. "Richard Tetley Glazebrook, 1854–1935." *Obituary Notices of Fellows of the Royal Society* 2:29–56.

Reid, C. 1996. *Hilbert*. New York: Springer.

Reid, E. G. 1932. *Applied Wing Theory*. New York: McGraw-Hill.

Reisch, G. A. 2005. *How the Cold War Transformed Philosophy of Science: To the Icy Slopes of Logic*. Cambridge: Cambridge University Press.

Reißner, H. 1910. "Die Seitensteuerung der Flugmaschinen." *Zeitschrift für Flugtechnik und Motorluftschiffahrt* 1:101–6, 117–23.

———. 1912a. "Einige Bemerkungen zur Seitenstabilität der Drachenflieger." *Zeitschrift für Flugtechnik und Motorluftschiffahrt* 3:39–43.

———. 1912b. "Wissenschaftliche Fortschritte der Flugtechnik." *Jahrbuch der Luftfahr* 2:343–57.

Relf, E. F. 1913. "Photographic Investigation of the Flow Round a Model Aerofoil." Reports and Memoranda No. 76.

———. 1924a. "An Electrical Method of Tracing Streamlines for the Two-Dimensional Motion of a Perfect Fluid." Reports and Memoranda No. 905.

———. 1924b. "An Electrical Method of Tracing Streamlines in the Two-Dimensional Motion of a Perfect Fluid." *Philosophical Magazine* 48:535–39.

Relf, E. F., F. Bramwell, and A. Fage. 1912. "On the Determination on the Whirling Arm of the Pressure-Velocity Constant for a Pitot Tube, and on the Absolute Measurement of Velocity in Aeronautical Work." Reports and Memoranda No. 71.

Ricardo, H. R. 1948. "Frederick William Lanchester, 1868–1946." *Obituary Notices of Fellows of the Royal Society* 5:757–66.

Riedler, A. 1908. "Über die Entwicklung des maschinentechnischen Studiums." *Zeitschrift des Vereins deutscher Ingenieure* 52:702–7.

Robertson, R. 1936. "Joseph Ernest Petavel, 1873–1936." *Obituary Notices of Fellows of the Royal Society* 2:183–203.

Roland, A. 1985. *Model Research: The National Advisory Committee for Aeronautics, 1915–1958.* 2 vols. Washington, DC: Scientific and Technical Information Branch, National Aeronautics and Space Administration.

Rotta, J. C. 1990. *Die Aerodynamische Versuchsanstalt in Göttingen, ein Werk Ludwig Prandtls.* Göttingen: Vandenhoeck und Ruprecht.

Rouse, H., and S. Ince. 1957. *History of Hydraulics.* New York: Dover.

Routh, E. J. 1877. *A Treatise on the Stability of a Given State of Motion, Particularly Steady Motion.* London: Macmillan.

———. 1884. *The Advanced Part of a Treatise in the Dynamics of a System of Rigid Bodies.* 4th ed. London: Macmillan.

Rowe, D. E. 1986. "'Jewish Mathematics' at Göttingen in the Era of Hilbert and Klein." *Isis* 77:422–49.

———. 1989. "Klein, Hilbert, and the Göttingen Mathematical Tradition." *Osiris* 5:86–213.

———. 1994. "The Philosophical Views of Klein and Hilbert." In Chikara, Mitsuo and Dauben, *Intersection of History and Mathematics,* 187–202.

———. 2001. "Felix Klein as Wissenschaftspolitiker." In Bottazzini and Dalmedico, *Changing Images in Mathematics,* 69–91.

———. 2004. "Making Mathematics in an Oral Culture: Göttingen in the Era of Klein and Hilbert." *Science in Context* 17:85–129.

Rowe, D. E., and J. McCleary, eds. 1989. *The History of Modern Mathematics.* Vol. 2, *Institutions and Applications.* Boston: Academic Press.

Rudwick, M. J. S. 1985. *The Great Devonian Controversy: The Shaping of Scientific Knowledge among Gentlemanly Specialists.* Chicago: University of Chicago Press.

Ruerup, R., ed. 1979. *Wissenschaft und Gesellschaft: Beiträge zur Geschichte der Technischen Universität Berlin, 1879 1979.* Berlin: Springer-Verlag.

Runge, C. 1911. "Über die Längsschwingungen der Flugmaschinen." *Zeitschrift für Flugtechnik und Motorluftschiffahrt* 2:193–96, 201–4.

———. 1912. "Über die Längsschwingungen von Fleugzeugen." *Zeitschrift für Flugtechnik und Motorluftschiffahrt* 3:38–39.

———. 1913. "The Mathematical Training of the Physicist at the University." In Hobson and Love, *Proceedings of the Fifth International Congress of Mathematicians,* 2:598–607.

Runge, C., and L. Prandtl. 1907. "Das Institut für angewandte Mathematik und Mechanik." *Zeitschrift für Mathematik und Physik* 54:263–80.

Runge, I. 1949. *Carl Runge und sein wissenschaftliches Werk.* Göttingen: Vandenhoeck und Ruprecht.

Rupke, N., ed. 2002. *Göttingen and the Development of the Natural Sciences*. Göttingen: Wallstein Verlag.

Russell, B. 1903. *The Principles of Mathematics*. Cambridge: Cambridge University Press.

Rutherford, D. E. 1959. *Fluid Dynamics*. Edinburgh: Oliver and Boyd.

Ryle, G. 1949. *The Concept of Mind*. London: Hutchinson.

Saffman, P. G. 1992. *Vortex Dynamics*. Cambridge: Cambridge University Press.

Sanderson, M. 1972. *The Universities and British Industry, 1850–1970*. London: Routledge and Kegan Paul.

———. 1988. "The English Civic Universities and the 'Industrial Spirit.'" *Historical Research* 61:90–104;

Sayers, W. H. 1919. "The Scientific Progress of Aviation during the War." *Aeroplane* 16: 1309–10.

———. 1922. "The Arrest of Aerodynamic Development." *Aeroplane* 22:138.

Schaffer, S. 1992."Late Victorian Metrology and Its Instrumentation: A Manufactory of Ohms." In Bud and Cozzens, *Invisible Connections*, 457–78.

———. 1994. "Rayleigh and the Establishment of Electrical Standards." *European Journal of Physics* 15:277–85.

———. 1995. "Accurate Measurement is an English Science." In Wise, *Values of Precision*, 135–72.

———. 1998. "Physics Laboratories and the Victorian Country House." In Smith and Agar, *Making Space for Science*, 149–80.

Scharlau, W., ed. 1990. *Mathematische Institute in Deutschland, 1800–1945*. Braunschweig: Vieweg.

Schickore, J., and F. Steinle, eds. 2006. *Revisiting Discovery and Justification: Historical and Philosophical Perspectives on the Context Distinction*. Dordrecht: Springer.

Schlichting, H. 1951. *Grenzschicht-Theorie*. Karlsruhe: Verlag G. Braun.

———. 1975. "An Account of the Scientific Life of Ludwig Prandtl." *Zeitschrift für Flugwissenschaft* 23:287–316.

Schneider, I., H. Trischler, and U. Wengenroth, eds. 2000. *Oszillation. Naturwissenschaftler und Ingenieure zwischen Forschung und Markt*. Munich: Oldenbourg.

Schopenhauer, A. 1883–86. *The World as Will and Idea*. Trans. into 3 vols. by R. B. Haldane and J. Kemp. London: Trübner.

Schroeder-Gudehus, B. 1973. "Challenge to Transnational Loyalties: International Scientific Organizations after the First World War." *Science Studies* 3:93–118.

Schubring, G. 1981. "The Conception of Pure Mathematics as an Instrument in the Professionalisation of Mathematics." In Mehrtens, Bos, and Schneider, *Social History of Nineteenth Century Mathematics*, 111–34.

———. 1989. "Pure and Applied Mathematics in Divergent Institutional Settings in Germany: The Role and Impact of Felix Klein." In Rowe and McCleary, *History of Modern Mathematics*, 170–220.

———. 1990. "Zur strukturellen Entwicklung der Mathematik an den deutschen Hochschulen 1800–1945." In Scharlau, *Mathematische Institute in Deutschland*, 264–79.

———. 1994. "Die deutsche mathematische Gemeinde." In Fauval, Flood, and Wilson, *Möbius und sein Band*, 394–406.

Schuster, A. 1921. "John William Strutt, Baron Rayleigh." *Proceedings of the Royal Society* A 98:i–l.

Sears, W. R. 1956. "Some Recent Developments in Airfoil Theory." *Journal of the Aeronautical Sciences* 23:490–99.

Serby, J. E. 1971. "Wimperis, Harry Egerton, 1876–1960." In Williams and Palmer, *Dictionary of National Biography*, 1063–64.

Seward, A. C., ed. 1917. *Science and the Nation: Essays by Cambridge Graduates with an Introduction by the right Hon. Lord Moulton, K.C.B., F.R.S.* Cambridge: Cambridge University Press.

Shapere, D. 1986. "External and Internal Factors in the Development of Science." *Science and Technology Studies* 4:1–9.

Shapin, S. 1996. *The Scientific Revolution*. Chicago: University of Chicago Press.

Shapin, S., and S. Schaffer. 1985. *Leviathan and the Air-Pump: Hobbes, Boyle and the Experimental Life*. Princeton: Princeton University Press.

———. 1999. "Response to Pinnick." *Social Studies of Science* 29:249–53; subsequent discussion, 253–59.

Shaw, H. 1919. *A Text-Book of Aeronautics*. London: Griffin.

Shinbrot, M. 1973. *Lectures on Fluid Mechanics*. New York: Gordon and Breach.

Siegmund-Schultze, R. 1993. "Hilda Geiringer-von Mises, Charlier Series, Ideology, and the Human Side of the Emancipation of Applied Mathematics at the University of Berlin during the 1920s." *Historia Mathematica* 20:364–81.

———. 2001. *Rockefeller and the Internationalization of Mathematics between the Two World Wars: Documents and Studies for the Social History of Mathematics in the 20th Century*. Basel: Birkhäuser.

———. 2004. "A Non-Conformist Longing for Unity in the Fractures of Modernity: Towards a Scientific Biography of Richard von Mises (1883–1953)." *Science in Context* 17:333–70.

———. 2007. "Philipp Frank, Richard von Mises, and the Frank-Mises." *Physics in Perspective* 9:26–57.

Sigurdsson, S. 1994. "Unification, Geometry and Ambivalence: Hilbert, Weyl and the Göttingen Community." In Gavroglu, Christianidis and Nicolaidis, *Trends in the Historiography of Science*, 355–67.

———. 1996. "Physics, Life, and Contingency: Born, Schrödinger, and Weyl in Exile." In Ash and Söllner, *Forced Migration and Scientific Change*, 48–70.

Simmons, L. F. G., and E. Over. 1924. "Note on the Application of the Vortex Theory of Aerofoils to the Prediction of Downwash" *Report and Memoranda No. 914*.

Skempton, A. W. 1970. "Alfred John Sutton Pippard, 1891–1969." *Biographical Memoirs of Fellows of the Royal Society* 16:463–78.

Skidelsky, R. 1983. *John Maynard Keynes: Hopes Betrayed, 1883–1920*. London: Macmillan.

———. 1992. *John Maynard Keynes: The Economist as Saviour, 1920–1937*. London: Macmillan.

Smith, A. M. O. 1975. "High-Lift Aerodynamics." *Journal of Aircraft* 12:501–30.

Smith, C., and J. Agar, eds. 1998. *Making Space for Science: Territorial Themes in the Shaping of Knowledge*. Houndmills: Macmillan.

Smith, C., and N. Wise. 1989. *Energy and Empire: A Biographical Study of Lord Kelvin*. Cambridge: Cambridge University Press.

Sneddon, I. N. 1957. *Elements of Partial Differential Equations*. New York: McGraw-Hill.

Snow, C. P. 1967. Foreword to *A Mathematicians Apology*, by G. H. Hardy. Cambridge: Cambridge University Press.

Sokal, A., and J. Bricmont. 1998. *Fashionable Nonsense: Postmodern Intellectuals' Abuse of Science*. New York: Picador.

Sommer, D. 1960. *Haldane of Cloan: His Life and Times, 1856–1928.* London: George Allen and Unwin.

Sommerfeld, A. 1935. "Zu L. Prandtls 60. Geburtstag am 4. Februar 1935." *Zeitschrift für angewandte Mathematik und Mechanik* 15:1–2.

Southwell, R. V. 1924. "Notes on the Stability of Laminar Shearing Motion in a Viscous Incompressible Fluid" *Philosophical Magazine* 48:540–553.

———. 1925a. "Aeronautical Problems of the Past and of the Future." *British Association for the Advancement of Science: Report of 93rd Meeting, Southampton,* 395–417.

———. 1925b. "Some Recent Work of the Aerodynamic Department. National Physical Laboratory." *Journal of the Royal Aeronautical Society* 29:146–67.

———. 1927. Review of *The Elements of Aerofoil and Airscrew Theory,* by H. Glauert. *Mathematical Gazette* 13:394–95.

———. 1931. "Aeronautical Progress, 1914–1930. The James Forrest Lecture." *Minutes of the Proceedings of the Institution of Civil Engineers* 230:333–80.

———. 1938. "The Changing Outlook of Engineering Science." *British Association for the Advancement of Science: Report of the Annual Meeting, Cambridge,* 163–79.

Southwell, R. V., and Letitia Chitty. 1930. "On the Problem of Hydrodynamic Stability, 1: Uniform Shearing Motion in a Viscous Fluid." *Philosophical Transactions of the Royal Society of London* A 229:205–53.

Southwell, R. V., H. B. Squire. 1932. "A Modification of Oseen's Approximate Equation for the Motion in Two Dimensions of a Viscous Incompressible Fluid." *Philosophical Transactions of the Royal Society of London* A 232:27–64.

Spalart, P. R. 1998. "Airplane Trailing Vortices." *Annual Review of Fluid Dynamics* 30:107–38.

———. 2008. "Induced Drag and Wake Kinetic Energy." *Aeronautical Journal* 112:54.

Spiers, E. M. 1980. *Haldane: An Army Reformer.* Edinburgh: Edinburgh University Press.

Stadler, F. 2001. *The Vienna Circle: Studies in the Origins, Development, and Influence of Logical Empiricism.* Vienna: Springer.

Staley, R. 1992. "Max Born and the German Physics Community: The Education of a Physicist." Ph.D. dissertation, University of Cambridge.

Stanton, T. E. 1910. "Report on the Experimental Equipment of the Aeronautical Department of the National Physical Laboratory." Reports and Memoranda No. 25.

Stanton, T. E., and L. Bairstow. 1910. "Report from the National Physical Laboratory on Experiments Made to Determine the Relative Efficiencies of Certain Designs of Rudders and Lifting Planes for Dirigibles." Reports and Memoranda No. 24.

Stanton, T. E., and J. R. Pannell. 1914. "Similarity of Motion in Relation to the Surface Friction of Fluids." *Philosophical Transactions of the Royal Society of London* A 214:199–224.

Steiner, M. 1998. *The Applicability of Mathematics as a Philosophical Problem.* Cambridge: Harvard University Press.

Stokes, G. G. 1880. "On the Steady Motion of Incompressible Fluids." In *Mathematical and Physical Papers,* 1:1–16. Cambridge: Cambridge University Press.

———. 1899. "Mathematical Proof of the Identity of the Stream-Lines Obtained by Means of a Viscous Film with Those of a Perfect Fluid Moving in Two Dimensions." *Report of the British Association for the Advancement of Science* 68:143–44. Reprinted in 1905, in *Mathematical and Physical Papers,* 278–82 (Cambridge: Cambridge University Press).

Streeter, V. L. 1948. *Fluid Dynamics.* New York: McGraw-Hill.

Strutt, R. J. 1924. *John Willam Strutt: Third Baron Rayleigh.* London: Arnold.

Stuewer, R. H. 1970. "Non-Einsteinian Interpretations of the Photoelectric Effect." In *Historical* and *Philosophical Perspectives of Science*, ed. R. Stuewer, 246–63. Minnesota Studies in the Philosophy of Science, vol. 5. Minneapolis: University of Minnesota Press.

Sturm, T., and G. Gigerenzer. 2006. "How Can We Use the Distinction between Discovery and Justification? On the Weaknesses of the Strong Programme in the Sociology of Science." In Schickore and Steinle, *Revisiting Discovery and Justification*, 133–58.

Sutton, O. G. 1954. *Mathematics in Action*. London: Bell.

———. 1965. *Mastery of the Air: An Account of the Science of Flight*. London: Hodder and Stoughton.

Tani, I. 1977. "History of Boundary-Layer Theory." *Annual Review of Fluid Mechanics* 9:87–111.

———. 1979. "The Wing Section Theory of Kutta and Zhukovski." In Müler, Roesner, and Schmidt, *Recent Developments in Theoretical and Experimental Fluid Mechanics*, 511–16.

Taylor, G. I. 1916a. "Pressure Distribution over the Wings of an Aeroplane in Flight." Reports and Memoranda No 287. Reprinted in 1963, in *Scientific Papers*, vol. 3, chap. 4 (Cambridge, Cambridge University Press).

———. 1916b. "Pressure Distribution Round a Cylinder." Reports and Memoranda No. 191.

———. 1921. "Scientific Method in Aeronautics." *Aeronautical Journal* 25:474–91.

———. 1926. "Note on the Connection between the Lift in an Aerofoil in a Wind and the Circulation round It." *Philosophical Transactions of the Royal Society* A 225:238–45.

———. 1966. "When Aeronautical Science Was Young." *Journal of the Royal Aeronautical Society* 70:108–13.

———. 1971. "Aeronautics before 1919." *Nature* 233:527–29.

Taylor, G. I., and J. W. Maccoll. 1935. "The Mechanics of Compressible Fluids." In Durand, *Aerodynamic Theory*, 3:209–50.

Temple, G., J. L. Nayler, and E. F. Relf. 1965. "Leonard Bairstow, 1880–1963." *Biographical Memoirs of Fellows of the Royal Society* 11:23–40.

Thomson, G. P. 1919. *Applied Aerodynamics*. London: Hodder and Stoughton.

Thomson, W. (Lord Kelvin). 1869. "On Vortex Motion." *Transactions of the Royal Society of Edinburgh* 25:217–60.

Thurston, Albert P. 1911. *Elementary Aeronautics or the Science and Practice of Aerial Machines*. London: Whittaker.

Thwaites, B. 1960. *Incompressible Aerodynamics*. Oxford: Clarendon Press.

Tietjens, O. 1929. *Hydro- und Aeromechanik nach Vorlesungen von L. Prandtl*. Vol. 1, *Gleichgewicht und reibungslose Bewegung*. Berlin: Springer.

———. 1931. *Hydro- und Aeromechanik nach Vorlesungen von L. Prandt.Vol. 2. Bewegung reibender Flüssigkeiten und technische Anwendungen*. Berlin: Springer.

———. 1957a. *Applied Hydro- and Aerodynamics: Based on the Lectures of L. Prandtl*. New York: Dover. (First American ed., 1934.)

———. 1957b. *Fundamentals of Hydro- and Aeromechanics: Based on the Lectures of L. Prandtl*. New York: Dover. (First American ed., 1934.)

Times. 1908. "Reviews of Technical Books." Engineering supplement, no. 193, November 4, 6.

Tobies, R. 1989. "On the Contribution of Mathematical Societies to Promoting Applications of Mathematics in Germany." In Rowe and McCleary, *History of Modern Mathematics*, 223–47.

———. 2002. "The Development of Göttingen into the Prussian Centre of Mathematics and the Exact Sciences." In Rupke, *Göttingen and the Development of the Natural Sciences*, 116–42.

Tokaty, G. A. 1971. *A History and Philosophy of Fluid Mechanics*. Henley-on-Thames: Foulis. Reprinted in 1994 (New York: Dover).

Töpfer, C. 1912. "Bemerkungen zu dem Aufsatz von H. Blasius 'Grenzschichten in Flüssigkeiten mit kleiner Reibung.'" *Zeitschrift für Mathematik und Physik* 60:397.

Toulmin, S. 1953. *The Philosophy of Science: An Introduction.* London: Hutchinson.

Trefftz, E. 1913. "Graphische Konstruktion Joukowskischer Tragflächen." *Zeitschrift für Flugtechnik und Motorluftschiffahrt* 4:130–31.

Trischler, H. 1992. *Luft- und Raumfahrtforschung in Deutschland, 1900–1970.* Frankfurt: Campus Verlag.

Truesdell, C. 1952. "A Program of Physical Research in Classical Mechanics." *Zeitschrift für angewandte Mathematik und Physik* 3:79–95.

———. 1955. "Experience, Theory and Experiment." In *Proceedings of the 6th Hydraulic Conference, 1955.* Bulletin 36, State University of Iowa Studies in Engineering, 3–18.

———. 1968. *Essays in the History of Mechanics.* Berlin: Springer.

Turner, F. M. 1980. "Public Science in Britain, 1880–1919." *Isis* 71:589–608.

Turner, G. L'E., ed. 1976. *Patronage of Science in the Nineteenth Century.* Leyden: Noordhoff.

Turner, P. K. 1911. "Aeronautical Research." *Aeroplane* 1:298.

Uebel, T. E. 1996. "Epistemology and Social History of Knowledge: The Prospect of Constructivist Naturalism." In Johannessen and Nordenstam, *Wittgenstein and the Philosophy of Culture,* 338–59.

———. 2000. "Logical Empiricism and the Sociology of Knowledge: The Case of Neurath and Frank." *Philosophy of Science* (Proceedings) 67:S138–50.

Unwin, William C. 1911. "Hydraulics." In *Encyclopaedia Britannica,* 11th ed., 14:35–110. New York: Encyclopaedia Britannica.

Van Dyke, M. 1964. "Lifting-Line Theory as a Singular Perturbation Problem." *Archiwum Mechaniki Stosowanej* 16:601–14.

———. 1994. "Nineteenth-Century Roots of the Boundary-Layer Idea." *Society for Industrial and Applied Mathematics (SIAM) Review* 36:415–24.

Varcoe, I. 1970. "Scientists, Government and Organised Research in Great Britain 1914–16: The Early History of the DSIR." *Minerva* 8:192–216.

Vaughan, D. 1996. *The Challenger Launch Decision: Risky Technology, Culture, and Deviance at NASA.* Chicago: University of Chicago Press.

Villamil, R. de. 1913. "The 'Sin² Law.'" *Aeronautics* 6:55–56.

Vincenti, W. 1990. *What Engineers Know and How They Know It.* Baltimore: Johns Hopkins University Press.

Vogel-Prandtl, J. 1993. *Ludwig Prandtl. Ein Lebensbild: Erinnerungen, Dokumente.* Mitteilungen aus dem Max-Planck-Institut für Strömungsforschung, no. 107. Göttingen.

Vogt, A. B. 2007. "In Memoriam Richard von Mises." Institut für Mathematik, Humboldt-Universität zu Berlin. Preprint 14.

Vorreiter, A. 1909. "Kritik der Drachenflieger." *Zeitschrift des Vereins deutscher Ingenieure* 53:1093–1102, 1140–48, 1572–77, 1749–64.

Walker, P. B. 1974. *Early Aviation at Farnborough: The History of the Royal Aircraft Establishment.* Vol. 2. London: MacDonald.

Warner, E. P. 1936. *Airplane Design: Performance.* New York: McGraw-Hill.

Warwick, A. 1989. "International Relativity: The Establishment of a Theoretical Discipline." *Studies in the History and Philosophy of Science* 20:139–49.

———. 1992. "Cambridge Mathematics and Cavendish Physics: Cunningham, Campbell and Einstein's Relativity, 1905–1911. Part I. The Uses of Theory." *Studies in the History and Philosophy of Science* 23:625–56.

———. 1993. "Cambridge Mathematics and Cavendish Physics: Cunningham, Campbell and Einstein's Relativity, 1905–1911. Part II. Comparing Traditions in Cambridge Physics." *Studies in the History and Philosophy of Science* 24:1–25.

———. 2003. *Masters of Theory: Cambridge and the Rise of Mathematical Physics.* Chicago: Chicago University Press.

Watson, H. W., and E. J. Routh. 1860. *Cambridge Senate–House Problems and Riders for the Year 1860 (With Solutions).* Cambridge: Macmillan.

Weaver, J. R. H., ed. 1937. *Dictionary of National Biography, 1922–1930.* Oxford: Oxford University Press.

Wegener, P. 1997. *What Makes Airplanes Fly? History, Science, and Applications of Aerodynamics.* 2nd ed. New York: Springer.

Wengenroth, U., ed. 1993. *Die TU München. Annäherungen an ihre Geschichte.* Munich: Technische Universität München.

Werskey, G. 1988. *The Visible College: A Collective Biography of British Scientists and Socialists of the 1930s.* London: Free Association Books.

Weyl, A. R. 1965. *Fokker: The Creative Years.* London: Putnam.

Wheaton, B. R. 1983. *The Tiger and the Shark: Empirical Roots of Wave-Particle Dualism.* Cambridge: Cambridge University Press.

Whittacker, E. 1951. *A History of the Theories of Aether and Electricity.* Vol.1. London and Edinburgh: Thomas Nelson and Sons. (Orig. pub. 1910.)

W.H.W. [W. H. White?]. 1912. Review of *Stability in Aviation. Nature* 88:406–7.

Wiener, M. J. 1992. *English Culture and the Decline of the Industrial Spirit, 1850–1980.* London: Penguin.

Wieselsberger, C. 1914a. "Beitrag zur Erklärung des Winkelfluges einiger Zugvögel." *Zeitschrift für Flugtechnik und Motorluftschiffahrt* 5:225–29.

———. 1914b. "Einige Ergebnissse der theoretischen und experimentellen Untersuchungen von Tragflächen." *Österreichische Flugzeitschrift* 8:33–38.

———. 1914c. "Mitteilungen aus der Göttinger Modellversuchsanstalt, 16: Der Luftwiderstand von Kugeln." *Zeitschrift für Flugtechnik und Motorluftschiffahrt* 5:140–45.

———. 1916. "Tragflächenuntersuchungen der englischen Versuchsanstalt in Teddington." *Zeitschrift für Flugtechnik und Motorluftschiffahrt* 7:19–21.

———. 1918a. "Dreideckeruntersuchungen." *Technische Berichte* 3:302–8.

———. 1918b. "Luftwiderstandmessungen an wirklichen Flugzeugteilen." *Technische Berichte* 3:275–79.

———. 1918c. "Untersuchung eines Rumpfkühlers." *Technische Berichte* 3:107–11.

Wiggs, R. 1971. *Concorde: The Case against Supersonic Transport.* London: Ballantine.

Wigner, E. P. 1967. "The Unreasonable Effectiveness of Mathematics in the Natural Sciences." In *Symmetries and Reflections,* by E.P. Wigner, 222–37. Bloomington: Indiana University Press.

Wilde, O. 1966. "Phrases and Philosophies for the Use of the Young" (1894). In *Complete Works of Oscar Wilde.* London: Collins.

Williams, E. T., and H. M. Palmer, eds. 1971. *The Dictionary of National Biography, 1951–1960.* Oxford: Oxford University Press.

Wilson, A. 1973. *The Concorde Fiasco.* Harmondsworth: Penguin Books.

Wilson, D. B. 1984. "A Physicist's Alternative to Materialism: The Religious Thought of George Gabriel Stokes." *Victorian Studies* 28:69–96.

Wimperis, H. E. 1926. "The Relationship of Physics to Aeronautical Research." *Journal of the Royal Aeronautical Society* 30:668–75.

———. 1928. "The Progress of Aeronautical Research in Britain." In *IV Congresso Internazionale di Navigazione Aerea*, 4:393–401. Rome: Tipografia del Senato.

Wise, M. Norton, ed. 1995. *The Values of Precision.* Princeton: Princeton University Press.

Wittgenstein, L. 1922. *Tractatus Logico-Philosophicus.* London: Routledge and Kegan Paul.

———. 1967. *Philosophical Investigations.* Oxford: Blackwell.

———. 1969. *Über Gewißheit. On Certainty.* Oxford: Blackwell.

Wolf, S. L. 2003. "Physicists in the 'Krieg der Geister': Wilhelm Wien's 'Proclamation.'" *Historical Studies in the Physical and Biological Sciences* 33:337–68.

Wolstenholme, J. 1867. *Mathematical Problems.* London: Macmillan.

Wright, H. T. 1912. "Aeroplanes from an Engineer's Point of View." *Aero* 6:374–80.

Wuest, W. 2000. "Ludwig Prandtl als Lehrer in Hannover und Göttingen 1901–1947." In Meier, *Ludwig Prandtl*, 173–204.

Zilsel, E. 2000. *The Social Origins of Modern Science.* Ed. D. Raven, W. Krohn, and R. S. Cohen. Boston: Kluwer.

Index